Springer Proceedings in Energy

Series Editors

Muhammad H. Rashid, Department of Electrical and Computer Engineering, Florida Polytechnic University, Lakeland, USA

Mohan Lal Kolhe, Faculty of Engineering and Science, University of Agder, Kristiansand, Norway

The series Springer Proceedings in Energy covers a broad range of multidisciplinary subjects in those research fields closely related to present and future forms of energy as a resource for human societies. Typically based on material presented at conferences, workshops and similar scientific meetings, volumes published in this series will constitute comprehensive state-of-the-art references on energy-related science and technology studies. The subjects of these conferences will fall typically within these broad categories:

- Energy Efficiency
- Fossil Fuels
- Nuclear Energy
- Policy, Economics, Management & Transport
- Renewable and Green Energy
- Systems, Storage and Harvesting
- Materials for Energy

eBook Volumes in the Springer Proceedings in Energy will be available online in the world's most extensive eBook collection, as part of the Springer Energy eBook Collection. To submit a proposal or for further inquiries, please contact the Springer Editor in your region:

Kamiya Khatter (India)
Email: kamiya.khatter@springer.com
Loyola D'Silva (All other countries)
Email: loyola.dsilva@springer.com

Anil Kumar · G. N. Tiwari · Brian Norton ·
Deepak Tiwari
Editors

Selected Proceedings of the 1st International Conference on Advanced Materials for Sustainable Innovation; IC-AMSI 2024; 28–30 August; New Delhi; India

Harnessing Sustainable Energy Technologies for Effective Decarbonization

Springer

Editors
Anil Kumar
Department of Mechanical Engineering
Delhi Technological University
North West, Delhi, India

Brian Norton
Tyndall National Institute
International Energy Research Centre (IERC)
Cork, Ireland

G. N. Tiwari
BAG Energy Research Society Complex
Varanasi, Uttar Pradesh, India

Deepak Tiwari
School of Engineering and Technology
Vivekananda Institute of Professional Studies—Technical Campus
Pitampura, Delhi, India

ISSN 2352-2534 ISSN 2352-2542 (electronic)
Springer Proceedings in Energy
ISBN 978-981-96-4494-0 ISBN 978-981-96-4492-6 (eBook)
https://doi.org/10.1007/978-981-96-4492-6

This Springer imprint is published by the registered company Springer Nature Singapore Pte Ltd.
The registered company address is: 152 Beach Road, #21-01/04 Gateway East, Singapore 189721, Singapore

Preface

It is our pleasure to present academicians and scholars with the proceedings of research papers scheduled for presentation at the International Conference on Advanced Materials for Sustainable Innovation (IC-AMSI, 2024) 28th–30th August 2024 in School of Engineering & Technology, Vivekananda Institute of Professional Studies—Technical Campus (VIPS-TC), AU-Block (Outer Ring Road) Pitampura, Delhi-110034, India. This is the first international conference organized by the Department of Applied Science and Humanities, School of Engineering & Technology at VIPS-TC. The industry and academia have received the conference.

The core organizing committee for organizing IC-AMSI, 2024 is:

- Chief Patron: Dr. S. C. Vats, Chairman VIPS-TC
- Patron: Mr. Vineet Vats, Vice Chairman, VIPS-TC

Conference Conveners
Dr. Sonal Khurana, Assistant Professor VSE&T, VIPS-TC, Dr. Vanita Bhardwaj, Assistant Professor VSE&T, VIPS-TC

Organizing Secretaries: Dr. Deepak Tiwari, Assistant Professor VSE&T, VIPS-TC, Dr. Navneet Kumar, Assistant Professor VSE&T, VIPS-TC, Dr. Srishti, Assistant Professor VSE&T, VIPS-TC

Conference Co-Conveners
Dr. Anand Kumar Singh, Assistant Professor VSE&T, VIPS-TC, Dr. Lakshika Chutani, Assistant Professor VSE&T, VIPS-TC

The proceeding comprises papers from leading academicians, research scholars, and industry experts. The conference papers cover important research areas and the latest trends in the industry. About 192 papers from 200+ authors were presented, and about 250 participants attended the conference. In all, high-quality papers have been selected for presentation during the conference. The main topics of the conference have been classified into the following categories:

Soft and Bio Materials, Composite Materials,
Smart and Functional Materials,
Polymeric Materials, Shape Memory Alloys,
Advanced Materials, Piezoelectric Materials,
Electronic and Photonic Materials, Materials Computational Methods,
Renewable Raw Materials, Green Hydrogen, Plasmonic,
Sensors and Device Fabrications, Sustainable Materials,
Nanomaterials, Additive Manufacturing,
Materials Processing,
Advanced Materials Processing,
Renewable Energy,
Powder Metallurgy,
Materials Engineering,
Metallurgical Engineering.

The technical advisory committee is pleased to mention that the papers have been received on all the topics. Such a voluminous proceeding is only possible with the generous support received from various quarters. The committee would like to record its deep sense of gratitude and appreciation for the persistent efforts of all the reviewers. We are grateful to all the papers' authors for contributing to enrich this international conference of IC-AMSI, 2024. We also acknowledge the excellent cooperation received from our colleagues and experts on the review panel for their painstaking efforts in reviewing the papers. We thank the sponsors and extend full credit to the publisher (Springer) for accepting the proposal to publish conference proceedings with the book title Selected Proceedings of the 1st International Conference on Advanced Materials for Sustainable Innovation; IC-AMSI 2024; 28–30 August; New Delhi; India under the series title Harnessing Sustainable Energy Technologies for Effective Decarbonization, Springer Proceedings in Energy. Their timely support was not only a strength but also an inspiration for the organizers.

We sincerely hope that the engineering fraternity will find this publication a valuable source of knowledge for the application of mechanical engineering.

North West, Delhi, India — Anil Kumar
Varanasi, India — G. N. Tiwari
Cork, Ireland — Brian Norton
Pitampura, India — Deepak Tiwari

Contents

1 **Recent Solutions: Combating Climate Change and Its Impact** 1
Kunsh Sabharwal, Goyam Jain, and Komal Aggarwal

2 **The Interplay of Technology and Community Engagement in Advancing Sustainable Practices** 25
Charvi Gupta, Ram Sudan, and Lakshika Chutani

3 **Analysing Poverty–Environment–Climate Change Nexus and Its Impact on Sustainable Development Goals Attainment** 35
Nidhi Sharma, Jayati Ahuja, Urvashi Chawla, Himanshi, and Mudita Nagpal

4 **Sustainable Nutrition: The Environmental Benefits of Vegan and Vegetarian Choices** 49
Nidhi Sharma, Oshi Sharma, Tanisha Sharma, and Mudita Nagpal

5 **Analyzing the Need and Viability of Sustainable Fashion as a Pathway to Decarbonization** 63
Nidhi Sharma, Mudita Nagpal, and Tanuja Nautiyal

6 **GreenShield—A Natural Language Processing Based Approach to Prevent Greenwashing and Attain Decarbonization** 77
Nidhi Sharma, Adithya Venkateswaran, Raghav Tiwari, and Mudita Nagpal

7 **Energy-Development Nexus Review: PV Potential in Asian Least Developed Countries (ALDCs)** 93
Shivani Singh and Rahul Dev

8 ARTIFICIAL Intelligence in Home Energy Management Systems: Enhancing Sustainability and Grid Stability in Smart Grids 109
Radhika Kavra, Gaurisha Malhotra, Khushboo Khattar, and Lakshika Chutani

9 Predictive Analysis of Air Quality Index (AQI) and Identification of Influential Factors Using Machine Learning Models 131
Safina Shokeen, Taranveer Singh, Shashank Nautiyal, and Gautam Sethi

10 Sustainable Solutions for Delhi's Air Pollution: A Data Driven Approach 143
Kinjal Bansal and Vishal Srivastava

11 Analyzing the Potential of Artificial Meat Production as a Sustainable Solution to Decarbonization 159
Mudita Nagpal, Tanisha Gupta, Krrishika Gupta, and Nidhi Sharma

12 Bio-Plastics: A Sustainable Alternative for a Fossil Fuel-Reliant Industry 173
Parshav Jain, Pranay Bisht, and Komal Aggarwal

13 Studies on Durability Properties of Green Concrete by Partial Replacement of Coarse Aggregate with Jhama Bricks 193
Harideep Kumar Danthala, V. Mallikarjuna Reddy, and C. Vivek Kumar

14 Transforming Renewable Energy: Discovering the Development of Piezoelectric Substance 207
Utsav Singhal, Ansh Chawla, Amit Kumar, and Vanita Bhardwaj

15 Performance Analysis of Energy Harvesting for Sustainable Satellite-Terrestrial Relay Communication System with Path Selection in Eavesdroppers Presence 215
Varun Jain, B. V. R. Reddy, and Ashish Payal

16 Solar Energy Harvesting Methods for Sustainable Development 225
Bharat Kumar Nandmehar, Manasvi, and Navneet Kumar

17 Integrated Cleaner and Sun Tracker for Solar Panels 239
Siddharth Jha, Sarang, Tejaswi, Toukeer Khan, Monika Bhardwaj, and Pratul Arvind

18 A Review on Reliability and Maintenance of Photovoltaic Thermal Systems 249
Ravi Kumar, Deepali Atheaya, and Anil Kumar

19 **Energy Saving Enhanced Brain Tumor Classification Accuracy in Intelligent Systems Using Deep Convolutional Neural Network** ... 257
Abhimanu Singh and Smita Jain

20 **Simulation and Modelling of Nano Grid System Using Solar PV and Hydro Power Plant** ... 267
Vandana Sondhiya and Kaustubh Dwivedi

21 **An Investigation of Mismatch Between Energy Yields of Building Integrated Photovoltaic Systems** ... 279
Mansi Singh and Anil Kumar Rai

22 **Hybrid PVT-TEG Collector Coupled with VAR System: A Comparative Study** ... 289
Abhishek Tiwari and Shruti Aggarwal

23 **A Sustainable Enhanced MPPT Approach with Whale Optimization Algorithm (WOA) Tuned PID Control for Solar Water Pumping** ... 301
Vishal Srivastava and Anuj Rawat

24 **Enhancing Evaporation and Condensation in Solar Desalination Systems Through Hybridization: A Mini-Review** ... 313
Rahul Patel, Rahul Dev, Ajaya Bharti, Dhananjay Singh, and Anil Kumar

25 **Mathematical Modeling for Exergy Analysis of a Sea Water Reverse Osmosis Desalination System** ... 325
Gaurav Aggarwal and Surendra Singh Kachhwaha

26 **Enviro-Economic Assessment of Some Designs of PCM Incorporated Solar Stills** ... 337
Aseem Dubey and Akhilesh Arora

27 **Nanomaterials and Their Applications: A Sustainable Approach to Complete Water Refinement** ... 349
Bhavvya Mehta and Shubhangi Madan

28 **Thermal Characterization of Beeswax/CuO Composites as Latent Heat Storage Media for Space Heating Applications** ... 363
Arun Kumar Sharma, Prashant Sharma, Bhupendra Gupta, Anil Kumar, and Prashant Baredar

29 **Smart Grid Management: Machine Learning for Load Balancing and Fault Detection** ... 373
Ramandeep Saha, Garima Singhal, Shubhankit Sadhukar, Ajay Kumar, Aniket Singh, and Mohd Atif Wahid

30 Effect of Inclination Angle on Heat Pipe (HP) Performance with Graphene Nanofluids in Electronic Cooling Applications 389
Aruna Veerasamy and T. Dharini

31 Enhancing Home Comfort and Efficiency: A Comprehensive Study of Smart Home Automation Systems 401
Himanshu Lohia, Nupur, Sunidhi Prasad, Rishi Singhal, Aniket Mehra, and Divya Agarwal

32 Estimation of Optimum Condensing Temperature for Three Stage Vapor Compression Refrigeration System Using Alternative Refrigerants 413
Saquib Khan, Vaibhav Jain, and Kanchan Mugil

33 Design and Experimental Analysis of Free Piston Stirling Cooling System: An Environment Friendly Cooling Technique 427
Ehsas Srivastava, Kavya Taneja, Vaibhav Jain, and Kanchan Mudgil

34 Turning Trash into Treasure: The Role of Plastic Waste Recycling in Decarbonization 439
Mudita Nagpal, Sania, Khushi Garg, and Nidhi Sharma

35 Smart Waste Segregation System for Urban Sustainability 455
Ajeet Pathak, Shivani, Preeti Goyal, Amit Das, and Swati Sharma

36 Consumer Insights on E-Waste, Energy, and the Circular Economy: A Case Study of Delhi 465
Urvashi Chawla, Jayati Ahuja, and Sonal Khurana

Chapter 1
Recent Solutions: Combating Climate Change and Its Impact

Kunsh Sabharwal, Goyam Jain, and Komal Aggarwal

Abstract Climate change states long-term shifts in temperature and weather patterns due to global warming chiefly caused by a disbalance in atmospheric greenhouse gaseous (GHGs) such as carbon dioxide (CO_2), water vapor, methane (CH_4), nitrous oxide (NO_x), and CFC. Climate scientists have claimed that human activities such as the burning of fossil fuels (coal, petroleum and gases), deforestation etc. generate GHGs that trap the Sun's heat and increase the temperature. The evidence of climate change has been observed through the temperature rise, variation in rainfall, drought, and flood patterns, increasing sea levels, desertification, landslides etc. are the main causes. In recent years various technologies have been initiated especially for the use of renewable energy sources, that reduces pollution from the environment and helps in climate change. This article reviews the recent renewable energy sources that are clean, accessible, sustainable, and reliable and will drastically cut down CO_2 emissions in the environment. We discuss recent technologies like plant growth system through soilless culture, nanotechnology, artificial intelligence in renewable resources and ocean energy converters, that play a crucial role in combating climate change. On the other hand, we also discuss some green technologies like transparent solar cells, panels and biohybrid chandeliers as emerging tools for sustainable development and climate control.

Keywords Climate change · Sustainable development · Renewable

---

K. Sabharwal · G. Jain · K. Aggarwal (✉)
School of Engineering and Technology, Vivekananda institute of professional studies–Technical campus, New Delhi, India
e-mail: komal.aggarwal@vips.edu

A. Kumar et al. (eds.), *Selected Proceedings of the 1st International Conference on Advanced Materials for Sustainable Innovation; IC-AMSI 2024; 28–30 August; New Delhi; India*, Springer Proceedings in Energy,
https://doi.org/10.1007/978-981-96-4492-6_1

1.1 Introduction

Climate change denotes enduring alterations in temperature and weather patterns resulting primarily from global warming induced by an imbalance in atmospheric greenhouse gases (GHGs) such as water vapor, carbon dioxide (CO_2), methane (CH_4), nitrous oxide (NO_x), and chloro-fluoro-carbons (CFCs) [1, 2]. Though the pressing issue of climate change stems from persistent shifts in global climate conditions, primarily attributed to the cumulative effect of greenhouse gas emissions, human activities which include the burning of fossil fuels along with widespread deforestation, are major contributors to this phenomenon, amplifying the greenhouse effect and consequent warming of the Earth's atmosphere. Scientists emphasize that human activities such as fossil fuel combustion (coal, petroleum, and gases) and deforestation emit GHGs that trap solar radiation, elevating global temperatures [3].

Since 1850, the Earth's temperature has risen by an average of 0.11 °F (0.06 °C) per decade, totalling an increase of approximately 2 °F. However, the pace of warming has accelerated significantly since 1982, reaching 0.36 °F (0.20 °C) per decade, more than three times faster than the long-term average. In 2023, global temperatures soared to unprecedented heights, marking it as the warmest year on record since 1850. It surpassed the 20th-century average of 57.0 °F (13.9 °C) by 2.12 °F (1.18 °C) and exceeded the pre-industrial average (1850–1900) by 2.43 °F (1.35 °C). Remarkably, the ten hottest years in recorded history have all occurred within the past decade (2014–2023) [4].

Fluctuations in the precipitation patterns can be seen due to high temperatures giving rise to greater amounts of evaporation and have caused desertification of land due to drying and death of various plants making the land furthermore infertile along with increased risks of soil erosion and landslides with patterns of droughts and floods which cannot be predicted. This is also contributing to the rise in sea-levels putting areas close to seas, oceans at great risk of being hit by cyclones and tsunamis, while posing a greater threat of being submerged one day [5]. The major contributor to Climate Change are the emissions from traditional energy sources which are constantly being burnt rigorously and being used without limits leading not only to Climate Change but also giving hint at the threat of exhaustion of these resources, leaving humanity with no possible methods to produce energy for the activities supporting life [6].

The idea of Sustainable Development, which is a holistic approach to growth that seeks to meet the needs of the present without compromising the ability of future generations to meet their own needs, should be taken into regard as soon as possible in order to evade any possible adversities in the future that may give rise to environmental damage and scarcity of resources. Sustainable Development emphasizes the balanced integration of economic development, social equity, and environmental protection. This concept promotes the efficient use of resources, reducing waste and pollution, and fostering innovation and resilient communities. By focusing on long-term impacts and ensuring inclusive participation, sustainable development aims to create a harmonious coexistence between human activities and the planet's ecosystems, ultimately contributing to the well-being and prosperity of both current and future populations [7, 8].

The move towards renewable sources of energy can prove to be a game changer not only in the context of environment protection and reducing environmental damage but also at the same time making sure that the energy generated by these cleaner and greener sources of energy are significantly much more in amount than the traditional sources [9, 10]. Using resources such as wind/air to power windmills for energy generation, water streams/rivers/ponds/lakes to power turbines to create electricity, alongside solar panels to generate power from the sun's rays are few of the traditional renewable sources of energy that can be used to our advantage. Use of biomass and composting to create biogas-fuels is also another viable option.

These already known techniques to form renewable energy paired with new cutting-edge technologies are the basis of this article, where we critically examine the orthodox methods and sources of energy (crude oil, petroleum gas etc.) while throwing light on the currently present renewable energy sources, which are clean, accessible, sustainable, dependable, and poised to significantly curtail major emissions. In this work we are focusing on recent approaches towards sustainable and renewable energy technologies like plant growth system through soilless culture, nanotechnology, and ocean energy converters, that can cut down the usage of fossil fuel. We are also discussing the role of artificial intelligence in sustainable energy goals. Moreover, we will also be drawing the attention towards innovative green technologies like transparent solar cells and panels, along with biohybrid chandeliers and buildings illustrating their potential role in advancing sustainable development agendas and help in the climate change.

1.2 Causes of Climate Change

There are a few natural but primarily human activities which are causing the average rise in the earth's temperature. Rapid climate change, characterized by swift and significant alterations in global climate patterns, is primarily driven by various human activities coupled with natural phenomenon. The causes of rapid climate change include:

1. **Greenhouse Gas Emissions**: Gases like Methane (CH_4), carbon dioxide (CO_2), and nitrous oxide (N_2O), are being continuously released due to human interferences with environment like deforestation, burning of fossil fuels, industrial processes, and agriculture. These intensify the greenhouse effect, thus leading to the trapping of heat in the atmosphere, resulting in climate change and accelerated rate of global warming [11].

2. **Deforestation**: Chopping down trees for agricultural expansion, logging and urbanization, reduces the Earth's capacity to absorb CO_2 through photosynthesis. This loss of carbon sinks contributes to increased atmospheric CO_2 levels, exacerbating climate change [12].
3. **Burning of Fossil Fuels**: The combustion of fossil fuels for tasks of energy production in industries and transportation raise levels of CO_2 in the atmosphere. This significant source of greenhouse gas emissions amplifies the warming of the planet and accelerates climate change [13].
4. **Feedback Mechanisms**: Certain feedback mechanisms, such as the melting of polar ice caps and permafrost, can trigger further warming. As ice caps disappear, less sunlight gets reflected back into space, leading to more heat absorption by the Earth's surface. Additionally, these melted ice caps and glaciers release stored methane and carbon dioxide (greenhouse gases), into the atmosphere, amplifying the highly negative greenhouse effect [14].
5. **Industrial Processes**: Industrial activities that include production of cement, manufacturing of chemicals, and mining, emit greenhouse gases in combination with other severely harmful pollutants that contribute to climate change. These emissions alter the composition of the atmosphere and impact global climate patterns [15].
6. **Land Use Changes**: Converting natural landscapes into residential areas, agricultural lands, and other human-modified environments disrupts ecosystems and affects the Earth's climate-regulating mechanisms. Land use changes can lead to altered precipitation patterns, temperature extremes, and biodiversity loss, influencing climate dynamics [16].
7. **Transportation Emissions**: The transportation sector, particularly road vehicles, airplanes, and ships, releases large amounts of CO_2 and other pollutants into the environment. The reliance on fossil fuels for powering vehicles for transportation contributes to rapid climate change by increasing greenhouse gas concentrations [17].
8. **Waste Management**: Improper waste disposal and management practices, such as landfill emissions and incineration of waste, release methane and CO_2 into the air. These emissions contribute to the greenhouse effect and accelerate climate change [18].
9. **The Ozone Layer**: The ozone layer too plays a crucial role in influencing climate change through its interaction with greenhouse gases and its impact on atmospheric dynamics. Here are some key points regarding the ozone layer's effect on climate change:

The ozone layer, located in the stratosphere, helps absorb and scatter a portion of the incoming ultraviolet (UV) radiation from the sun. This absorption process warms the stratosphere but also protects the Earth's surface from harmful UV radiation. Human activities, particularly the release of chlorofluorocarbons (CFCs) and other ozone-depleting substances, have led to the depletion of the ozone layer, especially over polar regions. The thinning of the ozone layer allows more UV radiation to reach the Earth's surface, impacting ecosystems, human health, and contributing to climate change [19].

Addressing these causes of rapid climate change through new and advanced strategies, such as the move towards to renewable sources of energy production, promoting sustainable land use paired with cutting down emissions from transportation and industries, and enhancing waste management, can prove to be game changing to counter the impacts of climate change and lead ourselves towards a more sustainable future.

1.2.1 Effects of Climate Change

The after-effects of prolonged climate change encompass a range of long-term consequences resulting from continuous alterations in global climate patterns over an extended period. These impacts extend across various dimensions of the environment, society, and economy shown in Fig. 1.1. Here is an elaboration on the key after-effects of prolonged climate change:

1. **Rising Sea Levels**: Prolonged climate change causes ice caps and glaciers to melt and increase the sea levels. This, in-turn, contributes to coastal erosion, heightened flooding risks, and the displacement of communities residing in low-lying areas [20].

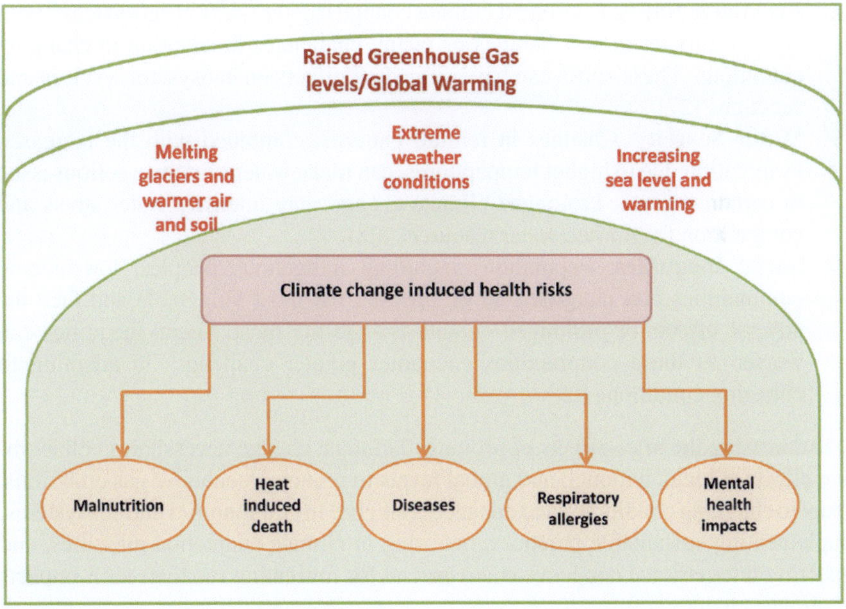

Fig. 1.1 Effects of climate change on the environment and human health [30]

2. **Extreme Weather Events**: Continued climate change intensifies occurrences of extreme weather events like heatwaves, droughts, hurricanes, and heavy rainfall. These events can inflict severe damage on not only human health but also infrastructure and agriculture [21].
3. **Biodiversity Loss**: Prolonged climate change disrupts ecosystems and habitats, resulting in loss of biodiversity. Several species struggle to adapt to the ever-changing climatic conditions, leading to population declines and potential extinctions [22].
4. **Food Insecurity**: Alterations in temperature and precipitation patterns impact agricultural productivity, potentially causing food shortages and insecurity. Prolonged climate change can disrupt food supply chains, exacerbating hunger and malnutrition in vulnerable populations [23].
5. **Health Impacts**: Prolonged exposure to changing climate conditions can have adverse effects on human health. Increased heatwaves can result in illnesses, while shifts in precipitation patterns can affect the quality of water and surge the spread of waterborne diseases [24].
6. **Economic Disruptions**: The long-term impacts of climate change can disrupt economies through infrastructure damage, loss of livelihoods in sectors like agriculture and fisheries, increased healthcare costs, and challenges to industries reliant on stable environmental conditions [25].
7. **Migration and Conflict**: Prolonged climate change can contribute to environmental displacement, forcing people to migrate due to sea-level rise, droughts, or other climate-related events. This can lead to heightened competition for resources and potential conflicts over land and water [26].
8. **Ecosystem Shifts**: Continued climate change triggers shift in ecosystems, with species moving to new habitats or facing challenges in adapting to changing conditions. These shifts can have cascading effects on ecosystem services and functions [27].
9. **Water Scarcity**: Changes in rainfall patterns, combined with the increased evaporation due to higher temperatures, can make water scarcity a serious issue in certain regions. Prolonged climate change may intensify water stress and competition for limited water resources [28].
10. **Social Inequities**: Populations including indigenous peoples, low-income communities and marginalized groups, are the most vulnerable and face the adverse effects of prolonged climate change the most. Social inequities can worsen as these communities encounter greater challenges in adapting to changing conditions [29].

Addressing the after-effects of prolonged climate change necessitates collaborative efforts at local, national, and global levels to reduce greenhouse gas emissions, adapt to changing conditions, and enhance resilience in communities and ecosystems. Implementing sustainable practices, investing in climate adaptation measures, and fostering international cooperation are crucial for mitigating the long-term impacts of climate change.

1.3 Solutions

1.3.1 Some Already Existing Techniques

In order to derive solutions to the above life threatening and environment polluting effects of climate change we need to pay attention and understand the currently present renewable energy sources that are non-polluting and help to resolve the issue of climate change. Some already existing and in use techniques are stated in Fig. 1.2 and mentioned.

1. **Wind Energy**: Wind is commonly regarded as a natural phenomenon with the capability to generate electricity by harnessing the kinetic energy of moving air. For many years, wind power has been employed to produce electricity. It is considered one of the most prevalent forms of renewable energy, thanks to its abundant and virtually inexhaustible supply [31].
2. **Hydro Electricity**: This type of energy leverages the force of moving water, such as that cascading over a waterfall, to generate electricity via a turbine. It serves as an alternative to fossil fuels by avoiding the emission of harmful gases into the atmosphere and is also capable of producing power in areas where conventional energy resources are limited [32].
3. **Solar Energy**: Solar energy stands as the most abundant form of renewable energy globally, offering a pollution-free and fully sustainable solution. By capturing the Sun's radiated light and heat through various technologies, solar energy can be converted into electricity. Installing solar panels in residential settings can reduce energy expenses, lower carbon emissions, and aid in combating climate change. Additionally, solar panels represent a valuable investment opportunity [33].
4. **Bio Energy**: Bioenergy is derived from living organisms through biological processes that convert stored chemical or potential energy into heat, light, or mechanical work. This form of energy has been utilized for many years, primarily in the form of biofuels such as ethanol and biodiesel [34].
5. **Geothermal Energy**: Geothermal energy is a reliable, clean, and cost-effective method for generating electricity by harnessing the Earth's internal heat. This sustainable and environmentally friendly energy source does not depend on fossil fuels. Instead, it utilizes naturally occurring hot water and steam found in various underground layers of soil and rock to produce electricity [35].

Apart from these, recent innovations which are constantly underway with technologies like Plant Growth Systems through soilless culture, Nanomaterials, Ocean Energy Converters and AI paired with existing technologies to improve their efficacy and output and help in combating climate change.

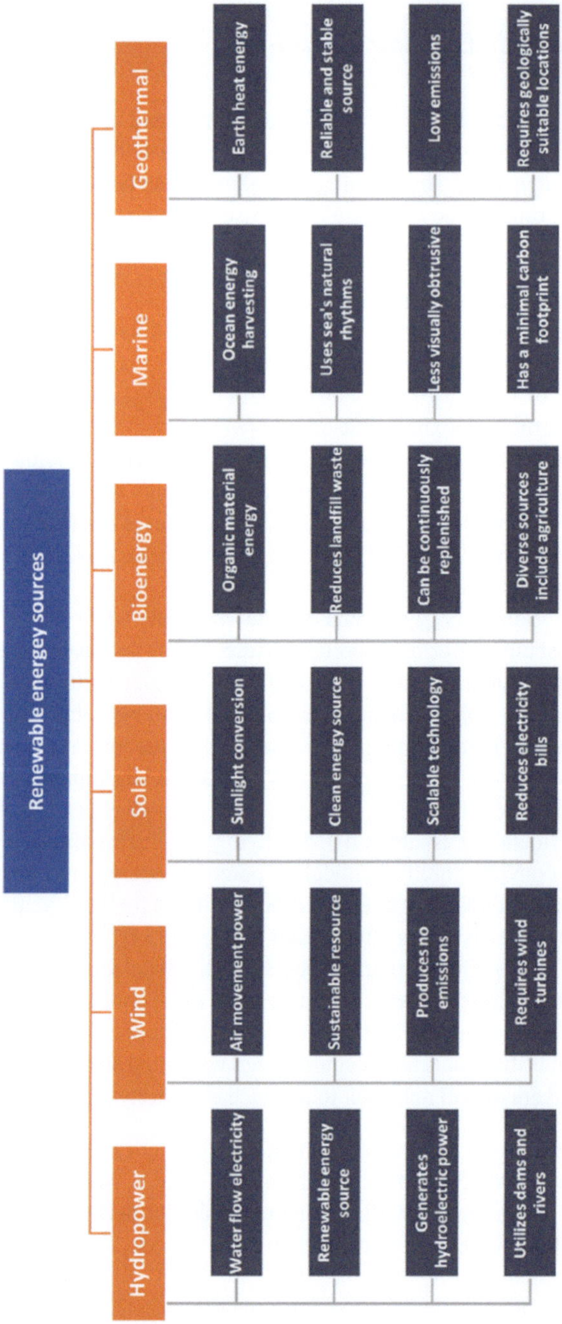

Fig. 1.2 Classification of various renewable energy sources [37]

1.3.2 Plant Growth System Through Soilless Culture

Variations in climatic factors can lead to various morphological changes in crops grown in open land ecosystems, including self-incompatibility, poor setting of seeds, male sterility, retardation in growth, embryo abortions, low nutrient content, yield loss, and even crop failure. Traditional agricultural systems, which often use large amounts of water on extensive arable land and rely heavily on agrochemicals, can lead to issues such as nutrient leaching, chemical residues in soil and water bodies, increased insect competition, weed proliferation, and soil erosion, as depicted in Fig. 3a. In contrast, controlled environment plant growth chambers provide substantial benefits. They require 80% less land and 90% less water than traditional agriculture, while more effectively managing nutrient runoff and translocation. These chambers are based on the technology of standard nutrient diffusion for plant growth, ensuring efficient use of nutrients like water, along with photosynthetic assimilation, and metabolic efficiency. They also regulate climatic factors like carbon sequestration, carbon footprint and greenhouse gas emissions, through a mechanistic model integrated with machine learning, which analyses critical and non-critical levels, dynamics, and influences of water, climate, and nutrients within the controlled environment [36].

The controlled environment plant growth chamber produces plants of outstanding physical and chemical quality shown in Fig. 3b. This system also plays an important role in combating the ill effects of climate change by making sure to conserve natural resources, and fostering research, employment, and industrial development. Though to better understand the effectiveness of climate change mitigation in controlled environments compared to natural ecosystems (with or without soil), more comparative studies are needed.

1.3.3 Nanotechnology

Nanotechnology has tremendous possibilities in the world of sequestration and reduction of greenhouse gases, along with uses in wastewater treatment, biofuel production and environmental remediation using structured materials such as functionalized nanomaterials, nanocomposites, nano catalysts, nano lubricants, nano zeolites, nano silica, and nano-coated metal organic frameworks, etc. using a sustainable approach shown in Fig. 4a and b. Nanotechnology offers a wide array of opportunities to create a more sustainable future for our planet in the face of climate change challenges [30]. Here are some key ways in which nanotechnology can contribute to sustainability:

1. **Energy Efficiency**: Nanotechnology enables the development of energy-efficient solutions by improving the performance of energy storage devices, enhancing the efficiency of solar cells, and optimizing insulation materials. These advancements help reduce energy consumption, thereby promoting renewable energy sources

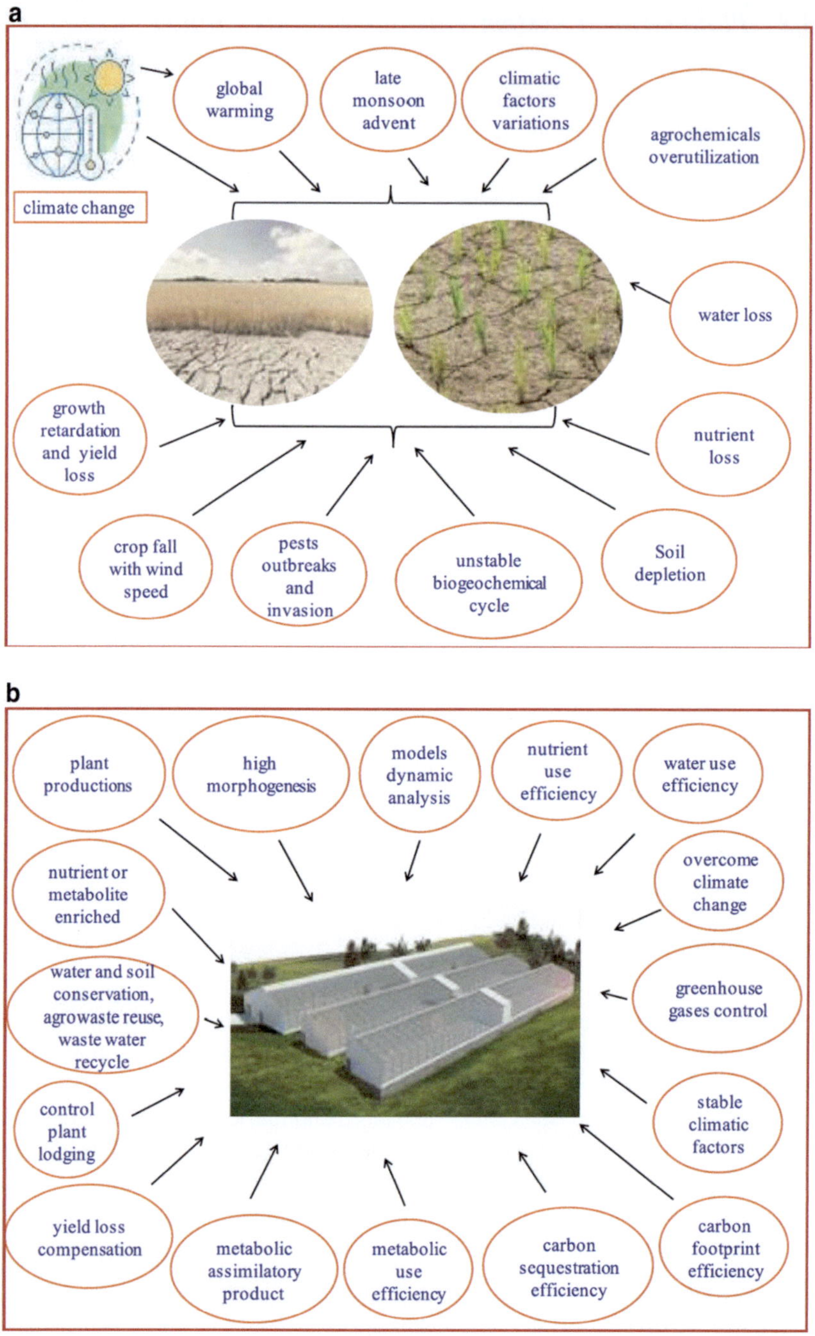

Fig. 1.3 **a** Stability of ecosystem in conventional environment agricultural system [36]. **b** Stability of ecosystem in controlled environment agricultural system [36]

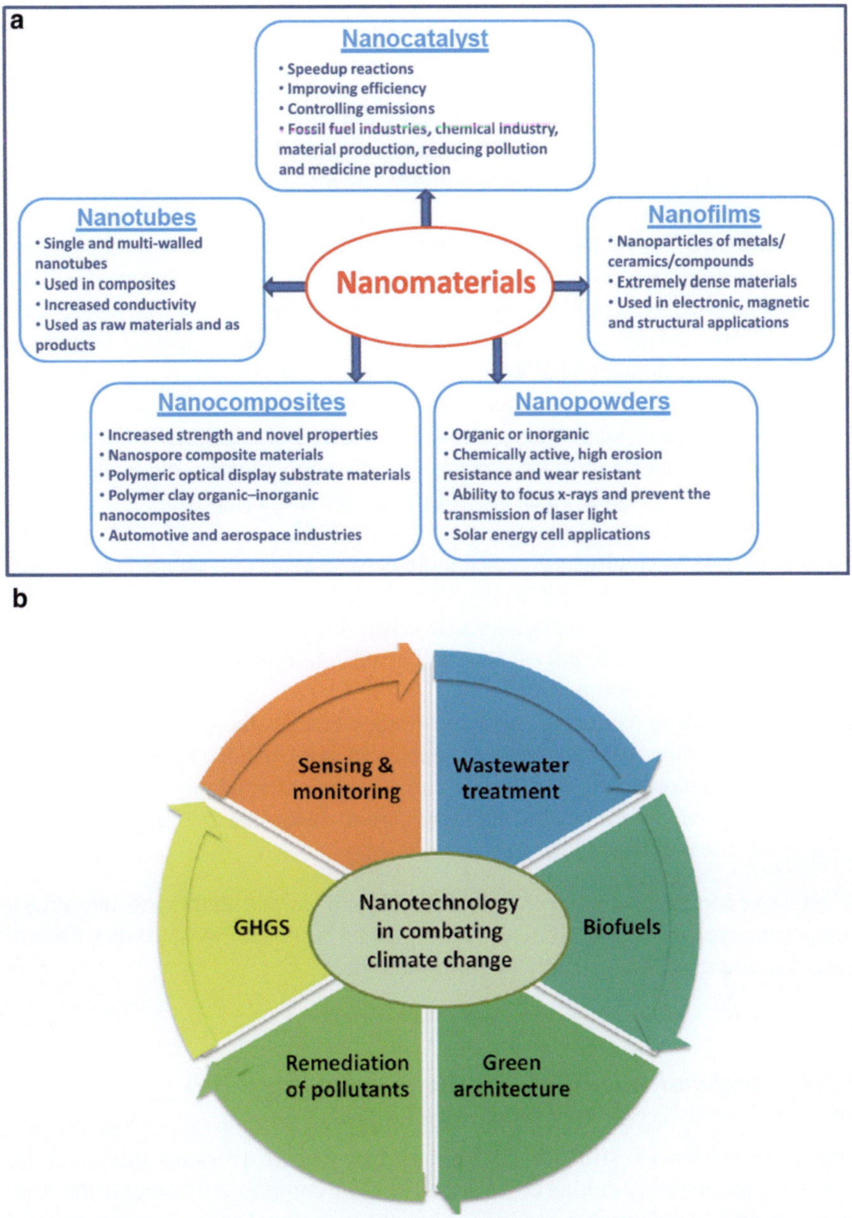

Fig. 1.4 **a** Different nanomaterials and their potential applications [30]. **b** Various applications of nanotechnology in combating climate change [30]

to mitigate the impact of climate change by lowering greenhouse gas emissions [30].

2. **Environmental Remediation**: Nanotechnology plays a crucial role in environmental cleanup and remediation by providing innovative materials and technologies for pollution control. Nanomaterials can effectively remove pollutants from air, water, and soil, contributing to improved environmental quality and ecosystem health [30].
3. **Greenhouse Gas Sequestration**: Nanotechnology offers solutions for capturing and storing greenhouse gases, such as carbon dioxide, methane, and nitrous oxide. Nanomaterials with high adsorption capacities can help mitigate climate change by reducing the emission of these gases into the atmosphere, and contributing to the efforts to combat global warming [30].
4. **Water Treatment**: Nanotechnology-based processes are used for water purification and treatment, addressing water scarcity and pollution issues. Nanomaterials like nano adsorbents and nano catalysts can efficiently remove contaminants from water sources, ensuring clean and safe water for various applications, including agriculture, industry, and human consumption [30].
5. **Sustainable Agriculture**: Nanotechnology plays a role in sustainable agriculture by offering solutions such as nano fertilizers, nano-pesticides, and precision farming technologies. These innovations help improve crop productivity, reduce environmental impact, and enhance soil health, contributing to food security and ecosystem sustainability [30].
6. **Green Building Materials**: Nanotechnology enables the development of eco-friendly building materials, such as nanocomposites, nanocoating's, and energy-efficient windows. These materials enhance the sustainability of buildings by improving energy efficiency, reducing carbon footprint, and promoting green architecture practices [30].

By harnessing the potential of nanotechnology in various sectors, we can address climate change challenges, promote environmental sustainability, and work towards a more resilient and eco-friendly future for our planet.

1.3.4 Artificial Intelligence in Renewable Resources

The incorporation of artificial intelligence has played a critical stimulant for improving productivity, cutting costs, and addressing complex difficulties in the innovative domain of sustainable and renewable energy technologies shown in Fig. 1.5 [37, 38].

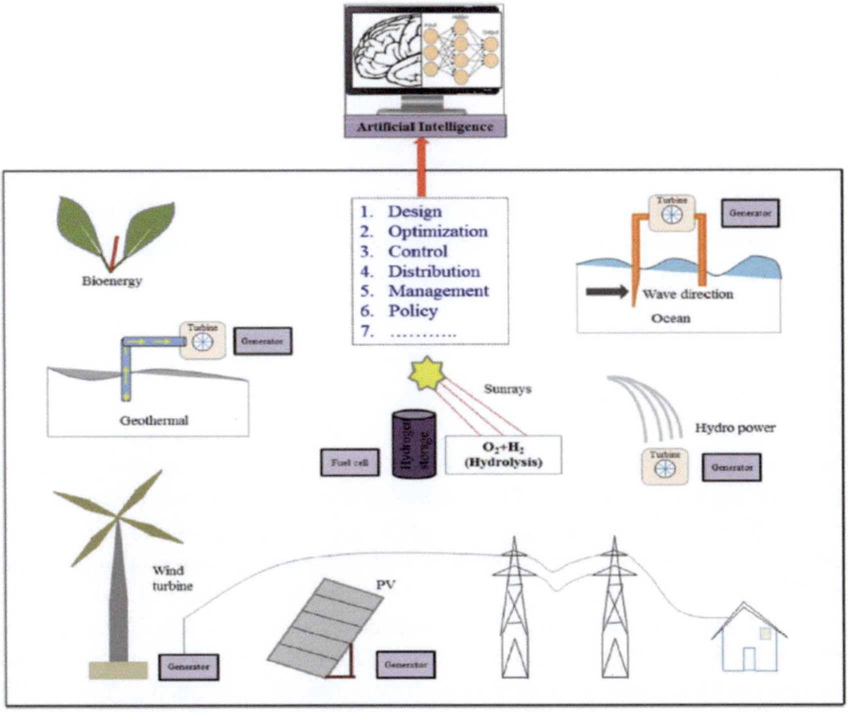

Fig. 1.5 Schematic representation of AI applications in renewable energy-based technologies [37]

Solar Energy

- AI simulation techniques have been used to predict solar radiation and optimize the performance of solar panels.
- AI-based predictive maintenance has been applied to improve the efficiency and reliability of solar energy systems.

Wind Energy

- AI and machine learning algorithms have been utilized to predict, monitor, and control wind turbines, enhancing their performance and output.
- Wind power prediction models using AI algorithms have been developed to optimize energy production and grid integration.

Geothermal Energy

- AI approaches have shown promise in analysing geothermal potential using surface and subsurface data, aiding in site selection and resource optimization.
- AI models have been used to address challenges in geothermal energy systems, improving efficiency and sustainability.

Hydro Energy

- AI technologies have been employed in hydroelectric power systems for predictive maintenance, safety improvements, and operational efficiency.
- Machine learning algorithms have been utilized to optimize the management and distribution of hydro energy resources.

These examples demonstrate the diverse applications of artificial intelligence in various renewable energy technologies, showcasing its potential to enhance performance, efficiency, and sustainability in the sector.

1.3.5 Ocean Energy Converters

The technology for field testing and evaluating the power outputs along with the quality of ocean energy converters is essential for enhancing their power generation capabilities and fostering the growth of the ocean energy converter industry. By examining the current state of research as well as the testing requirements for tidal and wave energy converters worldwide, a method for field testing and assessing the power and quality performance of these converters was developed. This approach involved proposing measurement requirements for tidal parameters, wave parameters, and electrical parameters, as well as investigating and constructing field testing systems for ocean energy converters, as illustrated in Fig. 1.6 [39]. Consequently, this technology is poised to play a pivotal role in advancing the industrialization of ocean energy converters and contributing to global climate control efforts.

The development of precise field testing and assessment methods allows for real-world evaluation of ocean energy converters under various environmental conditions. This ensures that the converters are optimized for maximum efficiency and reliability. Additionally, the construction of dedicated field-testing systems provides a controlled environment for rigorous testing, minimizing the risks associated with deploying unproven technologies in the open ocean. Overall, this comprehensive approach not only accelerates the maturation of ocean energy technologies but also supports the broader goal of reducing reliance on fossil fuels and mitigating climate change.

1.3.6 Biohybrid Chandeliers and Buildings

This chandelier not only illuminates the space but also fosters photosynthesis by supporting tiny microalgae, which absorb carbon dioxide and release breathable oxygen into the room. This organic process creates a novel symbiotic relationship between objects and people, wherein vital resources are continuously exchanged and waste products facilitate necessary metabolic processes. This dynamic mirrors biospheric systems, where waste is repurposed as a valuable resource rather than being discarded. The design of the bionic chandelier features glass leaves inhabited

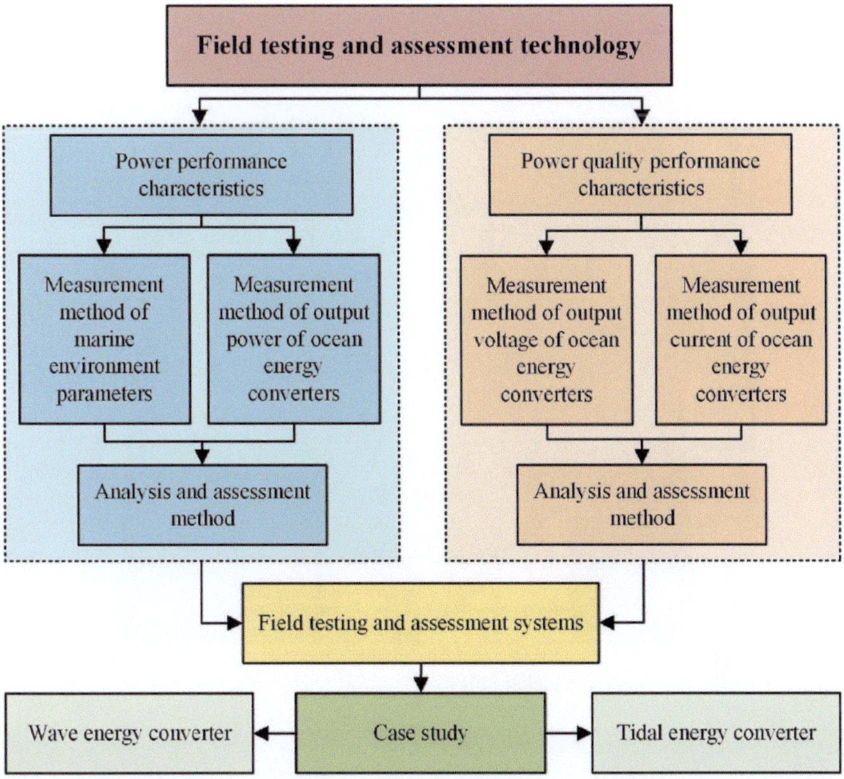

Fig. 1.6 Field testing and assessment technology for ocean energy converters [39]

by green algae, sustained by natural daylight. These leaves absorb carbon dioxide from the air and continuously emit oxygen. The Exhale bionic chandelier consists of a handmade, burnished metal structure supporting individual glass leaf modules of three different sizes. These modules are arranged in a radial pattern, replicating the natural form of plants, with a total of seventy modules, as illustrated in Fig. 1.7 [40].

Biohybrid Buildings introduce a pioneering façade system designed to cultivate microalgae for generating renewable heat and biomass, as illustrated in Fig. 1.8. The density of cells within the bioreactors varies according to the amount of available light and the harvesting strategy employed. Increased daylight promotes more algae growth, which in turn provides enhanced shading for the building. Biomass offers the advantage of flexible use for both generation of power and heat which can be stored with minimal energy loss. Additionally, cultivating microalgae using flat panel Photo-Bio-Reactors (PBRs) eliminates the need for extra land and remains largely unaffected by weather conditions [41].

Fig. 1.7 Exhale—the bionic chandelier [40]

Fig. 1.8 Lab created leaves filled with algae [41]

Fig. 1.9 Transparent solar panel windows [42]

1.3.7 Transparent Solar Panels

Photovoltaic glass represents one of the most advanced new solar panel technologies, poised to revolutionize the solar energy applications. Transparent solar panels can not only generate electricity from windows in homes, offices, cars, or even smartphones but they also allow normal visible light to pass through, selectively harnessing a portion of the solar spectrum that is invisible to the naked eye, as shown in Figs. 1.9 and 1.10. To achieve this remarkable technology, researchers have developed the Transparent Luminescent Solar Concentrator (TLSC) [42–44].

Fig. 1.10 Transparent solar panel [43]

1.4 Conclusion

As times and technologies advance at an unprecedented pace and species continue to evolve, the production of energy has become an urgent necessity. This imperative for energy reduction, however, carries significant responsibilities. If the methods of generating energy ail to adhere to principles of sustainable development, they risk becoming detrimental to humanity. Inadequate attention to sustainability in energy practices could precipitate numerous severe consequences, including environmental degradation, resource conflicts, and a rise in illnesses and diseases, potentially leading to catastrophic outcomes for our planet.

The reliance on non-renewable and traditional sources of energy–such as coal, oil, and natural gas–has become increasingly problematic. These sources not only deplete finite resources but also contribute to severe pollution and environmental harm. In contrast, the adoption of cleaner, greener, and safer renewable energy sources–such as wind, solar, hydroelectric power coupled with biomass and geothermal energy–offers a more sustainable alternative. Transitioning to renewable energy is crucial for addressing the urgent issue of pollution and ensuring that future generations inherit a world where they can breathe clean air and live healthy lives.

Renewable energy presents a viable solution for several pressing global challenges. It plays a critical role in mitigating climate change by reducing greenhouse gas emissions, which are a major driver of global warming. Additionally, renewable energy sources contribute to decreased air and water pollution, which translates into better public health outcomes and a higher quality of life. Economically, renewable energy can spur job creation, stimulate innovation, and foster energy independence, further enhancing its appeal as a sustainable choice.

The necessity of transitioning from non-renewable to renewable energy sources is both clear and compelling. Embracing renewable energy is a pivotal step in our collective effort to combat environmental degradation and address the urgent threats of climate change. By prioritizing renewable energy, we can work towards a more sustainable and equitable world, safeguarding the planet for generations to come (Table 1.1).

Table 1.1 Table of the remarks and finding of studies

S.no	Technology	Main findings	Remarks/impacts
1	Plant growth system through soilless culture	Controlled environment plant growth chambers to facilitate the development of plants to help combat climate change	Regulates climatic factors, greenhouse gas emissions, carbon sequestration, and carbon footprint through a mechanistic model combined with machine learning
2	Artificial intelligence	AI and machine learning algorithms, AI simulation techniques, etc	Tracking data patterns to predict further changes in climate along with identify the causes for the change
3	Nanotechnology	Nanostructured materials such as functionalized nanomaterials, nanocomposites, nano catalysts, nano lubricants, nano zeolites, nano silica, and nano-coated metal organic frameworks, etc	Has tremendous possibilities in the world of sequestration and reduction of greenhouse gases, along with uses in wastewater treatment, biofuel production and environmental remediation
4	Biohybrid chandeliers and buildings	The design of the bionic chandelier consists of glass leaves, doubling as a source for the purification of air while also fulfilling the purpose of the lamp. The green algae, when fully matured/ fermented can be taken out to be used as a fertilizer	Generates heat and biomass as renewable energy sources, alongside air purification and providing light. Lastly, but not the least, the algae can serve as much needed fertilizer for various other plants/gardens in the surroundings
5	Ocean energy converters	Converters that that trap air pockets by the use of oscillating water columns to drive a turbine along with oscillating body converters that utilize the energy from wave motion and last but not the least, overtopping converters that make use of height differences in the water levels during tides and waves generation	Advancement in the industrialization of ocean energy converters and helps in climate control worldwide
6	Transparent solar panels	Photovoltaic glass doubling as solar panels and windows	Generates electricity from windows in offices, homes, car's sunroof, or even smartphones

Acknowledgements The authors are thankful to VIPS-TC, New Delhi, India for the infrastructural support.

References

1. I. Hadjipaschalis, A. Poullikkas, V. Efthimiou, Overview of current and future energy storage technologies for electric power applications. Renew. Sustain. Energy Rev. **13**, 1513–1522 (2009). https://doi.org/10.1016/j.rser.2008.09.028
2. K. Abbass, M.Z. Qasim, H. Song, M. Murshed, H. Mahmood, I. Younis, A review of the global climate change impacts, adaptation, and sustainable mitigation measures, Environ. Sci. Pollut. Res. **29**, 42539–42559 (2022). https://doi.org/10.1007/s11356-022-19718-6
3. A. Kumar, S. Nagar, S. Anand, Climate change and existential threats, Glob. Clim. Chang. pp. 1–31 (2021). https://www.sciencedirect.com/science/article/abs/pii/B9780128229286000058
4. R. Lindsey, L. Dahlman, NOAA Climate.gov, Climate change: global temperature (2024), https://www.energy.gov/sites/default/files/2024-02/093.%20Rebecca%20Lindsey%20and%20Luann%20Dahlman%2C%20NOAA%2C%20Climate%20Change_%20Global%20Temperature.pdf
5. S. Prabhu, V. Chitale, *Decoding India's Changing Monsoon Patterns: A Tehsil-level Assessment*, (Council on Energy, Environment and Water, New Delhi, 2024), https://www.ceew.in/publications/decoding-changing-monsoon-rainfall-patterns-due-to-climate-change-in-india#:~:text=As%20Earth's%20surface%20temperatures%20rise,due%20to%20a%20warming%20climate.
6. S. Mahapatra, Biofuels and their sources of production: a review on cleaner sustainable alternative against conventional fuel, in the framework of the food and energy nexus. Energy Nexus **4**, 100036 (2021). https://doi.org/10.1016/j.nexus.2021.100036
7. United Nation, Climate Action, Climate Action Fast Fact, https://www.un.org/en/climatechange/science/key-findings#physical-science.
8. A. Hussaina, S.M. Arif, M. Aslam, Emerging renewable and sustainable energy technologies: state of the art. Renew. Sustain. Energy Rev. **71**, 12–28 (2017). https://doi.org/10.1016/j.rser.2016.12.033
9. A.G. Olabi, M.A. Abdelkareem, Renewable energy and climate change, Renew. Sustain. Energy Rev. **158**, 112111 (2022). https://doi.org/10.1016/j.rser.2022.112111
10. N.S. Rathore, N.L. Panwar, *Renewable energy sources for sustainable development* (New India Publishing Agency, New Delhi, India, 2007)
11. A. Mikhaylov, N. Moiseev, K. Aleshin, T. Burkhardt, Global climate change and greenhouse effect. Entrep. Sustain. Issues **7**, 2897–2913 (2020). https://doi.org/10.9770/jesi.2020.7.4(21)
12. P. Longobardi, A. Montenegro, H. Beltrami, M. Eby, Deforestation induced climate change: effects of spatial scale. PLoS ONE **11**(4), e0153357 (2016). https://doi.org/10.1371/journal.pone.0153357
13. D.J. Soeder, *Fossil Fuels and Climate Change. In: Fracking and the Environment*, (Springer, Cham, 2021). https://doi.org/10.1007/978-3-030-59121-2_9
14. P. Ceppi, F. Brient, M.D. Zelinka, D.L. Hartmann, Cloud feedback mechanisms and their representation in global climate models. WIREs Clim. Chang. **8**, e465 (2017). https://doi.org/10.1002/wcc.465
15. B.K. Sovacool, S. Griffiths, J. Kim, M. Bazilian, Climate change and industrial F-gases: a critical and systematic review of developments, sociotechnical systems and policy options for reducing synthetic greenhouse gas emissions. Renew. Sustain. Energy Rev. **141**, 110759 (2021). https://doi.org/10.1016/j.rser.2021.110759
16. D. Beillouin, R. Cardinael, D. Berre, A. Boyer, M. Corbeels, A. Fallot, F. Feder, J. Demenois, A global overview of studies about land management, land-use change, and climate change

effects on soil organic carbon. Glob. Change Biol. **28**, 1690–1702 (2022). https://doi.org/10.1111/gcb.15998
17. A.A.R.N. Avotra, A. Nawaz, Asymmetric impact of transportation on carbon emissions influencing SDGs of climate change. Chemosphere **324**, 138301 (2023). https://doi.org/10.1016/j.chemosphere.2023.138301
18. P.J. Castro, J.M.M. Aráujo, G. Martinho, A.B. Pereiro, Waste management strategies to mitigate the effects of fluorinated greenhouse gases on climate change. Appl. Sci. **11**, 4367 (2012). https://doi.org/10.3390/app11104367
19. P.W. Barnes et al., Ozone depletion, ultraviolet radiation, climate change and prospects for a sustainable future. Nat. Sustain. **2**, 569–579 (2019). https://doi.org/10.1038/s41893-019-0314-2
20. R. Gehrels, E. Garrett, Chapter 11–Rising sea levels as an indicator of global change, in *Climate Change Observed Impacts on Planet Earth*, Third edn, (2021) pp. 205–217. https://doi.org/10.1016/B978-0-12-821575-3.00011-6
21. B. Clarke1, F. Otto, R. Stuart-Smith, L. Harrington, Extreme weather impacts of climate change: an attribution perspective. Environ. Res. Clim. **1**, 012001 (2022). https://doi.org/10.1088/2752-5295/ac6e7d
22. S. Habibullah, B.H. Din, S.H. Tan, H. Zahid, Impact of climate change on biodiversity loss: global evidence. Environ. Sci. Pollut. Res. **29**, 1073–1086 (2022). https://doi.org/10.1007/s11356-021-15702-8
23. S. Dasgupta, E.J.Z. Robinson, Attributing changes in food insecurity to a changing climate change. Sci. Rep. **12**, 4709 (2022), https://www.nature.com/articles/s41598-022-08696-x
24. J. Paavola, Health impacts of climate change and health and social inequalities in the UK. Environ. Health **16**, 113 (2017). https://doi.org/10.1186/s12940-017-0328-z
25. J.D. Winne, G. Peersman, The adverse consequences of global harvest and weather disruptions on economic activity. Nat. Clim. Chang. **11**, 665–672 (2021), https://www.nature.com/articles/s41558-021-01102-w
26. K. Burrows, P.L. Kinney, Exploring the climate change, migration and conflict nexus. Int. J. Environ. Res. Public Health **13**, 443 (2016). https://doi.org/10.3390/ijerph13040443
27. G.M. Harris, S.E. Sesnie, D.R. Stewart, Climate change and ecosystem shifts in the southwestern United States. Sci. Rep. **13**, 19964 (2023). https://doi.org/10.1038/s41598-023-46371-x
28. S.N. Gosling, N.W. Arnell, A global assessment of the impact of climate change on water scarcity. Clim. Change **134**, 371–385 (2016). https://doi.org/10.1007/s10584-013-0853-x
29. S.N. Islam, J. Winkel, Climate change and social inequality, in *Department of Economic and Social Affairs*, (2017), p. 152, https://www.un.org/sites/un2.un.org/files/2020/08/1597341819.1157.pdf
30. N. Chautala, J. Saxena, R. Prasad, Nanotechnology as a sustainable approach for combating the environmental impacts of climate change. J. Agric. Food Res. **12**, 100541 (2023). https://doi.org/10.1016/j.jafr.2023.100541
31. International Renewable Energy Agency—Wind Energy, https://www.irena.org/Energy-Transition/Technology/Wind-energy
32. International Renewable Energy Agency—Hydropower, https://www.irena.org/Energy-Transition/Technology/Hydropower
33. International Renewable Energy Agency—Solar Energy, https://www.irena.org/Energy-Transition/Technology/Solar-energy
34. International Renewable Energy Agency—Bioenergy and Biofuels, https://www.irena.org/Energy-Transition/Technology/Bioenergy-and-biofuels
35. International Renewable Energy Agency—Geothermal Energy, https://www.irena.org/Energy-Transition/Technology/Geothermal-energy
36. A. Sharma, M. Hazarika, P. Heisnam, H. Pandey, V.S. Devadas, M. Wangsu, Controlled environment ecosystem: a plant growth system to combat climate change through soilless culture. Crop. Des. **3**, 100044 (2024). https://doi.org/10.1016/j.cropd.2023.100044

37. A. Bennagi, O. AlHousrya, D.T. Cotfas, P.A. Cotfas, Comprehensive study of the artificial intelligence applied in renewable energy. Energ. Strat. Rev. **54**, 101446 (2024). https://doi.org/10.1016/j.esr.2024.101446
38. S.K. Jha et al., Renewable energy: present research and future scope of artificial intelligence. Renew. Sustain. Energy Rev. **77**, 297–317 (2017). https://doi.org/10.1016/j.rser.2017.04.018
39. X. Wang, H. Xia, y. Guo, Y. Duan, M. Wang, Y. Liu, Research on field testing and assessment technology of ocean energy converters. Ocean. Eng. **285**, 115539 (2023). https://doi.org/10.1016/j.oceaneng.2023.115539
40. J. Melchiorri, Bionic Chandelier (2020). https://www.julianmelchiorri.com/Bionic-Chandelier
41. Arup, World-first façade system to cultivate micro-algae to generate heat and biomass as renewable energy sources, Biohybrid Buildings (2024) https://www.arup.com/projects/solar-leaf
42. Revolutionizing Renewable Energy: Transparent Solar Panels, https://www.anker.com/blogs/solar/transparent-solar-panels
43. R. Lunt, Michigan State University, Solar Magazine, Transparent Solar Panels: Reforming Future Energy Supply (2020), https://solarmagazine.com/solar-panels/transparent-solar-panels
44. T. Shiyani, Solar PV Panels—Recent Advances and Future Prospects, Transparent Solar PV Panels (2023), https://www.intechopen.com/chapters/86250.

Chapter 2
The Interplay of Technology and Community Engagement in Advancing Sustainable Practices

Charvi Gupta, Ram Sudan, and Lakshika Chutani

Abstract Environmental, Social, and Governance (ESG) refers to a framework designed to evaluate an organization's business operations and effectiveness regarding various sustainability and ethical considerations. In this paper we show that how technology and community involvement impact sustainable development-is a boon or a bane.

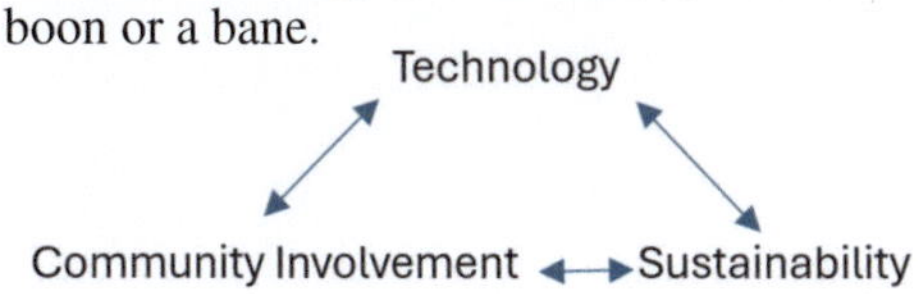

Keywords Environmental · Social · Governance framework · Sustainability

2.1 Literature Review

Technology has become an indispensable force shaping the trajectory of human progress. In the ever-evolving landscape of innovation, it serves as the catalyst for societal transformation and individual empowerment. From the intricate algorithms powering artificial intelligence to the seamless connectivity of a globalized world, technology intertwines with our daily lives, reshaping how we communicate work, and perceive reality. As we navigate this digital frontier, the ethical considerations surrounding technology's impact on privacy, equity and the environment become pivotal. Embracing the responsible and mindful development of technology ensures a future where innovation harmonizes with the values of humanity.

In the twenty-first century technology stands as the defining force propelling human civilization into new frontiers. From the advent of the internet to the era of

C. Gupta · R. Sudan · L. Chutani (✉)
School of Engineering & Technology, VIPS-TC, Delhi, India

C. Gupta
e-mail: Charvigupta2605@gmail.com

A. Kumar et al. (eds.), *Selected Proceedings of the 1st International Conference on Advanced Materials for Sustainable Innovation; IC-AMSI 2024; 28–30 August; New Delhi; India*, Springer Proceedings in Energy,
https://doi.org/10.1007/978-981-96-4492-6_2

artificial intelligence, our world is undergoing an unprecedented transformation. The ubiquity of smartphones, interconnected devices and the power of data have woven a digital fabric that shapes every facet of our lives. This transformative force has not only revolutionized communication and information dissemination but has also become the backbone of industries, economies and global interconnectedness.

2.1.1 The Dual Nature of Development

Yet as we are marvels of the technological progress a dichotomy emerges. The same tools that connect us globally can also exacerbate divides both digital and societal. The promise of automation and efficiency comes hand in hand with concerns about job displacement and ethical ramifications. Striking a balance between the rapid pace of technological advancement and the need for thoughtful regulation and ethical considerations becomes imperative.

2.1.2 The Ethical Imperative

The ethical dimensions of technology echo loudly in our collective conscience. From issues of privacy invasion to the environmental toll of manufacturing and disposing of electronic devices, navigating the ethical landscape of our tech-driven world requires a concerted effort. Responsible innovation, sustainable practices and a commitment to inclusivity are essential to ensuring that the benefits of technology are equitably distributed and the potential harms are mitigated.

2.1.3 Shaping a Human-Centric Future

As we stand on the cusp of further technological breakthroughs, the imperative is clear—to shape a future where technology serves humanity rather than dictates it. Embracing a human-centric approach involves fostering education and digital literacy, encouraging diversity in tech industries and designing technologies that prioritize human well-being. The narrative of technology is still unfolding and our choices today will determine whether it becomes a force for empowerment and equality or a source of division and inequality. In the grand tapestry of human history technology stands as both the architect and the canvas of our collective destiny.

2.2 Introduction

Achieving sustainable development demands a holistic approach that harmonizes economic growth, social equity and environmental preservation in line with the 17 UN Sustainable Development Goals (SDGs) [1]. These goals address a wide array of global challenges including poverty, inequality and climate change. With increasing occurrences of climate disasters such as floods, wildfires and hurricanes the need for urgent collective action to combat these crises has become more critical.

The involvement of local communities, government policies and technological advancements are key pillars in driving sustainable development. Communities are essential in adopting sustainable practices at the grassroots level—whether through waste management, use of renewable energy or sustainable farming. Governments in turn have a crucial role in providing the framework for sustainability by investing in green infrastructure while funding renewable energy projects and promoting electric vehicles which are becoming an increasingly viable alternative in reducing carbon emissions in the transportation sector [1].

Technological innovation, especially in rapidly developing countries like India is proving to be a game-changer in advancing sustainability. High-tech solutions such as blockchain for ensuring supply chain transparency, AI for optimizing resource consumption and the growing carbon markets are helping industries to reduce emissions and foster greener practices. The global transition towards greener energy systems has been further reinforced by the "UAE Consensus," a significant agreement that focuses on reducing reliance on fossil fuels and committing nearly 200 countries to achieve net-zero emissions by 2050 [1].

During past years a lot of research has been done in the field of sustainability where some of them may be found in [2, 3, 6, 7]. In [5] support of non government organisations on sustainability can been seen whereas in [4] community involvement for sustainable development as conceptual model is shown.

As the world continues to confront climate-related challenges, the path to achieve a sustainable future will hinge on international collaboration, government policies and the integration of cutting-edge technologies. Ensuring a balance between economic development, environmental protection and social well-being is the only way to create a resilient, sustainable world for future generations.

2.3 Survey

We conducted a survey for different age groups (14–22 years, 23–45 years and 45 + years) and obtained the following results regarding the interdependency of community involvement, sustainability and technology on each other.

2.3.1 *For Age Group 14–22*

Community Involvement and Sustainability

1. How frequently do you participate in community activities related to sustainability (example: clean up drives, trees planting and recycling programs) Figs. 2.1–2.12.
2. What barriers do you think exist in your community that hinder sustainable development? (Select all that apply)

Technology and Sustainability

1. How often do you use technology (like smartphones, tablets or computers) to learn about environmental issues?

Fig. 2.1 Community involvement and sustainability survey for age group 14–22

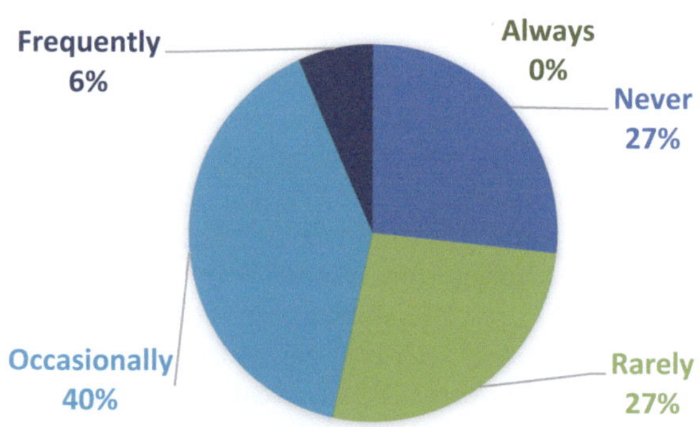

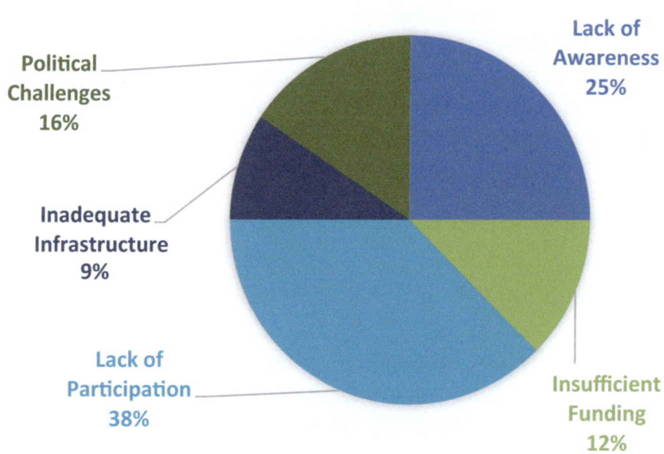

Fig. 2.2 Community involvement and sustainability survey for age group 14–22

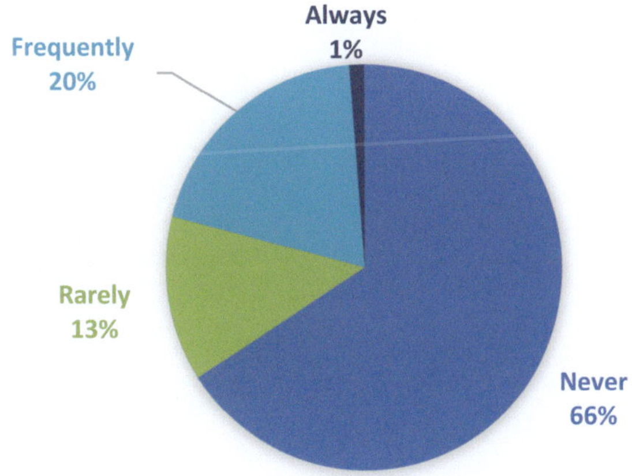

Fig. 2.3 Technology and sustainability survey for age group 14–22

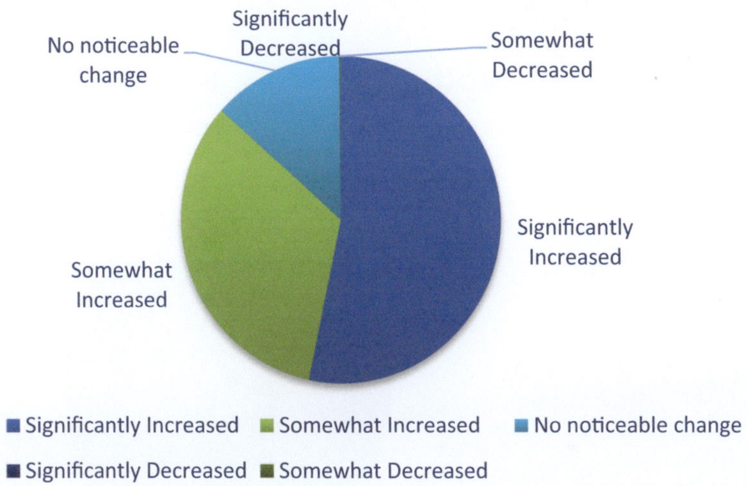

Fig. 2.4 Technology and sustainability survey for age group 14–22

2. In your opinion, how has social media influenced your awareness and actions toward sustainability?

2.3.2 For Age 23–45 Years

Community Involvement with sustainable development

1. How frequently do you participate in community activities related to sustainability (example: clean up drives, trees planting and recycling programs)?

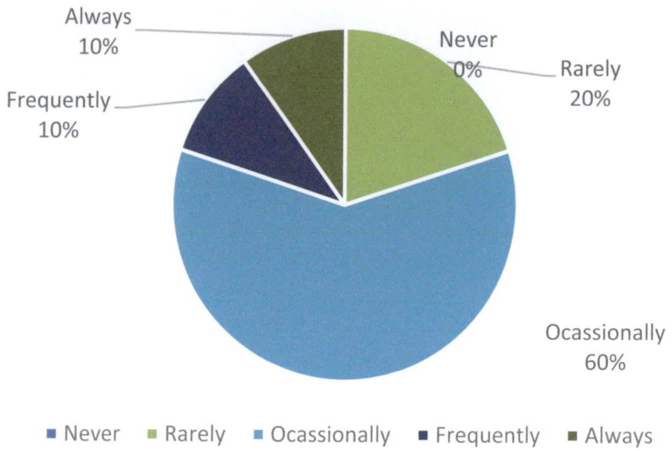

Fig. 2.5 Community involvement and sustainability survey for age group 23–45

Fig. 2.6 Community involvement and sustainability survey for age group 23-45

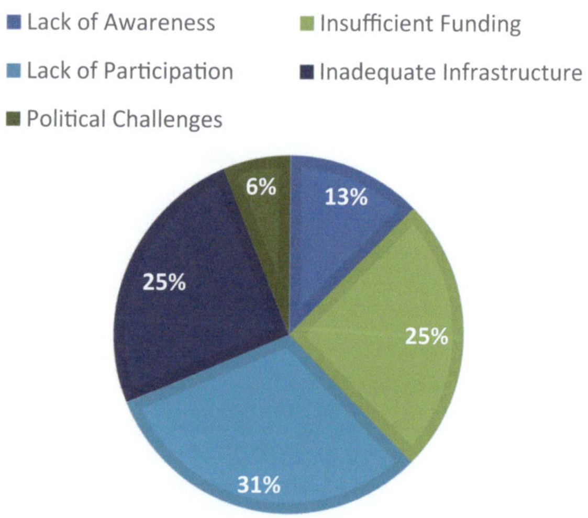

2. What barriers do you think exist in your community that hinder sustainable development? (Select all that apply)

Technology with Sustainable Development

1. How often do you use technology (like smartphones, tablets or computers) to learn about environmental issues?
2. In your opinion how has social media influenced your awareness and actions toward sustainability?

Fig. 2.7 Technology and sustainability survey for age group 23–45

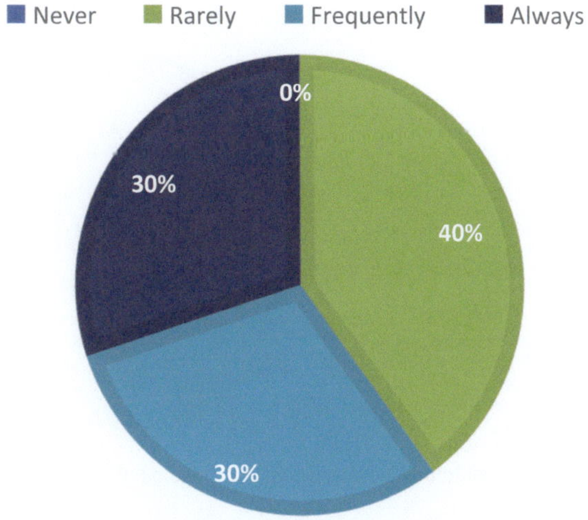

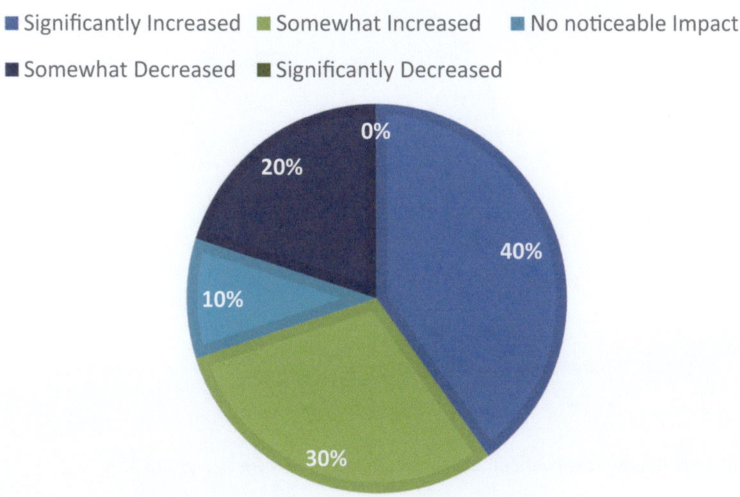

Fig. 2.8 Technology and sustainability survey for age group 23–45

2.3.3 For Age Group 45+

Community Involvement and Sustainability

1. How frequently do you participate in community activities related to sustainability (example: clean up drives, trees planting and recycling programs)?
2. What barriers do you think exist in your community that hinder sustainable development? (Select all that apply)

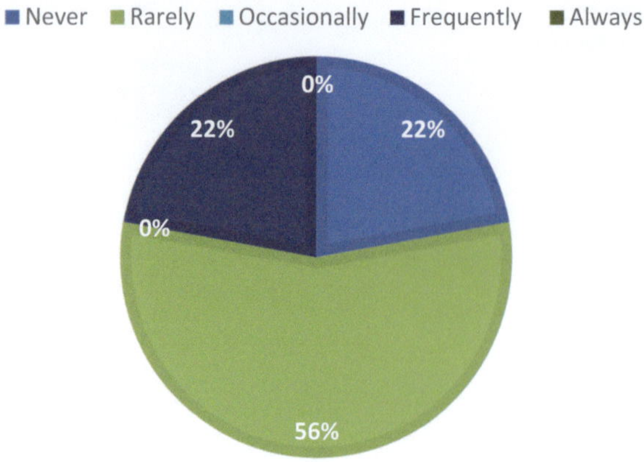

Fig. 2.9 Community involvement and sustainability survey for age group 45+

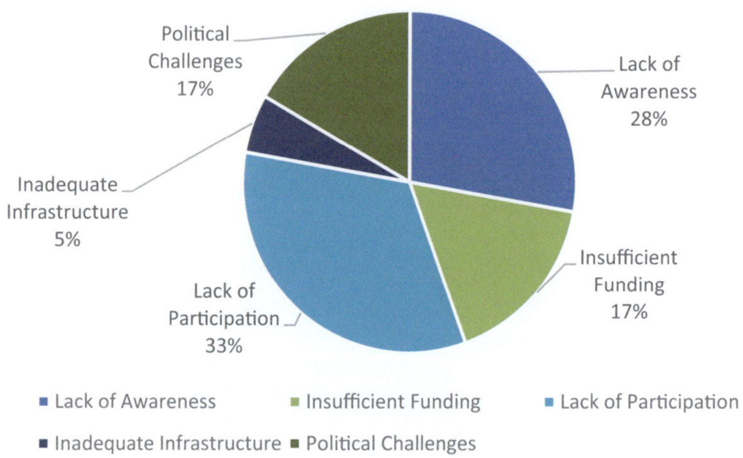

Fig. 2.10 Community involvement and sustainability survey for age group 45+

Technology and Sustainability

1. How often do you use technology (like smartphones, tablets or computers) to learn about environmental issues?
2. In your opinion, how has social media influenced your awareness and actions toward sustainability?

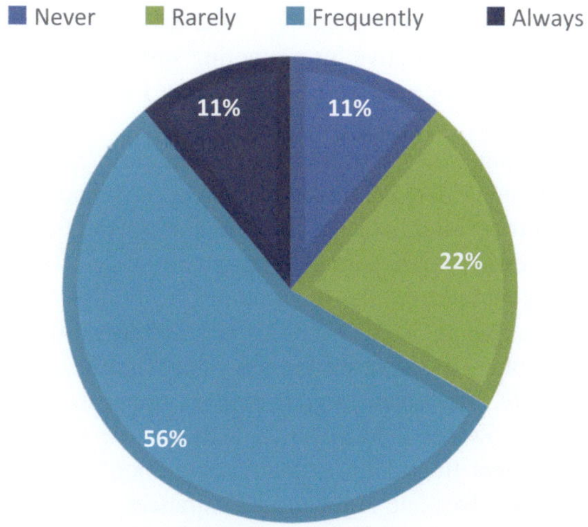

Fig. 2.11 Technology and sustainability survey for age group 45+

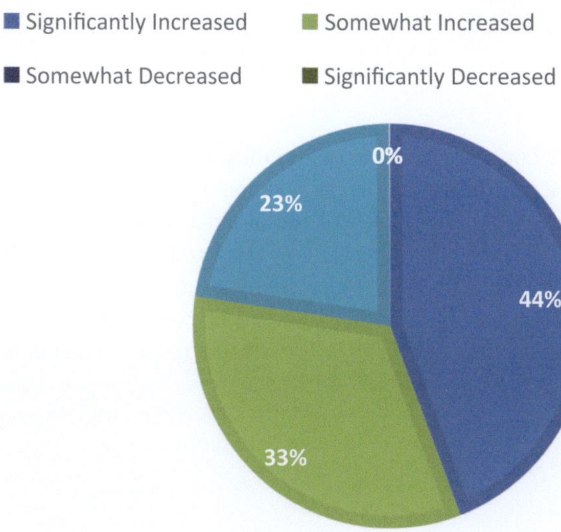

Fig. 2.12 Technology and sustainability survey for age group 45+

2.4 Analysis of Survey

The survey included questions involving community, sustainability and technology for the age group 14–22, 23–45 and 45 + years. According to the data people of age group 14–22 and 23–45 occasionally take part in sustainable practices such as clean up drives, trees plantation, recycling programs etc. Whereas 45 + year community rarely take part in such practices. The main cause that hinders the sustainable development is lack of participation, lack of awareness and insufficient funding. Moreover survey shows that age group 14–22 the educational workshops is the most effective community initiative for promoting sustainable development. For the age 23–45 its recycling programs and for age 45 + its clean up drives and as well as recycling programs.

Community agree that the community led initiatives are more effective than government led initiatives in promoting sustainable development. Also people of age group 14–22 perceive neutrally about the impact of community involvement or local environmental issue. Whereas 23–45 and 45 + age group perceive it positively.

According to the community technology has somewhat influenced their purchasing decisions regarding ecofriendly products and services. Advancements in technology like renewable energy play significant role in promoting sustainability in real life. Moreover, in their opinion social media has significantly increased their action towards sustainability. Most people use technology like smartphones and computers frequently to learn more about environmental issues. Therefore technology solves one of the main hindrances in sustainable development that is lack of awareness.

2.5 Conclusion

From the above diagrams we conclude that the community involvement and new technologies evolving over the time contribute to the sustainable development of the country. If the community come together and take initial small steps individually it will make a greater difference and help in achieving our sustainable goals. The new technological advancements such as artificial intelligence has helped to achieve the goal. The government initiatives such as installments of power plants, using renewable energy, solar panels etc. has significantly contributed to the sustainability goals. India's main goal is to achieve net zero by 2030.

References

1. Sustainable Development Newsletter (2023), http://assocham.org
2. D. Mitlin, Sustainable development: a guide to the literature. Environ. Urban. **4**(1), 111–124 (1992)
3. I. Zelenika, J. Pearce, Barriers to appropriate technology growth in sustainable development. Eur. J. Sustain. Dev. **4**(6) (2011).
4. Y. Li, C. Hunter, Community involvement for sustainable heritage tourism: a conceptual model. J. Cult. Herit. Manag. Sustain. Dev. **5**(3), 248–262 (2015)
5. S. Bass, Participation in sustainable development: the roles of non-governmental organisations, in *Presentation to the 20th Waigani Seminar, Environment and Development* (1993)
6. I. Alotaibi, M. Abido, M. Khalid, A. Savkin, A comprehensive review of recent advances in smart grids: a sustainable future with renewable energy resources. Energies **13**, 6269 (2020)
7. S. Sharda, M. Singh, K. Sharma, Demand side management through load shifting in IoT based HEMS: overview, challenges and opportunities. Sustain. Cities Soc. **65**, 102517 (2021)

Chapter 3
Analysing Poverty–Environment–Climate Change Nexus and Its Impact on Sustainable Development Goals Attainment

Nidhi Sharma, Jayati Ahuja, Urvashi Chawla, Himanshi, and Mudita Nagpal

Abstract A good environment is a prerequisite for the health and wellbeing of all. Rapid urbanization and migration toward big cities have resulted in the formation of urban slum clusters, particularly in metropolitan areas. These clusters, emerging in unorganized settlements, bring about environmental and social problems. This study focuses on four slum clusters in Delhi, utilizing general surveys and complexometric data analysis to identify key poverty indicators. These indicators are further analyzed with respect to rising pollution levels, contributing to climate change. Our work highlights the integration of climate change action with poverty reduction strategies to achieve the Sustainable Development Goals (SDGs) by developing an application using artificial intelligence and engineering tools. The proposed app streamlines the real-time information and donation processes by means of AI, ensuring that the resources reach those in greatest need. It further enhances transparency by providing real-time updates on project progress, impact assessments, and financial reports.

Keywords Poverty · Environment · Poverty-Environment nexus · Sustainable development goals · Feedback loops · Artificial intelligence

---

N. Sharma · J. Ahuja · U. Chawla (✉) · Himanshi · M. Nagpal
School of Engineering and Technology, VIPS-TC, Delhi, India
e-mail: 35817702723_cse@vipstc.edu.in

N. Sharma
e-mail: nidhisharma@vips.edu

J. Ahuja
e-mail: 06417702723_cse@vipstc.edu.in

Himanshi
e-mail: 05917702722_cse@vips.edu

M. Nagpal
e-mail: mudita.nagpal@vips.edu

A. Kumar et al. (eds.), *Selected Proceedings of the 1st International Conference on Advanced Materials for Sustainable Innovation; IC-AMSI 2024; 28–30 August; New Delhi; India*, Springer Proceedings in Energy,
https://doi.org/10.1007/978-981-96-4492-6_3

3.1 Introduction

Poverty is marked by extreme deprivation of well-being and encompasses multiple aspects. It involves low wages and the inability to access essential goods and services needed for a dignified life. Poverty includes inadequate health and education, limited access to clean water and sanitation, lack of physical security, lack of representation, and insufficient opportunities and capacity to improve one's circumstances.

While traditional poverty metrics typically rely on income levels, it's crucial to understand that poverty extends beyond economic constraints to encompass diverse social, cultural, and political factors. Economically, poverty is gauged by income thresholds established by governments or international bodies, identifying individuals or households falling below these benchmarks as living in poverty. Today, the assessment of poverty increasingly utilizes the Multidimensional Poverty Index (MPI), which incorporates 10 indicators across health, education, and living standards. This approach acknowledges that poverty encompasses a wide range of deprivations beyond income alone [1].

According to Multidimensional Poverty Index 2023, 4.2% of population in India is in severe multidimensional poverty and 18.7%of the population is vulnerable to multidimensional poverty and 21.9%is below the poverty line [1].

3.1.1 *Cause of Poverty*

Rapidly Increasing Population: India's population grows by 2.2% annually, adding about 17 million people each year, which increases the demand for resources.

Low Agricultural Productivity: Fragmented landholdings, lack of capital, traditional farming methods, and storage wastage result in low agricultural output.

Underutilized Resources: Underemployment and disguised unemployment, especially in farming, lead to low agricultural output and reduced living standards.

Low Rate of Economic Growth: Slow economic development widens the gap between demand for and availability of goods and services.

Social, Political, and Climatic Factors: Inheritance laws, the caste system, British colonization, traditions, and natural calamities (floods, earthquakes, cyclones) impede poverty eradication and economic development.

3.1.2 *Slums*

Slums are informal urban settlements with poor housing, inadequate infrastructure, and unsanitary conditions. Rapid urbanization drives people to cities in search of better opportunities, but the lack of affordable housing forces many into slums. Insufficient urban planning and social support systems exacerbate the situation, leading to growing slum populations and increased marginalization. The characteristics of a slum depend on its site, situation, and socio-cultural factors. The degree of deprivation varies among slums, which can lead to disparities in health outcomes even within the same city. The majority of people who end up in slums are employed in manufacturing, construction, household services, rag picking, and various of small-scale industries like selling fruits and vegetables [2].

3.1.3 Correlation between Poverty and Environment

The poverty-environment nexus highlights a reciprocal relationship where impoverished communities, heavily reliant on natural resources like agriculture, fishing, and forestry, often lack access to sustainable technologies. This dependency leads to unsustainable environmental practices such as deforestation and soil erosion, exacerbating issues like water resource depletion. Income inequality also increases carbon emissions and ecological footprints, with high-income countries mitigating pollution through better regulation and cleaner technologies, while middle-income countries face worsening environmental degradation due to resource and infrastructure disparities [3].

Conversely, environmental degradation and climate change intensify poverty by increasing the frequency and severity of extreme weather events such as floods, heatwaves, and storms. These environmental challenges further compromise public health through pollution and poor living conditions. The impact of such events is disproportionately felt in slum areas due to their inadequate infrastructure and limited resilience. This interplay creates a feedback loop where poverty and environmental degradation reinforce each other, perpetuating a cycle of vulnerability and hardship for affected communities (Fig. 3.1).

3.2 Related Work Done in the Field

Wealth inequality significantly affects environmental quality attributes. One explanation for this effect could be underprivileged people's exploitation of natural resources who lack alternative sources of income to survive. Rising inequality is associated with higher pollution levels, implying that poor areas and unequal societies in developing countries frequently overlook environmental issues. This negligence can be attributed to a lack of awareness and inadequate environmental policies in these areas [4].

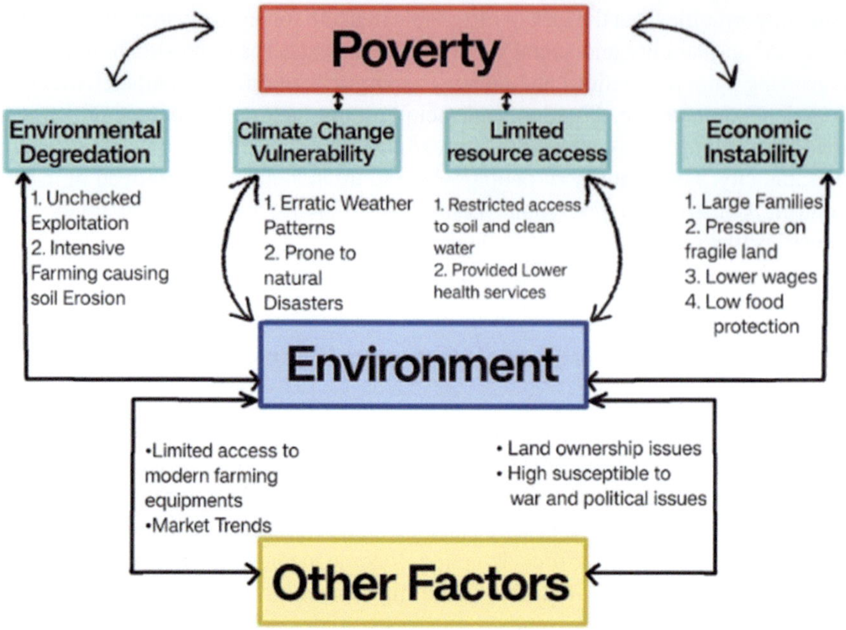

Fig. 3.1 Flowchart on correlation of environment and poverty

The complex interplay between poverty and the environment, as well as impoverished communities and natural resources, has sparked heated debate. Poor communities are frequently stuck in a cycle of poverty due to resource scarcity, forcing them to further harm the environment in order to meet their immediate needs [5].

The distribution of wealth is unequal, and impoverished people suffer disproportionately because of their extreme vulnerability in society. Income inequality makes it more difficult to enact environmental laws that are effective in reducing emissions. Higher income inequality frequently leads to the enactment of fewer environmental protection laws, which raises emissions [6].

The health of low-income households is at risk, and efforts to mitigate climate change are made more difficult by conventional fuels like coal and firewood, which also significantly increase indoor pollutants and greenhouse gas emissions. The difficulties low-income households encounter in the face of strict environmental regulations have been brought to light by the use of these fuels in China [7].

There is a correlation between poverty, energy access, and climate vulnerability; 14 indicators have been found to be important in this relationship. Together, these indicators—which cover infrastructure, education, awareness, energy, agriculture, and income—help create an institutional framework for sustainable development [8].

In addition to this work, our research focuses on understanding the poverty-environment nexus from an Indian perspective, providing a deep analysis of the same in the slum areas of a developing country like India. This study aims to shed light on the unique challenges and potential solutions within this context.

3.3 Methodology

3.3.1 Study Area

The Delhi NCR region faces significant challenges in quality of life, driven by high pollution levels, a dense migrant population, elevated mortality rates from pollution-related diseases, and insufficient green spaces [9]. Smog, a visible sign of worsening pollution, becomes particularly severe in winter due to low temperatures and stagnant air, combined with high NOx emissions from fossil fuel combustion [10]. Furthermore, 3.43% of Delhi's population is multidimensionally poor, reflecting significant socio-economic disparities within the region [11].

To understand the poverty-environment nexus in Delhi's slum areas, Haiderpur, Seemapuri, Paharganj, and Bhalswa Jahangirpuri slums have been taken into consideration. These slums are located in financially distressed regions of New Delhi. Slum formation is a relatively recent phenomenon in Delhi, primarily occurring after the growth of urban and industrial centers. As the city's industrial and commercial sectors expanded, areas such as Okhla, Shahdara, and other industrial belts emerged [13]. This urban expansion attracted a large influx of migrants, who established slum settlements (Fig. 3.2).

Studying such areas is crucial to delve into the poverty-environment nexus because:

1. Slums are among the most poverty-stricken areas in the city and have a high concentration of population below the poverty line.
2. The environment of the slums is directly affected by factors related to a lack of monetary assets.
3. The residents of these areas are among the most vulnerable populations globally, lacking essential services and infrastructure. The lack of proper infrastructure leads to environmental degradation within the slums.
4. Slum environments are commonly observed to have low education rates and healthcare access, which helps us explore these factors' effects on climate and environment.

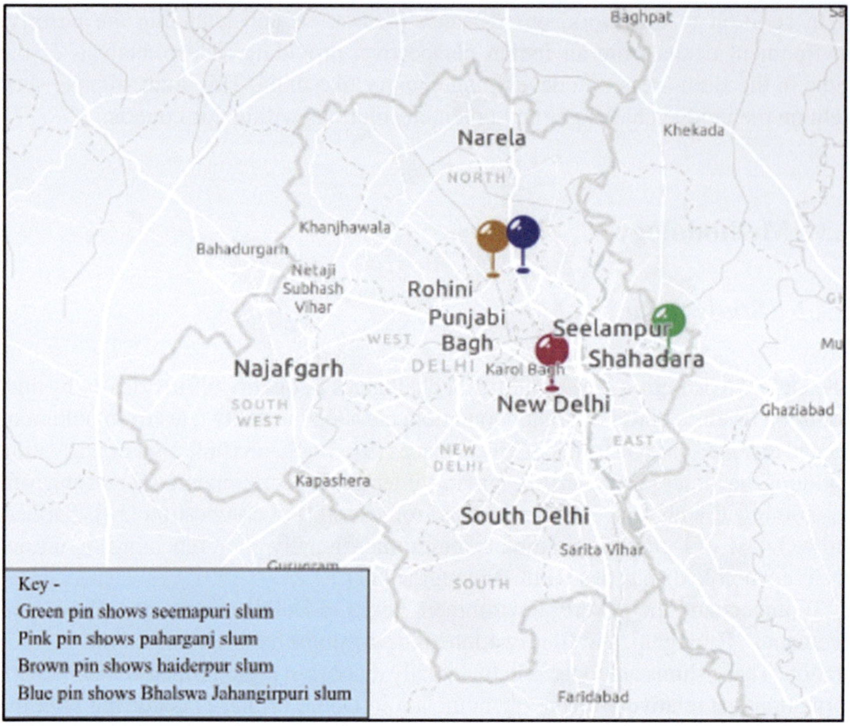

Fig. 3.2 Location of the slums taken into consideration on the map of Delhi (using https://www.arcgis.com/index.html)

3.3.2 Data Sets Used

Primary Data Set

The following data was obtained from the personal interview done in Haiderpur and overviews age, education and other parameters important for drawing conclusions (Tables 3.1, 3.2 and 3.3).

Secondary Data Sets

To successfully understand how poverty in the 3 slums- Paharganj slum, Seemapuri slum and Bhalswa Jahangirpuri slum-reflects on the environment and vice versa, poverty indicators such as the following were used:

a. AQI
b. Education
c. Sanitation
d. Water Quality of ground water

Table 3.1 Demographic information of responses of personal interview done in Haiderpur slum

1	Parameter	Total	Male	Female
2	No. of people interviewed	100	62	38
3	Educated till grade 8 (out of 50)	48	34	14
4	Educated till grade 12 (out of 50)	16	12	4
5	Educated till grade (out of 50)	12	10	2
6	Uneducated (out of 50)	24	6	18
7				
8	Working population (out of 50)	76	56	20
9	Unemployed population (out of 50)	24	6	18
10				
11	Ages 1–18 (out Of 50)	14	8	6
12	Age 19–60 (out of 50)	78	48	30
13	Ages 60 + (out of 50)	8	6	22
14				
15	History of chronic diseases (out of 50)	28	16	12
16				
17	Do own a health/ration card (out of 50)	46	22	24

3.3.3 Methods Used

To understand the poverty-environment nexus, different aspects were studied. These include the following:

Offline Survey Conducted

A visit to the Haiderpur slum in Delhi was organized and a survey interview of 100 individuals living in this area was personally conducted. The questionnaire for the interview focused on basic statistics of the population sample and problems faced in their living conditions. We learned about the unique challenges people face and their causes.

Reviewing Pre-Existing Data

The census provided by the Government of India and other sources were studied and reviewed in-depth, analyzing various poverty indicators that the three slums in the sample population may have in common (Fig. 3.3).

Table 3.2 Different parametric values for Delhi and the 3 slums that were studied (using www.academic.oup.com)

Parameter	Delhi	Bhalswa-Jahangirpuri	Paharganj	Seemapuri
AQI				
AQI-US	107	152	58	74
PM 2.5	36	57	35	23
PM10	119	197	53	68
Temperature	33	33	34	34
Humidity	71	71	67	67
Literacy rate				
Total literacy rate	86.21%	77.70%	84.24%	82.39%
Male literacy rate	90.94%	72.40%	78.28%	76.96%
Female literacy rate	80.76%	59.80%	71.21%	65.87%
Groundwater				
pH of groundwater	–	7.4	7	6.30
Nitrate N(mg/l) of groundwater	–	0.3	0.69	0.68
TDS(mg/L) of groundwater	–	346	768	337
Population				
Total slum population	1,67,87,941	1,97,148	1,74,613	5,39,914
Male slum population	89,87,326	1,06,388	92,984	2,86,091
Female slum population	78,00,615	90,760	81,659	2,53,823

Table 3.3 Different parametric values for sanitation in slums [15]

1	Type of facility used	Percentage of households (2019) (%)
2	Flush/pour system	99.31
3	Pit latrine with slab	0.00
4	Hanging toilet	0.26
5	No facilities/Open	0.01
6		
7	Facilities in sharing with other household	Percentage of household (2019)
8	Yes	77.32
9	No	22.68

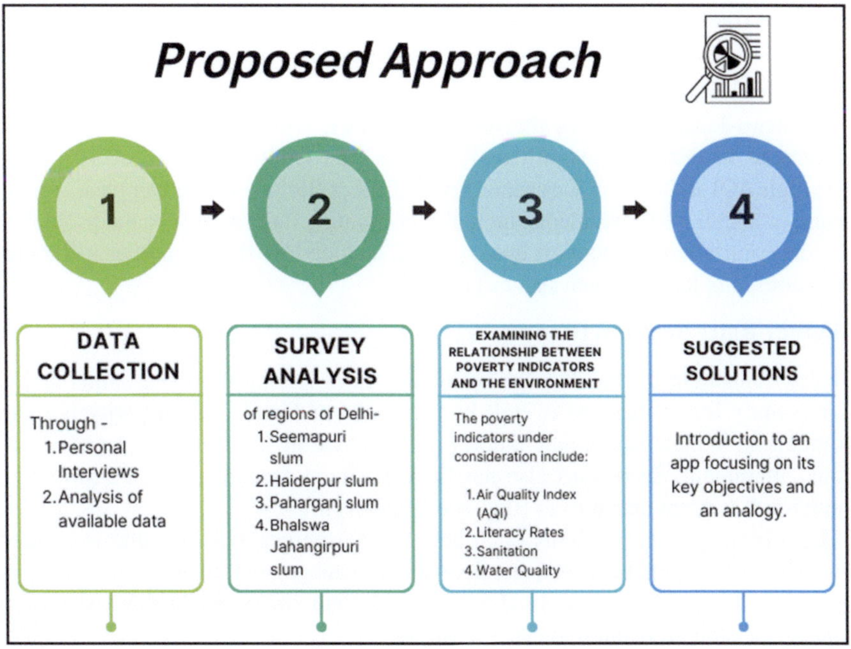

Fig. 3.3 Flowchart explaining the proposed approach

3.4 Result and Discussion

3.4.1 Case Studies and Regional Analysis

The NSS 69th Round Survey by the Delhi Government estimated the total number of slums as 6,343 and the total number of households therein was estimated at 10.20 lakhs. About 29% of the slums had 20–60 households, while the remaining 71% had more than 60 households each. On average, 161 households per slum were found ~~to be~~ in these slums [12].

Seemapuri has the highest population, followed by Bhalswa-Jahangirpuri, with Paharganj having the smallest population among the three slums. The high population density in Seemapuri highlights the need for substantial infrastructure and service provision [13]. Seemapuri's large population indicates a high-density settlement with substantial needs for housing, infrastructure, and services.

3.4.2 Poverty Indicators and Their Impact on Environment

AQI (www.aqi.in)

Bhalswa-Jahangirpuri has the worst air quality, with an AQI of 152 ppm and particulate matter levels significantly higher than Delhi's average of 107 ppm. The high AQI indicates a hazardous level of air pollution, which can have serious health implications for residents, including respiratory and cardiovascular problems. Factors contributing to poor air quality could include industrial activities, vehicular emissions, and lack of green spaces [14].

Literacy Rates in Slums (Census 2011)

Bhalswa-Jahangirpuri has the lowest literacy rates among the three slums and is significantly lower than the overall literacy rates of Delhi. The disparity is more pronounced in male and female literacy rates, indicating a substantial gap in educational attainment. Seemapuri also has literacy rates lower than the Delhi average, though not as pronounced as in Bhalswa-Jahangirpuri.

Low literacy rates in slums contribute to environmental issues due to limited awareness of sustainable practices, poor waste management, overuse of resources, reduced advocacy for environmental policies, and increased health risks from inadequate sanitation and hygiene education.

Sanitation (2021)

The study population in the mentioned rounds overwhelmingly had access to improved sanitation facilities, with over 95% of households reporting access to flush or pour-flush latrines. Very few households reported using open defecation or hanging toilets, indicating a high level of access to improved sanitation infrastructure [15].

Both Bhalswa and Seemapuri suffer from inadequate infrastructure, particularly in water and sanitation. These deficiencies lead to significant health problems, including skin rashes, eye irritation, and malnutrition, particularly affecting children [16].

Sanitation and health challenges in poverty-stricken areas profoundly impact the environment. Improper waste management practices contribute to environmental pollution, affecting air and soil quality. These interconnected issues not only degrade the immediate living conditions but also hinder sustainable development efforts, perpetuating socio-economic disparities in affected communities.

Water Quality of Ground Water (2021) (Cpcb.Nic.In)

Bhalswa-Jahangirpuri and Seemapuri generally have acceptable water quality parameters, with Bhalswa having the most balanced profile. Bhalswa-Jahangirpuri and Paharganj have ideal pH levels, while Seemapuri's water is slightly acidic, which may need attention. All three slums have nitrate levels well below the safe limit. Paharganj has high TDS levels, suggesting the need for water treatment to reduce mineral content.

Furthermore, in Seemapuri, the sample survey reveals that 50% of slum residents use water tankers, 39% use boring water for drinking, while only 10% have private taps in households. The quality of water from tankers is not up to the mark for safety and no filtration or pathogen-killing system is equipped. The area lacks the infrastructure needed for clean water supply and sanitation, resulting in dependence on polluted sources, as noticed by the high TDS.

This contaminated water can cause waterborne diseases, adversely affecting public health and putting additional strain on the already limited healthcare resources in these areas. Hence, poor groundwater quality in slum areas exacerbates environmental degradation and perpetuates poverty.

Problems Inferred from the Haiderpur Survey

The individuals raised following problems observed in their environment:

1. The foul smell and possible diseases from the open dumping ground and the absence of any allotted garbage collection trucks and transport were reported and can lead to poor AQI.
2. The population near the Haiderpur slum was not monitored by the concerned authorities properly, causing a spike in the number of crimes and substance abuse around the area. Many stated that the absence of streetlights may be contributing to the problem stated. These conditions highlight how inadequate infrastructure perpetuates safety concerns and social instability, impacting residents' well-being and reinforcing cycles of poverty.
3. Some of the interviewed women stated their distress about the frequent absence of clean drinking water and electricity in the slum, causing a lot of problems involving sanitation, jobs, education and more.
4. Lack of the Water cleanliness is a major contributor in the observation that 28% of the sample population had a history of chronic diseases. This indicates a direct link between environmental degradation, inadequate infrastructure, and health outcomes in impoverished communities.

3.5 Suggested Solutions

We have created an application to Ease Donations. This will help in creating awareness, crowd funding and ensure that the donations reach the right hands.

Application's Objectives and Goals

The application would broadly focus upon environmental degradation and pollution along with poverty, due to lack of resources in the lower classes. It strives to understand the data on such matters with the help of an Artificial Intelligence algorithm which highlights news based on certain keywords, which can be changed or adjusted.

Goal 1: The project would focus on making an application that collects data and recent news that focuses on the environment and environmental degradation caused by poverty in various parts of the world in a single application and arranges it based on the severity of the matter.

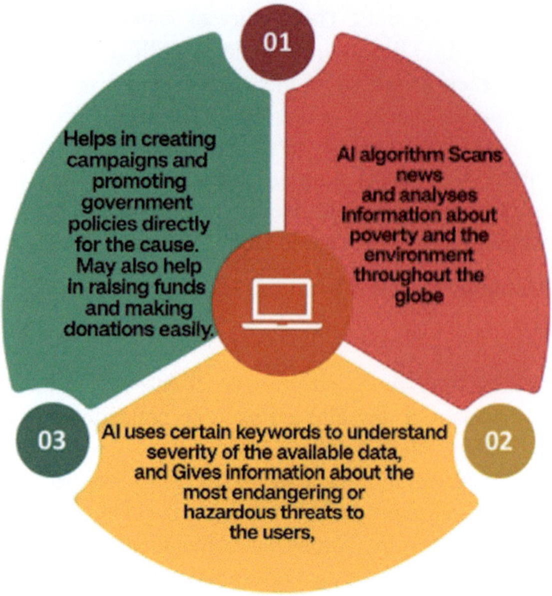

Fig. 3.4 Flowchart explaining proposed analogy for the app

Goal 2: It would create a community of users that are made aware of the prospects of nature and pollution and the struggle of others, especially in economically weaker regions.

Goal 3: It would not only help people in spreading awareness but also in taking action. The application would include direct links to do so and may help in raise more funds as an Artificial intelligence system understand emotional patterns and donating patterns of each user, focusing the news recommended to them accordingly (Fig. 3.4).

Analogy.

3.6 Limitation of the Study

For a better understanding of these dynamics over time, longitudinal studies are needed. The study may not include all relevant climate variables that could influence the poverty-environment nexus. Although the proposed AI-powered application shows promise, its practical implementation and scalability require further exploration. Technology infrastructure, algorithm biases, and data privacy challenges must be addressed.

3.7 Conclusion

The association between poverty and the environment is evident in the slums of Delhi and even highlights the urgent requirement for innovation and sustainable solutions to augment its conditions. Climate changes not only exacerbate the existing challenges such as pollution, poor sanitation, lack of green spaces, increase in spread of disease and more but also worsen them.

We can notice that the distribution of these side effects of this correlation between the two is not even and some poverty-stricken areas might be more affected than others, usually based on various factors like population, percentage of educated population and the location are few to state. The vice versa is also applicable. It is recognizable that improvement in either of the two will aid in solving the other.

The development of an AI-based application, as even proposed in this study, presents a viable solution for enhancing transparency and efficiency of aid distribution by removing unnecessary middlemen. It is important for involvement of technological solutions to solve this issue at hand.

3.8 Future Scope

Future research should emphasize investigating the implementation and scalability of AI-powered early warning systems in low-resource settings, assessing their impact on disaster preparedness and response in vulnerable communities. Additionally, exploring the adoption of sustainable agricultural practices in impoverished regions and their effectiveness in enhancing food security, along with evaluating the socioeconomic benefits of integrating traditional knowledge with modern sustainable techniques, is essential.

Examining the feasibility and impact of transitioning to renewable energy sources in impoverished areas and analyzing the role of green technologies in reducing energy poverty and promoting sustainable development should be prioritized.

Lastly, future research should focus on the role of climate finance mechanisms, including international funding and investment flows, in supporting climate-resilient development pathways and addressing adaptation and mitigation needs in vulnerable communities, alongside strengthening capacity-building efforts and educational programs focused on climate change adaptation, mitigation strategies, and sustainable development practices.

References

1. Oxford Poverty and Human Development Initiative, Multidimensional Poverty Index 2023: Unstacking Global Poverty: Data for High Impact Action, pp. 1–3 (2023)
2. M. Swaminathan, Aspects of urban poverty in Bombay. Environ. Urban. **7**(1), 133–144 (1995)

3. K.U. Ehigiamusoe, M.T. Majeed, E. Dogan, The nexus between poverty, inequality and environmental pollution: evidence across different income groups of countries. J. Clean. Prod. **341**, 130863 (2022). https://doi.org/10.1016/j.jclepro.2022.130863
4. S. Ullah, M.S. Awan, Environmental Kuznets curve and income inequality: Pooled mean group estimation for Asian developing countries. Forman J. Econ. Stud. (2019), https://www.fccollege.edu.pk/wp-content/uploads/Environmental-Kuznets-Curve-and-Income-Inequality-Pooled-Mean-Group-Estimation-for-Asian-Developing-Countries.pdf
5. J. Bucknall, C. Kraus, P. Pillai, Poverty and the environment, in *World Bank* (2000)
6. M.A. Baloch, Danish, S.U.-D. Khan, Z.S. Ulucak, A. Ahmad, Analyzing the relationship between poverty, income inequality, and CO_2 emission in Sub-Saharan African countries. Sci. Total. Environ. (2020), https://www.sciencedirect.com/science/article/abs/pii/S0048969720333878
7. M. Haroon, Energy poverty in the face of stringent environmental policies: an analysis of mitigating role of energy storage in China. J. Energy Storage (2024). https://doi.org/10.1016/j.est.2023.110396
8. R.N. Yadava, B. Sinha, Energy-poverty-climate vulnerability nexus: an approach to sustainable development for the poorest of poor. Environ. Dev. (2022)
9. N. Sharma, S. Taneja, A. Bhatt, Empirical analysis of life quality based on air pollution in states of India. J. Stat. Manag. Syst. **23**(7), 1213–1226 (2020). https://doi.org/10.1080/09720510.2020.1799579
10. N. Sharma, S. Taneja, B. Suri, S. Kaur, A. Bhatt, Photochemical smog in Delhi: impact, analysis and future trends. J. Chem. Health Risks **14**(2), 2935–2944 (2024)
11. NITI Aayog, National Multidimensional Poverty Index: A Progress Review 2023, Government of India (2023)
12. S. Ali, Urban slums in Delhi. Directorate of Economics and Statistics and Office of Chief Registrar (Births and Deaths), Government of NCT of Delhi (n.d.), https://des.delhi.gov.in/sites/default/files/urban_slums_in_delhi.pdf
13. Census India, Delhi population census 2011, Delhi religion, literacy, sex ratio (2011), https://www.censusindia.co.in/states/delhi
14. AQI, New Delhi air quality index (n.d.), https://www.aqi.in/dashboard/india/delhi/new-delhi
15. S. Kanungo, P. Chatterjee, J. Saha, T. Pan, N.D. Chakrabarty, S. Dutta, Water, sanitation, and hygiene practices in urban slums of Eastern India. J. Infect. Dis. **224**(15), S573–S583 (2021). https://doi.org/10.1093/infdis/jiab354
16. Central Pollution Control Board (CPCB), CPCB | Central Pollution Control Board (2023), https://cpcb.nic.in/nwmp-data-2021/

Chapter 4
Sustainable Nutrition: The Environmental Benefits of Vegan and Vegetarian Choices

Nidhi Sharma, Oshi Sharma, Tanisha Sharma, and Mudita Nagpal

Abstract The production of animal-based foods has a huge impact on the environment, especially in terms of greenhouse gas emissions, overgrazing of pasture land by livestock. This calls for the analysis of the environmental sustainability of whole diets, which includes plant-based foods versus meat-based foods and its health benefits which is yet to be fully explored. Fewer studies have been conducted on the regional impact of varying types of animal-derived foods consumed by people. The present study aims to look at the environmental impact of the diets of a specific group of consumers, including those who eat meat, vegetables, and vegan foods. We have surveyed the sample population across different age groups, regions and income groups segments and identified the different diet consumption patterns. Countrywide analysis of food consumptions has been conducted. Animal products, like milk and dairy, were the biggest contributors to the environmental footprint in both meat-eaters and vegetarians. The study shows that switching to a plant-based diet can greatly reduce the carbon footprint on a vegan or vegetarian diet. A model to show the calories and carbon footprint on the given food has been worked upon to guide consumers to plan healthy and environment friendly diets.

Keywords Sustainability · Greenhouse gas · Plant-based diet · Carbon footprint · Meat substitutes · Dietary consumption patterns · Vegan and vegetarian diets

N. Sharma · O. Sharma (✉) · T. Sharma · M. Nagpal
School of Engineering and Technology, VIPS-TC, Delhi, India
e-mail: 06617702723_cse@vipstc.edu.in

N. Sharma
e-mail: nidhisharma@vips.edu.in

T. Sharma
e-mail: 09617702723_cse@vipstc.edu.in

M. Nagpal
e-mail: mudita.nagpal@vips.edu

A. Kumar et al. (eds.), *Selected Proceedings of the 1st International Conference on Advanced Materials for Sustainable Innovation; IC-AMSI 2024; 28–30 August; New Delhi; India*, Springer Proceedings in Energy,
https://doi.org/10.1007/978-981-96-4492-6_4

4.1 Introduction

Veganism refers to the practice of averting oneself from including meat and dairy, along with derived products from one's diet and daily routine. Veganism offers a potential solution to global food scarcity due to reduced requirement in terms of resources and energy with a wider availability. SDG 12 (Responsible Consumption and Production), natural resources are more sustainably utilised in a vegan lifestyle. SDG 3 (Good Health and Well-being). Veganism allows the body to reach better health with lesser risks involved. SDG 13 (Climate Action). Farming of animals leads to degradation of environmental health, biodiversity loss and prevention of future dangers to habitats can be affiliated with veganism [1].

In recent times, a steady increase has been observed in adoption of veganism across the globe. This may also be tracked to the Go-Vegan Movement that has spiked in India as well. Adopting a vegan lifestyle helps greenhouse gas emissions, saves water and prevents deforestation. By sticking to a diet that excludes animal products, we can reduce our impact on the environment and combat climate change to promote a fair world.

4.1.1 Plant-Based Diets in India

India, as a nation, mostly presents its citizens as vegetarian rather than completely vegan. This stems from various connotations of what people perceive as health–promoting consumables. These may vary as:

- Regional/Geological Restrictions: Areas with less vegetational variety promote people to look for non-vegetarian alternatives.
- Cultural and Religious Perception: Indian diversity involves outlooks that instil beliefs about the stance of veganism and its benefits. For instance, followers of Buddhism and Jainism have advocated and practised a form of veganism in India for centuries, with the belief that humans should not inflict pain on other voiceless animals [2].
- Sensory: A dislike of particular aspects of food choices like texture, taste and colour among others.
- Financial: The extent of cost-efficiency has acted as a major influence,
- Justice and World Hunger: Availability and social concerns have been found to be another factor of consideration in dietary choices
- Social Networks: Veganism has gained popularity and spread into the world lately. With connections and more discussions being conducted, people have been gaining knowledge of benefits to be derived from the same. Thus, people have been shifting to this way of life.

4.2 Objective of the Study

Carbon footprints pertaining to the industries like meat, dairy and poultry among others have been particularly prevalent and rising in comparison to the vegan industry and so, it is important to consider creating an impact at any possible level to promote a smooth transition into veganism and similar options to help reduce the degradation of environment along with improving personal health with what our own soil has to offer. We have concluded analytical reports on different levels to highlight how carbon footprints pertaining to the industries like meat, dairy and poultry among others have been particularly prevalent and rising in comparison to the vegan industry. Our program can help suggest better vegan alternatives for a healthier and environment-friendly lifestyle as reflected by the results of our research comparing what vegetarian and non-vegetarian diets can contribute to individually. It is important to consider creating an impact at any possible level to promote a smooth transition into veganism and similar options to help reduce the degradation of the environment along with improving personal health with what our own soil has to.

4.3 Carbon Footprints and Involvement

Carbon Footprints measure human impact on the environment in terms of heat- trapping gases released like carbon dioxide for a quantitative measure of consequences of our day to day life. Our diet and how our food is grown influences not only us but our surroundings as well. It leads to depletion and deterioration of precious resources and effluent discharge (Fig. 4.1).

4.3.1 Greenhouse Gas Emissions

The FAO's data unveils that almost two out of every ten units of greenhouse gas emissions are released due to animal farming and processes involved. Some herbivores, while digesting plant constituents, break down the matter into harmful warming greenhouse gases, mainly methane, which decompose into worse products over time. The grass and crops grown for cattle may be grown by fertilisers or other chemicals for faster production, introducing gases like nitrous oxide, another strong climate-warming gas [3].

Researchers at University of Oxford claim a deduction of up to 73% in Carbon footprints traced to livestock and issues conjoined to animal products derived [4].

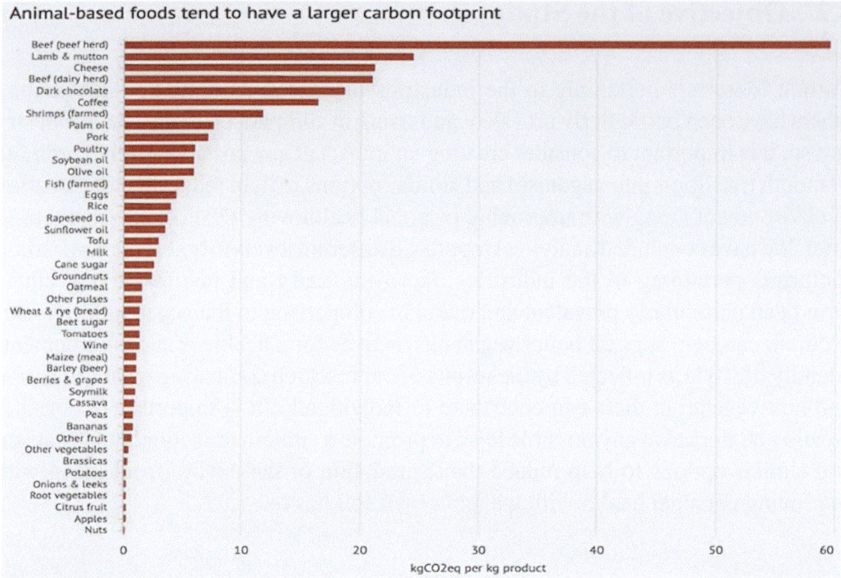

Fig. 4.1 Different animal products produce different carbon footprint levels. (*Source* Hannah Ritchie. Pablo Rosado.Max Roser.2022. "Environmental Impacts of Food Production" Published online at (OurWorldInData.org)

4.3.2 Water Resource Depletion

Water is required for enhancing production facilities of meat. Moreover, Farmers generally raise more cattles than feasible for a higher profit margin. This also brings up the extra requirements for livestock.

4.3.3 Energy Consumption

Significant energy is put to use in the meat industry. Emissions from manure management, rice cultivation, burning of crop residues and fuel use in agriculture must be recognised.

4.3.4 Waste Disposal

Waste management and disposal becomes a problem with the expansion of livestock rearing. It produces a high volume of biological waste rich in pathogens, enzymes and higher water content in raw materials [5]. Even though it is completely biodegradable it may pose a problem if improperly managed.

4.3.5 Land Resource in Great Demand

Deforestation is a concern due to the large expanse of land needed to raise animals for the market in profitable quantities. Furthermore, the forests cleared to clear more land for livestock reduces the carbon storage by 8%. This ultimately affects quality of air, water seepage and soil retention along with other impacts of deforestation. Longer breeding cycles and lifespan contribute to a higher demand for feedstock and land to meet the industry needs.

4.3.6 Animal Cruelty, Chemical Use and Biodiversity Risk

Biodiversity and habitat loss are profound with clearance of lands to feed the need for resources. This is further leading to the extinction of many species. Animal husbandries invite risks of deadly epidemics as these animals may become storage houses for various viruses and bacteria which multiply within the body. Coupled with the concept of trophic levels, these can be very efficient in spreading multiple diseases. Animal cruelty is observed with tightly packed areas that limit the animal's movement from transport to slaughter which is hard to deconstruct with these practices. Places that raise animals for agriculture are packed with harmful chemicals that are not ideal and often dangerous.

4.4 Related Work

A study on farms from 119 countries conducted by lead researcher Peter Scarborough, at Oxford University revealed that vegan diets resulted in a reduction of 75% in terms of greenhouse gases, water pollution and land utilization [6]. A pilot study conducted by human nutrition experts in Poland showed that vegan diets were responsible for 64% lower carbon footprint. This study also took into account the 60.9% reduction in land use and 62.9% less water footprints [7]. According to a study by researchers from the Stanford Woods Institute for the Environment, once a country makes a shift from imported to domestic products, the environment pays the price with its amenities being compromised [8]. The Good Food Institute studied the impact of meat consumption and found that bacteria from livestock being raised in large numbers was polluting agricultural areas and spreading viruses affecting the health of many meat consumers [9]. The UN's Intergovernmental Panel on Climate Change in 2022 touched upon how plant-based diets can reduce greenhouse gas emissions. The Center for Sustainable Systems at the University of Michigan in 2020 proposed that switching even 50% of animal produce for vegan products could result in a 35% decrease in dietary GHG emissions by 2030 [10].

4.5 Methodology

The study is carried out to compare factors and effects concerning the environment and human health in terms of dietary choices. Both indirect and direct surveys followed by a project application have been carried out.

For indirect analysis, we collected and analysed the data from records stating the trends of veganism and impact of the meat industry on different regions. Internationally, we have selected the USA and the UK as countries that are predominantly non vegetarian. Nationally, Rajasthan and West Bengal are being selected for comparative study as these states of India as one is predominantly regarded as vegetarian and other as Non vegetarian state. For direct survey, we successfully conducted a direct survey concerning information regarding people's diet within a sample population of 200 people. The responses have been put together to achieve a final outlook of majority and minority choices and expanding on the reasons for the same.

Proposed program: Compiling the results from the above segments, we have proposed a project for detection of calories and carbon footprints induced in the production of particular food choices. This allows us to introduce alternatives for the results obtained from the input by the user in accordance to better alternatives objectively (Fig. 4.2).

4.6 Indirect Analysis by Reviewing Pre-Existing Studies and Datasets

4.6.1 Comparative Analysis of Veganism as a Dietary Choice in USA and UK

See Table 4.1.

4.6.2 Comparative Analysis of Veganism as a Dietary Choice in Rajasthan and West Bengal

See Table 4.2.

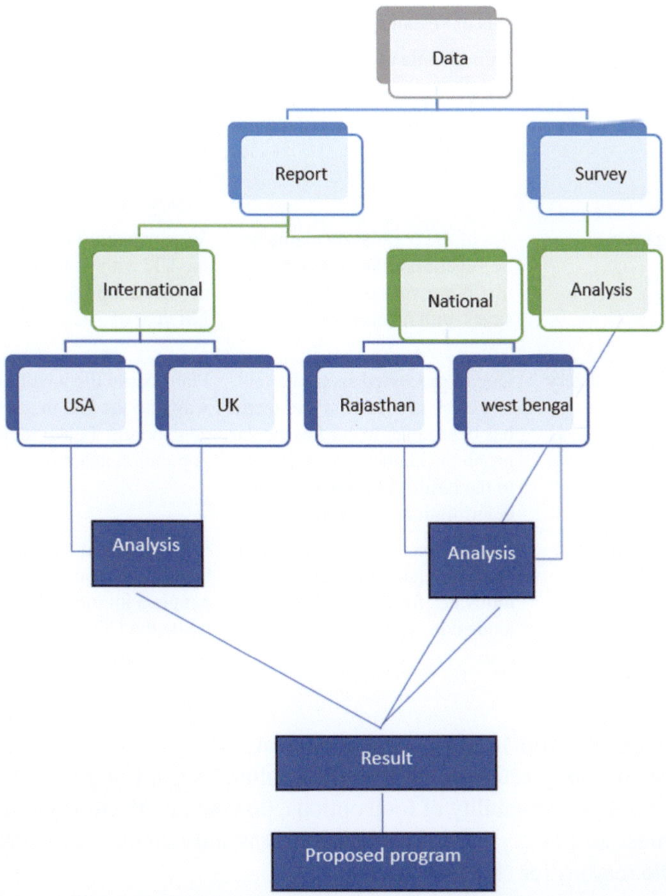

Fig. 4.2 Flow of the study and progression into conclusion

4.7 Direct Survey Analysis

To understand the personal preferences of people and the motivation behind the choices to analyse future prospects of veganism, direct surveys in the form of personal interviews have been conducted of a sample population (120 in number) of multiple age groups for reliable results.

4.7.1 Diets Followed

It is to be observed that almost 75% of the responses consist of people following a vegetarian diet along with a quarter of the subjects pertaining to meat- based diets

Table 4.1 Comparative analysis of veganism as a dietary choice in two leading countries [11–15]

Basic of distinction	United State of America	United Kingdom
Population constituents	USA is considered to be majorly the meat-consuming population in the world	The United Kingdom being a contributor the meat industry is now adopting rather swiftly to healthier and more sustainable dietary choices
State of veganism	5% of the population claims to be vegetarian. Veganism as a diet is not very popular in the states with quantity at around 4%	Veganism has been on the rise in the UK with an observable spike of 78% adopting the diet over the last year of 2023 itself
Consequences of dietary choices	Due to processed saturated fats and cholesterol in the diet there has been spike in cardiovascular problems, diabetes, cancer due to the harmful byproducts and deterioration of moratality rates	Plant-based diets will find their way into the UK much rapidly considering the willingness of the population adapt to alternatives
Reason for switching to plant-based diets	Health concerns and food trends have emerged as major motivates for diet of choice alongside animal welfare	Environmental and health benefits have motivated the people a lot into making a shift while the COVID pandemic has also pushed them for the same

varying in regularity with health as a major motivator alongside religious and cultural aspects due to strong religious beliefs. The cultural regard aligns with regional constraints with the availability of food options, flowing into the next generations as a cultural lifestyle. Further, animal welfare concerns and economic structure are also considerable reasons for choice.

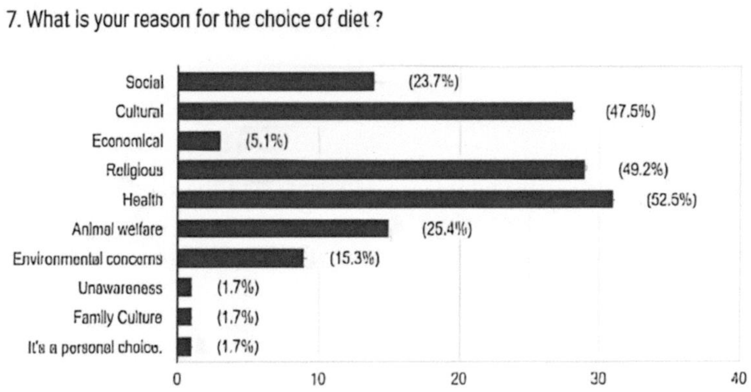

Table 4.2 Comparative analysis of dietary choices in two states of India [16–18]

Basic of distinction	Rajasthan	West Bengal
Population and preferences	About 75% of people follow a plant-based diet	98% of people are regular consumers of livestock
Religious factors and relevance	88.49% of Rajasthani follow Hinduism as religion of choice which adds to people following a plant-based diet with respect to religious boundaries	Religious connotations are different for this part of the subcontinent and not strict with the food intake of followers allowing them freedom of non-vegetarian options
Cultural and historical influence	The diet becomes a part of representing their culture and community	Meat and fish are vigorously served to deities like Kali and Durga. Such traditions have prevailed over years of history
Geographical restrictions	The hot and heavy climate of the state does not ideal conditions to raise animals coupled with water scarcity which diverts the focus to vegetation	Rivers are abundant in the state like the Ganges teeming with fish. This provides large water bodies and fertile lands to raise fish for most of the recipes
Major contributor to choice of diet	Faith is a big motivator in this state. However, the meat industry can find a place with advancements in technology and people altering historical culture and tradition to match other requirements	Heavy influence of regional constraints can be observed in West Bengal. It proves why a plant-based diet would deem unsuccessful in terms of convenience, economic loss and traditional strains

4.7.2 *Awareness and Willingness*

In terms of awareness, the response had been quite positive, with 80% having encountered veganism to be knowledgeable about the term and what it entails. With about 75% of the respondents agreeing that a vegetarian diet is a more proficient choice, 50% are considering making the shift towards it actively while 37.5% have chosen not to.

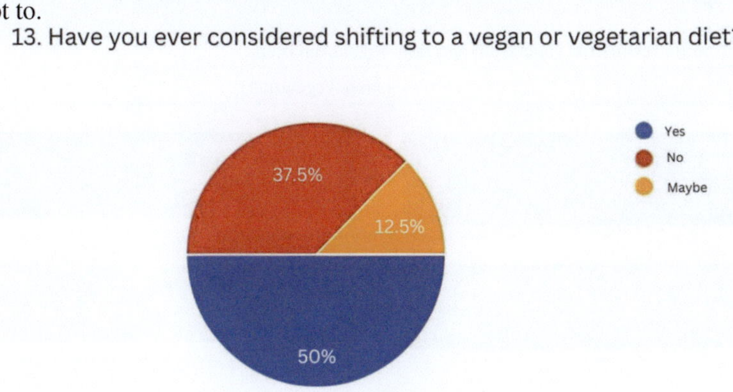

4.7.3 Challenges Faced

When questioned about the challenges towards following these diets, a little more than 50% of the survey attendees have claimed concerns about not receiving enough nutrients from such diets as their answer. It is indeed true that few nutritional requirements like Vitamin B12, Vitamin D, proteins, calcium and others may be hard to inculcate into the diet if not planned strategically. However, alternatives are easy to look into for these needs as shall be discussed later in the HEALTH ALTERNATIVES section.

4.7.4 Veganism in India: Present and Future

When questioned about the feasibility of a vegan diet in India, 62.7% of the participants suggest that veganism in India can be a successful sector in India over the next 25 years, about 42.4% have expressed that it can be a challenging process indicating difficulty in adapting to the changes. On the other hand, a total of 47.4% consider the change to be somewhat easy considering the willingness.

4.8 Proposed Program

In order to look for solutions, the initial problem must be dealt with as a priority. For a better lifestyle, it becomes very important to neutralise threats with healthier alternatives. Keeping this in mind, a program has been created to allow the user to predict the calories in their food items and the carbon footprints caused by it. This will make it easier to weigh out their options in terms of their meals. We have used the project to input vegetarian and non-vegetarian ingredients to provide a clear comparison between the outputs of the categories (Table 4.3).

Table 4.3 Program results for vegetarian and non-vegetarian diets

Vegetarian	Non-Vegetarian
1. Enter food name : Rice Food: Rice Calories (per 100g): 130 kcal Carbon Footprint (per 100g): 2.70 kg CO2	1. Enter food name : Beef Food: Beef Calories (per 100g): 250 kcal Carbon Footprint (per 100g): 27.00 kg C
2. Enter food name : Carrot Food: Carrot Calories (per 100g): 41 kcal Carbon Footprint (per 100g): 0.02 kg C	2. Enter food name : Chicken Food: Chicken Calories (per 100g): 239 kcal Carbon Footprint (per 100g): 6.90 kg CO2

The results imply and further elucidate the fact that non-vegetarian food products are significantly higher in calorie count and carbon emissions when put against safer vegetarian produce in the same amounts. The logic of the code proceeds as follows:

1. Take the name of the food item as input from the user.
2. Search for the given food item in a predefined array of food items.
3. If found, the food's details are displayed as calorie intake and carbon footprint released by the same.
4. If not found, the user is informed.
5. The code repeats itself and runs until the user enters "quit" to exit the program.

4.9 Future Prospects

A 2023 study from Nature Communications found that halving meat production could potentially decrease land and agricultural emissions by 31% by 2050 while also supporting food security. Science Advances report from this year, found that a global transformation to a vegan diet by 2050, would limit global warming up to 1.5 °C possibly. By 2030, with more influence of veganism the fall in air pollution could halt an estimated 236,000 premature deaths per year [10]. This entails that with appropriate and responsible measures our carbon footprints can be limited to a safer range if we make careful choices for our diet.

4.10 Health Alternatives

While a non-vegetarian diet provides some nutrients more than plant-based diets like proteins and some vitamins, it also introduces risks like poor heart health, cancer risk, hormonal imbalance, higher antibiotic inefficiency, low energy diffusion, risk of diseases and pandemics, obesity, diabetes, hypertension and hypercholesterolemia, pneumonia, early death and more. A vegan diet can reduce these problems due to lower body weight, blood pressure regulation, safer cholesterol levels, and vitamins observed in vegetarians and vegans [19, 20] (Table 4.4).

4.11 Conclusion

Livestock rearing and the whole process of meat production plays a huge role in the society's carbon footprints which is quite significant in terms of the numbers.

Veganism, if adopted in the lifestyle can counter the issue of increasing carbon emission in the atmosphere to a greater extent, it has gained popularity in the recent years for varying reasons as discussed above whether it be for health or environmental concerns, following latest trends, regional allowances, and traditional cultural

Table 4.4 Vegan alternatives for respective nutrients [21]

S.No	Vegan Alternatives	Calories per 100 g (Approx)
Protein		
1	Tofu	76 cal
2	Tempeh	193 cal
3	Edamame	121 cal
4	Black beans	132 cal
5	Chia seeds	486 cal
6	Hemp seeds	553 cal
7	Seitan	370 cal
Zinc		
1	Lentils	116 cal
2	Chickpeas	164 cal
3	Beans	127 cal
4	Sunflower seeds	584 cal
Iron		
1	Pulses	116 cal
2	Wholemeal bread and flour	247, 340 cal
3	Breakfast cereals fortified with iron	360 cal
4	Dark green, leafy vegetables, such as watercress, broccoli and spring greens, nuts, dried fruits, such as apricots, prunes and figs	Dark green, leafy vegetables, such as watercress: 11 cal, broccoli: 34 cal and spring greens: 32 cal, nuts: 607 cal, dried fruits: 243 cal, such as apricots, prunes and figs
Vitamin B12		
1	Unsweetened soya drinks fortified with vitamins B12	33 cal
2	Yeast extract, such as Marmite	185 cal
3	Nutritional yeast flakes which are fortified with vitamin B12	325 cal
Omega		
1	Ground linseed (flaxseed) oil	884 cal
2	Vegetable (rapeseed) oil	884 cal
3	Chia seeds	486 cal
4	Shelled hemp seeds	553 cal

faith. However, there is still a long way to go to make a more sustainable future for ourselves and our planet. It can only happen when society is sensitized about the benefits of vegan diet in terms of Environmental concern and sustainable development of the society. As indicated in our program, an Artificial Intelligence application can play a key role in designing the vegan meal plan as per the individual requirements and suggest the alternative options rich in nutrients and also caution the public regarding the diet restriction as per the health requirements. Community building and partnership can play a key role in achieving that objective.

References

1. Veganism, Sustainable Development Goals—Resource Centre. RELX SDG Resource Centre. https://sdgresources.relx.com/veganism
2. G. Salehi, E. DÃaz, R. Redondo, Forty-five years of research on vegetarianism and veganism: a systematic and comprehensive literature review of quantitative studies. Heliyon **9**(5) (2023)
3. D. Dunne, What is the climate impact of eating meat and dairy? Carbon Brief (2020), https://interactive.carbonbrief.org/what-is-the-climate-impact-of-eating-meat-and-dairy
4. How Much Does Being Vegan Reduce My Carbon Footprint By? Zero Smart (2022)
5. K. Jayathilakan, K. Sultana, K. Radhakrishna, A.S. Bawa, Utilisation of byproducts and waste materials from meat, poultry and fish processing industries: a review. J. Food Sci. Technol. **49**(3), 278–293 (2012). https://doi.org/10.1007/s13197-011-0290-7
6. P.C. Scarborough, M. Cobiac, Vegans, vegetarians, fish-eaters and meat-eaters in the UK show discrepant environmental impacts. Nat Food **4**, 565–574 (2023). https://doi.org/10.1038/s43016-023-00795-w
7. R. Góralska-Walczak, K. Kopczyńska, R. Kazimierczak, L. Stefanovic, Environmental indicators of vegan and vegetarian diets: a pilot study in a group of young adult female consumers in Poland" of researchgate **16**(1), 249 (2023). https://doi.org/10.3390/su16010249
8. Cassandra Brooks (2011) *Meat's Environmental Impact*, Stanford Woods Institute for the Environment, https://woods.stanford.edu/news/meats-environmental-impact
9. I. Emery, Meat's Sustainability Problem. POL-FS-ENVIRON ANIMAL AG-2018–0824 (2018), https://gfi.org/images/uploads/2018/10/AnimalAgEnvironment.pdf
10. J. Norris, *Going Vegan Reduces Your Environmental Impact*, Vegan Outreach (2024), https://veganoutreach.org/environment/
11. Statista Research Department, Veganism and vegetarianism in the United States—statistics and facts. Statista (2023)
12. The Vegan Society. Worldwide growth of veganism, https://www.vegansociety.com/news/media/statistics/worldwide
13. D. Snyder, *Eating Meat in America | The Pursuit*, (University of Michigan School of Public Health, 2018). https://sph.umich.edu/pursuit/2018posts/eating-meat-in-america.html
14. N. Wunsch, Veganism and vegetarianism in the United Kingdom—statistics and facts. Statista (2024)
15. A. Mridul, *5% of Brits are Vegan—But Their Understanding of Climate Language is Low*. (Green Queen, 2024), https://www.greenqueen.com.hk/number-of-vegans-in-uk-survey-climate-change-language/
16. S. Khamrai, Why does Rajasthan have the highest number of vegetarians in India? Quora (2018), https://www.quora.com/Why-does-Rajasthan-have-the-highest-number-of-vegetarians-in-India
17. Rajasthan Hindu Muslim Population, Population Census (2011), https://www.census2011.co.in/data/religion/state/8-rajasthan.html
18. Times of India, West Bengal second in non-veg consumption, Kolkata News (2016)
19. "Meat Consumption: Trends and Health Implications" Johns Hopkins Center for a Livable Future, https://clf.jhsph.edu/projects/technical-and-scientific-resource-meatless-monday/meatless-monday-resources/meatless-monday-resourcesmeat-consumption-trends-and-health-implications

20. J.H. O'Keefe, E.L. O'Keefe, C.J. Lavie, L. Cordain, Debunking the vegan myth: the case for a plant-forward omnivorous whole-foods diet. Sci. Direct **74**, 2–8 (2022). https://doi.org/10.1016/j.pcad.2022.08.001
21. NHS, The Vegan Diet (2022), https://www.nhs.uk/live-well/eat-well/how-to-eat-a-balanced-diet/the-vegan-diet/

Chapter 5
Analyzing the Need and Viability of Sustainable Fashion as a Pathway to Decarbonization

Nidhi Sharma, Mudita Nagpal, and Tanuja Nautiyal

Abstract Environmental cost due to textile industry is rising with ever growing demand in apparel sector. The textile industry, ranked as the sixth most environmentally damaging globally, is facing heightened scrutiny due to its expanding role in the fashion sector. The surge in consumerism, coupled with the rapid growth of the apparel industry, has led to the entry of both domestic and international clothing brands in India. This surge is notably driven by the prevalence of fast fashion trends and the increasing purchasing power of individual households. Fast fashion operates on large-scale production of inexpensive garments designed for short-term use, contributing to significant environmental repercussions. The consequences manifest in various forms, including substantial environmental impact, escalating waste generation from discarded clothing, discharge of harmful dyes, excessive water resource exploitation, and a growing carbon footprint. This paper delves into the environmental impact attributable to the textile industry, exploring sustainability practices within the textile and fashion sectors. Emphasis is placed on utilizing renewable feedstocks as raw materials and leveraging cutting-edge technology for proactive production planning. Additionally, the study compares per capita textile waste across different countries, shedding light on the factors contributing to India's comparatively lower per capita textile waste generation. Furthermore, the article presents case studies highlighting emerging eco-friendly fabric alternatives such as hemp, bamboo, and recycled fabrics. The conclusion suggests that sustainable fashion not only mitigates adverse environmental effects but also paves the way for developmental and innovative opportunities through the exploration of new products and designs.

N. Sharma · M. Nagpal (✉)
School of Engineering & Technology, Vivekananda Institute of Professional Studies-Technical Campus, Delhi, India
e-mail: mudita.nagpal@vips.edu

T. Nautiyal
Dr. Akhilesh Das Gupta Institute of Technology & Management, Delhi, India

A. Kumar et al. (eds.), *Selected Proceedings of the 1st International Conference on Advanced Materials for Sustainable Innovation; IC-AMSI 2024; 28–30 August; New Delhi; India*, Springer Proceedings in Energy,
https://doi.org/10.1007/978-981-96-4492-6_5

Keywords Sustainable · Textile · Waste generation · Renewable feedstock · Circular economy · Carbon footprints

5.1 Introduction

Textile and apparel industries have shown exponential growth owing to the latest trend of fast fashion which is being followed globally also. In the year 2019–2020, the Indian Textile and Apparel sector held a valuation of USD 108.5 billion, constituting 2% of the overall Gross domestic product, 12% of value of exports, and 7% of the value of industry output. During the period from 2010 to 2018, the industry experienced a compound growth rate of 13.8% [1]. By 2023, the valuation of the Indian textile industry has reached USD 226 billion (Fig. 5.1). Conventionally also this industry is considered to be an employee generator with more than 18 million people directly and 20 million people indirectly employed and contributing 7% to industrial production as of 2021 [2].

The Indian textile industry is mainly cotton based, with 65 percent of the cloth manufacture in the country being accounted for by cotton fabric. The production of raw cotton varies year to year depending upon weather conditions resulting in price fluctuations and thus affecting the industry [3].

In addition to its role as a manufacturing center, India is emerging as a significant consumer of textile products consumer, reaching a value of USD 75 billion in domestic consumption. This consumption is anticipated to rise between 2015 and 2025 due to economic growth, bringing an additional 118 million households into upper class in India. The surge in consumption is evident in the overall Indian spending on clothing, which escalated from USD 24.65 billion in 2010 to USD 69.3 billion in 2018. The income growths and shifts in patterns of consumption in India have fueled a thriving retail sector in the country. India is among the top five global exporters of numerous categories of natural fibres, yarns etc. In 2021, the Textile Industry accounted for 2% of India's GDP and constituted 12% of the country's total exports for that year [4].

The expansion in both production and consumption, however, results in a rise in the volume of textile waste generated. In India, textile waste is predominantly

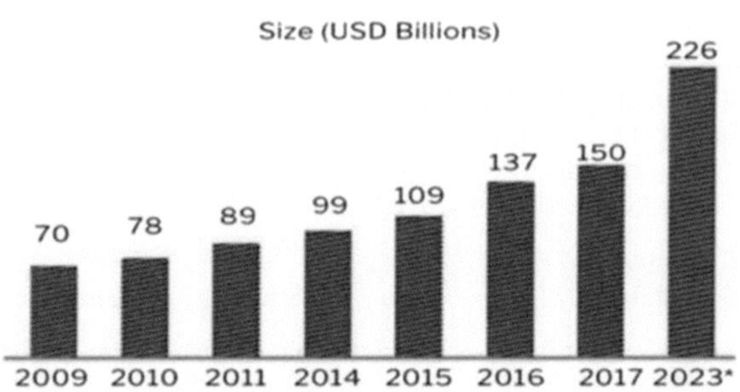

Fig. 5.1 Growth in the Indian apparel and textile industry [1]

generated through three streams: (1) pre-consumer waste, which includes the waste produced before the final products reach the market, including waste from spinning, trimmings from fabric and garments and fabric left unsold; (2) domestic post-consumer waste, involving products thrown out by consumers in domestic market; and (3) the imported waste stream, which comprises used clothes and rags brought to Indian market from other countries [5]. The consequences of the growing textile waste include environmental degradation and strain on waste management systems. Addressing the mounting waste crisis can be achieved through the recycling of textile waste and moving towards more sustainable options for fabrics. This paper delves into the environmental impact attributable to the textile industry, exploring sustainability practices within the textile and fashion sectors. Furthermore, the article presents case studies highlighting emerging eco-friendly fabric alternatives such as hemp, bamboo, and recycled fabrics.

5.2 Methodology

This paper explores the base of Indian textile industry vis a vis its burgeoning market in India and how it influences the country's economy. Furthermore, we have touched upon the consequences faced by the environment due to the apparel and textile industry and the way its unsustainable growth has impacted water resources and resulting in an increase carbon footprint. The various methods by which sustainability can be introduced in apparel industries have been discussed and reviewed upon. Moreover, the research presents the growth in the per capita textile waste generated in India in the last decade. Further, a comparison between the per capita textile waste among various nations has been made and the factors that contribute to India's relatively lower per capita textile waste generation have been highlighted. Lastly, sustainable clothing practices and renewable feedstock as raw material has been worked upon in study. The empirical study is based on the statistical data of Ministry of Textiles, government of India and other statutory bodies related with it. The objective of this research paper is to suggest the ways to make apparel sector sustainable and profit making as well by using environmentally conscious fashion rather than conventional synthetic fast fashion which causes water pollution, solid waste pollution as well as contribute towards micro plastic pollution.

5.3 Textile Industry and Pollution It Causes

5.3.1 Fast Fashion and Overconsumption

Today, the lack of adequate access to energy supplies and efficient transformation systems remains Fast fashion refers to the quick turnover of clothing collections to meet the latest trends. This leads to accelerated production cycles and a constant influx of new items into the market. Fast fashion brands often prioritize low production costs, achieved through outsourcing manufacturing to countries with lower labor costs and using inexpensive materials. The focus on trends and low-cost production results in garments with a short lifespan [6]. Items may go out of style quickly, leading consumers to discard them in favor of the latest trends. According to a report by Mckinsey, Apparel companies have strategically managed their supply chains and minimized the costs. Thus, the price of clothing has risen very slowly as compared to other consumer goods which has led to drastic rise in apparel sales [7] (Fig. 5.2).

Overconsumption in the fashion industry is characterized by consumers purchasing a high volume of clothing items, often impulsively and driven by changing trends. The combination of low prices and rapid turnover encourages a "wear and discard" mentality, where clothing is treated as disposable rather than durable goods. Constantly changing fashion trends drive consumers to buy new items regularly, even if their existing wardrobe is still in good condition [8].

The disposable nature of fast fashion results in a massive amount of textile waste. The waste produced by the textile industry includes liquid and air pollutants, and solid wastes. The lengthy process sequence and high resource consumption in the textile manufacturing process leads to significant waste production. Fast fashion and overconsumption contribute to the depletion of natural resources, such as water, energy, and raw materials used in clothing production [9].

5.3.2 Environmental Impact

A significant impact on the environment is caused by the fashion industry across various levels of its supply chain, including production of raw material, manufacturing, distribution, and disposal (Fig. 5.3). Some key environmental concerns associated with various stages of textile production include:

Raw material production

Conventional cotton farming often involves the use of synthetic fertilizers and pesticides, leading to water and soil pollution. The leather industry is associated with deforestation, as well as the use of toxic chemicals in tanning processes [10].

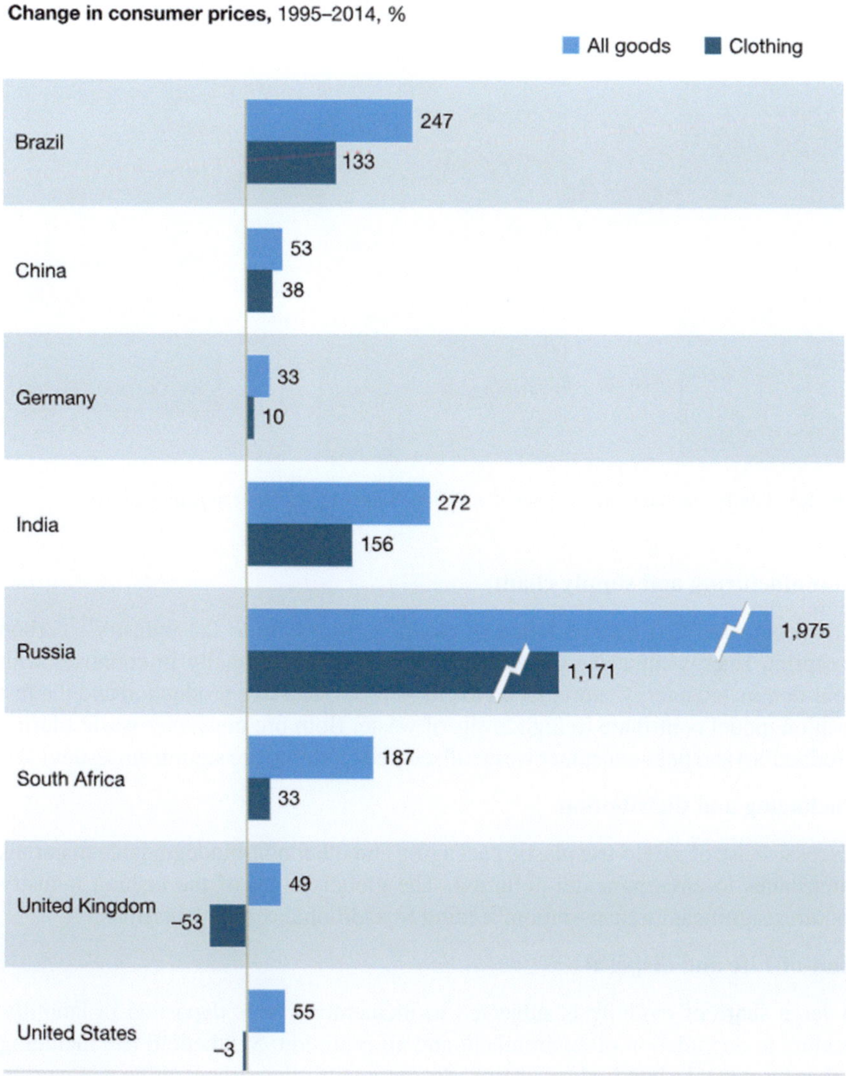

Fig. 5.2 Comparison of rise in clothing prices with other consumer goods in various countries [7]

Textile production

Large water quantities are required in textile dyeing and finishing processes. This releases toxins into water bodies, leading to pollution. The energy-intensive nature of textile manufacturing, particularly in synthetic fiber production results in emissions of greenhouse gases [11].

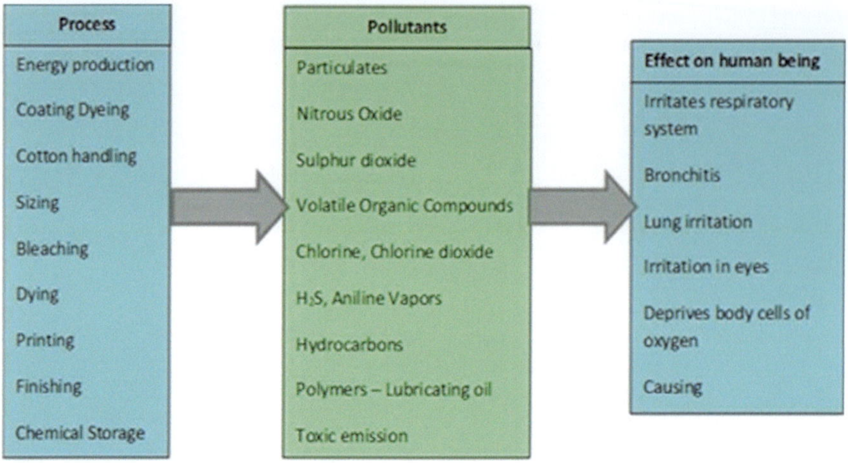

Fig. 5.3 Environmental pollution caused at various stages by textile & apparel industry

Manufacturing and supply chain

The production and transportation of clothing contribute to the industry's carbon footprint. Energy-intensive manufacturing processes, especially in countries with coal-dominated energy sources, add to emissions. Excessive production and the fast fashion model contribute to high levels of waste. Both pre-consumer waste (during production) and post-consumer waste (discarded clothing) are significant issues [12].

Packaging and distribution

Excessive use of single-use plastic packaging and other non-biodegradable materials contributes to environmental pollution. The global nature of the fashion industry requires significant transportation, leading to additional carbon emissions.

End-of-Life and disposal

A large share of clothing is subjected to incineration or is deposited in landfills, leading to degradation of environment and air pollution. Synthetic fibers including polyester, can take hundreds of years to decompose, further exacerbating waste problems. The biggest issues about environment which are connected to the textile sector are typically those caused by water body contamination brought on by untreated effluent discharge. Use of synthetic fiber causes micro plastic pollution [13, 14].

5.4 Growth in the Per Capita Textile Waste in India and a Comparison with Different Countries

In 2011, it was estimated that 100 ktons of textile waste was generated in India annually with most of it being produced from household sources [15]. With the population of India as 121,08,55,000 in 2011 [16], the per capita textile waste produced was 0.82 kg. After a decade, the textile waste generated in India has increased to 7793 ktons in 2021 [5]. The population of India has also observed a tremendous increase in a decade and risen to 136,30,06,000 [16]. In 2021, the per capita textile waste produced in India was 5.7 kg, which is approximately the seven times of the amount generated in 2011.

For year 2011, in a comparison of the per capita textile waste produced in India with other countries, it has been observed that the India's per capita textile waste is much less compared to other countries (Fig. 5.4). Further, the data for the per capita textile waste generated in various countries in 2018 is presented in Fig. 5.5. For India, the data for 2018 has not been reported from any sources. However, the per capita textile waste produced in India in 2021 was 5.7 kg which is lower than amount generated by many countries in 2018. This low per capita textile waste generation in India can be attributed to many factors like low income, joint family system where clothes are shared among family, emotional attachment with a particular garment if it belongs to someone close and reusing the garments as wipes for cleaning purposes [17].

Although India's per capita textile generation is lower than many countries, it still needs to work on its recycling capacity. In the year 2021, out of the 7793 tons textile waste produced, 59% of the waste was recycled, 19% was downcycled by secondary industries and still 17% ended up in landfills while 5% was incinerated [5]. Thus, India needs to strengthen its recycling industry, invest in sustainable options and move towards circularity in textile waste which involves the systematic development of a regenerative system in which textiles are circulated until their maximum value is preserved for the longest time through recycle/reuse (Fig. 5.6).

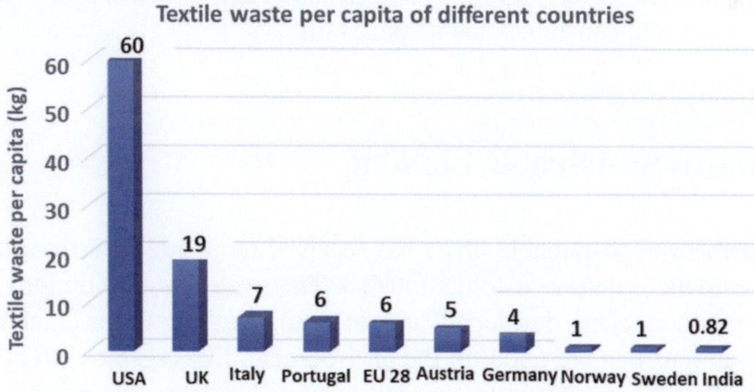

Fig. 5.4 Per capita textile waste generation in different countries (2011) [15]

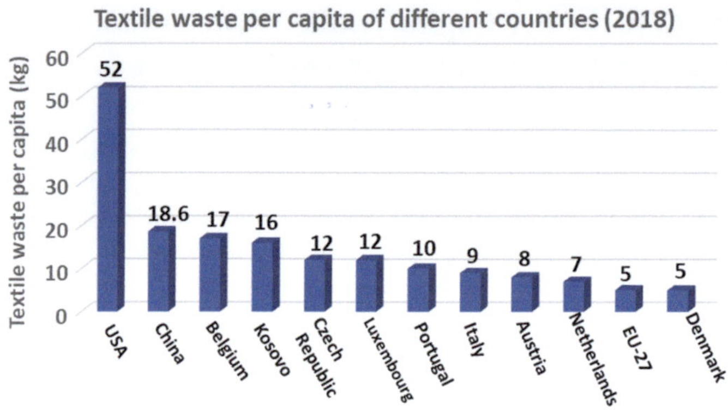

Fig. 5.5 Per capita textile waste generation in different countries (2018) [18]

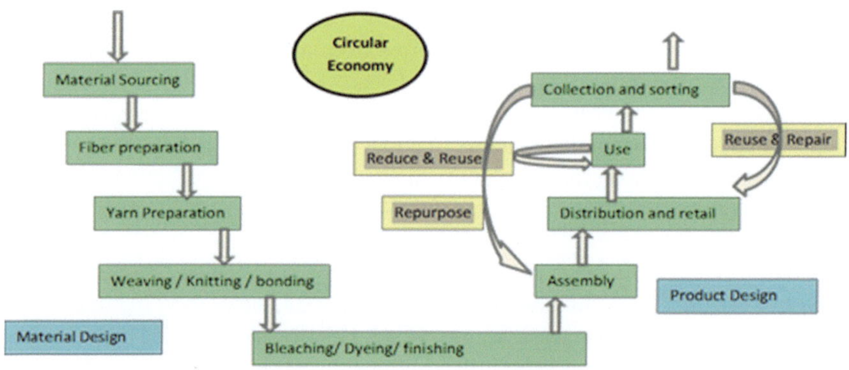

Fig. 5.6 Circular economy approach towards sustainable fashion

In the following sections, sustainable clothing will be discussed in detail. The goal of sustainable clothing is to addressing the social and environmental challenges related to the fashion sector, promoting a responsible and ethical approach to the way clothing is produced, consumed and disposed off.

5.5 Towards Sustainable Clothing

The development of sustainable fibers has received more attention from the textile industry due to the growing customer interest. There are now new options available that have been researched, developed, and released. Clothing that is created sustainably uses materials that are good for the environment, like organic cotton, bamboo, and Tencel, as well as methods that minimize waste and water usage.

Sustainable clothing is frequently created using natural, organic, or recycled materials since they have a lesser environmental impact than synthetic ones [19]. Sustainable clothing offers many advantages:

- **Water conservation**: is Many conventional garment production techniques consume a lot of water, which can put a burden on the nearby water supply. To safeguard this precious resource, sustainable clothing production techniques frequently utilize less water or recycled water.
- **Energy efficiency**: The creation of sustainable clothing frequently employs renewable energy sources and techniques which are energy-efficient. This can assist in lessening greenhouse gas emissions and the overall environmental effect of the fashion sector.
- **Reducing waste**: Steps to limit waste, like employing eco-friendly colors and recycling or repurposing materials, are frequently included in sustainable clothing production methods.
- **Reducing carbon footprint**: Local production of sustainable apparel both reduces the product's carbon footprint and supports local economy.

5.5.1 Sustainable Fashion

The term "sustainable fashion" was introduced in the 1960s due to increasing awareness among textile consumers about the environmental repercussions of clothing production and urged the industry to adopt more eco-friendly practices [20]. Slow fashion, particularly sustainable fashion, is aimed at empowering workers at every supply chain level, employ methods such as upcycling, recycling, and introducing organic, degradable and renewable materials into its processes [21]. Sustainable fashion is not just about the fabric or style but an approach towards cloth and accessory manufacturing which aims to reduce the adverse impact on Environment and foster Social Responsibility by using Eco-friendly raw material and green manufacturing process and embracing the Model of Circular Economy.

5.5.2 Conscious Fashion

The Conscious Fashion is essentially related to conscious consumerism [22]. Conscious consumerism works on encouraging consumers to rent clothing, buy used/pre-loved clothing or invest in higher-quality, longer- lasting apparel. Conscious consumerism approach works on styling the apparel to be timeless, stylish and affordable. Vintage and Heirlooms are examples of conscious fashion. If the garment carries an emotional value, the user will adapt the garment to suit new need when an heirloom garment goes out of fashion. The longevity of the garment will be ensured and thus help in making a fine balance between trend and sustainability.

5.5.3 *Waterless Fashion*

Sustainable Waterless textile dyeing is an emerging area of dyeing known by the name Air dyeing or Supercritical Carbon dioxide assisted dyeing ($scCO_2$). Generally, CO_2 is used as a fluid because in comparison to other fluids, it has lower pressure (Pc = 74 bar) and critical temperature (Tc = 31.1 °C). Moreover, it is non-flammable, nontoxic and easily available. Its liquid-like densities facilitate in dissolving hydrophobic dyes. Furthermore, its reduced diffusion characteristics and viscosities can result in shorter time required for dyeing in comparison to water [23].

Over the recent years, American firms namely, Colorzen and AirDye® have pioneered waterless textile dyeing. AirDye® practices placing fabrics into printing machines where colors formulated from paper are transferred to polyester fabric. Depending on the fabric, AirDye® technology consumes 95% less water, 86% less energy, and emits 84% fewer greenhouse gases compared to conventional dyeing methods. Its notable benefits include reduced hazardous waste emissions and minimized water usage. However, this technology is limited to synthetic materials like polyester, as natural fibers such as cotton cannot undergo the air-dye process. AirDye® is currently utilized by numerous textile converters across Europe, Asia, and North America [24–28].

5.6 Case Studies of Sustainable Fibers

5.6.1 *Hemp Fibre*

Fabric derived from the fibers of the cannabis plant is known as hemp fabric. It possesses durability, lightness, and antimicrobial characteristics [29]. In comparison to cotton, hemp cultivation demands less water and fewer pesticides [30]. The fibre is produced by the hemp stem. After being harvested, hemp must go through retting and separation. The full length of the hemp plant can produce fibres that range in length from 0.91 m (3 ft) to 4.6 m (15 ft). The fibres may be divided into smaller pieces later on in the operation [31]. In a study by Lamberti & Sarkar, 100% woven cotton was compared with 100% hemp fabric. It was observed that hemp performed equally well as cotton in terms of flammability, colour fastness, elongation, oil stain release and breaking and tearing strength [32].

5.6.2 *Lyocell*

Lyocell fibers are produced from sustainably managed eucalyptus tree plantations, utilizing wood pulp. Eucalyptus trees are chosen for their ability to thrive with reduced water and pesticide usage, distinguishing them from other crops commonly employed in textile production [33]. Lyocell is composed of numerous highly hydrophilic crystalline nano-fibrils arranged in a precise pattern. It stands out due to its exceptional absorption capacity, distinctive nano-fibril arrangement, and smooth surface. In a study by Shen and group, the environmental impact of lyocell was compared with viscose, modal and cotton. In the study, Lyocell is recognized for its sustainable features, given that its production involves lower water, energy, and chemical consumption in comparison to traditional fiber production. Further, the production results in low SO_2 and CO_2 emissions which leads to low terrestrial acidification and ecotoxicity [34].

5.6.3 *Econyl*

Econyl stands out as a regenerated nylon fiber produced from post-consumer waste like industrial plastic, discarded fishing nets, and fabric remnants by an Italian company, Aquafil [35]. To manufacture ECONYL, waste materials like reclaimed fishing nets undergo initial sorting and shredding at pre-treatment facilities. The shredded material is then subjected to the ECONYL process, moving to a regeneration plant where it enters large chemical reactors. Through a process involving de and repolymerization, these reactors break down the components of the material and regenerate polyamide-6 (PA6). The final product is then transformed into yarn [36].

The introduction of ECONYL in 2012 marked the beginning of a 'closed-loop' system, where end-of-life products made from polyamide-6 are reclaimed and converted into a more environmentally sustainable raw material, all while maintaining the quality of the final products [37]. From a life cycle environmental perspective, an ECONYL filament produces 50% fewer greenhouse gas emissions compared to an equivalent petrochemical filament. It has been estimated that when 10,000 tons raw materials are recycled into ECONYL, it results in the preservation of 70,000 crude oil barrels and helps in preventing the emission of 57,000 tons CO_2, when contrasted with conventional production techniques [38].

5.7 Conclusions

A significant impact on various aspects of society, economy, and the environment is caused by the textile industry. While textile industry contributes significantly towards the GDP of the countries and generate employment, it is also known for its high consumption of resources, including water, energy, and raw materials. Textile production generates substantial waste, including leftover fabric scraps, water pollutants, and chemicals used in various processes. Over the past decade, the per capita textile waste generated in India has gone up by seven times.

By embracing sustainability, the fashion industry can reduce the negative environmental impacts on the environment and lead to new prospects for development and innovation. Encouraging the use of emerging raw materials and techniques for fabric production, sustainable clothing made of environment friendly fabrics like hemp, lyocell, econyl which is made in an ethical and environmentally responsible manner, can stimulate innovation in the fashion sector. For instance, businesses may start adopting more environmentally friendly materials like organic cotton or bamboo, as well as investing in technology like waterless dyeing or 3D printing, to lessen their influence on the environment.

Addressing the mounting waste crisis can be achieved through the recycling of textile waste. It is important to comprehend the nature of waste to establish a circular economy which results in the reintroduction of waste into supply chain at the global level. The future of sustainable fashion business is bright given the surge in environmentally conscious consumers, government initiatives, and e- commerce.

References

1. K. Zheng, Ministry-of-Textiles, Annual report 2021–2022. NewDelhi: Ministry of Textiles, Government of India (2022)
2. M.N. Pervez, M.I.H. Mondal, Y. Cai, Y. Zhao, V. Naddeo, The textile institute book series, in *Textile waste management and environmental concerns* (2021) pp. 719–739
3. G. Yogananda, Bonfring. Int. J. Ind. Eng. Manag. Sci. **5**, 1 (2015)
4. The State of Fashion 2024: Finding pockets of growth as uncertainty reigns, (Mckinsey & Company), https://www.mckinsey.com/industries/retail/our-insights/state-of-fashion. Accessed Jan 2024
5. Wealth in Waste: India's Potential to Lead Circular Textile Sourcing. Fashion for Good, https://fashionforgood.com/our_news/wealth-in-waste-indias-potential-to-lead-circular-textile-sourcing/. Accessed Jan 2024
6. R. Rathinamoorthy, *Sustainable Innovations in Recycled Textiles*, (2018), pp. 19–52
7. Style that's sustainable: a new fast-fashion formula. Sustainability (Mckinsey, 2016). Accessed Jan 2024
8. H. Park, M.Y. Lee, W. Koo, J. Glob. Fash. Mark. **8**(4), 298–312 (2017)
9. E. Hur, T. Cassidy, Int. J. Fash. Des. Technol. Educ. **12**(2), 208–217 (2019)
10. Y. Chang, H.L. Chen, S. Francis, Application of recycled post-consumer fibers. Fam Consum Sci Res J **27**(3), 320–340 (1999)
11. T. Karthik, D. Gopalakrishnan, Environmental analysis of textile value chain: an overview, in *Roadmap to Sustainable Textiles and Clothing: Environmental and Social Aspects of Textiles and Clothing Supply Chain* (2014), pp.153–188
12. R.B. Chavan, Indian textile industry-environmental issues. NISCAIR-CSIR **26**, 11–26 (2001)
13. K.K.L. Moon, C.S.Y. Lai, E.Y.N. Lam, J.M. Chang, J. Text. Inst. **106**(9), 939–952 (2015)
14. Keith Slater "Environmental impact of textiles: production, processes and protection, in *Textile Institute* (Woodhead Publishing, Manchester, England, 2023)
15. A. Aggarwal, Env. Sci. Indian J. **17**, 180 (2021)
16. Population projection for India and States (2011–2036). National Commission on Population Ministry of Health and Family Welfare, https://main.mohfw.gov.in/sites/default/files/Population%20Projection%20Report%202011-2036%20-%20upload_compressed_0.pdf. Accessed Jan 2024
17. N. Bairagi, J. Text. Sci. Eng. **8**(1) (2017)
18. Are Asian and African textile hubs sustainability ready? Fibre2Fashion (2021), https://www.fibre2fashion.com/industry-article/9196/are-asian-and-african-textile-hubs-sustainability-ready. Accessed Jan 2024

19. M.D. Stănescu, Environ. Sci. Pollut. Res. 28(3) (2021)
20. S. Jung, B. Jin, Int. J. Consum. Stud. **38**(5), 510–519 (2014)
21. A. Johnston, *The First Steps towards Considerate Design Incorporating Cradle to Cradle Principles*, (London College of Fashion, London, 2012), https://innovatingsustainablefashion.files.wordpress.com/2012/07/cradle-to-cradle_copenhagen_final_small.pdf. Accessed Jan 2024
22. C. Marino, S. Bellini, Designing a conscious fashion experiences: strategies for generation Z, in *Beyond all limits. Proceeding of International Conference on Sustainability in Architecture, Planning, and Design*, (DADI_PRESS, 2022), pp. 632–638
23. I. Mahmud, S. Kaiser, Recent progress in waterless textile dyeing. J. Text. Sci. Eng **10**(6), 1–3 (2020)
24. M.M. Bomgardner, These new textile dyeing methods could make fashion more sustainable. Chem. Eng. News **96**(29), 28–33 (2018)
25. V. Dhanabalan, L. Sukanya, K.V. Lokesh, Air-dyeing technology-a review. Text Today (2015)
26. K. Deepak, Dyeing without water. Fibre2Fashion.com (2011), https://www.fibre2fashion.com/industry-article/5676/dyeing-without-water. Accessed Jan 202
27. T. Huang, X. Kong, H. Cui, T. Zhang, W. Li, P. Yu, J. Lin, Waterless dyeing of zipper tape using pilot-scale horizontal supercritical carbon dioxide equipment and its green and efficient production. J. Clean. Prod. **233**, 1097–1105 (2019)
28. K.M. Babu Air-flow dyeing-an eco-friendly water preserving fabric dyeing technology. Textile School (2019)
29. B.A. Khan, P. Warner, H. Wang, Antibacterial properties of hemp and other natural fibre plants: a review. BioResources **9**(2), 3642–3659 (2014)
30. H. Zhang, Z. Zhong, L. Feng, Advances in the performance and application of hemp fiber. Int. J. Simul. Syst. Sci. Technol. **17**, 1–18 (2016)
31. A. Shahzad, A study in physical and mechanical properties of hemp fibres. Adv. Mater. Sci. Eng. (2013)
32. D.D. Lamberti, A.K. Sarkar, Hemp fiber for furnishing applications. IOP Conf. Ser. Mater. Sci. Eng. (IOP Publishing, 2017) **254**(19), 192009
33. R. Medar, G. Mahale, Int. J. Curr. Microbiol. App. Sci, **9**(6), 3851–3867 (2020)
34. L. Shen, E. Worrell, M.K. Patel, Resources. Conserv. Recycl. **55**(2), 260–274 (2010)
35. M. Pohjakallio, Secondary plastic products—examples and market trends, in *Plastic Waste and Recycling*, (Academic Press, 2020), pp. 467–479
36. M. Perella, New fabrics make recycling possible, but are they suitable for high street. The Guardian (2015), https://www.theguardian.com/sustainable-business/sustainable-fashion-blog/2015/jan/22/fabric-recycling-closed-loop-process-high-street-fashion. Accessed Jan 2024
37. E. Wankowicz, Sustainable fibre for sustainable fashion supply chains: where the journey to sustainability begins, in *ICIL Conference* **13**, 342–352 (2016)
38. R. Nayak, L.V.T. Nguyen, T. Panwar, L. Jajpura, Sustainable technologies and processes adapted by fashion brands, in *Sustainable Technologies for Fashion and Textiles* (Woodhead Publishing, 2020), pp. 233–248

Chapter 6
GreenShield—A Natural Language Processing Based Approach to Prevent Greenwashing and Attain Decarbonization

Nidhi Sharma, Adithya Venkateswaran, Raghav Tiwari, and Mudita Nagpal

Abstract The global rise in demand for environment friendly products has led companies to adopt sustainability in the supply chain of consumer goods. However, few companies have resorted to unethical means and malpractices such as false advertising to increase the profit margin and thus deceive the public. This phenomenon, known as "greenwashing" involves misleading the public by making unsubstantiated claims about a product's environmental benefits. Greenwashing not only erodes consumer trust but also hinders genuine efforts towards sustainability and can lead to increased consumption of products that are not genuinely sustainable, exacerbating the problem of overconsumption and waste. In recent years, there has been a shift towards the digital era with the rise in technologies like artificial intelligence, data science, natural language processing, and many more. Researchers across the globe have been trying to solve this problem of greenwashing by utilizing these technologies. This paper aims to integrate the current technologies and research findings to devise a solution that effectively tackles greenwashing. A literature review has been conducted on more than 40 papers to determine the gaps in their findings and analyze them to conceptualize a framework using artificial intelligence and machine learning. The framework uses natural language processing to analyze the data provided by companies in their sustainability reports and compares the data with advertisements to verify their credibility. This tool helps prevent fraudulent malpractices and wrongful

N. Sharma · A. Venkateswaran (✉) · R. Tiwari · M. Nagpal
School of Engineering and Technology, VIPS-TC, Delhi, India
e-mail: 02317702723_cse@vipstc.edu.in

N. Sharma
e-mail: nidhisharma@vips.edu

R. Tiwari
e-mail: 05217702723_cse@vipstc.edu.in

M. Nagpal
e-mail: mudita.nagpal@vips.edu

A. Kumar et al. (eds.), *Selected Proceedings of the 1st International Conference on Advanced Materials for Sustainable Innovation; IC-AMSI 2024; 28–30 August; New Delhi; India*, Springer Proceedings in Energy,
https://doi.org/10.1007/978-981-96-4492-6_6

information dissemination to the general public by the companies involved. This, in turn, will ensure optimum resource utilization and curbing greenwashing practices, therefore, aiding in decarbonization and upholding consumer trust.

Keywords Greenwashing · Sustainability reporting · Artificial intelligence · Natural language processing · Transformers · Decarbonization

6.1 Introduction

Companies traditionally used financial reports to communicate their performance. However, growing concerns about sustainability led to the creation of sustainability reports. These reports go beyond just financial results and consider how a company impacts the environment, society, and the economy [1].

Many companies and organizations practice sustainability reporting on a global scale from the early 1990s when the environmental concerns and sustainability demands associated with the various sectors of society started to become more relevant [2]. Since then, several reporting methods have been associated with sustainability including inter alia, corporate social responsibility (CSR) reporting, greenhouse gas emissions (GHG) reporting, UN Sustainable Development Goals (SDG) reporting, and triple bottom line reporting [3].

The rise in demand for environment-friendly products resulted in using "green" ideas to sell their products to people who care about the environment. To attract these "green" customers, some companies make claims which may sound good for the environment but are ambiguous and even untrue [4]. This is called greenwashing. The campaign by Ford's "It Isn't Easy Being Green" is a well-known example of greenwashing. This campaign advertised the Escape SUV as environmentally friendly. Still, ironically enough a group of scientists said it had the worst fuel efficiency and released the most harmful emissions compared to other car companies. [5].

Greenwashing has seen a considerable rise in all three pillars of ESG (Environment, Social, and Governance) in the banking sector over the past decade according to a report published by the European Banking Authority in 2023 (Fig. 6.1) [6].

The 12th United Nation Sustainable development goal promotes buying products that are sustainable and environment-friendly in order to achieve responsible consumption and production [8]. However, despite the efforts made by the people, achieving this has become impossible with continued greenwashing practices. Therefore, it is essential to eliminate it.

Over the past decade, advancements in technologies like artificial intelligence and machine learning have led to development of numerous applications in the field of Environmental awareness and protection [9]. While a substantial amount of research has been undertaken to explore the links between artificial intelligence, greenwashing, and sustainability reporting, these studies often lack concrete implementation and definitive conclusions.

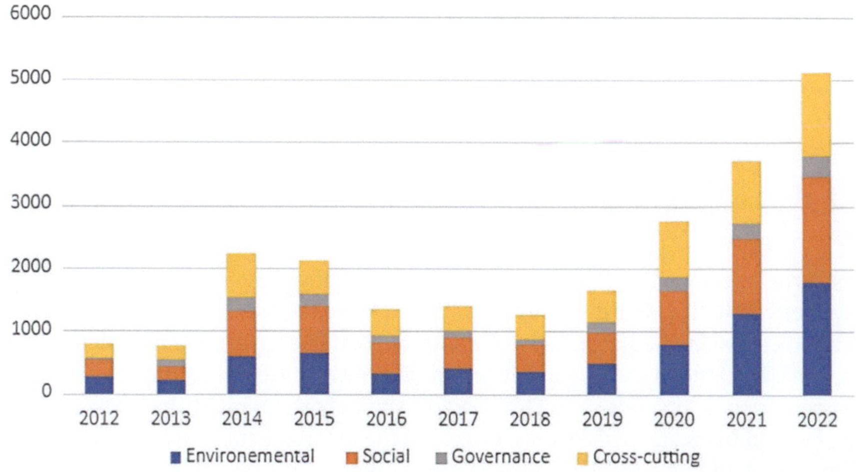

Fig. 6.1 Alleged cases of greenwashing on ESG topics (*Source* RepRisk as cited in EBA report on greenwashing) [6, 7]

De Villiers et al. [10] showed how generative AI models like ChatGPT can be used to improve the speed and efficiency of making reports but fail to account for the transparency and reliability of the information provided by AI [10].

Lien [11] in his thesis uses a web scraper to extract information from websites and detect greenwashing but does not account for video advertisements [11] which are generally more popular [12].

This study aims to propose a framework that is free of existing loopholes and can be easily implemented. It uses natural language processing and decision-making machine learning algorithms to compare the green claims made by companies in their advertisements with those stated in sustainability reports like CSR and ESG to verify their credibility and hold the offenders accountable. This paper fills the gap in current research by developing a solution that can be readily implemented with minimal costs, includes all forms of advertisement, and addresses the reliability issues of artificial intelligence, as mentioned by De Villiers and his team [10].

6.2 Related Work

6.2.1 Sustainability Reporting

Organizations like KPMG define sustainability reports as "reports that include both numbers and descriptions about a company's financial, social, and environmental performance, all given equal importance" [13]. Similarly, the World Business Council for Sustainable Development (WBCSD) describes them as "public reports companies use to show stakeholders their position and activities related to the economy, environment, and society" [14].

6.2.2 Greenwashing

The term "greenwashing" was coined in the late 1980s, by environmentalist named Jay Westervelt [15]. But it's being widely used since the late 1990s, after appearing in a book about marketing and the environment [16]. The term's origin is often traced back to a 1980s hotel campaign that encouraged towel reuse, presenting it as an environment friendly practice [17]. This example shows how seemingly harmless actions can be used to create a misleading image having environmental consequences. The definition, by Lyon and Maxwell [18] clearly defined greenwashing as the "selective disclosure of positive information about a company's environmental or social performance, without full disclosure of negative information on these dimensions" [18]. De Freitas et al. [19] identified three main ways people who study this stuff define greenwashing; Hiding the Bad Stuff, Saying One Thing, Doing Another and Looking Good, Not Being Good [19].

6.2.3 Impact of Greenwashing on Decarbonization

The growing popularity of eco-friendly products has created a situation where companies can trick consumers by appearing more environment friendly than they are to be more competitive [20].

As companies are constantly competing for attention and money, Greenwashing can be a tool for survival [19]. With more and more people worried about climate change and increasing carbon footprints, many consumers are looking for companies that care about the environment [21]. Companies use greenwashing to attract these "green" consumers and get good word-of-mouth advertising [22]. Even if a company only pretends to be socially responsible, it can still appear more attractive to certain groups of people, leading to more customers and sales [23].

One of the glaring examples where greenwashing is rampant is Voluntary carbon market (VCM) where some companies gain carbon credits by unscrupulous manner without reducing the real emissions and also making money in the process [24].

In the short term, greenwashing might benefit existing stakeholders by making the company more profitable [25]. But this comes with a cost of hurting society as a whole in the longer run by generating waste and increasing carbon emissions.

If companies keep on making vague and misleading claims about being "green," consumers eventually stop caring about these claims altogether [26, 27]. The whole idea of a product being "green" loses its meaning and in the end, everyone loses: consumers get tricked, companies struggle, and the environment suffers.

Another reason why consumers doubt green advertising is that environmental issues can be complicated. It can be hard for the average person to understand the science behind these issues. Companies can take advantage of this by making claims that sound good but are misleading [28].

By understanding why companies greenwash, we can see the problem more clearly. Companies might be tempted to use greenwashing to get ahead in the short term, but the long-term consequences can be significant.

6.2.4 *Overview of Artificial Intelligence Technologies in Environment Studies*

Information technology is becoming increasingly important for managing and understanding environmental issues. These technologies are used for planning, predicting, monitoring, and controlling environmental processes, and they can be applied at different scales and timeframes.

A new field called Environmental Informatics is emerging to address these needs. This field combines areas like Artificial Intelligence (AI), Geographic Information Systems (GIS), modeling and simulation, and user interfaces. AI has proven particularly useful in environmental management [29]. For example, AI systems can advise emergency response teams on handling accidents [30] or help predict fish stocks [31] and model water quality [32].

Recent research in AI focuses on developing Knowledge-Based Systems (KBS). When applied to environmental issues, KBS are called different things, such as Decision Support Systems (DSS), Environmental Decision Support Systems (EDSS), or Multiple Objective Decision Support Systems (MODSS) [7].

While there are many definitions of DSS, a simple way to understand them is as computer programs designed to help people make decisions about complex problems. These programs are flexible and can adapt to different situations. They are particularly useful for problems that are poorly defined or unstructured, often faced in strategic management. By providing relevant information and analysis, DSS can significantly improve decision-making processes [33].

Some examples include OASIS, which helps in optimizing water flow and making informed decisions about water resource allocation [34], and STORMCAST which helps forecast storms over the Scandinavian Peninsula [35].

6.2.5 Practical Examples of Implementation of AI in Improving Sustainability Reports and Eliminating Greenwashing

A journal article from De Villiers and his team demonstrated how generative AI can be used to improve the quality of sustainability reports and speed up the process at the same time [8].

The conference paper by Kazuya Uekedo and his colleagues compared companies' CSR reports with their corporate values in an attempt to catch greenwashing using machine learning and data mining [36].

In another conference paper, S. Kim and his team developed a prediction model for detecting greenwashing using ESG-BERT in KOSPI 200 Enterprises. They focused on a BERT-based metric for greenwashing to address limitations in inherent assessment methodologies [37].

A research article by Yukiro Nakao and the team aimed at analyzing images and photographs in sustainability reports to disclose corporate reporting strategies using image recognition with deep learning [38].

Wayne Moodaley and Arnesh Telukdarie came up with a conceptual framework for training large language models to detect false and misleading claims in sustainability disclosures [39].

In an article, Hyewon Kang and Jinho Kim analyzed the text information in Corporate Sustainability Reports using natural language processing and noticed disparities in the balance of positive and negative information that was disclosed [40].

In a discussion paper by Sebastian Schmidt and his colleagues, they correlate the US metal industry with atmospheric SO_2 concentrations through Sentinel-5P satellite data using spatial regression analysis and use the data to catch companies involved in greenwashing [41].

In a thesis by Lien, he attempted to automate the approach to detect greenwashing using machine learning. He used a web scraper to extract information about companies, preprocessed it, and used it to train machine learning models [9].

Smeuninx and the team analyzed the readability of sustainability reports using natural language processing and corpus linguistics to make them more accessible to the general public [42].

Marcelo Gutierrez-Bustamante and Leonardo Espinosa-Leal aimed at the automation of evaluation of sustainability reports through natural language processing and text mining and evaluated the extent of affinity of Nordic companies [43].

Wei Wu and colleagues propose a novel environmental reporting system (BI-ESRS) using unsupervised neural network-enabled spatial–temporal analytics that leverages both blockchain and Internet of Things (IoT) technologies. This system automates the collection of environmental data, ensuring reports are reliable and the data trail can be easily tracked [44].

Carolina Rodriguez and team [45] conducted a comprehensive study to identify red flags for "greenwashing" by Colombian companies across different industries. They used a quantitative approach to analyze data at a single point in time. Three machine learning algorithms (decision tree, random forest, and logistic regression) were employed, along with a special technique (double machine learning) to pinpoint what factors cause greenwashing and which variables are most significant [45].

6.2.6 Research Gaps

Carolina Rodriguez and team [45] conducted a comprehensive study to identify the factors aiding in identifying Greenwashing practices in companies across various industrial sectors in Colombia. However, their paper fails to evaluate long-term environmental practices due to a lack of longitudinal data. The study also lacks relevance across various situations owing to the absence of case studies or comparative analysis [45].

Yuri Nakao and the team utilized image analysis in sustainability reports to disclose corporate reporting strategies. However, the study did not include the reader's interpretations and therefore fails to highlight the reader's perception and image evaluation [38].

The conceptual framework given by Wayne Moodaley and the team uses large language models that are trained beforehand to detect false claims in sustainability disclosures. However, the training data set used may be limited. The given framework is also not suitable for identifying green claims that relate to various forms of greenwashing [39].

In the study led by Sebastian Schmidt and his colleagues, a correlation was found between the US metal industry and the atmospheric SO_2 concentrations using spatial regression analysis along with using web text data to catch the companies making false claims. However, the high proportion of companies without a website (40.6%) inhibits a more thorough analysis [41].

Wei Wu and colleagues devised a framework of smart reporting systems using technologies such as IoT and Blockchain to enhance the viability of ESG disclosure. However, the authenticity and credibility of ESG disclosures remain in doubt due to non-transparency which requires a lot of annual work [44].

6.3 Research Methodology and Discussions

6.3.1 Overview

First, advertisements are extracted from websites. This includes audio, video, and static ads and cannot be directly used in its raw form. Hence, the data is processed. If any green claims are identified, the data from advertisements undergoes feature extraction to remove unwanted content and retain only the green claims. The company is identified and its sustainability reports are retrieved from the web and similarly

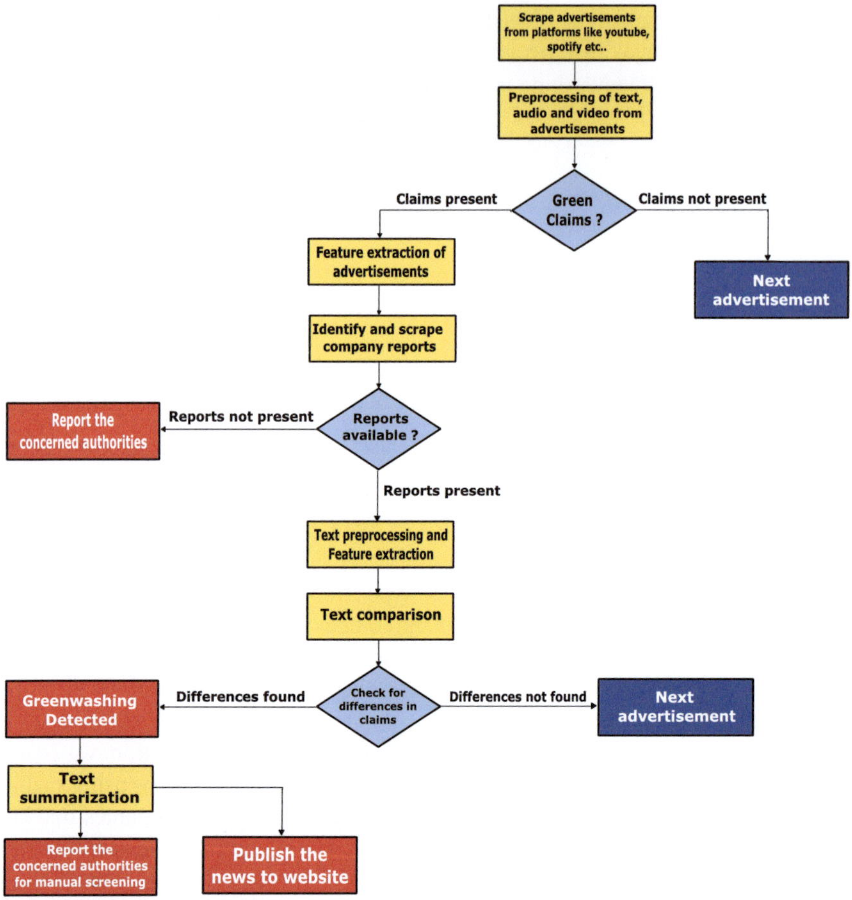

Fig. 6.2 Method overview

processed and subjected to feature extraction. If company reports are not found online, authorities are notified to conduct an investigation. The green claims in the advertisements are then compared to those in the company reports. If discrepancies are found, a summary of these differences is provided, and the authorities are notified and the public is informed through websites (Fig. 6.2).

6.3.2 Materials and Methods Used

Data collection. Video, audio, and static advertisements are collected from websites like YouTube, Spotify, etc. and company sustainability reports like CSR from the company's websites using web scraping. Web scraping utilizes the HyperText

Transfer Protocol (HTTP) to communicate with web servers and automatically retrieve the necessary data. Tools such as Selenium, Beautiful Soup, and Scrapy can be used for this purpose.

Data preprocessing. The processing of data involves two steps: Text extraction and text preprocessing.

1. Text preprocessing is a method to make the data more suitable for analysis. This requires mainly 3 steps. First, the text is split into smaller units called tokens. This process is called tokenization. The text is then converted to lowercase, punctuation and stop words removed, and finally, it is lemmatized. Common libraries used for this purpose include spaCy and NLTK.
2. The text from sustainability reports usually in pdf format can be extracted using libraries like PyMuPDF, which is then further processed.
3. Static advertisements are analyzed using image preprocessing. The images are resized and normalized to maintain consistency and improve efficiency. Next, the text from images is extracted using optical character recognition which can be performed using a tool called tesseract. This textual data then undergoes text preprocessing.
4. Audio processing involves effectively transforming audio advertisements into text. It involves loading the audio files using LibROSA and then normalizing them using methods like converting them into mono and reducing background noises. The normalized audio is then converted into text using speech-to-text which can be implemented using libraries like Google Cloud text-to-speech.
5. For video advertisements frames are extracted, resized, and normalized using libraries like OpenCV. Key frames containing important information are identified which are then sent for image preprocessing. Audio is extracted using tools like MoviePy which is then sent for audio pre-processing.

Feature Extraction. This process transforms raw data into a numerical format that can be efficiently used by large language models without compromising the actual information and is an essential step for comparing texts. This form can be achieved via embeddings, which involves converting tokens obtained from text preprocessing into dense vector representations to capture the context and meaning. This data then undergoes Named Entity Recognition (NER) and part-of-speech tagging to extract the green claims. Feature extraction is performed using transformers. Transformers are neural networks based on multi-head attention mechanisms. Transformers track sequential data to learn the context and meaning [46]. Some popular pre-trained transformers include BERT, GPT, and RoBERTa.

Text Comparison. It involves comparing and contrasting the environmental claims companies make in their advertisements versus those found in their reports using transformers like BERT. Semantic similarity is a method that utilizes parameters from feature extraction and measures the degree to which two texts carry the same meaning. If claims appear in both documents and are semantically similar, they will not be deemed as greenwashing. However, if claims are present only in advertisements but not in reports, or if similar claims have differing meanings, this will be considered greenwashing. For example, if a company asserts that it is fully carbon–neutral but the report states it will achieve carbon–neutral status in the future, this would be classified as greenwashing. Additionally, if negative environmental observations are mentioned in the report but not in the advertisement, this will be deemed greenwashing. To classify extracted environmental claims as pro-green or anti-green, each claim is evaluated using a combined scoring approach. This includes assigning keyword scores based on the environmental impact of specific terms, such as $+2$ for "reduced carbon footprint" and -2 for "deforestation." Sentiment analysis assesses the emotional tone of the sentences, with values like $+0.6$ for positive statements, such as "We reduced carbon footprint by 12% last year," and -0.7 for negative statements, like "Water consumption increased by 20%." Semantic similarity measures how closely the claim's meaning aligns with environmental concepts. The combined score is calculated as follows:

$$TotalScore = \left(\sum_{i=0}^{n} KeywordScore_i + SentimentScore \right) \times SemanticScore \quad (6.1)$$

where, $KeywordScore_i$ represents the score for each of the n keywords present in the sentence. Keyword scores are manually assigned during labeling data for training the model. Sentiment score and Semantic similarity range between -1 to 1. A total score greater than zero indicates a pro-green claim, while a total score less than zero indicates an anti-green claim.

Text Summarization. The differences in claims between the two documents can be summarized to provide a quick overview. Since claims are mostly not too long, hence summarization can be simply performed by using a loop and appending each claim to an empty list.

```
def summarize_differences(differences):
  summary = []
  for claim1, claim2 in differences:
    summary.append(f"Claim in ad: {claim1}\nClaim in
report: {claim2}\n)
  return "\n".join(summary)
```

[Sample code for summarization in python].

Decision-making Algorithms. A decision tree is a decision-making algorithm that operates by posing a simple question and then branches into subsequent sub-trees based on the answer [47]. Initially, it starts with the root node or the root attribute, which in this case is the presence of green claims. This is determined using pre-trained transformers like GPT and fine-tuning it using sample texts. From there, it branches into a decision node, where it checks for the availability of reports, and leads to two leaf nodes which indicates the appropriate actions to take. The tree further branches into another decision node, which analyzes the differences between company reports and advertisements. Finally, this decision node branches into two

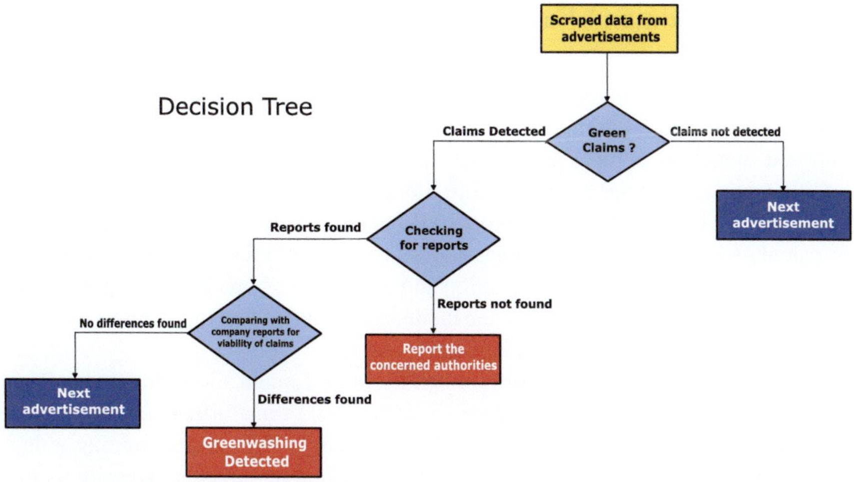

Fig. 6.3 Decision tree algorithm

leaf nodes: one for proceeding with the next advertisement and the other for notifying relevant authorities and publishing the results (Fig. 6.3).

Deployment and maintenance. Before deployment, the model needs to be fine-tuned using sample texts that contain different words used in green claims as mentioned in Table 6.1, to avoid errors and provide accurate results. Also, this approach works only if several advertisements are covered at the same time. Hence the model needs to be scaled to perform actions on multiple inputs simultaneously. This can be accomplished using services like AWS lambda, Kubernetes, etc. To implement this approach, we need two websites. One will be a dashboard that will notify authorities about potential greenwashing attempts. The second will be similar to a news website to inform the general public. These websites can be developed using a variety of technologies, including but not limited to Flask, Angular, and FastAPI.

Table 6.1 Sample keywords for training the model [48]

Environment/mother nature	renewable energy
Ozone safe/friendly	Planting trees
Sustainable/eco-friendly	Non-toxic/clean/green
Plastic	100% Natural
Carbon–neutral/footprint/offsets	Reduced/increased water usage
Greenhouse emissions	Good for the planet
Recycle/renewable/reuse	Degradable/compostable

6.4 Conclusion and Future Scope

In this study, we have conducted thorough research on greenwashing and proposed an Artificial intelligence-based model for detecting greenwashing. The model we have proposed works by comparing various advertisements and company reports such as CSR reports, financial reports, etc. to identify irregularities and differences. The reports and advertisements are preprocessed and fed into decision-making algorithms for calculated decisions and responses. This model can therefore be of the utmost importance in achieving decarbonization and aiding our efforts in building towards a better environment. However, since the model is still under development, its accuracy is a topic that needs to be discussed in later stages.

This paper presents opportunities for further enhancement. Given the scale of data involved, implementing more efficient algorithms and transformers could expedite processes and extend coverage to more advertisements along with avoiding redundant screenings of ads, whether on the same or different websites. Algorithms that directly skip to the environment section can improve efficiency. Developing methods to include page and paragraph numbers for claims in reports, along with timestamps for video and audio advertisements, would assist authorities in manual investigations.

6.4.1 Limitations

This paper presents itself with certain limitations. This approach requires covering all existing advertisements to eliminate greenwashing which will be time-consuming. Semantic limitations arise from the varied language usage in advertisements and thus, will require a lot of fine-tuning to work with different languages. Also, offline advertisements that are published in newspapers and weekly magazines cannot be analyzed using this approach. The web is a large space with huge amounts of data. Therefore, large amounts of computing power and energy are required to make the model fast, wide-ranging, and accurate which can negatively impact the environment.

The major limitation to this approach however is the cooperation that is required from the governments. Malpractices such as bribery and forging hamper the integrity of official documents and reports. The proposed model should be followed by an appropriate action plan to make it beneficial to the community.

6.4.2 Suggestions

- Laws should mandate that companies can only advertise themselves as environmentally friendly if they submit a report.
- The laws should be made more stringent to prevent unfair practices by ensuring necessary actions are taken and the entire process should be made transparent.

References

1. D. Claus-Heinrich, *Assessing the quality of sustainability reporting: an alternative methodological approach,* (Institute for Sustainable Management, Switzerland, 2005)
2. Murphy, David F. ORCID: https://orcid.org/0000-0001-8416-5627 and Bendell, Jem ORCID: https://orcid.org/0000-0003-0765-4413 (1997) In the company of partners: business, environmental groups and sustainable development post-Rio. Policy Press, Bristol, UK. United Nations. (n.d.).
3. W. Moodaley, A. Telukdarie, Greenwashing, sustainability reporting, and artificial intelligence: a systematic literature review. Sustainability **15**, 1481 (2023). https://doi.org/10.3390/su15021481
4. A. Gillespie, *'The 1990s', The Long Road to Sustainability: The Past, Present, and Future of International Environmental Law and Policy*, online edn, (Oxford Academic, Oxford, 2018). https://doi.org/10.1093/oso/9780198819516.003.0007
5. D. Friedman, D. MacKenzie, *Automaker rankings 2004* (UCS Publications, Cambridge, MA, 2004)
6. Anon, EBA progress report on greenwashing (2023), https://www.eba.europa.eu/sites/default/files/document_library/Publications/Reports/2023/1055934/EBA%20progress%20report%20on%20greewnwashing.pdf
7. Anon Reprisk, ESG Data science, www.reprisk.com
8. THE 17 GOALS, Sustainable Development (n.d.), https://sdgs.un.org/goals. Accessed 11 Jul 2024
9. U. Cortés, M. Sànchez-Marrè, L. Ceccaroni, I. Rodríguez-Roda, M. Poch, Artificial intelligence and environmental decision support systems. Appl. Intell. **13**, 77–91 (2000). https://doi.org/10.1023/A:1008331413864
10. C. De Villiers, R. Dimes, M. Molinari, How will AI text generation and processing impact sustainability reporting? Critical analysis, a conceptual framework, and avenues for future research. Sustain. Account. Manag. Policy J. Forthcom. (2023). https://doi.org/10.1108/SAMPJ-02-2023-0097
11. A.S. Lien, *Machine Learning to Detect Corporate Greenwashing* (Master Thesis, University of Stavanger, 2023), https://hdl.handle.net/11250/3090545
12. QuickFrame, Why you should choose video over static ads. QuickFrame (2024), https://quickframe.com/blog/video-vs-static/. Accessed 12 Jul 2024
13. KPMG/WIMM. KPMG international study of corporate sustainability reporting 2002, (Graduate Business School, Amsterdam, 2002) p. 72
14. World Business Council for Sustainable Development (WBCSD). Sustainable development reporting: striking the balance, (WBCSD: Geneva, Switzerland, 2002).p. 73
15. P. Romero, Beware of green marketing, warns Greenpeace exec. ABS-CBN (2008), www.abs-cbnnews.com/special-report/09/16/08/beware-green-marketing-warns-greenpeace-exec
16. J. Greer, K. Bruno, Greenwas2h: the reality behind corporate environmentalism. Multinatl. Monit. **30**(9) (1996)
17. K. Becker-Olsen, S. Potucek, Greenwashing, in *Encyclopedia of Corporate Social Responsibility* eds. by S.O. Idowu, N. Capaldi, L. Zu, A.D. Gupta (Springer, Heidelberg, Berlin, 2013)
18. T.P. Lyon, J.W. Maxwell, Greenwash: corporate environmental disclosure under threat of audit. J. Econ. Manag. Strat. **20**, 3–41 (2011). https://doi.org/10.1111/j.1530-9134.2010.00282.x
19. S.V. De Freitas Netto, M.F.F. Sobral, A.R.B. Ribeiro, G.R. de Luz Soares, Concepts and forms of greenwashing: a systematic review. Environ. Sci. Eur. **32**(19), 1–12 (2020)
20. B. Parguel, F. Benoît-Moreau, F. Larceneux, How sustainability ratings might deter 'greenwashing': a closer look at ethical corporate communication. J. Bus. Ethics **102**, 15–28 (2011). https://doi.org/10.1007/s10551-011-0901-2
21. F. Testa, O. Boiral, F. Iraldo, Internalisation of environmental practices and institutional complexity: can stakeholders pressures encourage greenwashing? J. Bus. Ethics **147**, 287–307 (2018). https://doi.org/10.1007/s10551-015-2960-2

22. J. Lu, G. Chen, D. Frierson, Response of the zonal mean atmospheric circulation to El Nino versus global warming. J. Clim. **21**, 5835 (2008). https://doi.org/10.1175/2008JCLI2200.1
23. W.S. Laufer, Social accountability and corporate greenwashing. J. Bus. Ethics **43**, 253–261 (2003). https://doi.org/10.1023/A:1022962719299
24. Anon ISDA Linklaters, Navigating the Risks of Greenwashing in the Voluntary Carbon Market (2024). https://www.isda.org/a/I9wgE/Navigating-the-Risks-of-Greenwashing-in-the-Voluntary-Carbon-Market.pdf
25. D. Jamali, C. Karam, Corporate social responsibility in developing countries as an emerging field of study. Int. J. Manag. Rev. **20**, 32–61 (2018). https://doi.org/10.1111/ijmr.12112
26. J.F. Solomon, C. Rhianon Pel Edgley, The abandoned mandatory OFR: a lost opportunity for SER? Soc. Responsib. J. **4**, 324–348 (2008). https://doi.org/10.1108/17471110810892848
27. R. Mayer, D. Scammon, C. Zick, Poisoning the well: do environmental claims strain consumer credulity? Adv. Consum. Res. **20** (1993)
28. M. Zimmer, T.Stafford, M. Stafford, Green issues: dimensions of environmental concern. J. Bus. Res. **71** (1994)
29. F.J. Radermacher, W.F. Riekert, B. Page, L.M. Hilty, Trends in environmental information processing, in *Thirteenth World Computer Congress 94* eds. by K. Brunnstein, E. Raubold, vol 2, pp. 597–604 (1994)
30. N.M. Avouris, Co-operating knowledge-based systems for environmental decision-support. Knowl.-Based Syst. **8**(1), 39–53 (1995)
31. L. Sazonova, G. Osipov, Intelligent system for fish stock prediction and allowable catch evaluation, in *Workshop Binding Environmental Sciences and Artificial Intelligence* (*BESAI 98*) eds. by U. Cort´es, M. S‘anchez-Marr‘e, pp. 161–176 (1998)
32. J. Wright, L. Wiggins, R. Jain, T. John Kim eds. *Expert Systems in Environmental Planning* (Springer-Verlag, 1993) ISBN 3–540–56063–7
33. BfG, Decision Support Systems (DSS) for River Basin Management. Bundesanstalt für Gewässerkunde (German Federal Institute of Hydrology) Report No. 4/2000, Koblenz, Germany (2000)
34. SFWMG (South Florida Water Management District) OASIS: The South Florida Water Management District's Operation Artificial Intelligence Program (1987)
35. G. Hartvigsen, D. Johansen, Cooperation in a distributed artificial intelligence environment—the stormcast application. Eng. Appl. Artif. Intell. **3**, 229–237 (1990)
36. K. Uekado, L. Feng, M. Suzuki, H. Ohwada, Classification of CSR using latent Dirichlet allocation and analysis of the relationship between CSR and corporate value, in *Knowledge Management and Acquisition for Intelligent Systems. PKAW 2018* eds by K. Yoshida, M. Lee, Lecture Notes in Computer Science, vol 11016. (Springer, Cham, 2018). https://doi.org/10.1007/978-3-319-97289-3_21
37. S. Kim, Y. Shin, S. Park, S. Joel, S.T. Kim, J.H. Oh, detecting greenwashing in sustainability disclosures: a prediction model for KOSPI 200 enterprises using ESG-BERT, in *2023 IEEE International Conference on Big Data (BigData)*, (Sorrento, Italy, 2023), pp. 6198–6200. https://doi.org/10.1109/BigData59044.2023.10386798
38. Y. Nakao, A. Ishino, K. Kokubu, H. Okada, Exploring visual communication in corporate sustainability reporting: using image recognition with deep learning. Corp. Soc. Responsib. Environ. Manag. (2024). https://doi.org/10.1002/csr.2735
39. W. Moodaley, A. Telukdarie, A conceptual framework for subdomain specific pre-training of large language models for green claim detection. Eur. J. Sustain. Dev. **12**(4), 319 (2023). https://doi.org/10.14207/ejsd.2023.v12n4p319
40. H. Kang, J. Kim, Analyzing and visualizing text information in corporate sustainability reports using natural language processing methods. Appl. Sci. **12**, 5614 (2022). https://doi.org/10.3390/app12115614
41. S. Schmidt, J. Kinne, S. Lautenbach, T. Blaschke, D. Lenz, B. Resch, Greenwashing in the US metal industry? A novel approach combining SO2 concentrations from satellite data, a plant-level firm database and web text mining. Sci. Total. Environ. **835**, 155512 (2022), ISSN 0048–9697. https://doi.org/10.1016/j.scitotenv.2022.155512. https://www.sciencedirect.com/science/article/pii/S0048969722026080

42. N. Smeuninx, B. De Clerck, W. Aerts, Measuring the readability of sustainability reports: a corpus-based analysis through standard formulae and NLP. Int. J. Bus. Commun. **57**(1), 52–85 (2020). https://doi.org/10.1177/2329488416675456
43. M. Gutierrez-Bustamante, L. Espinosa-Leal (2022) Natural language processing methods for scoring sustainability reports—a study of Nordic listed companies. Sustainability **14**, 9165 (2022). https://doi.org/10.3390/su14159165
44. W. Wu, W. Chen, Y. Fu, Y. Jiang, G.Q. Huang, Unsupervised neural network-enabled spatial-temporal analytics for data authenticity under environmental smart reporting system. Comput. Ind. **141**, 103700 (2022), ISSN 0166–3615. https://doi.org/10.1016/j.compind.2022.103700. (https://www.sciencedirect.com/science/article/pii/S0166361522000975)
45. C. Henao-Rodríguez, J.P. Lis-Gutiérrez, H.D. Angulo-Bustinza, Unveiling greenwashing in Colombian manufacturing: amachine learning approach. Res. Glob. **8**, 100196 (2024), ISSN 2590–051X. https://doi.org/10.1016/j.resglo.2024.100196. (https://www.sciencedirect.com/science/article/pii/S2590051X24000054)
46. A. Vaswani, N. Shazeer, N. Parmar, J. Uszkoreit, L. Jones, A.N. Gomez, L. Kaiser, I. Polosukhin, Attention is all you need (2017). https://arxiv.org/abs/1706.03762
47. Decision Tree Algorithm in Machine Learning—JavatPoint (n.d.), www.javatpoint.com. https://www.javatpoint.com/machine-learning-decision-tree-classification-algorithm. Accessed 11 Jul 2024
48. Central Consumer Protection Authority of India, Guidelines for Prevention and Regulation of Greenwashing. F.No.-CCPA/28/2023 (2023), https://consumeraffairs.nic.in/sites/default/files/file-uploads/latestnews/Draft%20Guidline%20with%20approval.pdf

Chapter 7
Energy-Development Nexus Review: PV Potential in Asian Least Developed Countries (ALDCs)

Shivani Singh and Rahul Dev

Abstract Energy indirectly affects the productivity of the industrial and economic sectors, which is a major factor in any country's economic health. In the least developed Asian countries, most people still struggle to meet their basic energy needs. Without addressing inequality in the least developed Asian countries, a World Bank report states that universal access to renewable energy will remain unattainable. The answer is to switch to renewable energy sources, such as wind, solar, and biofuels. Among all renewable energy sources, photovoltaic (PV) technologies have shown to be a viable source of inexpensive, clean energy. This research delves into a thorough discussion of economic and geographical condition of Asian least developed countries. Consequently, in order to maintain the development of solar technology capacity, a thorough analysis of the renewable energy policies of the least developed Asian countries is required, along with a discussion of the policy's shortcomings. The location of solar energy potential in the least developed Asian nations—Bangladesh, Bhutan, Afghanistan, Myanmar, Republic of Yemen, Timor-Leste, Nepal, Cambodia, and Lao People's Democratic Republic—is provided in depth by this study.

Keywords Renewable power generation · Photovoltaic · SDG 7 · Solar PV potential

S. Singh · R. Dev (✉)
Department of Mechanical Engineering, Motilal Nehru National Institute of Technology Allahabad, Prayagraj, Uttar Pradesh, India
e-mail: rahuldsurya@mnnit.ac.in

S. Singh
e-mail: shivani.2022rme10@mnnit.ac.in

A. Kumar et al. (eds.), *Selected Proceedings of the 1st International Conference on Advanced Materials for Sustainable Innovation; IC-AMSI 2024; 28–30 August; New Delhi; India*, Springer Proceedings in Energy,
https://doi.org/10.1007/978-981-96-4492-6_7

7.1 Introduction

An increase in energy consumption is often necessary for a country's socioeconomic progress to result in higher living standards and a decline in poverty. However, the GDP of a nation, which makes up roughly 90% of this indicator, is typically used to estimate the level of living in that nation [34]. Renewably sourced energy has been used historically, and its importance is now acknowledged because fossil fuels are becoming scarce and environmental issues are becoming more pressing [3, 39]. According to several studies [22, 24, 32, 38], using renewable energy sources can help reduce pollution and carbon dioxide (CO_2) levels. The International Renewable Energy Agency (IRENA) [28] states that in order to meet the goals of the 2015 Paris Climate Agreement, which limit global average temperature increases to 2 °C, the proportion of renewable energy in the world's energy mix must increase from 16% in 2020 to 77% by 2050 [21] .But it's important to remember that the least developed countries have some of the lowest rates of grid power availability, which means they depend heavily on imported energy. According to Sadat and Sabory [35], a country's ability to succeed depends on its ability to become energy independent, which lessens reliance on outside energy sources. A robust relationship has been observed between socioeconomic prosperity, stability at the national level, and the attainment of energy independence via the encouragement of domestic source diversification.

7.2 Least Developed Nations

"Low-income nations with significant structural barriers to sustainable development are known as least developed countries, or LDCs. Their human capital is scarce, and they are extremely susceptible to environmental and economic shocks".

7.2.1 Categorization of LDCs

Three criteria i.e., GNI per capita, HAI and EVI are used to categorize any nation as least developed countries (LDCs). The value of these parameter—Gross National Income (GNI) per capita must be ≤$1,088, Human Assets Index (HAI) must be ≤60 and Economic Vulnerability Index (EVI) must be ≥36. Out of all three criteria, if meeting the two criteria designated as the status of Least developed countries (LDCs). The Committee for Development (CDP) reviews the list of least developed nations (LDCs) every three years, and at the moment, 45 countries are included. The following regions comprise the distribution of these 45 LDCs: "Africa (33): Angola, Benin, Burkina Faso, Burundi, Central African Republic, Chad, Comoros, Democratic Republic of the Congo, Djibouti, Eritrea, Ethiopia, Gambia, Guinea,

Table 7.1 Geographical location data of Asian least developed countries (ALDC) [6]

Country name	Latitude	Longitude (°E)	Altitude (m)	Gross National Income (GNI) per capita ≤ $1,088 (figures as per 2024)	Human Asset Index (HAI) ≤ 60 (figures as per 2024)	Economic Vulnerability Index (EVI) ≥ 36 (figures as per 2024)
Afghanistan	34.53°N	69.17	1,809	$436.96	37.50	49.4
Bangladesh	23.71°N	90.41	22	$2684	26.68	77.26
Cambodia	12.56°N	104.99	126	$1590	29.98	74.98
Lao people's democratic republic	19.85°N	102.49	2,880	$2502	26.91	74.17
Myanmar	21.91°N	95.95	5,881	$1381	24.35	78.52
Nepal	27.70°N	85.32	1,289	$1300	24.57	77.36
Timor-Leste	8.87°S	125.72	2,963	$2313	38.80	61.01
Republic of Yemen	15.55°N	48.51	3,667	$350	62.45	38.86

Guinea-Bissau, Lesotho, Liberia, Madagascar, Malawi, Mali, Mauritania, Mozambique, Niger, Rwanda, Sao Tome and Principe, Senegal, Sierra Leone, Somalia, South Sudan, Sudan, Togo, Uganda, United Republic of Tanzania, and Zambia. Asia (8): Yemen, Timor-Leste, Bangladesh, Cambodia, Lao People's Democratic Republic, Myanmar, Nepal, and Afghanistan. Haiti in the Caribbean (1). Three Pacific Islands: Tuvalu, Kiribati, and the Solomon Islands".

7.2.2 Asian Least Developed Countries

Asian Least developed nation comprises of "Afghanistan, Bangladesh, Bhutan, Cambodia, Lao People's Democratic Republic, Myanmar, Nepal, Timor-Leste & Republic of Yemen". The geographical and economic condition of Asian least developed countries is mentioned in Table 7.1.

7.2.2.1 Afghanistan

Afghanistan has a great deal of potential for using solar, wind, hydro, biomass, and geothermal energy, among other renewable energy sources. These have the potential to make Afghanistan energy self-sufficient; of these, solar energy with a capacity of 222,000 MW and wind energy with a capacity of 66,000 MW are particularly promising [30, 31]. Afghanistan is anticipated to have 64.68 million people living there by 2050, based on UN estimates and a 2.4% annual growth rate [5]. Afghanistan is currently experiencing a noticeable rise in its energy use. Precise estimates of

Table 7.2 Solar PV potential location in Afghanistan [7]

S. No	Parameters	Location (maximum potential)	Location (minimum potential)
1	Specific photovoltaic power output (SPPO) (kWh/kWp)	Ghazni (5.65)	Kunduz (4.49)
2	Direct normal irradiation (DNI) (kWh/m^2)	Ghazni (7.16)	Kunduz (4.67)
3	Global horizontal irradiation (GHI) (kWh/m^2)	Hilmand (6.13)	Kunduz (4.98)
4	Diffuse horizontal irradiation (DHI) (kWh/m^2)	Nimroz (2.46)	Ghor (1.71)
5	Global tilted irradiation (GTI) (kWh/m^2)	Zabul and Kandahar (6.97)	Kunduz (5.63)
6	Air temperature (T_a)	Nimroz (24.9 °C)	Bamyan (7.7 °C)

electricity demand were given in the 2012 Afghanistan electricity Sector Master Plan (APSMP). According to World Bank data, net demand increased by 9.8% annually on average from 2,800 GWh in 2012 to 15,909 GWh by 2032 [36]. As of right now, renewable energy makes up about 3% of Afghanistan's installed capacity—hydropower excluded. By 2032, this percentage is expected to rise to 10% of the total demand. Afghanistan's solar energy potential (see Table 7.2) is projected at 222 GW, with more than 300 bright days annually.

7.2.2.2 Bangladesh

At the moment, Bangladesh produces 723.26 MW of power using renewable energy sources. Of this total, solar power accounts for 67.61%, hydroelectric power for 31.80%, and other power sources such as wind, biogas, and biomass account for 0.58%. Over 6 million solar photovoltaic (PV) systems that have been built generate 489 MW of solar energy [10, 29]. There are 6,408,721 renewable energy plants in operation in the nation. However, according to Hossain et al. [27], over 42% of rural areas still do not have access to electricity. With enough of sunlight accessible for the most of the year, Bangladesh's advantageous geographic location makes it perfect for the production of solar energy [27]. According to Ahmed et al. [2], solar radiation peaks in March and April and troughs in December and January. When fog, rain, and clouds are taken out of the equation, solar energy availability is between 4.0 and 6.5 kWh/m^2/day, with an average of 6 to 9 h of sunny daylight per day for around 300 days a year (see Table 7.3). The need for solar electricity can be met by this abundant solar radiation [1, 37].

Table 7.3 Solar PV Potential Location in Bangladesh [8]

S. No	Parameters	Location (maximum potential)	Location (minimum potential)
1	SPPO (kWh/kWp)	Chittagong (4.25)	Rangpur (3.85)
2	DNI (kWh/m^2)	Chittagong (3.84)	Rangpur (2.95)
3	GHI (kWh/m^2)	Chittagong (4.97)	Rangpur (4.53)
4	DHI (kWh/m^2)	Khulna (2.63)	Sylhet (2.51)
5	GTI (kWh/m^2)	Chittagong (5.36)	Rangpur (4.87)
6	T_a	Khulna (26.6 °C)	Rangpur (25.3 °C)

Table 7.4 Solar PV potential location in Cambodia [9, 15]

S. No	Parameters	Location (maximum potential)	Location (minimum potential)
1	SPPO (kWh/kWp)	Kâmpóng Cham (4.36)	Krong Preah Sihanouk (3.88)
2	DNI (kWh/m^2)	Kâmpóng Chhnang (4.21)	Kep (3.32)
3	GHI (kWh/m^2)	Kâmpóng Cham (5.49)	Krong Preah Sihanouk (4.83)
4	DHI (kWh/m^2)	Siemréab and Svay Rieng (2.54)	Kaôh Kong (2.45)
5	GTI (kWh/m^2)	Kâmpóng Cham (5.62)	Krong Preah Sihanouk (4.94)
6	T_a	Kâmpóng Cham (28.9 °C)	Chin (27.1 °C)

7.2.2.3 Cambodia

Photovoltaic (PV) systems are expanding in Cambodia as a result of both individual initiatives and auction schemes. It is anticipated that the industry sectors and government of Cambodia will have installed about 425 MW of PV capacity by 2021, mostly through ground-mounted systems unrelated to auction schemes. In order to carry out a 100 MW auction plan, the government and the Asian Development Bank have partnered since 2019. Furthermore, PV systems, which account about 455 MW at various levels of completion, are still being developed without previous requests [23]. Solar PV potential of Cambodia is shown in Table 7.4.

7.2.2.4 Lao People's Democratic Republic

The "long-range energy alternatives planning system (LEAP) model" (Lao PDR Energy Outlook 2020) was used by the Ministry of Energy and Mines (MEM) and ERIA to create a thorough review of Lao energy prospects and future forecasts up to 2040. Imports and hydropower coal are Laos's primary energy sources. However, because hydropower output quickly decreases in the summer, the government is compelled to purchase electricity. The Lao's PV installed capacity as of 2021 is around 34 MW. To diversify the electricity generation, the Lao PDR government has begun awarding projects related to photovoltaics. Additionally, Wealth Power Group

Table 7.5 Solar PV potential location in Lao people's democratic republic [9, 10]

S. No	Parameters	Location (maximum potential)	Location (minimum potential)
1	SPPO (kWh/kWp)	Champasak (4.23)	Phôngsali (3.69)
2	DNI (kWh/m^2)	Attapu (3.97)	Houaphan (3.02)
3	GHI (kWh/m^2)	Champasak (5.23)	Phôngsali (4.34)
4	DHI (kWh/m^2)	Champasak (2.48)	Xékong (2.37)
5	GTI (kWh/m^2)	Champasak (5.43)	Phôngsali (4.60)
6	T_a	Champasak (27.9 °C)	Houaphan (23.1 °C)

Table 7.6 Solar PV potential location in Myanmar [12]

S. No	Parameters	Location (maximum potential)	Location (minimum potential)
1	SPPO (kWh/kWp)	Magway (4.47)	Kachin (4.09)
2	DNI (kWh/m^2)	Magway (4.54)	Kachin (3.71)
3	GHI (kWh/m^2)	Magway (5.32)	Kachin (4.70)
4	DHI (kWh/m^2)	Tanintharyi (2.46)	Naypyitaw (2.26)
5	GTI (kWh/m^2)	Magway (5.76)	Kachin (5.17)
6	T_a	Mandalay (28.8 °C)	Chin (25.6 °C)

has shown interest to construct a big photovoltaic system with a capacity of 580 MW [11]. PV potential of Lao People's Democratic Republic is shown in Table 7.5.

7.2.2.5 Myanmar

During the past 10 years, the ongoing political and economic unrest has posed significant challenges for the growth of solar photovoltaic (PV) in Myanmar (see Table 7.6). Solar power generation capacity showed a gradual increase, despite notable breakthroughs, rising from an initial level of about 30 MW in 2016 to approximately 90 MW in 2019. Solar grow Renewables and China Machinery Engineering Corp., two Chinese firms, now hold most of the PV installation contracts in Myanmar. However, keeping in mind the present political chaos, some developers are reevaluating their projects and may even be thinking about withdrawing. The limited availability of funds is a result of financial institutions cautious lending practices in the current uncertain environment.

7.2.2.6 Nepal

Nepal now uses 267 kilowatt-hours (kWh) of electricity per person annually [13]. Table 7.7 illustrates the enormous potential that Nepal has for solar PV. The goal

Table 7.7 Solar PV potential location in republic of Nepal [14]

S. No	Parameters	Location (maximum potential)	Location (minimum potential)
1	SPPO (kWh/kWp)	Western (5.84)	Central (4.55)
2	DNI (kWh/m^2)	Western (7.35)	Central (4.29)
3	GHI (kWh/m^2)	Western (6.09)	Central (4.94)
4	DHI (kWh/m^2)	Eastern (2.59)	Far-Western (2.47)
5	GTI (kWh/m^2)	Western (6.99)	Central (5.51)
6	T_a	Central and Eastern (25.6 °C)	Far-Western (24.6 °C)

set forth by Nepal's Ministry of Energy, Water Resources, and Irrigation is for the country's per capita energy consumption to reach 1500 kWh by 2030.

7.2.2.7 Timor-Leste

Timor-Leste is a tropical country with the ability to utilize a wide variety of renewable energy sources, such as solar (see Table 7.8), wind, hydro, and bioenergy. These resources have the potential to greatly improve the nation's energy supply in the future if they are handled well. Timor-Leste has a net export of crude oil, but its lack of distillation facilities and heavy reliance on imported energy for transportation and electricity result in high yearly energy expenses. For example, the government's yearly gasoline spending in 2007 was approximately US $26 million, and it continued to rise in the following years. The nation can produce solar energy over its whole territory because of its average annual solar radiation, which varies from 14.85 to 22.33 MJ/m^2 per day (Timor-Leste Plan 2011–2030). Leveraging this potential, Timor-Leste has installed 1,228 solar energy units in remote household areas [15].

Table 7.8 Solar PV potential location in republic of Timor-Leste [16]

S. No	Parameters	Location (maximum potential)	Location (minimum potential)
1	SPPO (kWh/kWp)	Dili (4.85)	Aileu (4.48)
2	DNI (kWh/m^2)	Ambeno (5.61)	Ermera (5.03)
3	GHI (kWh/m^2)	Ambeno (5.94)	Aileui (5.53)
4	DHI (kWh/m^2)	Lautém (2.35)	Ainaro (2.13)
5	GTI (kWh/m^2)	Ambeno (6.08)	Aileui (5.63)
6	T_a	Liquiçá, Ermera and Bobonaro (28.3 °C)	Aileu (26.4 °C)

Table 7.9 Solar PV potential location in republic of Yemen [17]

S. No	Parameters	Location (maximum potential)	Location (minimum potential)
1	SPPO (kWh/kWp)	San'a' (5.74)	Al Hudaydah (4.81)
2	DNI (kWh/m^2)	Hadramawt (7.41)	Al Hudaydah (4.63)
3	GHI (kWh/m^2)	Shabwah (6.85)	Al Hudaydah (5.93)
4	DHI (kWh/m^2)	Hajjah and Al Hudaydah (2.78)	Amanat Al Asimah (2.1)
5	GTI (kWh/m^2)	Hadramawt (7.30)	'Adan (6.37)
6	T_a	Al Hudaydah (31.7 °C)	Amanat Al Asimah (20 °C)

7.2.2.8 Republic of Yemen

Since 2014, Yemen has been the scene of foreign military involvement amid a domestic armed struggle. The majority of people have been cut off from the public energy infrastructure during the conflict. Still, the nation has turned to decentralized solar energy because no other options were available, leading to an unprecedented deployment of solar (home) systems [19]. Yemen is the most economically, educationally, commercially, and tourismally disadvantaged country in the Middle East, largely due to the lack of electricity [33, 34]. Hence, a comprehensive solution to the electricity problem is essential to promote national prosperity and lessen its challenges. Because Yemen has an abundance of renewable energy sources, including biomass, geothermal, solar (see Table 7.9), and wind power, its people are fortunately able to handle this issue on their own [4, 25, 26]. Potentially enough ability to generate electricity and close the current energy gap is offered by the abundance of renewable resources. The utilization of these renewable resources can also be a useful means of achieving the Sustainable Development Goals (SDGs) and offer a number of long-term benefits, including the generation of jobs, improvement of energy security, encouragement of business ventures, progress of environmental welfare, and reduction of global warming [20].

7.3 Results and Discussion

Asian least developed countries have a huge potential for installing the solar PV power in the region. But due to lack of government commitment, policymakers, industry peoples etc. real potential is not harnessed in proper manner. Figure 7.1 shows the huge potential for photovoltaic power in Asian least developed countries.

From the Fig. 7.2, it is clearly shown that the Specific photovoltaic power output has great potential for Asian least developed countries.

From Fig. 7.3, It is clearly seen that the Global horizontal irradiation is maximum in the Yemen region. But from above review discussion of Yemen, country have not explored much in solar Photovoltaic.

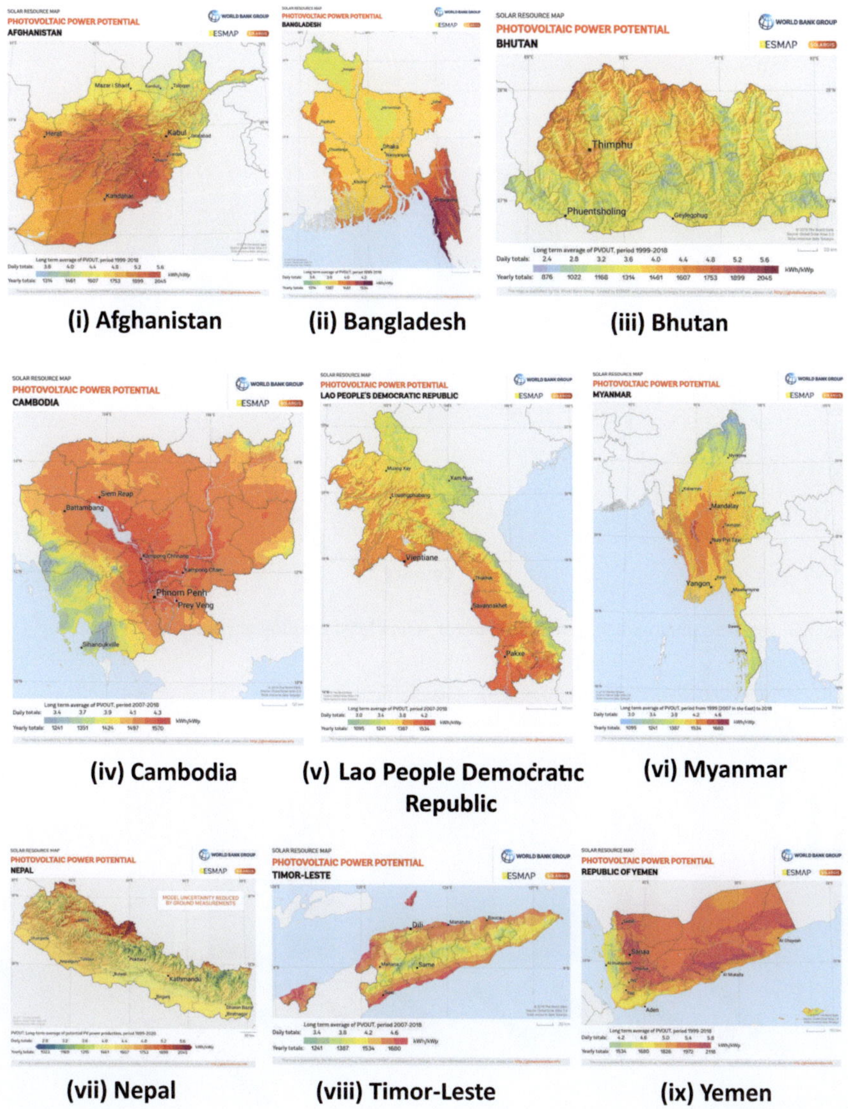

Fig. 7.1 Photovoltaic power potential [18] (i) Afghanistan (ii) Bangladesh (iii) Bhutan (iv) Cambodia (v) Lao People's democratic republic (vi) Myanmar (vii) Nepal (viii) Timor-Leste (ix) Yemen

From Fig. 7.4, it is clearly seen that the Diffuse horizontal irradiation for Asian least developed countries having a great potential for power generation.

Optimum tilt of PV for Asian least developed countries can be seen in Fig. 7.5. Solar PV output greatly influenced by the parameter such as placing of PV module at optimum angle.

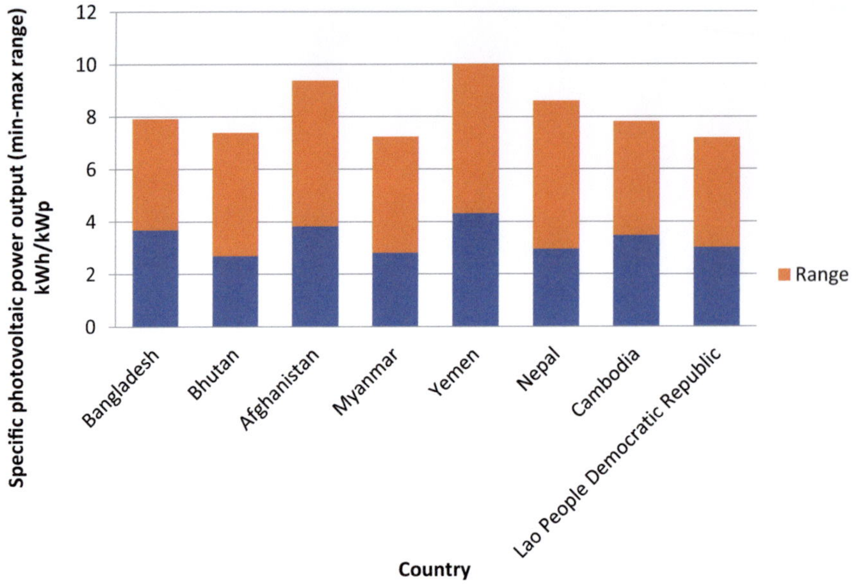

Fig. 7.2 Specific photovoltaic power output for Asian least developed countries

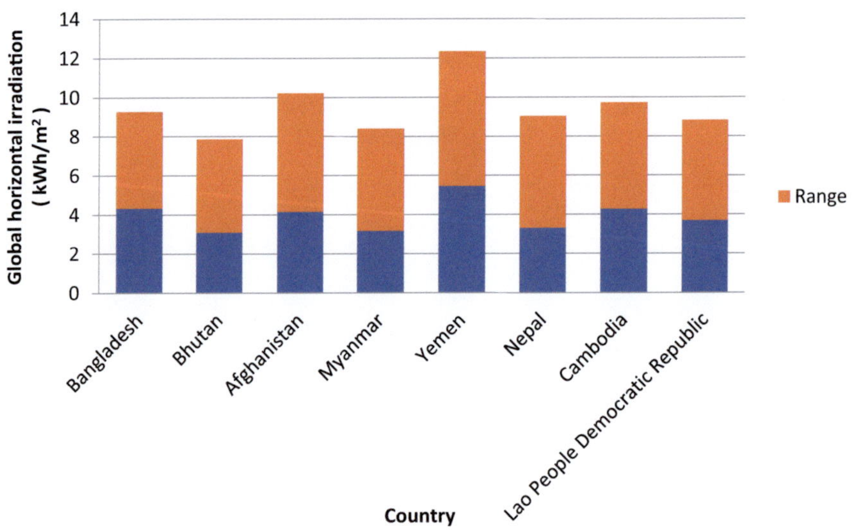

Fig. 7.3 Global horizontal irradiation for Asian least developed countries

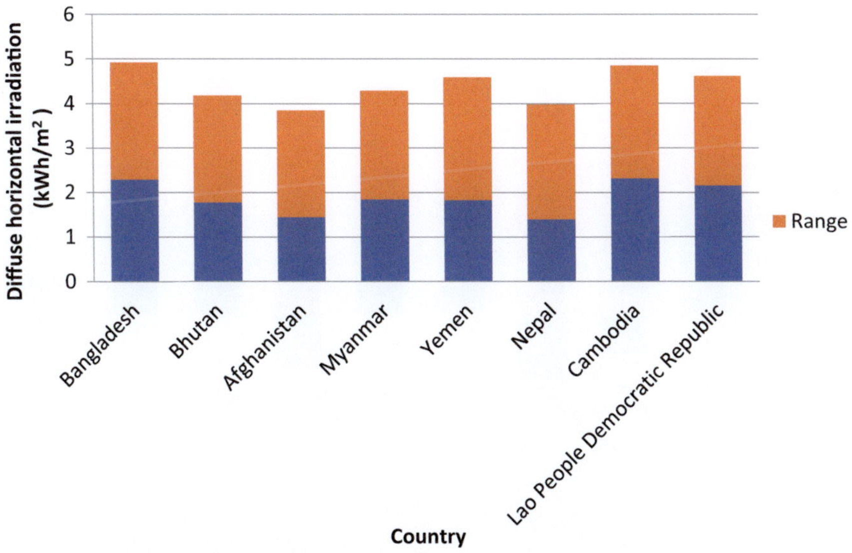

Fig. 7.4 Diffuse horizontal irradiation for Asian least developed countries

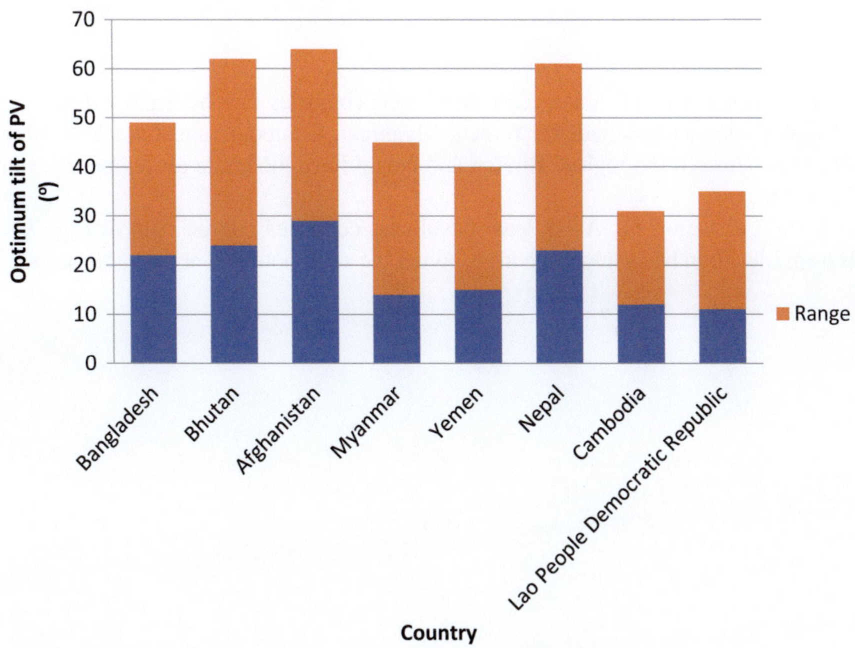

Fig. 7.5 Optimum tilt of PV for Asian least developed countries

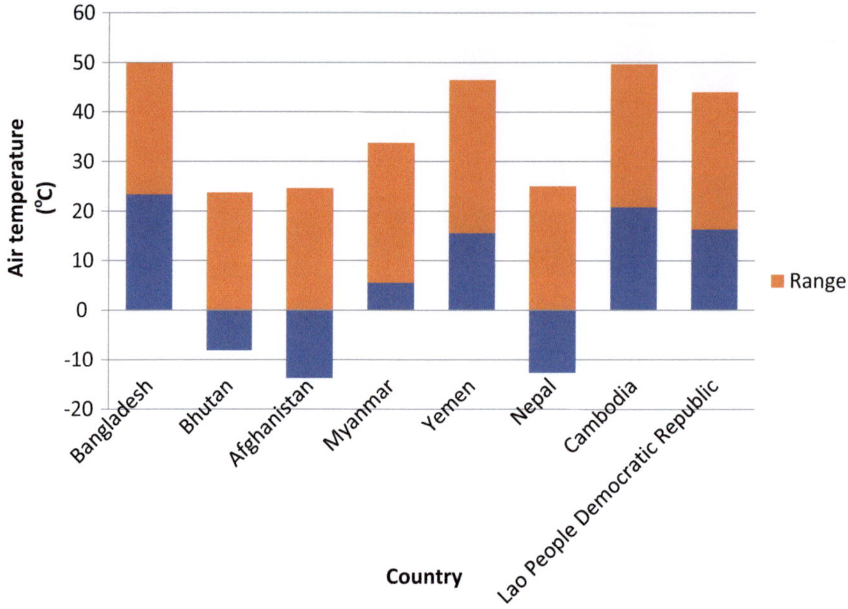

Fig. 7.6 Air temperature for Asian least developed countries

Air temperature for Asian least developed countries can be seen in Fig. 7.6. Countries such as Bangladesh, Yemen, Myanmar, Cambodia and Laos have high air temperature in the region. Bhutan and Nepal have moderate air temperature in the region.

Terrain elevation for Asian least developed countries can be seen in Fig. 7.7. Terrain Elevation has a huge role in receiving the solar potential at the earth surface.

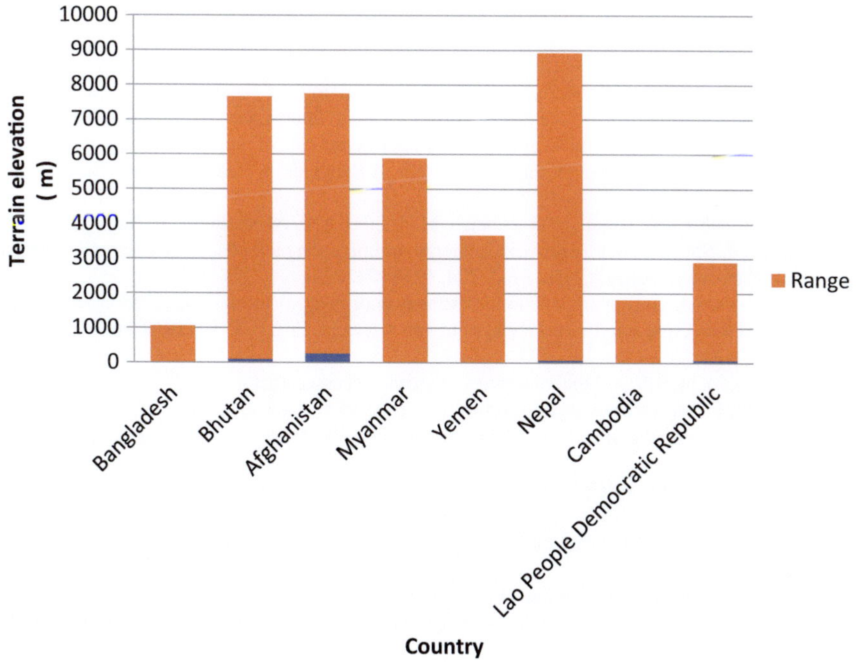

Fig. 7.7 Terrain elevation for Asian least developed countries

7.4 Conclusion

- Asian least developed countries are located in the torrid zone (23.5 °N to 23.5 °S), where maximum solar radiation is available. These nations have shown interest in installing PV systems to meet their future energy demands. Asian least developed countries have significant potential for solar PV power installation.
- The solar energy potential varies by country, with Afghanistan (4.67 to 7.16 kWh/m2), Bangladesh (2.95 to 4.97 kWh/m2), Nepal (4.29 to 7.35 kWh/m2), Myanmar (3.71 to 5.32 kWh/m2), Cambodia (4.83 to 5.49 kWh/m2), Yemen (4.63 to 7.34 kWh/m2), and Timor-Leste (5.03 to 5.94 kWh/m2).
- To achieve socioeconomic development, the enabling infrastructure, policies and regulations, skills, and institutional capacity in ALDCs must be upgraded.

References

1. M.D. Ahiduzzaman, A.S. Islam, Greenhouse gas emission and renewable energy sources for sustainable development in Bangladesh. Renew. Sustain. Energy Rev. **15**(9), 4659–4666 (2011)
2. S. Ahmed, M.T. Islam, M.A. Karim, N.M. Karim, Exploitation of renewable energy for sustainable development and overcoming power crisis in Bangladesh. Renew. Energy **72**, 223–235 (2014)
3. O.A. Al-Shahri, F.B. Ismail, M.A. Hannan, M.H. Lipu, A.Q. Al-Shetwi, R.A. Begum, N.F. Al-Muhsen, E. Soujeri, Solar photovoltaic energy optimization methods, challenges and issues: a comprehensive review. J. Clean. Prod. **284**, 125465 (2021)
4. A.Q. Al-Shetwi, Design and economic evaluation of electrification of small villages in rural area in Yemen using stand-alone PV system. Int. J. Renew. Energy Res. (IJRER) **6**(1), 289–298 (2016)
5. Anon (2023), https://worldpopulationreview.com. Accessed 18 Jun 2023
6. Anon (2023), https://www.un.org/ohrlls/content/list-ldcs. Accessed 10 Dec 2023
7. Anon (2023), https://globalsolaratlas.info/map?c=34.055761,67.7042,6&r=AFG. Accessed 11 Dec 2023
8. Anon (2023), https://globalsolaratlas.info/map?c=23.647356,90.3458,6&r=BGD. Accessed 12 Dec 2023
9. Anon (2023), https://globalsolaratlas.info/map?c=18.26151,103.8583,6&r=LAO. Accessed 14 Dec 2023
10. Anon (2023), https://live7.bmd.gov.bd/. Accessed 15 Dec 2023
11. Anon (2024), https://www.irena.org/publications/2021/Aug/Renewableenergystatistics-2021. Accessed 5 Jan 2024
12. Anon (2024), https://globalsolaratlas.info/map?c=19.436959,96.67085,5&r=MMR. Accessed on 17 Feb 2024
13. Anon (2024), https://globalsolaratlas.info/map?c=28.415608,84.12915,7&r=NPL. Accessed 18 Mar 2024
14. Anon (2024), https://globalsolaratlas.info/map?c=-8.813139,125.6917,8&r=TLS. Accessed on 30 Mar 2024
15. Anon (2024), https://globalsolaratlas.info/map?c=15.583037,48.175,6&r=YEM. Accessed on 31 Mar 2024
16. Anon (2024), https://thehimalayantimes.com/business/electricity-consumption-increases-to-267-units. Accessed on 26 Apr 2024
17. Anon (2024), https://www.iea.org/policies/5844-the-national-energy-security-strategy-and-policy. Accessed on 4 May 2024
18. Anon (2024), https://www.unep.org/resources/un-environment-annual-report-2017. Accessed on 10 May 2024
19. D. Ansari, C. Kemfert, H. Al-Kuhlani, *Yemen's Solar Revolution: Developments, Challenges, Opportunities*, (Politikberatung kompakt, DIW Berlin, 2019)
20. A. Armin Razmjoo, A. Sumper, A. Davarpanah, Energy sustainability analysis based on SDGs for developing countries. Energy Sources Part A Recover. Util. Environ. Eff. **42**(9), 1041–1056 (2020)
21. S.K. Baidya, M.L. Shrestha, M.M. Sheikh, Trends in daily climatic extremes of temperature and precipitation in Nepal. J. Hydrol. Meteorol. **5**(1), 38–51 (2008)
22. V. Belessiotis, E. Delyannis, The history of renewable energies for water desalination. Desalination **128**(2), 147–159 (2000)
23. L.W. Beurskens, M. Hekkenberg, P. Vethman, Renewable energy projections as published in the national renewable energy action plans of the European member states. ECN and EEA (2011)
24. M.K. Firozjaei, O. Nematollahi, N. Mijani, S.N. Shorabeh, H.K. Firozjaei, A. Toomanian, An integrated GIS-based ordered weighted averaging analysis for solar energy evaluation in Iran: current conditions and future planning. Renew. Energy **136**, 1130–1146 (2019)

25. S.M.B. Gadhi, M.A. Mukbel, A review of renewable energy activities in Yemen. Renew. Energy **14**(1–4), 459–465 (1998)
26. A. Ghalib Alkholidi, Renewable energy solution for electrical power sector in Yemen. Int. J. Renew. Energy Res. **3**(4), 803–811 (2013)
27. M.F. Hossain, S. Hossain, M.J. Uddin, Renewable energy: Prospects and trends in Bangladesh. Renew. Sustain. Energy Rev. **70**, 44–49 (2017)
28. IRENA, *World Energy Transitions Outlook 2023: 1.5°C Pathway* (International Renewable Energy Agency, Abu Dhabi, 2023)
29. A. Islam, E.S. Chan, Y.H. Taufiq-Yap, M.A.H. Mondal, M. Moniruzzaman, M. Mridha, Energy security in Bangladesh perspective—an assessment and implication. Renew. Sustain. Energy Rev. **32**, 154–171 (2014)
30. G.A. Ludin, M.A. Amin, A. Aminzay, T. Senjyu, Theoretical potential and utilization of renewable energy in Afghanistan. Aims Energy **5**(1), 1–19 (2016)
31. G.A. Ludin, H. Matayoshi, M.S.S. Danish, A. Yona, T. Senjyu, Hybrid PV/wind/diesel based distributed generation for an off-grid rural village in Afghanistan. J. Energy Power Eng. **11**(2) (2017)
32. S. Moonchai, N. Chutsagulprom, Short-term forecasting of renewable energy consumption: augmentation of a modified grey model with a Kalman filter. Appl. Soft Comput. **87**, 105994 (2020)
33. S. Nathaniel, S.A.R. Khan, The nexus between urbanization, renewable energy, trade, and ecological footprint in ASEAN countries. J. Clean. Prod. **272**, 122709 (2020)
34. S. Nathaniel, O. Anyanwu, M. Shah, Renewable energy, urbanization, and ecological footprint in the Middle East and North Africa region. Environ. Sci. Pollut. Res. **27**(13), 14601–14613 (2020)
35. S.M.W. Sadat, N.R. Sabory, Afghanistan renewable energy sector's human resources estimation until 2032, in *Repa Proceeding Series*, pp. 96–101 (2020)
36. World Bank Group. Afghanistan Renewable Energy Development Issues and Options. World Bank (2018)
37. U. Zafar, T.U. Rashid, A.A. Khosa, M.S. Khalil, M. Rashid, An overview of implemented renewable energy policy of Pakistan. Renew. Sustain. Energy Rev. **82**, 654–665 (2018)
38. S. Zeng, Y. Liu, C. Liu, X. Nan, A review of renewable energy investment in the BRICS countries: history, models, problems and solutions. Renew. Sustain. Energy Rev. **74**, 860–872 (2017)
39. S. Zereshkian, D. Mansoury, A study on the feasibility of using solar radiation energy and ocean thermal energy conversion to supply electricity for offshore oil and gas fields in the Caspian Sea. Renew. Energy **163**, 66–77 (2021)

Chapter 8
ARTIFICIAL Intelligence in Home Energy Management Systems: Enhancing Sustainability and Grid Stability in Smart Grids

Radhika Kavra, Gaurisha Malhotra, Khushboo Khattar, and Lakshika Chutani

Abstract The rapid advancements in power systems have facilitated the introduction of smart grids (SGs), which enable bidirectional communication between utilities and consumers. A critical component of SGs is the Home Energy Management System (HEMS), which plays a significant role in Demand-Side Management (DSM) to ensure grid stability. Traditional optimization techniques used in HEMS are being increasingly complemented and replaced by Artificial Intelligence (AI)-based methods due to their superior performance. This review paper explores various AI-driven optimization techniques, like Deep Learning (DL) and Machine Learning (ML) algorithms, and their applications in HEMS. It discusses the architecture of HEMS, including smart controllers, smart meters, Renewable Energy Sources (RES), Energy Storage Systems (ESS), and Electric Vehicles (EVs), and highlights the importance of communication technologies in these systems. The paper also examines different DSM programs and optimization strategies for scheduling appliances, providing a detailed literature survey on meta-heuristic, ML, and DL approaches. The findings indicate that AI-based techniques significantly enhance energy efficiency and user comfort, marking a shift towards more intelligent and sustainable energy management solutions. This review aims to give a clear overview of the current situation and future possibilities of AI in HEMS.

Keywords Artificial intelligence (AI) · Home energy management systems (HEMS) · Smart grid · Machine learning (ML) · Deep learning (DL) · Optimization techniques · Demand-side management

---

R. Kavra (✉) · G. Malhotra · K. Khattar · L. Chutani
School of Engineering and Technology, Vivekananda Institute of Professional Studies—Technical Campus, Delhi, India
e-mail: radhika.kavra@vips.edu

A. Kumar et al. (eds.), *Selected Proceedings of the 1st International Conference on Advanced Materials for Sustainable Innovation; IC-AMSI 2024; 28–30 August; New Delhi; India*, Springer Proceedings in Energy,
https://doi.org/10.1007/978-981-96-4492-6_8

8.1 Introduction

8.1.1 Background of Home Energy Management Systems (HEMS)

HEMS have emerged as pivotal tools in managing and optimizing energy consumption within residential settings. These systems leverage advanced technologies to monitor, control, and reduce household energy usage, thus contributing to both cost saving and environmental sustainability. Over the years, HEMS have evolved from simple energy monitoring devices to sophisticated systems that integrate various components such as smart meters, sensors, and home automation devices [1, 2].

8.1.2 Importance of AI in Energy Optimization

The incorporation of AI into HEMS marks an essential advancement in the field of energy management. AI techniques, including machine learning algorithms, deep learning methods, and optimization algorithms, enable more accurate energy consumption predictions, dynamic load management, and efficient integration of renewable energy sources [3, 4]. AI-driven HEMS can analyse a large amount of data is used to find patterns and improve energy use and make real-time adjustments, thereby enhancing overall energy efficiency and reliability. This not only helps in reducing energy costs but also plays a crucial role in minimizing carbon footprints and promoting sustainable living [5–7].

8.1.3 Scope and Objectives of the Review

The goal of this review is to present a thorough examination of the state of AI applications in HEMS today, emphasizing the different AI methods used and how they are really put to use.

The objectives are listed as follows:

1. Identify significant advancement in modern energy systems by smart grids.
2. Examine real-world case studies demonstrating the impact of AI on energy optimization.
3. Identify the challenges and limitations associated with AI-driven HEMS.
4. Discuss future directions and emerging trends that could shape the evolution of AI in home energy management.
5. Explore different machine learning and deep learning methodologies used in HEMS.

8.2 Smart Grids (SGs) and AI

SGs represent a significant advancement in modern energy systems, facilitating a bi-directional flow of information between utilities and consumers as shown in Fig. 8.1. This integration of communication and information technology with the traditional power grid enhances the efficiency, reliability, and sustainability of electricity services [5].

AI in SGs: AI plays an extra ordinary role in optimizing the functionality and efficiency of SGs. By leveraging AI technologies, SGs can be used in the following ways:

- Energy Demand Prediction: AI algorithms, specifically ML models, evaluate and analyse historical energy consumption data to predict future demand. This foreseen capability allows utilities to manage load distribution efficiently, reducing the risk of overloading and enhancing grid stability.
- Renewable Energy Integration: AI facilitates the integration of renewable energy sources, such as solar and wind, into the smart grid. ML models optimize the operation of these sources by forecasting production based on weather patterns and historical data. This ensures a balanced and efficient energy supply.

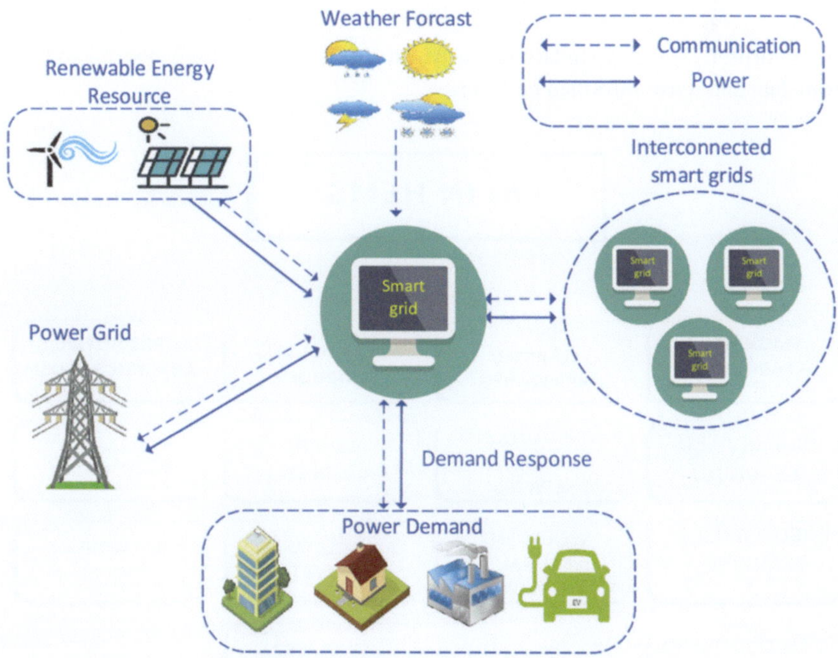

Fig. 8.1 Structure of smart grids

- Automated Grid Operations: AI-driven automation in smart grids includes real-time monitoring, fault detection, and self-healing mechanisms. These capabilities improve the resilience and reliability of the grid, minimizing downtime and operational costs.

Therefore, the incorporation of AI into SGs is transformative, offering enhanced efficiency, reliability, and sustainability [8]. By predicting energy demand, integrating renewable sources, and automating operations, AI ensures that smart grids meet the dynamic needs of modern energy consumers while promoting environmental sustainability [5].

8.3 AI Techniques in HEMS

The integration of AI in HEMS involves a variety of advanced techniques as shown in Fig. 8.2, aims for optimizing energy consumption and enhancing efficiency.

8.3.1 Machine Learning (ML) Algorithms

ML algorithms [1] are foundational to AI-driven HEMS, enabling systems to learn from data and make informed decisions.

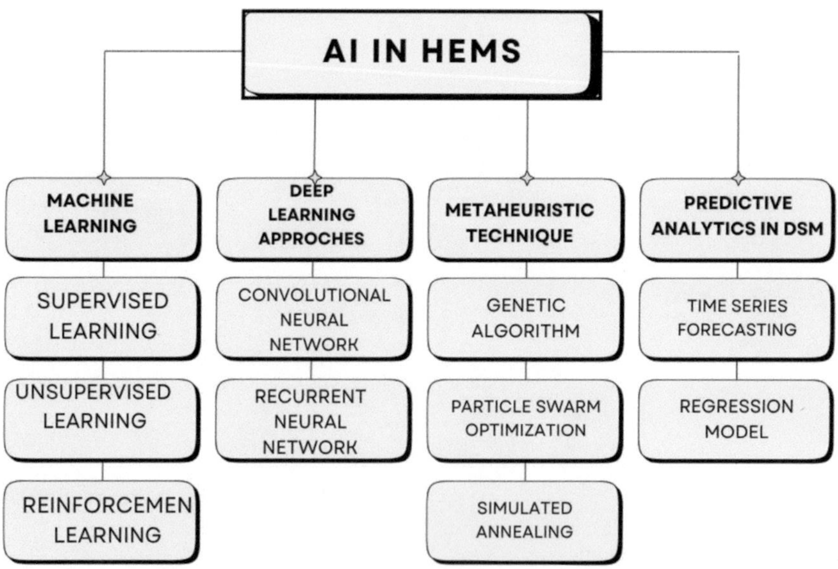

Fig. 8.2 Shows flowchart of AI-techniques in HEMS

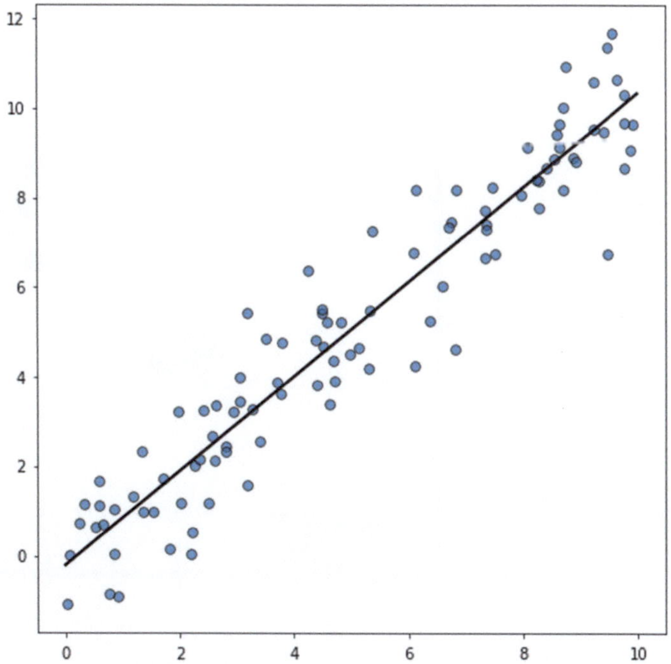

Fig. 8.3 A simple scatter plot with a regression line

Supervised Learning (SL): SL involves training models on labelled datasets to predict outcomes. Classification techniques, such as decision trees and support vector machines (SVM), are used for categorizing energy usage patterns. Regression techniques, including linear regression, demonstrated in Fig. 8.3, and polynomial regression, predict continuous energy consumption values [1].

Equation for Linear Regression:

$$y = \beta o + \beta \iota x \tag{8.1}$$

Unsupervised Learning (USL): USL algorithms identify patterns and anomalies without labelled data. Clustering techniques, as shown in Fig. 8.4, such as k-means and hierarchical clustering, group similar energy consumption behaviour. Anomaly detection methods, like Isolation Forests, identify unusual energy usage that may indicate inefficiencies or faults [1].

Reinforcement Learning (RL): RL involves training an agent to make decisions by maximizing cumulative rewards. Figure 8.5 shows the RL agent-environment interaction. Policy-based methods, such as policy gradient algorithms, directly optimize the policy. Value-based methods, like Q-learning, given by Eq. (8.2) [9], estimate the value of actions to derive optimal policies [2, 9].

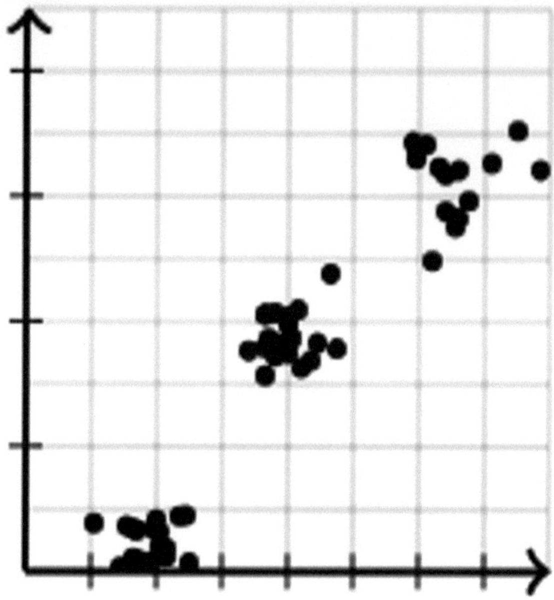

Fig. 8.4 A clustering diagram showing grouped data points

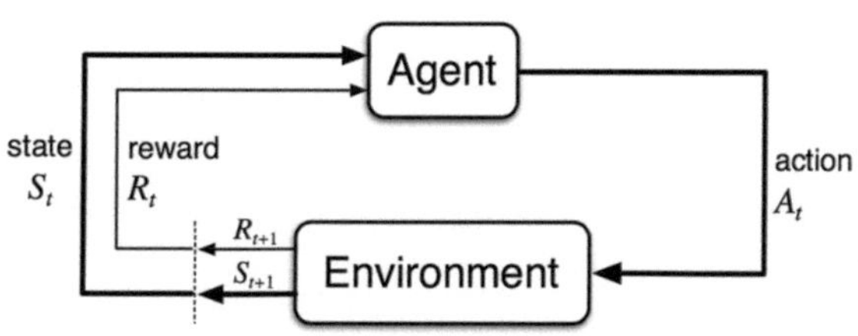

Fig. 8.5 A diagram showing the RL agent-environment interaction

$$Q(s, a) = Q(s, a) + \left(r + \max_{a''} Q(s', a') - Q(s, a)\right) \tag{8.2}$$

8.3.2 Deep Learning (DL) Approaches

DL methods are particularly effective in handling complex and high-dimensional data in HEMS.

8.3.2.1 Convolutional Neural Networks (CNNs)

CNNs are usually used for tasks like recognizing images, but they can also be adapted to predict energy demand. They do this by turning time-series data (data that changes

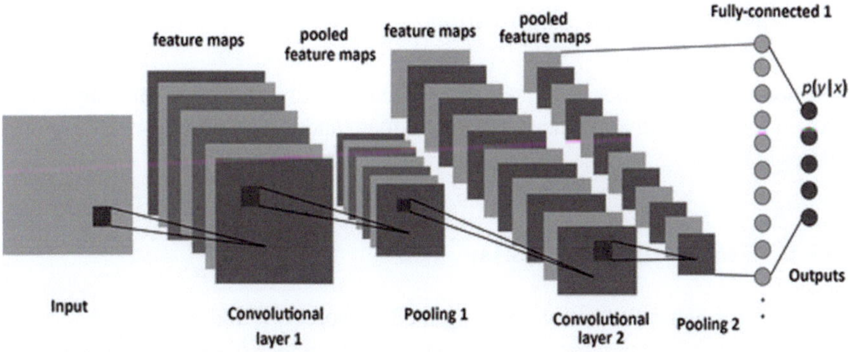

Fig. 8.6 Architecture of a convolutional neural network (CNN)

over time) into a form that looks like images. This helps CNNs spot patterns in energy use and pick out important details, such as trends or unusual energy consumption [6]. Figure 8.6 shows how CNNs work. The convolutional layers find important features in the data, the pooling layers reduce the size of the data to make the process faster, and the fully connected layers make the final predictions about energy use [10].

8.3.2.2 Recurrent Neural Networks (RNNs)

RNNs, especially Long Short-Term Memory (LSTM) networks, are designed for data that unfolds over time, like energy usage data. LSTMs are good at remembering information over long periods, which helps them predict future energy usage by looking at past patterns [6, 7]. Figure 8.7 shows how RNNs work by illustrating how information moves through different time steps. It features LSTM cells, which help keep important information to make better predictions about future energy needs.

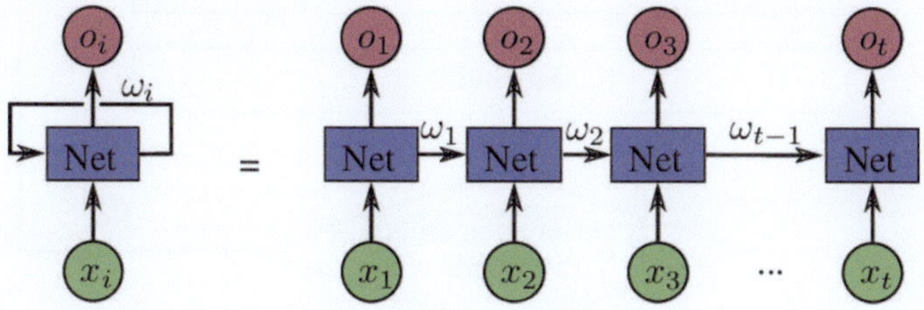

Fig. 8.7 RNN Architecture: flow of information through time steps

8.3.3 Meta Heuristic Techniques

Meta heuristic techniques in HEMS are utilized to optimize energy consumption by efficiently scheduling appliances, balancing load, and integrating renewable energy sources [11].

8.3.3.1 Genetic Algorithms (GAs)

GAs work for natural selection, picking the best solutions and making them better over time where fitness function for selection is given by Eq. (8.3). In HEMS, GAs help find the best schedules for using energy. They do this by creating different options and then improving them through three main steps: selection, crossover, and mutation [2, 12, 13]. Figure 8.8 shows the Genetic Algorithm process.

$$f(x) = \sum\nolimits_{i=1}^{n} ci.xi \tag{8.3}$$

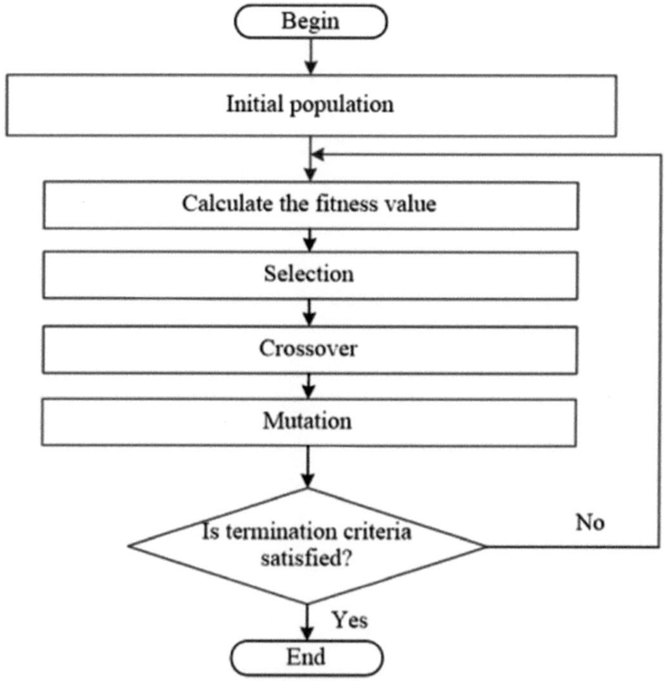

Fig. 8.8 Flowchart of the genetic algorithm process, showing selection, crossover, and mutation steps

8.3.3.2 Particle Swarm Optimization (PSO)

PSO is inspired by how birds and fish move in groups. In energy management systems, PSO helps find the best solution to reduce energy use and costs. Each "particle" (or solution) moves through the problem space, influenced by its best previous position and the overall best position, leading to better energy management while keeping users comfortable [2, 12, 14, 15]. Figure 8.9 helps in explaination this process by showing how particles move in the solution space, adjusting their position based on their previous best and the global best, ultimately converging on the optimal solution for energy savings.

Equation for Velocity Update:

$$v_i(t+1) = wv_i(t) = c_1r_1(p_i - x_i) + c_2r_2(g - x_i) \quad (8.4)$$

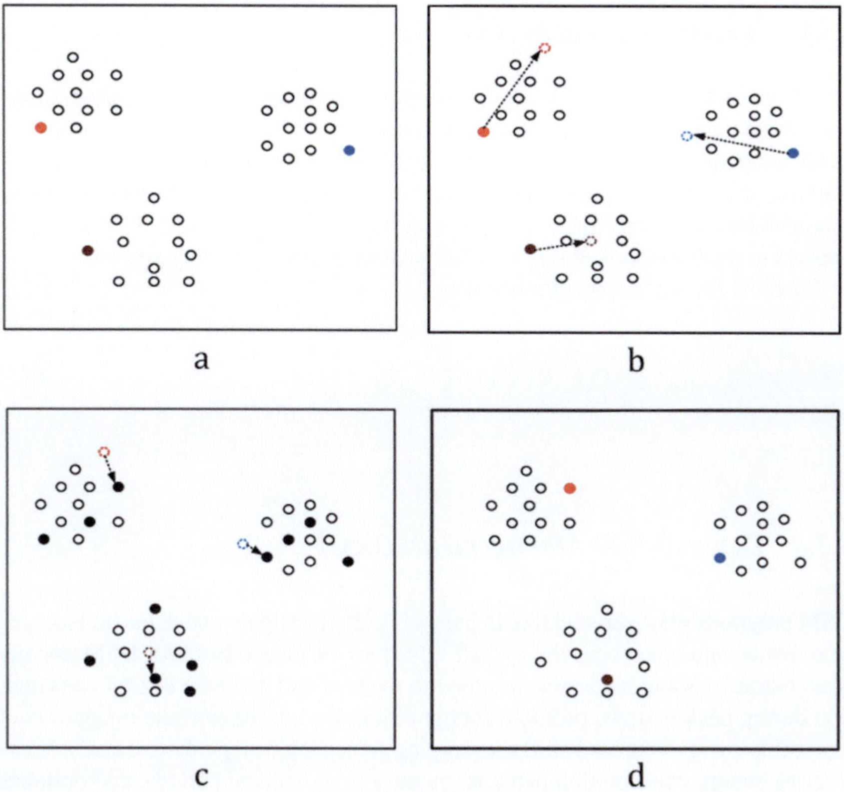

Fig. 8.9 Diagram showing particle movement in solution space

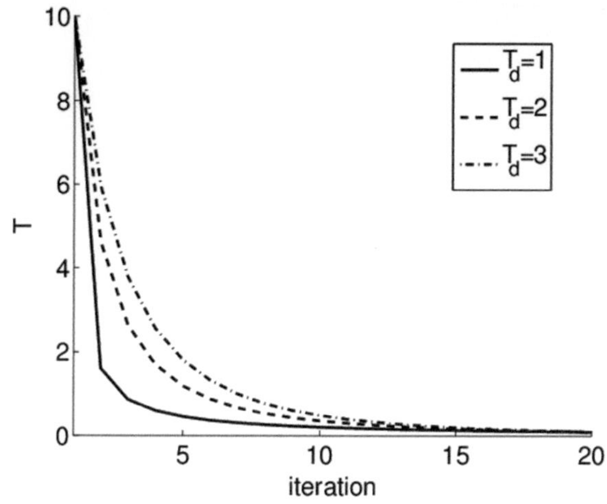

Fig. 8.10 Cooling schedule graph: temperature decreases over iterations

8.3.3.3 Simulated Annealing (SA)

In AI-based HEMS, the SA algorithm explores various energy configurations. Initially, at higher "temperatures," the system explores a wide range of solutions, but as the temperature lowers, exploration narrows, leading to more efficient energy use. Rapid cooling can result in suboptimal solutions, so a slow, controlled cooling process is crucial for AI-driven HEMS to achieve optimal energy efficiency [11]. Figure 8.10 appears to represent the cooling schedule in a simulated annealing process.

Equation for Acceptance Probability:

$$P\left(E, E', T\right) = \begin{cases} 1 & \textit{if } E' < E \\ e^{-\frac{E'-E}{T}} & \textit{if } E' \geq E \end{cases} \tag{8.5}$$

8.3.4 *Demand-Side Management (DSM)*

DSM programs play a crucial role in balancing energy supply and demand, reducing peak loads, and enhancing the overall efficiency of power systems. DSM encompasses strategies and measures designed to manage and decrease energy consumption during peak periods, promote energy efficiency, and support the integration of renewable energy sources [7]. By leveraging AI techniques, predictive analysis can forecast energy consumption patterns, identify peak demand periods, and optimize energy usage. Here's how predictive analysis is integrated into DSM:

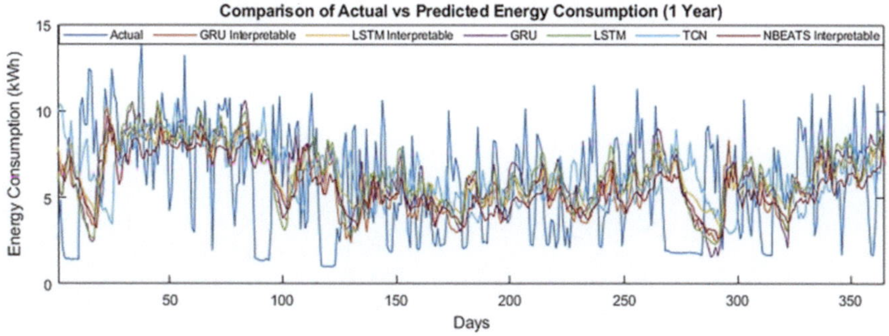

Fig. 8.11 A time series plot showing actual versus predicted energy consumption

Energy storage optimization

Predictive models can optimize the usage of energy storage systems (like batteries) by forecasting when to store excess energy and when to release it. This ensures that stored energy is used efficiently, reducing costs and improving reliability.

Time series forecasting

Time series forecasting involves techniques like ARIMA (Auto Regressive Integrated Moving Average) which is mathematically given by Eq. (8.5) [16] and Prophet. These models predict future energy consumption based on past trends. Figure 8.11 illustrates time series graph showing actual and predicted energy consumption.

$$y_t = c + \phi_1 y_{t-1} + \theta_t \in_{t-1} + \in_t \tag{8.6}$$

Regression Models

Regression models, both linear and nonlinear, are used to predict energy consumption based on various factors such as weather conditions and occupancy patterns. Polynomial regression can model complex relationships between variables and is given by Eq. (8.6) as well as Fig. 8.12 shows that how well the ML model predicts new data using training data.

$$y = \beta o + \beta_1 x + \beta_2 x^2 + \ldots . \beta_n x^n \tag{8.7}$$

Fault Detection and Prevention

Predictive analytics can spot possible problems or inefficiencies in the power grid by looking at past performance data and monitoring it in real time. This helps catch issues early, allowing for timely maintenance and reducing the chance of system breakdowns.

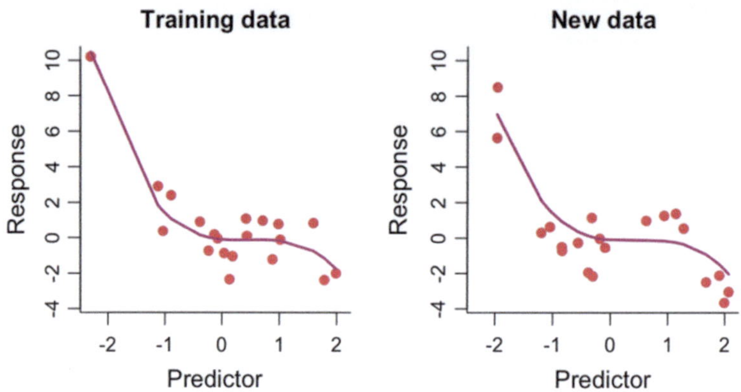

Fig. 8.12 A plot showing the fit of a polynomial regression model to the data

8.4 Qualitative Proof of AI in HEMS

The integration of AI into HEMS has led to significant advancements in various applications, enhancing energy efficiency, cost savings, and sustainability [8, 17]. This section provides a detailed analysis of AI applications in residential energy management, demand response strategies, and renewable energy integration, supported by examples and factual data.

8.4.1 Residential Energy Management

Smart thermostats leverage AI algorithms to optimize home heating and cooling systems, adjusting temperatures based on user behaviour and preferences. These devices use machine learning to predict when occupants are home and automatically adjust settings to maximize comfort and energy efficiency.

Example: The Nest Learning Thermostat uses AI to learn users' schedules and preferences, resulting in substantial energy savings. Also, AI techniques, such as time series forecasting and regression models, are employed to predict household energy consumption patterns. Accurate load forecasting allows for better energy management and planning, reducing the risk of overloading the grid and enabling utilities to optimize energy distribution.

Example: Utilizing ARIMA models, utilities can predict daily energy consumption with high accuracy, improving load balancing and reducing peak demand. Research indicates that such models can enhance prediction accuracy by up to 20%.

8.4.2 Demand Response Strategies

Real-time pricing (RTP) models use AI to dynamically adjust electricity prices based on demand and supply condition. By providing consumers with price signals that reflect real-time energy costs, RTP encourages users to shift their energy usage to off-peak times, reducing overall demand and stabilizing the grid [2, 15].

Example: AI-driven RTP models have been implemented in several pilot programs, resulting in up to 20% reductions in peak demand and significant cost savings for consumers.

Automated Demand Response (ADR) systems utilize AI to automatically adjust or curtail energy usage during peak demand periods. These systems can control various household appliances, such as HVAC systems and water heaters, to reduce load without compromising user comfort.

Example: The Open ADR standard has been widely adopted, enabling seamless communication between utilities and smart devices for efficient demand response. Studies have shown that ADR can lead to a 15% reduction in peak load.

8.4.3 Renewable Energy Integration

Solar Energy Management: AI algorithms are employed to optimize the performance of solar panels by predicting energy production based on weather forecasts and historical data. Machine learning models can adjust the angle of solar panels, manage battery storage, and optimize energy usage to maximize efficiency.

Example: AI-driven solar management systems have demonstrated up to 20% improvements in energy production efficiency, significantly enhancing the viability of solar energy.

Wind Energy Optimization: AI techniques are used in wind energy systems for predictive maintenance and energy prediction. Machine learning models analyse data from sensors to predict potential failures, enabling proactive maintenance and reducing downtime.

Additionally, AI can predict wind energy production based on weather patterns, optimizing the integration of wind energy into the grid.

Example: Implementing AI-based predictive maintenance has reduced operational costs by up to 25% and increased turbine availability by up to 10%.

8.5 AI in HEMS: Balancing Challenges

The integration of AI into HEMS offers significant advantages but also introduces substantial challenges. This section delves into critical issues related to data privacy and security, scalability and integration, model accuracy and reliability, leveraging technical terms and supported by recent research findings.

8.5.1 Data Privacy and Security

The implementation of AI in HEMS necessitates the collection and analysis of extensive sensitive consumer data, such as energy usage patterns. Protecting this data from breaches and unauthorized access is essential. AI systems must adhere to data protection regulations like the General Data Protection Regulation (GDPR) and utilize advanced encryption techniques to secure data.

- **Technical Terms**: Data anonymization, encryption algorithms, GDPR compliance
- **Recent Research**: A 2023 study published in the *Journal of Energy Informatics* revealed that 65% of consumers express concerns about data privacy in AI-driven HEMS. This underscores the necessity for robust encryption and anonymization techniques to maintain consumer trust.

8.5.2 Scalability and Integration

A significant challenge in deploying AI-driven HEMS is integrating AI with existing infrastructure. Ensuring that AI algorithms can efficiently scale and seamlessly work with various hardware and software components in homes is crucial. This integration demands compatibility with diverse devices and communication protocols, such as Zigbee, Z-Wave, and Wi-Fi.

- **Technical Terms**: Interoperability, communication protocols, scalability
- **Recent Research**: A 2022 report in the IEEE Internet of Things Journal highlighted the interoperability challenges in smart homes, noting that 40% of integration failures stem from compatibility issues between AI systems and existing HEMS devices.

8.5.3 Model Accuracy and Reliability

The accuracy and reliability of AI models in HEMS are vital for effective energy management. These models [18, 19] must be resilient to data variations and capable of handling real-time predictions and adjustments. Challenges include balancing model complexity with computational efficiency and ensuring models are robust against data anomalies and noise.

- **Technical Terms**: Model validation, overfitting, real-time processing, anomaly detection
- **Recent Research**: A 2023 study in Applied Energy demonstrated that advanced validation techniques and real-time anomaly detection can enhance the reliability of AI models in HEMS by up to 30%, reducing the occurrence of incorrect energy predictions.

8.5.4 Challenges in Smart Grids

Integrating AI into smart grids presents challenges such as managing vast data volumes, ensuring cybersecurity [5], integrating with legacy systems, and navigating regulatory and ethical issues. Solutions include utilizing advanced data processing and cleaning techniques, implementing robust security measures, adopting modular designs for compatibility with existing infrastructure, leveraging cloud computing for scalability, and developing specialized training programs. Addressing these challenges effectively can enhance grid efficiency and reliability [1].

8.5.5 Meta Heuristic Techniques–Pros and Cons

Meta heuristic techniques, like genetic algorithms and particle swarm optimization, offer advantages in energy management such as flexibility in handling complex, non-linear problems and the ability to find near-optimal solutions efficiently. They can adapt to various constraints and objectives, making them suitable for dynamic environments. However, their limitations include potentially high computational costs, difficulty in guaranteeing global optimality, and sensitivity to parameter settings. Balancing these factors is crucial for effective implementation in energy management.

8.5.6 Implementing Demand-Side Management

Implementing Demand-Side Management (DSM) [7] involves strategies to influence and reduce energy consumption during peak periods. Key aspects include:

- **Incentives**: Offering rebates or financial incentives to encourage energy-saving behaviours.

- **Time-of-Use Pricing**: Implementing variable pricing to shift usage to off-peak times.
- **Smart Appliances**: Promoting energy-efficient appliances that can be remotely controlled.
- **Behavorial Programs**: Educating consumers and providing tools to monitor and manage their energy use.

DSM helps balance supply and demand, reduce energy costs, and enhance grid stability.

8.6 Future Directions in AI for HEMS: Advancements in AI Algorithm

8.6.1 *Next-Generation Machine Learning Techniques*

The future of HEMS lies in the development and application of next-generation machine learning techniques. Advanced algorithms such as Federated Learning as shown in Fig. 8.13, Transfer Learning, and Explainable AI (XAI) are set to revolutionize HEMS by offering more accurate predictions, better generalization across different datasets, and enhanced transparency. Federated Learning allows models to be trained across multiple decentralized devices without sharing raw data, improving privacy and scalability. Transfer Learning facilitates the application of pre-trained models to new, related tasks, reducing the need for extensive data. XAI ensures that AI decisions are interpretable, fostering trust and compliance with regulatory requirements [20, 21].

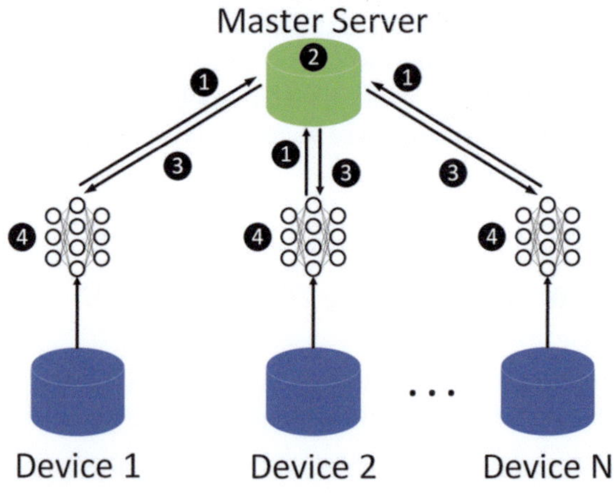

Fig. 8.13 Illustrating federated learning in HEMS, showing decentralized data training and aggregated model updates

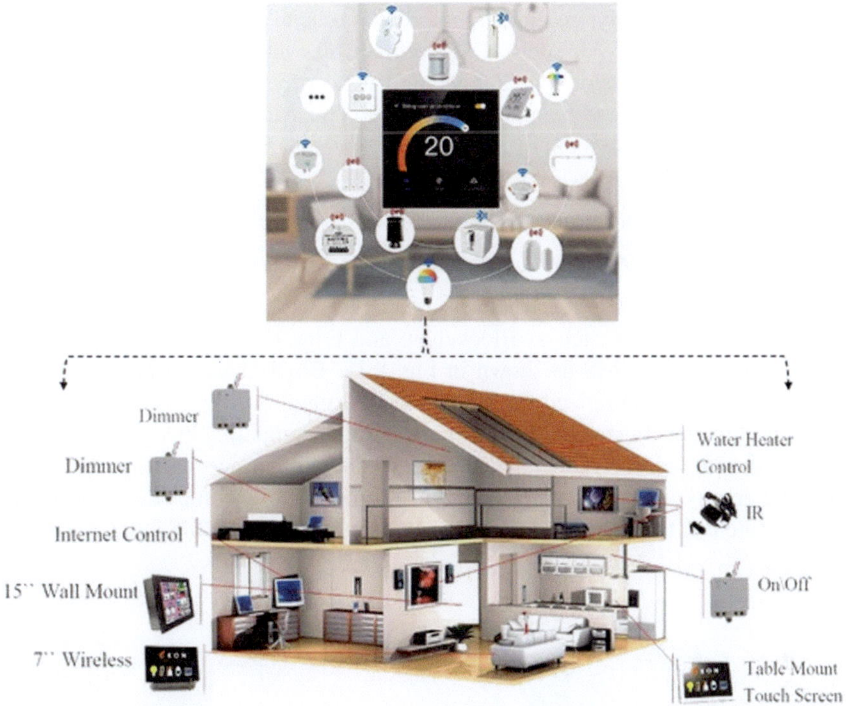

Fig. 8.14 Showing the interaction between IoT devices and AI algorithms in a smart home setup

8.6.2 Integration with IoT and Smart Devices

The integration of AI with the Internet of Things (IoT) is a critical advancement for HEMS. IoT devices, such as smart meters, sensors, and home automation systems, generate vast amounts of data that AI algorithms can analyse to optimize energy consumption in real-time. This synergy enables seamless communication between devices, facilitating automated control and energy-saving actions [22]. IoT devices in a smart home collect data, which AI algorithms analyze to automate tasks like adjusting lights, temperature, and security for convenience and efficiency as shown in Fig. 8.14. Recent research [23] highlights the benefits of AI-IoT integration in achieving significant energy savings and improving user comfort through intelligent automation and predictive-maintenance.

8.6.3 Policy and Regulatory Implications

As AI becomes more embedded in HEMS, robust policy and regulatory frameworks are essential to ensure safe, ethical, and fair use of technology. Developing standards for data privacy, security, and interoperability is crucial. Policies must address the ethical implications of AI, such as algorithmic bias and accountability [24, 25]. The European Commission's guidelines on trustworthy AI and the IEEE's global initiative on ethical AI are examples of efforts to create comprehensive frameworks that can guide the deployment of AI in energy systems [17, 26].

8.6.4 Blockchain, Edge Computing, and Beyond

Emerging technologies such as Blockchain and Edge Computing [27] are poised to further enhance HEMS as shown in Fig. 8.15. Blockchain can provide secure, transparent, and decentralized energy transaction platforms, enabling peer-to-peer energy trading and enhancing grid resilience. Edge Computing brings computation closer to the data source, reducing latency and bandwidth usage, which is critical for real-time energy management. Recent advancements in Quantum Computing also hold the potential to solve complex optimization problems in HEMS more efficiently.

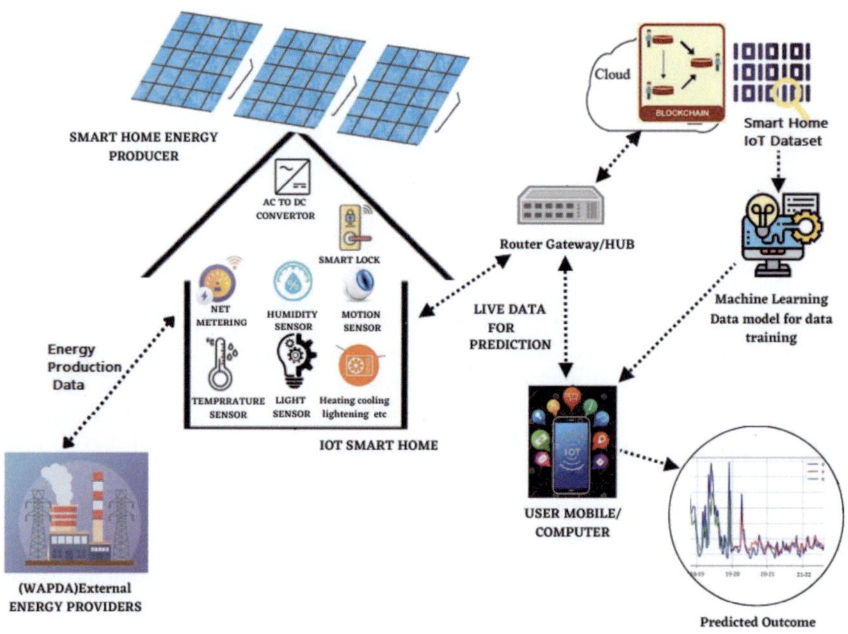

Fig. 8.15 Showing the application of blockchain in energy trading and Edge Computing in real-time energy management

8.7 Conclusion

8.7.1 Summary of Key Findings

Smart Grids: AI-driven smart grids represent a future where energy distribution is more efficient, reliable, and responsive to real-time demands.The synergy between AI and smart grids enables better demand forecasting, fault detection, and dynamic load management, ultimately leading to improved grid stability and reduced operational costs.

Metaheuristic Techniques: Metaheuristic algorithms, such as genetic algorithms, particle swarm optimization, and ant colony optimization, have proven effective in solving complex optimization problems in HEMS. These techniques offer robust solutions for energy scheduling, load balancing, and resource allocation, ensuring optimal energy usage and cost savings.

Demand-Side Management (DSM): AI's role in demand-side management is crucial for achieving energy efficiency goals. By implementing DSM strategies, AI systems can predict and respond to energy consumption patterns, engage consumers in energy-saving behaviors, and integrate renewable energy sources seamlessly. These strategies not only enhance energy efficiency but also contribute to a more sustainable and resilient energy system.

Emerging Technologies: Blockchain and Edge Computing promise further enhancements.

Efficiency: AI techniques, including supervised and unsupervised learning, reinforcement learning, and deep learning, significantly improve energy efficiency and system performance.

8.7.2 Implications for Stakeholders

Consumers: Experience enhanced energy efficiency and cost savings through intelligent automation and personalized energy management.

Utilities: Benefit from improved grid stability and reduced peak demand via predictive analytics and automated demand response.

Policymakers: Must establish robust regulatory frameworks for ethical AI deployment, data privacy, and interoperability standards.

Researchers and Developers: Have opportunities to innovate AI algorithms, integrate with IoT and smart devices, and explore emerging technologies to advance HEMS.

8.7.3 Recommendations for Future Research

Advanced AI Algorithms: Develop next-generation techniques like Federated Learning and Explainable AI to improve accuracy, scalability, and transparency.

Emerging Technologies: Investigate Blockchain for secure energy transactions and Edge Computing for real-time data processing.

8.8 Ethical and Regulatory Frameworks

Create comprehensive policies to address ethical issues, ensure data privacy, and promote responsible AI use.

Case Studies and Pilot Programs: Conduct real-world studies to validate AI applications in diverse residential settings, assess their impact, and refine methodologies.

8.9 Disclosure

Funding The author did not receive support from any organization for the submitted work.

Data Availabillity Authors can confirm that all relevant data are inculded in the article.

References

1. O.A. Omitaomu, H. Niu, Artificial intelligence techniques in smart grid: a survey. Smart Cities **4**(2), 548–568 (2021)
2. U. Zafar, S. Bayhan, A. Sanfilippo, Home energy management system concepts, configurations, and technologies for the smart grid. IEEE Access **8**, 119271–119286 (2020)
3. L. Tai, M. Liu, Deep-learning in mobile robotics—from perception to control systems: a survey on why and why not (2016)
4. Y. Shi, X. Xu, Deep federated adaptation: an adaptive residential load forecasting approach with federated learning. Sensors **22**(9), 3264 (2022)
5. I. Alotaibi, M. Abido, M. Khalid, A. Savkin, A comprehensive review of recent advances in smart grids: a sustainable future with renewable energy resources. Energies **13**, 6269 (2020)
6. J. Yu, A. de Antonio, E. Villalba-Mora, Deep learning (CNN, RNN) applications for smart homes: a systematic review. Computers **11**(2), 26 (2022)
7. S. Sharda, M. Singh, K. Sharma, Demand side management through load shifting in IoT based HEMS: overview, challenges and opportunities. Sustain. Cities Soc. **65**, 102517 (2021)
8. A. Almusaed, I. Yitmen, A. Almssad, Enhancing smart home design with AI models: a case study of living spaces implementation review. Energies **16**(6), 2636 (2023)
9. A.A. Amer, K. Shaban, A.M. Massoud, DRL-HEMS: deep reinforcement learning agent for demand response in home energy management systems considering customers and operators perspectives. IEEE Trans. Smart Grid **14**(1), 239–250 (2023)

10. E. Aslanci, K. Coşkun, Using hierarchies in reinforcement learning framework with non-stationary environments. RG **2**(2), 18773–06888 (2017)
11. S. Radhika, A. Chaparala, Optimization using evolutionary metaheuristic techniques: a brief review. Braz. J. Oper. Prod. Manag. **15**(1), 44–53 (2018)
12. N. Qayyum, A. Amin, U. Jamil, A. Mahmood, Techniques for home energy management: a review, in *Proceedings of the International Conference on Emerging Technologies* (ICOMET) (2019)
13. J. Jeyaranjani, D. Devaraj, Improved genetic algorithm for optimal demand response in smart grid. Sustain. Comput. Inform. Syst. **35**, 100710 (2022)
14. K. Parvin, M.A. Hannan, A.Q. Al-Shetwi, P.J. Ker, M.F. Roslan, T.M.I. Mahlia, Fuzzy based particle swarm optimization for modeling home appliances towards energy saving and cost reduction under demand response consideration. IEEE Access **8**, 210784–210799 (2020)
15. B.N. Silva, M. Khan, K. Han, Futuristic sustainable energy management in smart environments: a review of peak load shaving and demand response strategies, challenges, and opportunities. Sustainability **12**(14), 5561 (2020)
16. B. Nepal, M. Yamaha, A. Yokoe, T. Yamaji, Electricity load forecasting using clustering and ARIMA model for energy management in buildings. Jpn. Arch. Rev. **3**(1), 62–76 (2020)
17. M. Albadr, S. Tiun, M. Ayob, F. Al-Dhief, Genetic algorithm based on natural selection theory for optimization problems. Symmetry **12**, 1–31 (2020)
18. C. Strecha, Multi-view stereo as an inverse inference problem. Thesis (2007)
19. F. Iqbal, A. Altaf, Z. Waris, D.G. Aray, M.A.L. Flores, I.D.L.T. Díez, I. Ashraf, Blockchain-modeled edge-computing-based smart home monitoring system with energy usage prediction. Sensors **23**(11), 5263 (2023)
20. I. Behravan, S. Zahiri, S. Razavi, R. Trasarti, Clustering a big mobility dataset using an automatic swarm intelligence-based clustering method. J. Electr. Comput. Eng. Innov. **6**, 243–262 (2018)
21. A.K. Shaikh, A. Nazir, I. Khan et al., Short term energy consumption forecasting using neural basis expansion analysis for interpretable time series. Sci. Rep. **12**, 22562 (2022)
22. M. Khan, M.I.A. Sajjad, M. Tahir, A. Haseeb, IoT application for energy management in smart homes, in *Proceedings of the Engineering Conference*, vol 43 (2022)
23. H. Xu, W. Yu, D. Griffith, N. Golmie, A survey on industrial internet of things: a cyber-physical systems perspective. IEEE Access **6**, 78238–78259 (2018)
24. H.R.O. Rocha, I.H. Honorato, R. Fiorotti, W.C. Celeste, L.J. Silvestre, J.A.L. Silva, An Artificial Intelligence based scheduling algorithm for demand-side energy management in smart homes. Appl. Energy **282**(PA) (2021)
25. S. Albelwi, A. Mahmood, A framework for designing the architectures of deep convolutional neural networks. Entropy **19**(6), 242 (2017)
26. Y. Shi, T. Hoang, A. Savkin, C.-T. Lin, J.G. Zhu, H.V. Poor, Distributed model predictive control for joint coordination of demand response and optimal power flow with renewables in smart grid. Appl. Energy **290**, 116701 (2021)
27. Z. Qing, X. Lin, Y. Hao, J. Cao, Energy-aware scheduling in edge computing with energy internet. IEEE Access **8**, 229052–229065 (2020)

Chapter 9
Predictive Analysis of Air Quality Index (AQI) and Identification of Influential Factors Using Machine Learning Models

Safina Shokeen, Taranveer Singh, Shashank Nautiyal, and Gautam Sethi

Abstract Air pollution is the presence of dangerous compounds in the air that have a negative impact on health, resulting in millions of premature deaths every year. The high concentration of particulate matter is a significant problem. Understanding the causes of air pollution is therefore critical for avoiding and managing it. This article provides a full assessment as well as a comparative analysis of several Machine Learning models for forecasting air pollution particle matter. Furthermore, the study intends to examine major work in forecasting air quality using ML models. Furthermore, Support Vector Regression was used to analyze the link between training data and projected outcomes for continuous variables.

Keywords Air quality index · Prediction · AQI bucket · Air pollution · Environment · Air quality classification · Machine learning models · Support vector regression

9.1 Introduction

Air pollution is a widespread global problem caused by the burning of fossil fuels, industrial operations, automobile emissions, agricultural activities, waste disposal, and certain natural causes. Since India has experienced some of the worst and most widespread deterioration in air quality over the past 20 years, air pollution has become a major concern for regulatory agencies. Severe pollution outbreaks are highly likely during the post-monsoon season, particularly in the Indo Gangetic Plains (IGP).

S. Shokeen · T. Singh (✉) · S. Nautiyal · G. Sethi
School of Engineering and Technology, Vivekananda Institute of Professional Studies—Technical Campus, Delhi, India
e-mail: singhtaranveer094@gmail.com

S. Shokeen
e-mail: safina.shokeen@vips.edu

A. Kumar et al. (eds.), *Selected Proceedings of the 1st International Conference on Advanced Materials for Sustainable Innovation; IC-AMSI 2024; 28–30 August; New Delhi; India*, Springer Proceedings in Energy,
https://doi.org/10.1007/978-981-96-4492-6_9

An alarming example occurred throughout the course of a week in early November 2017, when Delhi's PM2.5 particle levels surpassed environmental health criteria by 11 times (or 25 times) the WHO norms, resulting in an environmental health emergency. The issue of air pollution requires immediate attention and extensive measures to protect human health and the environment. The public is informed about potential health risks and given the ability to take measures via accurate prediction and forecasting of air quality, which is why it is so important. A commonly used metric to assess the intensity of air pollution is the Air Quality Index (AQI). It is a significant tool in the transmission of Air quality data as well as identification and quantification of air pollution, with higher AQI values depicting the more severe health concerns. In India, The quality of Air is majorly affected by numerous pollutants [1], and each one adds up differently to the backdrop of air constituents. Simultaneously, the greater prevalence of nitrogen dioxide (mostly produced by combustion processes) lead to a higher AQI due to an increase in respiratory illnesses as well as worsening opacity from the atmosphere. Additionally, atmosphere pollution from sulphur dioxide, a polluter generated during burning of fossil fuels also connects to AQI as it induces respiratory problems [2] and acid rain The combined effect of these pollutants on AQI reflects the intricacies of air quality dynamics in India [3], necessitating holistic interventions for sustained environmental health benefits.

Recently, the importance of forecasting time serial variations of pollutants using machine learning algorithms in air quality prediction has increased. Climate prediction models can help to create useful strategies for the evaluation as well as projection of potential enhancements in the air quality index [4]. The data-driven, algorithm-based ML techniques have revolutionized the manner in which scientists evaluate the impact of various air pollutants on human health. The details regarding the complete processes used in these sophisticated approaches to evaluate and understand complex sets of data for air pollution, considering diverse air pollutants and their potential impacts on human health are available elsewhere [5].

The literature review focused on ambient air quality, paying special attention to pollutants which factor into the Air Quality Index (AQI). So, the major aspect of research is to study each parameter in detail so that CO, NO_2, SO_2 and PM2. 5, and PM10 [6]. In addition to this, Machine Learning models have also been a hot topic since they are time-saving and provide robust data processed. In the following, the authors utilize Support Vector Regression (SVR) for air quality prediction because it can represent complex relationships among input attributes and AQ indicators. SVR, a type of Support Vector Machine (SVM, was preferred because SVR can represent nonlinear relationship between variables and overcome outliers, which is good powerful for modelling the complex patterns in air quality dataset document. This study takes advantages of the ability to capture complex data patterns in SVR-based air quality prediction, so was proposed as a base work.

Table 9.1 AQI range with specific concentration of pollutants

AQI category (range)	PM10	PM2.5	NO_2	O_3	CO	SO_2	NH_3	Pb
Good (0–50)	0–50	0–30	0–40	0–50	0–1.0	0–40	0–200	0–0.5
Moderate (51–100)	51–100	31–60	41–80	51–100	1.1–2	41–80	201–400	0.5–1
Poor (101–200)	101–250	61–90	81–180	101–168	2.1–1	81–380	401–800	1.1–2
Very unhealthy (201–300)	251–350	91–120	181–280	169–208	10–17	381–800	801–1200	2.1–3
Severe (301–400)	351–430	121–250	281–400	209–748	17–34	801–1600	1200–1800	3.1–3.5
Hazardous (401–500)	430 +	250 +	400 +	748 +	34 +	1600 +	1800 +	3.5 +

9.2 Pollutants Contributing to AQI

This, by considering eight important pollutants—PM10, PM2.5, NO_2, SO_2, CO, O_3, NH_3, and Pb—the Air Quality Index follows a step-by-step approach to measure the quality of air [6]. National Ambient Air Quality guidelines provide health criteria that govern the classification of short-term pollutant concentrations into six categories ranging from "Good" to "Severe" [7]. This minute analysis not only gives an indication of the status of the air, but it also creates public awareness about possible health consequences [5]. Table 9.1 illustrates the relationship between the AQI range and the myriad of pollutants contributing to that range.

Each pollutant has a linear function contributing to the AQI while some concentration breakpoints have health risks. The levels of PM2.5 range 51 (31 µg/m3), to 100 (60 µg/m3) and 75 (45 µg/m3). The total AQI also provides a holistic assessment of air quality if the highest sub-index amongst contaminants is attained.

9.3 Machine Learning Prediction Models

Computational algorithms referred to as machine learning (ML) models are those that, without explicit programming, identify patterns in data to forecast or make judgments. Such models can be used effectively to solve any number of air pollution monitoring and forecast problems [8]. Figure 9.1 illustrates the fundamental architecture of a Machine Learning model for air pollution, illustrating how data can be used to determine the proper algorithm for making relevant decisions.

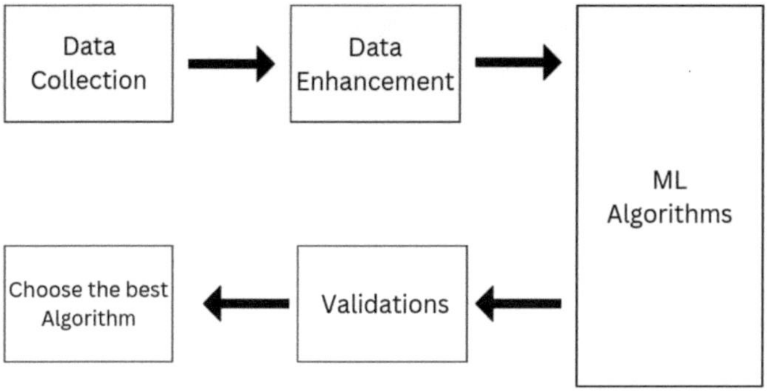

Fig. 9.1 Basic architecture of machine learning model [2]

Machine learning algorithms can forecast air quality by analyzing historical data, meteorological conditions, and pollutant emissions. They also contribute to determining the source of air pollution via monitoring stations and sensor networks [9]. Table 9.2 lists machine learning models for predicting air quality, including Random Forest, Gradient Boosting Regression, Support Vector Regression, Decision Tree Regression, and K-nearest neighbour, each with advantages and disadvantages for applications such as urban planning, healthcare management, and environmental impact assessment.

We used Support Vector Regression (SVR) for predictive analysis because it is good at handling nonlinear connections and outliers, which are prevalent in air quality data. It allows SVR to capture complex data patterns while unlike linear models, it is better.

This is suited modeling the complicated interactions between pollutants and AQI. Support Vector regression or SVR is one approach to machine learning that makes use of kernel functions to translate the input data into a higher-dimensional space and then find the best fitting hyperplane that predicts the continuous values of the AQI, which reduces the prediction errors given by SVR. A margin of tolerance, known as epsilon, further optimizes the model, and this in fact did make the model very successful in predicting AQI values based on pollutant concentrations and climatic variables.

To learn to predict AQI using machine learning algorithms like SVR, we will have to conduct the following procedures:

1. **Data Acquisition**: Gather information concerning pollutants, such as PM2.5, PM10, NO_2, SO_2, CO, and O_3, as well as meteorological conditions.
2. **Pre-processing of Data**: Clean and normalize the data so that it becomes uniform.
3. **Feature Selection**: Identify and maximize the most critical characteristics based on the factors that influence AQI.
4. Training the SVR model with historical data to identify correlation between contaminant and air quality index.

Table 9.2 Machine learning models for air quality

Model	Pros	Cons	Application
Random forest [10]	a. Comprehensive data analysis from multiple locations b. Comparative analysis enhances understanding of modeling approaches	a. Limited generalizability to other regions b. Dependency on historical data c. Complexity of some models may hinder interpretation	a. Assisting urban planning efforts b. Supporting healthcare management c. Driving technological innovation in air quality monitoring
Gradient boosting regression [11]	a. Comprehensive data source b. Identification of pollutants c. Robust model evaluation	a. Limited outlier detection b. Moderately strong correlations	a. Air quality forecasting b. Environmental impact assessment c. Development of automated systems
Support vector regression [10]	a. Integration of advanced machine learning techniques b. Practical application for public health protection	a. Scalability issues with SVR model	a Informing policy decisions
Decision tree regression [12]	a. Accurate prediction model b. Improved model performance	a. Model complexity	a. Early warning systems b. Public health management
K-nearest neighbour [13]	a. Simple to Implement b. No training phase	a. Computationally expensive b. Sensitive to feature scaling c. Memory intensive	a. Localized air quality assessment b. Early warning systems c. Sensor network optimization

5. **Predicting**: Map levels of pollution into AQI values for fresh data using the learned model.

9.4 Related Work

The section's main purpose is to provide a summary of the research and effort that has been put into monitoring air quality using various machine learning models over the last several years. Table 9.3 summarizes recent studies that use machine learning models to forecast air quality, demonstrating varying methodologies and conclusions across multiple contaminants and prediction strategies.

Table 9.3 An overview of articles with ML models to predict air quality

References	Year	ML Model	Particulate	Features
[16]	2018	Artificial neural network mode, random forest, decision tree model, least squares support vector machine model, deep belief network	CO, Pb, NO_2, O_3	Reviewed literature and compares research on air quality evaluation using big data analytics and machine learning techniques, emphasizing the importance of real-time monitoring in future smart cities
[10]	2019	Support vector regression and random forest regression	NO_x	Two prediction models were developed for distinct scenarios, enhancing accuracy in forecasting air indicators and offering guidance for urban air quality modeling and analysis
[19]	2020	Support vector machine, random forest, extreme gradient boosting, LightGBM and CatBoost	SO_2, CO, O_3, NO_2, PM10, and PM2.5	Combining random forest algorithm with sensor data and public weather data yields the best results for predicting AQI levels and rankings in many circumstances
[18]	2021	Artificial neural network	SO_2, CO, PM10, O_3, NOx, H_2S	The absolute air quality index found to be a good predictor of air quality model stability. Also, the results demonstrated that NARX outperformed a neural network, ANN, and ANFIS-based AQHI model for high-dimensional data evaluation
[11]	2023	Random forest, gradient boosting, lasso regression and the stacked regressor	PM10, PM2.5, NO_2, SO_2, CO, O_3	Stacked generalization machine learning models efficiently and precisely forecast air quality index and grade while mitigating overfitting concerns compared to individual models
[14]	2023	Multilinear regression, KNN, and HGBoost regression models	PM2.5	The HGBoost3 model outperformed existing PM2.5 models by adding the spatiotemporal link between air pollutants and meteorological factors to forecast air quality, resulting in lower error and higher R2 values

(continued)

Table 9.3 (continued)

References	Year	ML Model	Particulate	Features
[15]	2024	Linear regression, random forest regression	CO, NO_2, SO_2, PM2.5, and PM10,	Satellite remote sensing can reduce spatial uncertainty, but measurement results are mostly limited to the upper atmosphere, which has high measurement sensitivity. This study utilizes remotely sensed data to model the global Air Quality Index for contaminants including CO, NO2, SO2, PM2.5, and PM10
[17]	2024	Fuzzy time series with deep learning approach	SO_2, CO, O_3, NO_2, PM10, and PM2.5	The experimental findings show that the predicted background leads to stable forecasting and that the proposed model beats current models in predicting effects

In 2018, Gaganjot et al. [16] evaluated different big data and machine learning-based air quality forecasting approaches. The paper underlines the significance of real-time monitoring in future smart cities and identifies research problems and requirements in this field.

Huixiang et al. [10] predicted AQI in Beijing using SVR and RFR regression models. Furthermore, the experimental findings demonstrated that the RFR-based model outperformed other models in predicting NOx concentration.

Furthermore, in 2021, Kanchan et al. [18] discovered that most techniques lack the flexibility to adapt a new contaminant. Furthermore, [14] in 2023 emphasizes the need of adding spatiotemporal linkages between air pollutants and climatic aspects utilizing Multilinear regression, KNN, and HGBoost regression models for PM2.5 prediction. The HGBoost3 model's greater performance emphasizes the need of using advanced approaches to improve forecasting accuracy, as demonstrated by reduced error rates and higher R2 values. In 2024, Tania et al. [15] Linear regression and random forest regression models were used to predict numerous contaminants, with a focus on the importance of satellite remote sensing in reducing geographical uncertaintyThe paper thus uses remote sensing data to simulate the Air Quality Index for all pollutants over the whole Earth, thereby solving the problem of sparsity of measurement in very sensitive upper atmosphere regions.

Subraninan [17] integrated Fuzzy Time Series with Deep Learning Approach in the same year. The study highlighted the stability of forecasting while emphasizing the application to the future prediction of air quality, giving useful insights and approaches for the future research and development in this sector. These articles, put together, depict the changing landscape of ML models. The normalized difference is then added to the Base AQI to calculate the final AQI value. Furthermore, Fig. 9.2 depicts the most often used phrases while assessing various machine learning models and methodologies for air quality prediction.

Fig. 9.2 Most used keywords

9.5 AQI Bucket List Predictor

An AQI bucket list predictor would help classify AQI levels into specific buckets, thereby further aiding in public health management and also making individuals aware of the possible health risks associated with different levels of air pollution. Linear Interpolation method is usually used to find the threshold value of every pollutant for the Air Quality Index (AQI) that ties the level of concentration of a pollutant to its associated AQI value. Equation 9.1 represents the conversion of concentrations of pollutants to AQI values, thereby standardizing the measurement of air quality across different pollutants.

$$\text{Ppollutant sub} - \text{index} = \text{Base AQI} + \left[(\text{X} - \text{Given Value}) * 100/(\text{Breakpoint})\right] \tag{9.1}$$

where,

The pollutant sub-index reflects how air quality is said to be on a certain level of pollutant concentration.

The base AQI is the value of AQI corresponding to the lower bound of the breakpoint range. These values are used as reference points in calculations and are, in general, provided as part of AQI tables or standards.

X: the concentration of the pollutant being measured. Measured Concentration of particulate matter, (PM2.5 and PM10), nitrogen diox-ide, NO_2, ozone, O_3, carbon monoxide, CO, sulfur dioxide, SO_2, ammonia, NH_3. The value given is the concentration of the pollutant being measured and Breakpoint is the range of pollutant concentration corresponding to different AQI levels. In simple words, it defines the upper as well as the lower bounds within which the pollutant concentration falls to determine the AQI level. Equation 9.1 calculates the AQI with interpolation between the base AQI and the AQI value associated with the breakpoint range. This is achieved through calculation of the difference between the given pollutant concentration (X), subtracting the breakpoint, multiplying it by 100, and then dividing by the range of the breakpoint.

9.6 Results and Discussion

We have relied upon data provided by The Central Pollution Control Board (CPCB) https://www.data.gov.in/resource/real-time-air-quality-index-various-monitoring stations that include detailed air quality data, which has been sourced from various places across India. The data for the study were extracted from the Central Pollution Control Board (CPCB), which provides monitoring services of air quality throughout India. It is from this source that we used data regarding the concentration of pollutants for our analysis: PM2.5, PM10, NO_2, SO_2, CO, and O_3. It embraces a wide range of pollutants which includes PM10, PM2.5, NO_2, SO_2, CO, and O_3, which will provide insightful details about their concentration and impacts on air quality. This dataset is critical in our research work, for it gives us the opportunity to track trends and even estimate levels of pollution, which could be used as guidelines for furthering enhancements about air quality, particularly for the research purposes we are focused on. This dataset is so enriched that, with data sourced from reliable monitoring stations, we can extract in-depth information that lays the foundations for understanding the dynamics of air pollution in different regions of India, thereby helping us better understand and manage environmental health issues in the context of our research objectives.

Figure 9.3 shows the plot between predicted AQI values vs actual AQI values using the SVR Model. This is a robust algorithm of ML, usually applied for regression tasks, especially when the interactions in the data are complex and nonlinear. In this experiment, we used SVR to predict the AQI based on different features. During training, SVR will seek to optimize a hyperplane that minimizes errors of the predictions. The choice of kernel, regularization coefficient C and other parameters that are either inherent or specific to the chosen kernel are crucial in fine-tuning the model toward the best possible performance. After training, the quality of the model's prediction on the training set was measured using metrics like Mean Absolute Error (MAE) and Root Mean Squared Error (RMSE). The RMSE of pre-diction on the training set was around 99.53 units on average. The RMSE of the testing set also suggests deviation with estimates of approximately 92.72 units. Such results demonstrate how good the model is in prediction and may even induce more optimization and feature engineering to improve its performance and generalizability.

Most importantly, influential outcome about the effect of different pollutants on air quality, focusing on identified trends and patterns. We review how good a prediction of Python-based an approach for AQI bucket levels will be achieved in terms of accuracy and key influencing factors.

Figure 9.4 further shows the increase in AQI value with each and every pollutant present in the atmospheres. It explains that even the slight increase in benzene

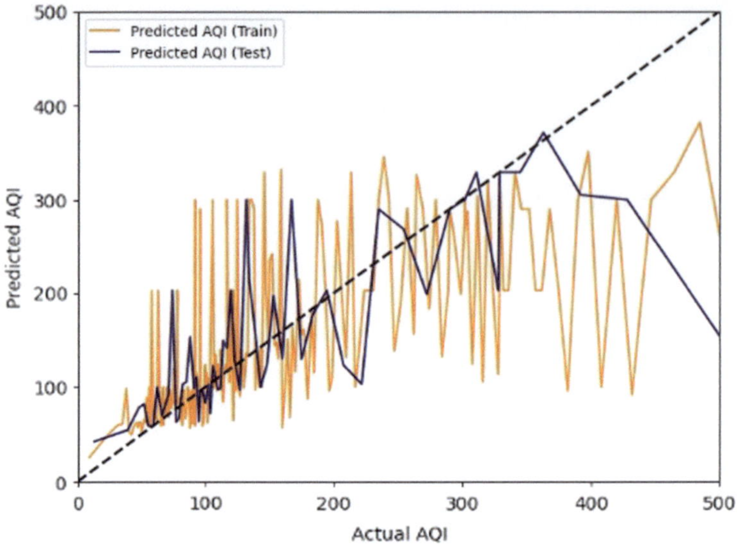

Fig. 9.3 Support vector regression: predicted vs actual AQI

concentration can lead to significant spikes in AQI levels due to its high toxicity and adverse health effects, particularly on respiratory and cardiovascular systems. Toluene concentrations exhibit a moderate impact on AQI levels, with noticeable increases correlating with higher AQI readings, primarily affecting respiratory health and contributing to smog formation. PM10 concentrations have a direct and substantial impact on AQI levels, with higher concentrations leading to severe air pollution episodes, respiratory issues, and reduced visibility. CO concentrations exhibit a strong correlation with AQI levels, particularly in urban areas with heavy traffic, posing significant health risks due to its affinity for hemoglobin and potential for causing carbon monoxide poisoning. Elevated NOx levels contribute to AQI spikes, primarily due to the formation of NO_2 and secondary particulate matter, exacerbating respiratory problems and playing a role in the formation of ground-level ozone. Even small changes in PM2.5 concentrations can significantly impact AQI levels, with higher concentrations posing serious health risks, including respiratory and cardiovascular problems, and reduced visibility in affected areas. Benzene, a volatile organic compound (VOC), is highly toxic and can cause significant spikes in AQI even at low concentrations. Its involvement in photochemical reactions can lead to the formation of ground-level ozone, contributing to smog and poor air quality. Benzene spikes often occur in areas with high industrial emissions or vehicular traffic, leading to sharp AQI increases.

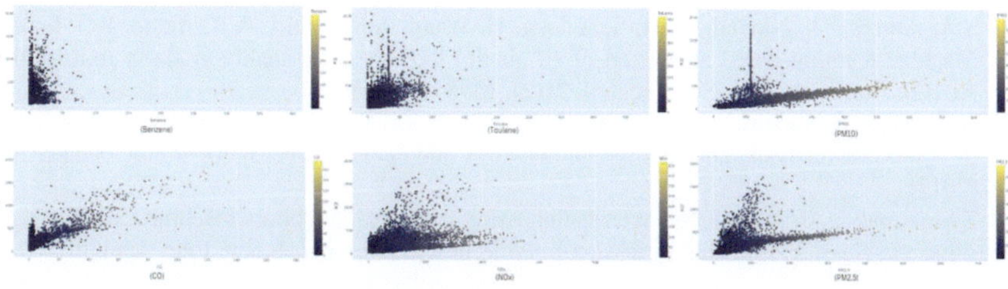

Fig. 9.4 Pollutants contributing to increase in AQI

9.7 Conclusion

Various Machine Learning modes were reviewed that investigated the AQI value. Further, the Support Vector Regression model was used to analyze the difference between the predicted and actual values of AQI based on the data obtained from CPCB. Also, the model was used to determine which pollutant has a greater impact in increase of AQI value and hence therefore crucial for the environment. Future work holds the further improvement in the collected data by using different machine learning models.

References

1. Kanchan, A.K. Gorian, P. Goyal, A review on air quality indexing system. Asian J. Atmos. Environ. **9**(2), 101–113 (2015)
2. K.K. Mokoena, C.J. Ethan, Y. Yu, A.T. Quachie, Interaction effects of air pollution and climatic factors on circulatory and respiratory mortality in Xi'an, China between 2014 and 2016. Int. J. Environ. Res. Public Health **17**(23), 9027 (2020)
3. R. Sharma, R. Kumar, D.K. Sharma, L.H. Son, I. Priyadarshini, B.T. Pham, S. Rai, Inferring air pollution from air quality index by different geographical areas: case study in India. Air Qual. Atmos. Health **12**, 1347–1357 (2019)
4. M.T. Udristioiu, Y.E. Mghouchi, H. Yildizhan, Prediction, modelling, and forecasting of PM and AQI using hybrid machine learning. J. Clean. Prod. **421**, 138496 (2023)
5. J.T. Spickett, H.L. Brown, K. Rumchev, Climate change and air quality: the potential impact on health. Asia Pac. J. Public Health **23**(2_suppl), 37S-45S (2011)
6. Y. Wang, L. Huang, C. Huang, J. Hu, M. Wang, High-resolution modeling for criteria air pollutants and the associated air quality index in a metropolitan city. Environ. Int. **172**, 10775 (2023)
7. S.A. Horn, P.K. Dasgupta, The air quality index (AQI) in historical and analytical perspective a tutorial review. Talanta **267**, 125260 (2023)
8. I.H. Sarker, Machine learning: algorithms, real-world applications and research directions. SN computer science **2**(3), 160 (2021)
9. S. Mansour, N. Nasser, L. Karim, A. Ali, Wireless sensor network-based air quality monitoring system, in *2014 International Conference on Computing, Networking and Communications (ICNC)*, (IEEE, 2014) pp. 545–550
10. H. Liu, Q. Li, D. Yu, Y. Gu, Air quality index and air pollutant concentration prediction based on machine learning algorithms. Appl. Sci. **9**(19), 4069 (2019)

11. S.A. Aram, E.A. Nketiah, B.M. Saalidong, H. Wang, A.R. Afitiri, A.B. Akoto, P.O. Lartey, Machine learning-based prediction of air quality index and air quality grade: a comparative analysis. Int. J. Environ. Sci. Technol. **21**(2), 1345–1360 (2024)
12. A. Jamal, R.N. Nodehi, Predicting air quality index based on meteorological data: a comparison of regression analysis, artificial neural networks and decision tree. J. Air Pollut. Health **2**(1) (2017)
13. T.M. Amado, Air quality characterization using k-nearest neighbors machine learning algorithm via classification and regression training in R. J. Comput. Innov. Eng. Appl. **2**(2), 1–7 (2018)
14. A. Mathew, P.R. Gokul, P. Raja Shekar, K.S. Arunab, H. Ghassan Abdo, H. Almohamad, A. Abdullah Al Dughairi, Air quality analysis and PM2.5 modelling using machine learning techniques: a study of Hyderabad city in India. Cogent Eng. **10**(1), 2243743 (2023)
15. T.S. Anggraini, H. Irie, A.D. Sakti, K. Wikantika, Machine learning-based global air quality index development using remote sensing and ground-based stations. Environ. Adv. **15**, 100456 (2024)
16. G.K. Kang, J.Z. Gao, S. Chiao, S. Lu, G. Xie, Air quality prediction: big data and machine learning approaches. Int. J. Environ. Sci. Dev. **9**(1), 8–16 (2018).
17. A. Unnikrishnan, S. Rajeswari, Optimal prediction of air quality index in metropolitan cities using fuzzy time series with deep learning approach. Ecol. Eng. Environ. Technol. (EEET) **25**(1) (2024)
18. O. Taylan, A.S. Alkabaa, M. Alamoudi, A. Basahel, M. Balubaid, M. Andejany, H. Alidrisi, Air quality modeling for sustainable clean environment using anfis and machine learning approaches. Atmosphere **12**(6), 713 (2021)
19. D.Q. Duong, Q.M. Le, T.L. Nguyen-Tai, D. Bo, D. Nguyen, M.S. Dao, B.T. Nguyen, Multi-source machine learning for AQI estimation, in *2020 IEEE International Conference on Big Data* (*Big Data*) (IEEE, 2020) pp. 4567–4576
20. D. Yao, J. Yang, X. Zhan, An improved random forest algorithm for class-imbalanced data classification and its application in PAD risk factors analysis. Open Electr. Electron. Eng. J. 7(1) (2013)

Chapter 10
Sustainable Solutions for Delhi's Air Pollution: A Data Driven Approach

Kinjal Bansal and Vishal Srivastava

Abstract The city of Delhi has become an important subject of research in the global discussion surrounding air pollution. The city experienced a significant deterioration in air quality between 2015 and 2020, and this study seeks to identify the complex factors that have contributed to this trend. The research employs a comprehensive approach that goes beyond the surface-level causes of pollution. It examines the role of rapid urbanization, industrial growth, vehicular emissions, and agricultural practices in contributing to pollution levels while recognizing the intricate interconnectedness of these factors. To address the issue, the study advocates for a holistic approach that considers the specific sources of emissions. By utilizing advanced analytical tools such as machine learning algorithms and satellite data integration, the research aims to provide accurate and actionable insights to inform targeted policy interventions. Furthermore, the study critically evaluates prior research and identifies gaps in existing methodologies. By introducing innovative approaches and leveraging advanced analytical tools, the study aims to offer a nuanced understanding of pollution trends and dynamics. In summary, the research underscores the need to adopt comprehensive, data-driven strategies to combat air pollution in Delhi and other urban centers worldwide. It highlights the impact of pollution on public health, ecosystems, and environmental sustainability and calls for collective action to create clean and healthy environments.

K. Bansal · V. Srivastava (✉)
School of Engineering & Technology, Vivekananda Institute of Professional Studies—Technical Campus, AU-Block (Outer Ring Road) Pitampura, Delhi 110034, India
e-mail: vishal.srivastava@vips.edu

K. Bansal
e-mail: 09617711922_ds@vips.edu

A. Kumar et al. (eds.), *Selected Proceedings of the 1st International Conference on Advanced Materials for Sustainable Innovation; IC-AMSI 2024; 28–30 August; New Delhi; India*, Springer Proceedings in Energy,
https://doi.org/10.1007/978-981-96-4492-6_10

10.1 Introduction

There have been so many researches carried out on human interactions with their physical environment regarding how human activities impact biotic interactions between organisms and microorganisms, on one hand, and the hydrosphere, lithosphere, and atmosphere constituting our environment on the other hand. A very important topic is pollution. This refers to the introduction of harmful substances into the environment. These harmful substances may exist in either solid, liquid, or gas forms. The level of these pollutants has, more often than not, been elevated, and the effects create a bad quality environment [2]. To some extent, human activities influencing technological developments, societal development, and provision of services have concurrently impacted bad environmental effects. The industrial revolution drastically heightened the degree at which pollution was causing an adverse health effect, given the massive production of large amounts of pollutants launched into the air. Global environmental pollution has social, economic, legislative, and lifestyle dimensions in a complex interlocking set [5]. It is further compounded by these historical and contemporary trends of urbanization and industrialization, currently occurring at unprecedented scales throughout the entire world [4]. While all anthropogenic air pollution in turn arises from fossil fuel combustion, the total rank of relative contributions from these sources is dominated by, and determined by, traffic. One of the grave public health hazards worldwide is anthropogenic air pollution, contributing to approximately 9 million deaths each year [1]. There are a range of adverse health effects due to air pollution, varying with the concentration, and individuals, especially more vulnerable and sensitive, are affected even by low levels of pollution [3].

Climate change and global warming are likely to exacerbate these issues, accompanied by increased hospitalization among the elderly and vulnerable populations. Fine and ultrafine particulate matter, which can penetrate deep into the alveoli and more easily translocate through the bloodstream, appear to be associated with worsening conditions [6]. The most severe air pollution problems are often observed in large cities, where vehicle emissions are considered a major source of poor air quality. In addition, industrial accidents may pose a lethal threat to nearby populations due to the release of chemical pollutants. Developing countries contribute to the problem further through overcrowding [7], uncontrolled urbanization, and industrial expansion, leading to a significant deterioration in air quality, especially in countries with socio-economic imbalances and limited knowledge about sustainable environmental management. Those with low incomes rely on wood or solid fuels for domestic use and are more vulnerable to severely polluted indoor air. While pollution affects both rural and urban areas in India, it can be attributed to rapid industrialization, urbanization, and an increased reliance on motorcycles as major contributors in rural areas. Moreover, household air pollution from biomass burning for heating and cooking is a significant source of pollution in India and Nepal [8]. There is spatial heterogeneity in India, with higher PM2.5 levels observed in the colder northern states compared to the southern states, which have a more tropical climate and less need for indoor heating. These indoor periods lead to significant health implications, primarily for women and children [9].

These alarming levels of air pollution represent a significant public health concern, underscoring the pressing need to address this issue across India, particularly in New Delhi. Air pollution affects climate change, public health, and individual well-being in various regions worldwide. Exposure to particulate matter can lead to respiratory and cardiovascular conditions, reproductive and central nervous system dysfunction [10, 11], and cancer. Additionally, other harmful pollutants, such as ozone, nitrogen oxides, sulfur dioxide, VOCs, dioxins, and polycyclic aromatic hydrocarbons, contribute to haze formation and adverse health effects. The severity of this global problem is exacerbated in India for two key reasons. First, the cost of air pollution in the country has exceeded the guidelines recommended by the World Health Organization for over a decade, exceeding nearly 8% of India's GDP. Second, despite India's high primary school enrollment rates, the country faces a significant learning deficit, with many children lacking essential skills [12]. This downward trend in national educational productivity raises concerns about the potential impact of air pollution on academic performance and the future capabilities of children. In the Indian context, New Delhi is at the center of this crisis, given its extreme air pollution and public health challenges. The economic costs of air pollution in India are high and are further compounded by weak environmental regulations and industrial resistance to strengthening such rules. The urgency for action against air pollution in India is further underscored by the economic loss due to premature deaths and morbidity, which approximates nearly 8% of the country's GDP. Control Board, Indian cities remain shrouded under clouds of toxic pollutants. That requires a broad overall understanding of the specific problems that India, particularly Delhi, faces [15]. The monumental situation in India as well as the country's huge learning deficit prove simultaneously that policies and interventions should be evidence-based about immediate health crises owing to air pollution and long-term educational challenges.

10.2 Sources of Air Pollution

Air pollution is one of the severe problems in India, especially in cities. One such city is Delhi, which has even more severe problems in terms of its geography: it is flanked by mountains and comprises nineteen districts. During winter, the air is cold and gets stuck at the bottom, creating a sort of lid for the pollutants. For all these reasons, Delhi remains in the top three list of the most polluted cities of the world due to construction activities, factories, thermal power stations, vehicular traffic, and contaminated gasoline [13]. What is worse, their cooking fuel sources use cow dung and biomass. The air pollution crisis in India needs holistic efforts in terms of sources like vehicle emissions, industrial pollution, dust from construction areas, power generation, and household cooking practices. The knowledge of the sources of pollution is therefore crucial so that the right regulation can be done accordingly and a better system of air quality management can be established. Population growth coupled with fast urbanization and industrialization heightens pollution levels in metropolitan cities, tier II cities, towns, and villages [14]. The government has tried hard to hold its stand by bringing in the odd–even vehicle use days and installing smog towers, but the operational effectiveness of all these has not yet been fully known. Besides all this, the steady deterioration of the air quality situation in megacities has gradually reached other regions as well, necessitating deep changes in the degree to control pollution [16]. Stricter pollution levels severely affect the health, and the increase in spending on health infrastructure is a consequence; labor productivity is reduced, as well as agricultural productivity. Particulate matter (PM) is one of the major health hazards in India, mainly because more than 1.3 billion Indians stay in areas that have crossed the benchmark provided by WHO [23]. While part of that reduction in life expectancy would come from decreasing 5.3 years for every annum of increased particulate pollution, the latter actually exceeds the other risks of cardiovascular diseases and child/maternal malnutrition. India contributes significantly to increased pollution, with the Northern Plains constituting the most affected region [22].

For the Delhi data set, considering the period of 2015–2020, the EDA analysis reveals that vehicular and industrial emissions are the dominant factors in pollution levels before the COVID-19 pandemic [21]. The Figs. 10.1 and 10.2, represent the density of particulate matter; it is observable to what extent these sources have contributed to air quality. The most common source of pollution remains vehicular and industrial, both before and after lockdowns, and hence, it is perceived as a major source of this kind of pollution despite the declining density after the lockdown [19]. Thus, still much work is to be done in relation to such sources of pollution.

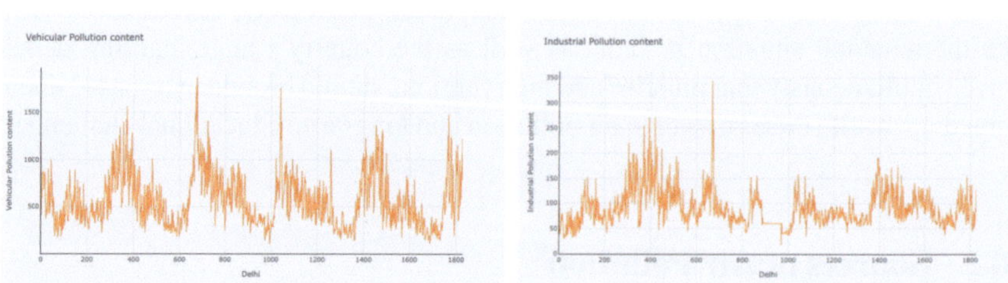

Fig. 10.1 Intensity of major causes of pollution: before COVID

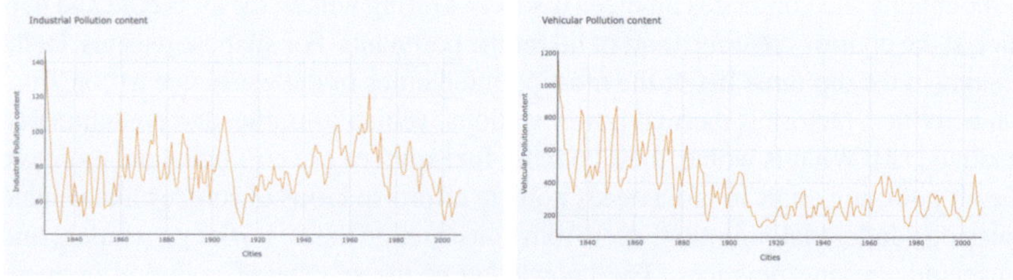

Fig. 10.2 Intensity of major causes of pollution: after COVID

10.2.1 Major Air Pollutants

India's air quality is a complex issue, with various pollutants contributing and interacting in intricate ways. These include Particulate Matters (PM1, PM2.5, PM10), NO_X, SOx, CO, Ozone, and Pb. Data from the National Air Quality Program reveals that PM2.5 and PM10 are consistently exceeding the National Ambient Air Quality Standard (NAAQS) in urban areas of the Indo-Gangetic Plain. This poses a challenge for policymakers who need to reduce emissions and ensure compliance with air quality standards. However, the levels of NO_X, SOx, and Ozone generally align with NAAQS guidelines, offering a more nuanced perspective on air quality management.

Road traffic emissions contribute significantly to air pollution in urban areas, as does road dust. In cities like Delhi, Mumbai, and Kolkata, road dust contributes 37%, 30%, and 61%, respectively, to PM emissions. The complex interaction between pollutants and their sources highlights the challenges India faces in maintaining air quality standards amidst rapid urbanization and industrialization.

When we focus on specific urban areas, we can see that road transport is the main cause of PM2.5 pollution. This is particularly evident in cities like Bengaluru (41%), Chennai (34%), Surat (42%), and Indore (47%). This breakdown highlights the different sources of pollution in each area, and emphasizes the need for customized policies to tackle the problem.

Looking more closely at the national averages for SO_2, NO_2, and PM10, we can see some interesting trends. There has been a decrease in SO_2 levels over the years, which suggests that pollution control measures have been effective. NO_2 levels have remained stable, with a slight decrease in recent years, despite an increase in emissions from vehicles. PM10 levels have fluctuated over time, while PM2.5 levels have remained stable. This could be due to various interventions, such as advances in vehicle technology and other pollution control measures, as documented by the Central Pollution Control Board (CPCB) in 2020.This nuanced understanding further emphasizes the need for a multifaceted approach to address air quality challenges in India, incorporating insights from diverse studies and acknowledging the impact of intervention measures on pollutant concentrations over time.

10.3 Exploratory Data Analyis

In the exploratory data analysis (EDA) of Delhi's AQI dataset, we conducted a comprehensive investigation into the distribution of AQI levels across various pollutant categories. The AQI, representing air quality, was segmented into six categories: very poor, poor, moderate, severe, satisfactory, and good [20]. To understand the underlying factors contributing to these AQI categories, we examined the levels of specific pollutants such as PM2.5, PM10, NOx, NO_2, and NH_3. Through graphical representations, we visualized the distribution of AQI categories based on the levels of these pollutants. Each pollutant's concentration was mapped against the corresponding AQI category, revealing distinct patterns and correlations. As per Figs. 10.3, 10.4, 10.5, 10.6, 10.7 this analysis provided insights into the major causes of pollution in Delhi, shedding light on the pollutants driving air quality deterioration. By segmenting AQI levels according to pollutant concentrations, we identified key contributors to poor air quality, facilitating a targeted approach to pollution control measures.

In addition to analyzing Delhi's AQI dataset, we conducted an EDA to investigate the impact of COVID-19 on AQI levels across India. By comparing AQI data before and after the onset of the pandemic, we discerned notable shifts in air quality patterns.

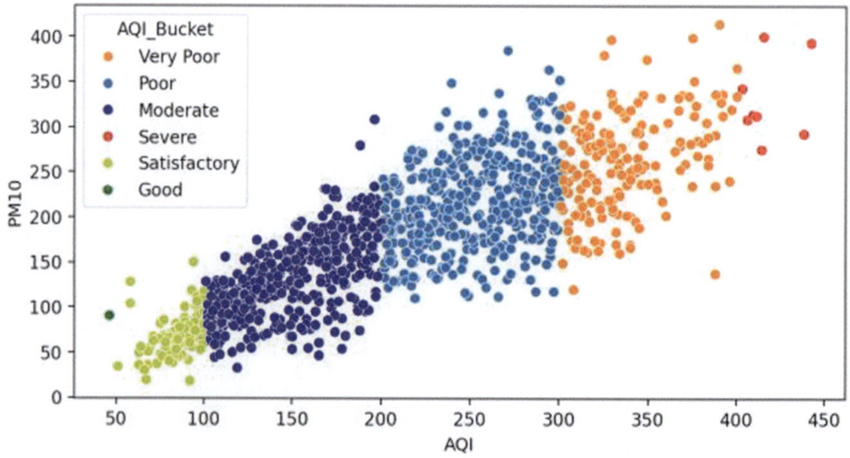

Fig. 10.3 AQI classification as per PM10 levels

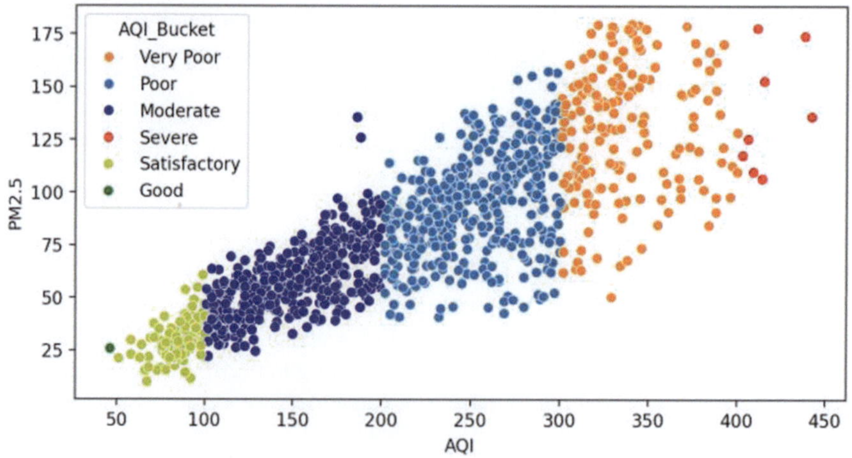

Fig. 10.4 AQI classification as per PM2.5 levels

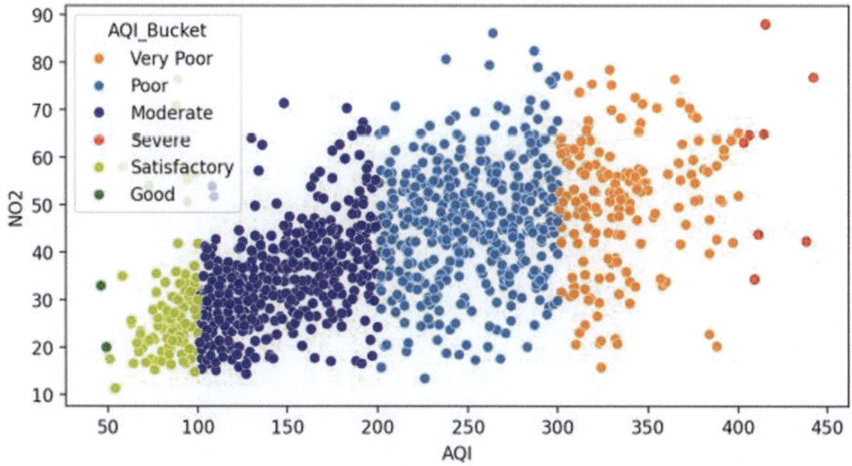

Fig. 10.5 AQI classification as per NO_2

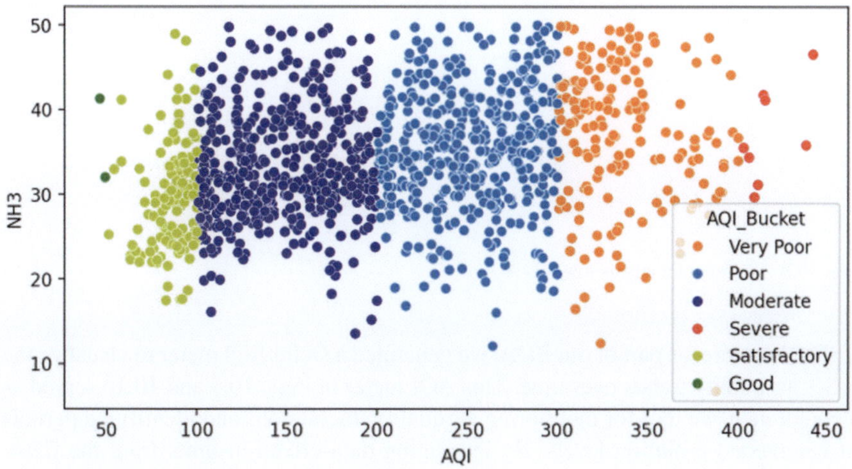

Fig. 10.6 AQI classification as per NH_3

This analysis provided valuable insights into the before-and-after effects of COVID-19 on AQI levels, revealing changes in pollution levels and their implications for public health and environmental sustainability.

Moreover, our EDA delved into people's satisfaction levels regarding AQI before and after COVID-19. In Fig. 10.8, by examining public sentiment and perception towards air quality, we gained a deeper understanding of how environmental factors influence societal well-being and quality of life [36]. This analysis offered insights into the evolving attitudes towards air pollution amidst the pandemic, informing future mitigation strategies and public awareness campaigns.

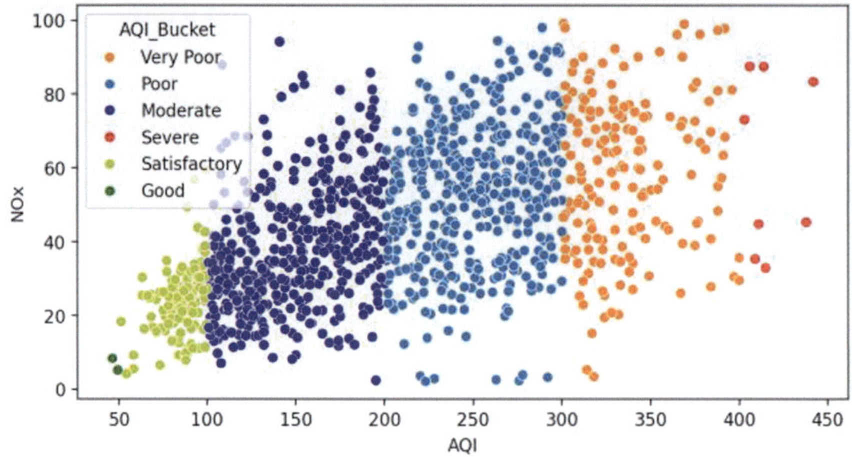

Fig. 10.7 AQI classification as per NO_X levels

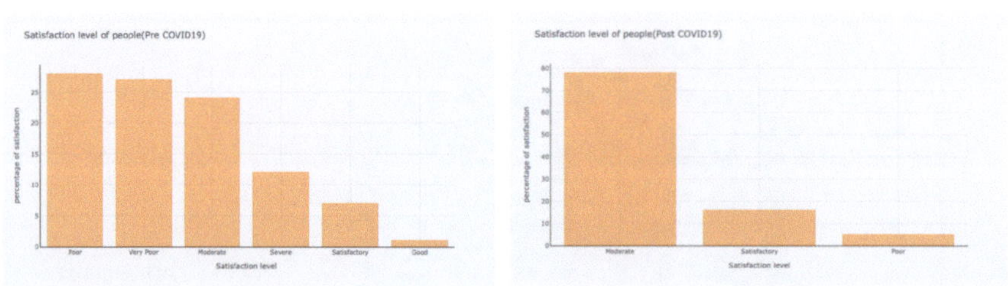

Fig. 10.8 Sentiment analysis towards AQI

Furthermore, as part of our EDA, we generated a Delhi AQI meter to visualize the city's air quality trends over time. This AQI meter in Figs. 10.9 and 10.10 served as a comprehensive tool for monitoring air quality fluctuations and identifying periods of heightened pollution levels. By leveraging data-driven insights from the EDA, we developed a nuanced understanding of Delhi's air quality dynamics, enabling informed decision-making and targeted interventions to address pollution challenges effectively.

10.4 Predicting Delhi's AQI Using Catboost Regression: Machine Learning Algorithm

The prediction of Delhi's Air Quality Index (AQI) using CatBoost regression stands as a pivotal component of this study, showcasing the efficacy of machine learning techniques in addressing complex environmental challenges [30]. Leveraging a

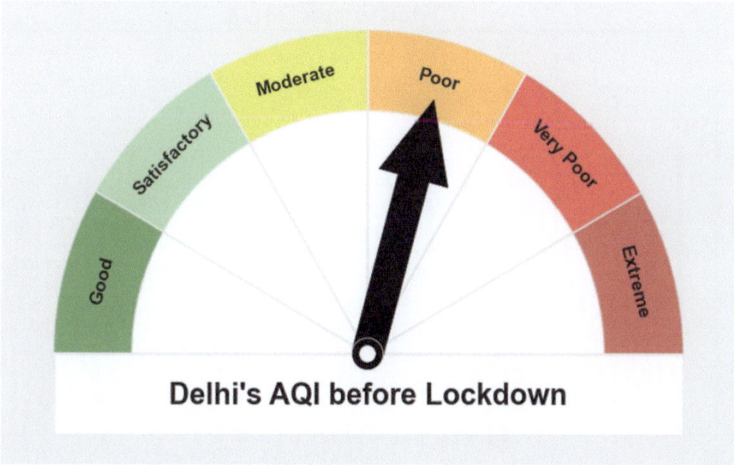

Fig. 10.9 Delhi's AQI status before COVID

Fig. 10.10 Delhi's AQI status after COVID

comprehensive dataset spanning from 2015 to 2020, which encompasses the tumultuous period marked by the peak years of the COVID-19 pandemic in India, this research delves into the intricate dynamics of air quality in the region. The dataset encapsulates various pollutants known to be major contributors to pollution in Delhi, including NO_2, SO_2, CO, O_3, PM2.5, and PM10 [26]. By harnessing this rich dataset, the CatBoost regression model was meticulously trained and rigorously tested using an 80–20 split, enabling robust evaluation of its predictive capabilities. Notably, the Root Mean Squared Error (RMSE) of the model, a crucial metric for regression models, was calculated to be 31.9994, signifying the average deviation of predicted

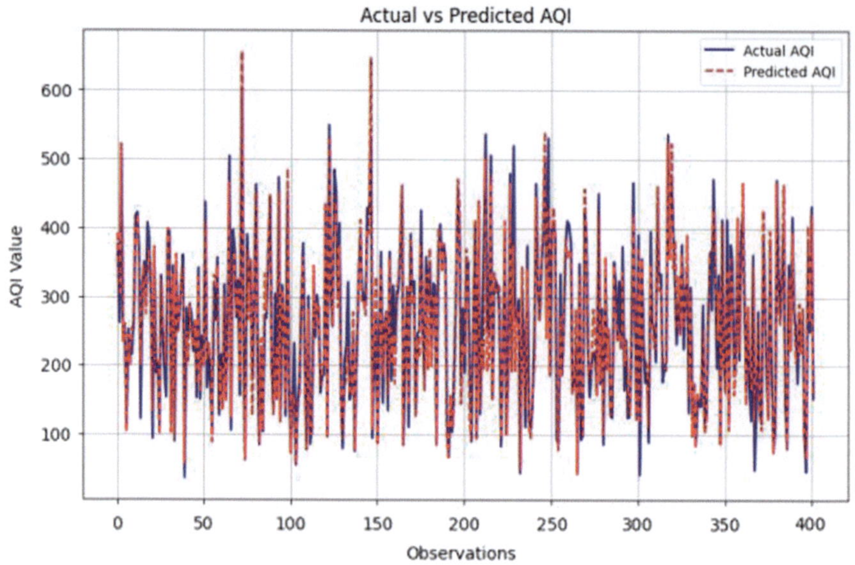

Fig. 10.11 Actual vs predicted AQI using

AQI values from the actual values. The graph in Fig. 10.11 depicting the alignment between actual AQI values (depicted in blue) and corresponding predicted AQI values (represented in red) offers profound insights into the model's performance across the spectrum of AQI levels. This visual representation not only highlights areas of accurate prediction but also underscores potential areas for refinement, thereby enhancing the model's predictive accuracy and reliability. Furthermore, by discretizing AQI values into distinct classes and computing classification metrics such as accuracy and precision, a comprehensive assessment of the model's performance was achieved as shown in Fig. 10.12. These metrics provide quantitative insights into the model's ability to accurately classify AQI levels, thereby informing stakeholders and policymakers about the reliability and effectiveness of the predictive model in real-world applications. In essence, the CatBoost regression model emerges as a potent tool for predicting Delhi's AQI, offering promising avenues for mitigating air pollution and safeguarding public health in the region, while also contributing to the broader discourse on environmental sustainability and sustainable development.

10.5 Major Policy Initiatives by Indian Government

The Government of India has undertaken a range of multifaceted measures to tackle air pollution. This includes engaging with diverse stakeholders, promoting public involvement, developing innovative clean air technologies, enhancing monitoring systems, and forging strategic global collaborations. Alongside these efforts, the

Fig. 10.12 Accuracy, precision, and confusion matrix on AQI classification machine learning technique [27]

```
Accuracy: 0.7935323383084577
Precision: 0.800755613407696
Confusion Matrix:
[[  1   3   2   0   0]
 [  0  22   8   0   0]
 [  0   6  73   0  18]
 [  0   0   1 137  17]
 [  0   0  17  11  86]]
```

government has introduced new schemes and programs, such as the Clean Fuel Research Initiative, National Mission for Cleaner Coal Utilization, transition to cleaner fuels, specialized missions on clean fuel, promotion of indigenous clean air technologies, and support for start-ups in this domain. These programs are characterized by strong synergies between central and state governments, public–private partnerships, and collaborations with international organizations and foundations. The key policies addressing air pollution mitigation and progress towards clean air encompass the following:

1. The 1981 Air Act provides an important legislative framework for India, establishing the rationale and enabling measures to regulate and prevent air pollution [17]. Under this Act, the National Ambient Air Quality Standards were set, which define benchmarks for various air pollutants to maintain air quality.
2. The National Air Quality Monitoring Programme is a significant initiative that systematically monitors air quality across an extensive network of stations in India, collecting data on key pollutants [18].
3. The National Air Quality Index is a summary tool that expresses air quality information in a simplified manner, using different segments to convey the health impacts, making it comprehensible even to the general public.
4. The Delhi and NCR Comprehensive Action Plan, formulated in 2018, outlines specific timelines and responsibilities for actions to prevent, control, and mitigate air pollution in this urban region, addressing its unique challenges and improving air quality.
5. The Graded Response Action Plan is an action plan devised by the Government of Delhi and the NCR to prevent, control, and reduce air pollution through a structured and graded approach, with 42 action points that are dynamic and adjustable based on changing pollution levels.
6. The Jawaharlal Nehru National Solar Mission emphasizes the promotion of solar energy, which can reduce dependence on conventional sources and, in turn, decrease air pollution levels.
7. The Environment Pollution Authority is an apex body with the mandate to implement environmental laws and coordinate efforts to tackle pollution, playing a key role in policymaking, implementation, and enforcement [20].

8. Under the National Clean Air Program established in 2019, India has set a target of reducing particulate pollution by 20–30% in 102 cities by 2024. The program has since been expanded to 131 cities, aiming to reduce particulate pollution by 40% by 2025–26. If successful, this could potentially add 2.1 years to the average life expectancy of Indians living in these cities, and 7.9 months for the rest of the country.
9. The Pradhan Mantri Ujjwala Yojana, conceived to curb indoor air pollution, provides free-of-cost LPG connections among women of low-income families, which would encourage cleaner cooking fuels.
10. The Ministry of Environment in India has been taking several steps towards combating air pollution. It has undertaken the National Clean Air Programme, which aims for reducing pollutants by 2024. The Central Pollution Control Board monitors the pollutants released into the air through monitoring stations installed at 804 locations across the states and cities of 344 cities. In line with the country's Swachh Bharat initiative, India had skipped Bharat Stage-V and transitioned directly from Bharat Stage-IV to Bharat Stage-VI emission norms in 2019.

Establishment of the National Green Tribunal (NGT) 2010 dealing with environmental issues, which provided much-needed impetus to the environmental consciousness of India. The Supreme Court itself is considered to be an environmental activist in that it declared that the right to life under Article 21 includes the right to a healthy environment [29]. The judiciary has made several efforts that led to keeping the environmental consciousness alive in the country, like reconstituting the Environment Protection Authority, the 'Polluter Pays' principle, freezing production licenses for harmful pesticides, relocating hazardous industries, and banning barium-containing crackers during Diwali. However, despite initiating the policy, the whole landscape is somehow disjointed and needs to be more pervasively and collectively enforced. Importantly, judicial activism cannot replace effective executive action.

10.6 Conclusion

In conclusion, the present study started with a holistic assessment of air quality in India, focusing only on Delhi, which today is the world's most polluted city. This was achieved by careful studies of data on air quality between the years 2015 and 2020, where, based on the trends and fluctuations, some significant conclusions could be drawn about significant particulate matter responsible for pollution. Many of the pollutants are declining significantly because of government policies and direct impacts related to the COVID-19 pandemic. Still, in such a scenario, the overall state of air quality in the region remains of concern-it seems the chronic nature of the pollution problem is being maintained. Our exploratory data analysis was thus able to unravel the complex intricacy of air pollution, wherein multiples of interplaying variables of several pollutants-Particulate Matters (PM1, PM2.5, PM10), NO_X, SOx,

CO, Ozone, and Pb-converge. Further, while analyzing sources of pollution, which comprises vehicular emissions, industrial activities, and other anthropogenic factors, an all-inclusive understanding of the problem was constituted. Recently, the authors also investigated the potential application of predictive machine learning models for estimating particulate matter concentrations. These authors again applied advanced techniques such as CatBoost regression to test whether type of model can still capture trends and predict values. ML in air quality management now gives guidance for future development with decisions that can be proactively taken and targeted interventions as well.

As we transition from analysis to action, a set of strategic solutions emerges to address the persistent air pollution crisis in India:

1. India's air pollution data accessibility and usability have significant room for improvement [33]. While data are available from central and state pollution control boards, the process is cumbersome, and the data are not always in user-friendly formats for analysis. Compared to other nations with smaller air quality monitoring networks, India lacks readily accessible, open, and well-structured public government air quality datasets. In many cases, the data, particularly in PDF format, require extensive pre-processing before becoming useful. Enhancing the accessibility of such datasets through better APIs and timely availability, as well as empowering diverse skill sets among Indian citizens, can enable more active engagement in addressing the nation's pressing public health crisis caused by air pollution.
2. Establishing control points in urban areas could significantly reduce pollutant emissions. Sustainability matrices could be developed to monitor the frequency of emission regulation.
3. The proliferation of private vehicles is a major contributor to air pollution in Indian cities. Consequently, policymakers should consider implementing regulations that restrict private vehicle ownership. The deterioration of air quality is also linked to the age of vehicles, indicating that older, high-emitting vehicles should be gradually phased out over a 10-year period. Furthermore, vehicle and industrial emissions should be subject to strict threshold limits, and violators of these regulations should face appropriate penalties.
4. Implementing infrastructural changes, such as building infrastructure to reduce traffic congestion in polluted areas and providing efficient public transport options like bus rapid transit or comprehensive public transit systems, improving walking and cycling facilities, and relocating point sources away from urban centers, would significantly reduce emissions.
5. Technological improvements, such as hybrid or fuel cell vehicles, as well as changes in fuel types like ultra-low sulphur fuels, CNG, methanol in Brazil, or hydrogen fuel in Japan, could reduce emission levels. Additionally, the decrease in sulphur content of fuels has led to declining SO_X emissions, and this trend is expected to further reduce air pollutant emissions.Emission trading, commonly referred to as cap and trade, is one of the control measures implemented in urban cities that are presently used in the US: the present concentrations of pollutants.

6. Carefully designed policies, technological advancements, and strategic land-use planning can be integrated to develop effective control strategies that enable the mitigation of emissions [15]. Implementing stricter emission standards, utilizing cleaner fuels, improving engine development, manufacturing environmentally-friendly vehicles, and employing post-emission treatment technologies can significantly reduce pollution levels in urban areas. Concrete policy interventions can also be implemented to limit exposure to pollutants in emissions. Relocating industries to the outskirts of cities represents a viable approach to address this issue.
7. The Indian government has launched the bold Smart City initiative, which leverages IoT technology to transform cities and prioritize clean air as a key objective. All stakeholders involved in measuring, controlling, and mitigating air pollution–including residents and the CPCB–need to more vigorously embrace these new technologies and forge stronger connections with smart city projects.

In brief, the quest for clean air in India has to be delivered hand in hand by the governments, industries, businesses, the stakeholders, and the citizens. The embedding of data-driven insights through exploratory data analysis and machine learning models would only enhance the accuracy and effectiveness of the control strategies formulated. These solutions will pave a way on this long journey ahead towards a cleaner and healthier environment for future generations in the years to come, born from the present time.

References

1. https://www.indiatoday.in/opinion-columns/story/why-air-pollution-in-india-is-a-public-health-crisis-and-demands-urgent-action-opinion-2347548-2023-03-16
2. https://aqli.epic.uchicago.edu/country-spotlight/india/
3. https://www.worldbank.org/en/country/india/publication/catalyzing-clean-air-in-india
4. https://www.nature.com/articles/s41598-023-39471-1
5. https://doi.org/10.3389/fpubh.2020.00014/full
6. https://www.sciencedirect.com/science/article/abs/pii/S0305750X21001686
7. https://www.teriin.org/article/air-pollution-india-major-issues-and-challenges
8. https://core.ac.uk/reader/234686750
9. https://aaqr.org/articles/aaqr-21-08-oa-0204
10. https://www.thinkglobalhealth.org/article/indias-pollution-problem
11. https://digitalcommons.du.edu/cgi/viewcontent.cgi?article=1293&context=irbe
12. https://www.ncbi.nlm.nih.gov/pmc/articles/PMC6489448/
13. https://www.orfonline.org/research/indias-air-pollution-challenge
14. https://www.ncaer.org/wp-content/uploads/2022/09/The-Solvable-Challenge-of-Air-Pollution-in-India_Michael-Greenstone-Santosh-Harish-Rohini-Pande-Anant-Sudarshan.pdf
15. https://archive.nyu.edu/jspui/bitstream/2451/42303/2/Air%20pollution%20in%20India.pdf
16. https://www.researchgate.net/publication/357838513_Air_Pollution_in_India_A_Critical_Assessment_and_Suggestive_Pathways_for_Clean_Air_CSIR-NIScPR_Discussion_Paper_Series
17. https://www.sciencedirect.com/science/article/pii/S2667010021004054
18. https://doi.org/10.1186/s12302-021-00575-y

19. https://www.ncbi.nlm.nih.gov/pmc/articles/PMC9107909/
20. https://www.researchgate.net/publication/292643438_Air_Quality_Index_-_A_Comparative_Study_for_Assessing_the_Status_of_Air_Quality
21. https://www.airnow.gov/aqi/aqi-basics/
22. https://cpcb.nic.in
23. WHO. Air Pollution, http://www.who.int/airpollution/en/. Accessed 5 Oct 2019
24. I.C. Eze, E. Schaffner, E. Fischer, T. Schikowski, M. Adam, M. Imboden et al., Long-term air pollution exposure and diabetes in a population-based Swiss cohort. Environ. Int. **70**, 95–105 (2014). https://doi.org/10.1016/j.envint.2014.05.014
25. R. Kelishadi, P. Poursafa, Air pollution and non-respiratory health hazards for children. Arch. Med. Sci. **6**, 483–95 (2010). https://doi.org/10.5114/aoms.2010.14458
26. USGCRP. Global climate change impacts in the United States, in *Climate Change Impacts by Sectors: Ecosystems* eds. by T.R. Karl, J.M. Melillo, T.C. Peterson (Cambridge University Press, New York, NY: United States Global Change Research Program, 2009)
27. P.M.Manucci, M. Franchini, Health effects of ambient air pollution in developing countries. Int. J. Environ. Res. Public Health. **14**:1048 (2017). https://doi.org/10.3390/ijerph14091048
28. Burden of Disease from Ambient and Household Air Pollution, http://who.int/phe/health_topics/outdoorair/databases/en/. Accessed 15 Aug 2017
29. D. Hashim, P. Boffetta, Occupational and environmental exposures and cancers in developing countries. Ann. Glob. Health **80**, 393–411 (2014). https://doi.org/10.1016/j.aogh.2014.10.002
30. Y. Guo, H. Zeng, R. Zheng, S. Li, G. Pereira, Q. Liu et al., The burden of lung cancer mortality attributable to fine particles in China. Total Environ. Sci. **579**:1460–6 (2017). https://doi.org/10.1016/j.scitotenv.2016.11.147
31. Q. Hou, X.Q. An, Y. Wang, J.P. Guo, An evaluation of resident exposure to respirable particulate matter and health economic loss in Beijing during Beijing 2008 olympic games. Sci. Total Environ. **408**:4026–32 (2010). https://doi.org/10.1016/j.scitotenv.2009.12.030
32. A. Kankaria, B. Nongkynrih, S. Gupta, Indoor air pollution in india: implications on health and its control. Indian J. Comm. Med. **39**:203–7. https://doi.org/10.4103/0970-0218.143019
33. I. Parajuli, H. Lee, K.R. Shrestha, Indoor air quality and ventilation assessment of rural mountainous households of Nepal. Int. J. Sust. Built Env. **5**:301–11 (2016). https://doi.org/10.1016/j.ijsbe.2016.08.003
34. T. Saud, R. Gautam, T.K. Mandal, R. Gadi, D.P. Singh, S.K. Sharma, Emission estimates of organic and elemental carbon from household biomass fuel used over the Indo-Gangetic Plain (IGP), India. Atmos. Environ. **61**:212–20 (2012). https://doi.org/10.1016/j.atmosenv.2012.07.030
35. D.P. Singh, R. Gadi, T.K. Mandal, T. Saud, M. Saxena, S.K. Sharma, Emissions estimates of PAH from biomass fuels used in rural sector of Indo-Gangetic Plains of India. Atmos. Environ. **68**:120–6 (2013). https://doi.org/10.1016/j.atmosenv.2012.11.042
36. M. Dherani, D. Pope, M. Mascarenhas, K.R. Smith, M.B.N. Weber, Indoor air pollution from unprocessed solid fuel use and pneumonia risk in children aged under five years: a systematic review and meta-analysis. Bull. World Health Organ. **86**:390–4 (2008). https://doi.org/10.2471/BLT.07.044529

Chapter 11
Analyzing the Potential of Artificial Meat Production as a Sustainable Solution to Decarbonization

Mudita Nagpal, Tanisha Gupta, Krrishika Gupta, and Nidhi Sharma

Abstract Artificial meat production is a reliable and efficient way of producing meat, which could contribute to global food security by reducing dependency on traditional livestock farming as it obviates the slaughtering and raising of animals for a daily diet. This involves growing meat in a controlled lab environment, offering numerous benefits such as reduced exposure to pesticides and bacteria, environmental sustainability, addressing ethical, cultural concerns, and a rise in global demand for protein rich food. The current state of artificial meat production procedures involves various methodologies like tissue engineering, 3D/4D bioprinting, biophotonics, and cloning. The intersection of biotechnology and artificial intelligence (AI) has brought about transformative advancements in the production of artificial meat. Biotechnological methods provide the foundation for cellular agriculture, while AI enhances the precision, speed, and scalability of the production process and allows for the creation of unique and innovative meat alternatives. Between 2024 and 2031, substantial growth is anticipated in the artificial meat market, driven by strategic insights and innovative approaches. The article highlights the favorable results achieved through various projects in different regions, demonstrating the potential of cultured meat production to effectively adapt and bring about beneficial changes. Further, the challenges and future perspectives associated with artificial meat production have also been discussed.

Keywords Artificial meat · Artificial intelligence · Biotechnology · Sustainability · Methodologies · Market scope

---

M. Nagpal · T. Gupta (✉) · K. Gupta · N. Sharma
Department of Applied Sciences, Vivekananda Institute of Professional Studies-Technical Campus, Delhi, India
e-mail: gtanisha599@gmail.com

A. Kumar et al. (eds.), *Selected Proceedings of the 1st International Conference on Advanced Materials for Sustainable Innovation; IC-AMSI 2024; 28–30 August; New Delhi; India*, Springer Proceedings in Energy,
https://doi.org/10.1007/978-981-96-4492-6_11

11.1 Introduction

The world is on the verge of an artificial meat production revolution, which could change the food industry around the globe and also offer an ethical and sustainable alternative to traditional livestock farming. This novel idea involves growing meat in controlled laboratory environments that have some advantages over ordinary methods. In this regard, artificial meat production helps reduce our dependence on traditional livestock farming by eliminating raising and slaughtering animals. Thus, addressing this shift satisfies vital concerns not only related to ethics but also culture, in addition to responding to rising global demand for foods rich in protein [1]. The advent of artificial meat production is timely as the planet grapples with food security issues and searches for sustainable responses to feed a burgeoning population. Among other things, one of the most interesting aspects of artificial meat is its potential to improve global food security [2]. Generally, conventional livestock rearing practices are land-intensive and require substantial amounts of water and feed; moreover, they often have negative environmental implications such as deforestation, greenhouse gas emissions, and pollution [3]. 7.1 billion tons of carbon dioxide equivalent came from animal agriculture in 2013, according to FAO [4]. Conversely, growing flesh under laboratory conditions can be more efficient and environmentally sustainable than this. This shortens the time that meat is exposed to pesticides and bacteria, resulting in a safer and cleaner product [5].

The current status of making artificial meats embraces several advanced technologies. For instance, 3D and 4D bioprinting are used to create complex meat structures by layering cells in precise patterns [6]. Biophotonics, which studies the interaction of light with biological matter, improves monitoring and control throughout the growth process so as to guarantee appropriate conditions for meat development [7]. Integrating biotechnology with artificial intelligence (AI) is a major driver of progress in artificial meat production. Cellular agriculture relies on key biotechnology methods to establish its base, while AI enhances precision, speed, and scalability. Growth conditions can be optimized by AI algorithms, while potential problems may be predicted before they occur or reduced efficiently, hence increasing cost-effective measures in production [8]. This combination of AI and biotechnology not only fast tracks the development of artificial meats but also leads to innovative alternative products that are distinct from any existing ones since they cater to specific dietary requirements dictated by personal preferences. It is expected that between 2024 and 2031 [9], the artificial meat market will experience an enormous increase in size. Different projects in various parts of the world predict this expansion to be driven by strategic insights and innovative approaches that have shown promising results.

This article examines different methods employed in the production of lab-grown meat, highlighting its potential as a positive environmental force. It contains information on how AI has been integrated into the process of making meat products. In addition, it incorporates several case studies, addressing such issues as costs, regulations, and consumer acceptance. The article further underscores cultured meat's ability to adapt as well as transform the global food industry.

11.2 Cultured Meat Production

Cultured meat production offers a feasible, non-butchering alternative to traditional meat, providing numerous social and environmental benefits. To create different kinds of meat, including beef, fish, and chicken, among others. This approach uses genetic engineering and tissue technology. Cultured meat can be classified into two types: in-vitro meat and genetically modified organisms (GMOs). A range of techniques can be used to produce in-vitro, such as scaffolding, self-organization, 3D/4D bioprinting, biophotonics, and nanotechnology (Fig. 11.1) [10].

11.2.1 Genetically Modified Organisms (GMO)

Cultured meat also includes cloned animals and genetically modified organisms. In GMOs, genetic engineering is used to alter their genes, incorporating DNA from other organisms [11]. Animal cloning is an intricate process aimed at producing species with identical genetic traits as their parents. Numerous animals, including

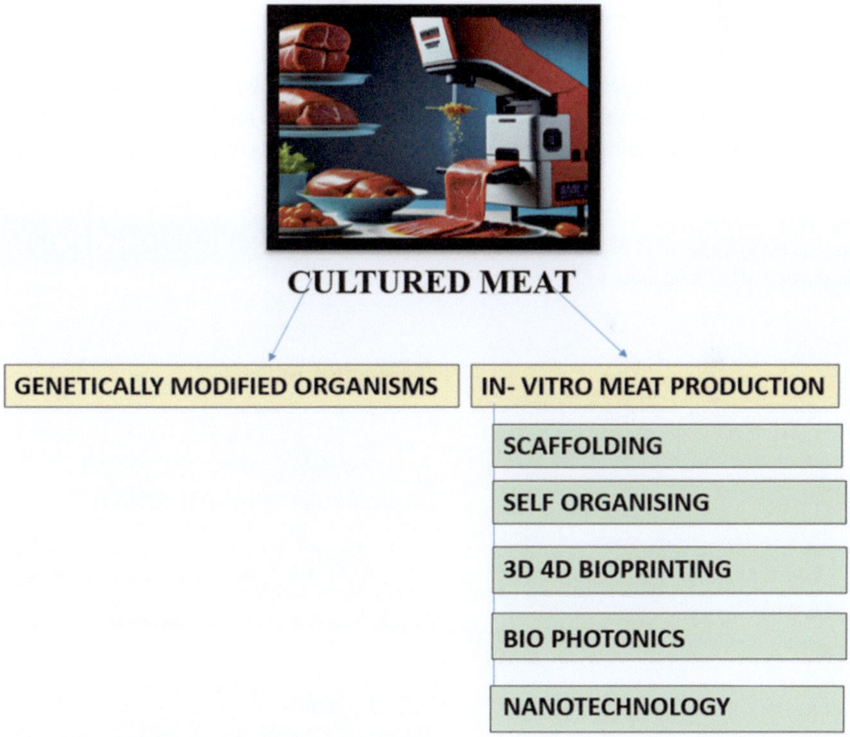

Fig. 11.1 Methods for cultured meat production

goats, pigs, and lambs, have been cloned up to this point but have never been eaten. Certain flaws in the cloning process include malformations, big or aberrant offspring syndrome, and even premature death [12, 13].

11.2.2 In Vitro Production Methods

This method was designed to overcome difficulties such as diffusion barriers and to create more speculative possibilities. It is a four-stage process (Fig. 11.2). The first stage involves collecting cells via biopsy from either living or deceased animals. Only connective and muscle tissue are extracted and, after that, put together to develop further. Cells are nourished, developed, multiplied, and kept in an isolation tube or a particular glass-based laboratory culture media throughout the second phase. They are given various nutrients, sugars, amino acids, and salts for the proper formation of the desired meat. In the third phase, the successfully raised cells are transferred to the cultivation bioreactor for further growth. After that, cells change into connective and muscle cells. After this, in about two to eight weeks, they form a great number of cells, which results in the formation of a chunk of meat. The fourth phase is called harvesting, as the cells have been formed into adequate-density meat [14]. There are four in vitro production methods.

Scaffolding: Firstly, the tissue sample of the required animal is collected, and then, to isolate the satellite cells, the chemical breakdown of the tissue is done. Myoblasts of the tissue are then cultured in an appropriate medium, and for growth, a plant-based serum is used. Using a bioreactor, stem cells are withdrawn from the

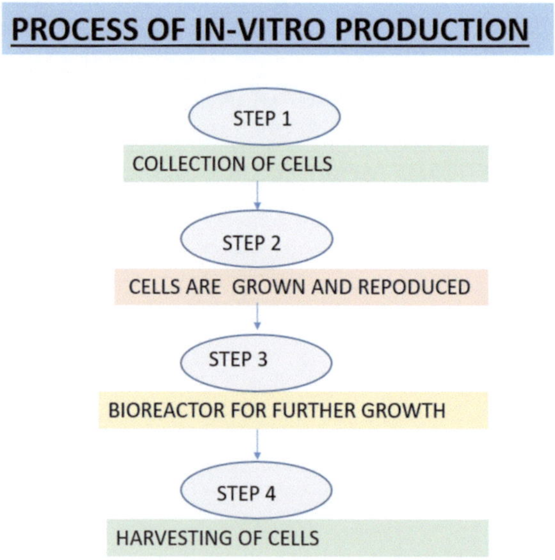

Fig. 11.2 Four phase process for in-vitro production of cultured meat

tissue. In the bioreactor, the cells first multiply, then differentiate. The exchange of gasses occurs when the medium is run via a porous scaffold, resulting in edible meat [15].

Self-organizing: This technique can be used to produce highly structured, self-organizing meat such as steak, as the muscle tissue of the cell can form itself [16]. The meat produced through this process is exactly like conventional meat, but due to the absence of blood vessels, if left unchecked, the meat can turn poisonous or be harmful to health [17].

3D/4D bioprinting: Bioprinting, another name for three- or four-dimensional (3D) organs, is based on traditional printing principles. It is an effective technique as it creates similar tissues and organs for organ therapeutic and regenerative tissue applications. The bioproduct models are created using computer assistance. The cells are first sprayed onto gels. The cells combine throughout the culturing process for the bio-product formation, which possesses the fundamental cell structure needed to transport blood. It's a precise procedure using similar technology. Tissues and target organs exhibit sensitivity to temperature and humidity, thus these are two characteristics that must be managed. It is employed to restore cardiovascular, bone, and muscle tissues [18]. World's first rib-eye steak was recently successfully grown in 2021 by Aleph Farms and The Technion, Israel Institute of Technology, utilizing 3D bioprinting. Steak is supposed to be juicy and soft and has fat similar to that of ordinary meat [19].

Biophotonics: It is the interaction of light photons with biological samples and is a recently developed process in which laser light is used to bind particles together. Desired structures are formed from the optical matter produced by the laser; in order for the material to mix and form a new solid structure, it is deposited and kept under light. Compared to previous methods, this technology makes it easier to fuse and infuse traits like fat, making it suitable for the production of meat. Up to now, this technique has been used to manufacture hamster ovaries and red blood cell [10, 20].

Nanotechnology: It is a technology that can increase the shelf life of meat without reducing nutrients. In nanotechnology, a molecule sized robot is designed that can create and assemble the molecules precisely at an atomic level. It is a promising solution to the current problems as it is affordable and produces edible meat [21, 22].

11.3 Sustainability & Environmental Impact of Cultured Meat

There must be an estimated increase of 60–110% in global food production by 2050 to feed 9–10 billion of the population [23]. Animals are an integral part of the ecosystem. Since meat production relies on butchery, it is a matter of concern. The annual consumption of various forms of meat has increased tremendously over the years (Fig. 11.3). A multi-faceted and holistic approach is required to face the big challenge of sustainability in the meat sector. Sustainability can be attained by

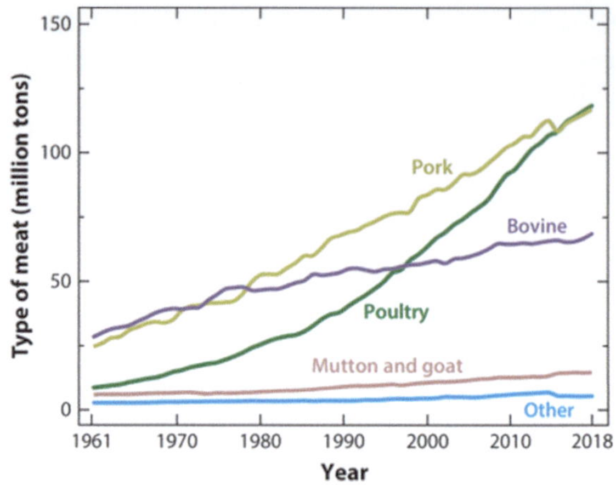

Fig. 11.3 Trend for the annual global consumption of various forms of meat [26]

utilizing tools such as precision livestock farming, digitalizing farming systems, livestock markets, and applying remote sensing and AI to manage production and mitigate environmental impact and GHG emissions [24].

A renewable form of meat production, and maybe the best substitute, is the development of artificial meat production. This may help people consume meat without guilt or fear [10]. There are some other factors that help in identifying if cultured meat improves sustainability and environmental impacts. An important factor is the cost-effectiveness and resource efficiency of the cultured meat production process. Furthermore, the production of a unit of cultured meat must have a lower total environmental impact than meat produced conventionally. Legal and regulatory clearance should also be obtained for the production technology. Ultimately, it should be the consumer's choice to switch from conventional meat to cultured meat. [25].

A third of the nitrogen emissions caused by humans today are from the production of meat; livestock alone emits 65 Tg N yr−1, which has an impact on the ecosystem. Thirty percent of water resources, seventy seven percent of all agricultural land, and twelve to twenty percent of greenhouse gas emissions caused by humans are attributed to animal agriculture. Numerous studies show that compared to conventionally produced meat, the total environmental effect of cultured meat is lower [27, 28] (Fig. 11.4). Mattick and his group conducted a life cycle analysis for the in vitro production of biomass for producing cultured meat. Results reflected that the greenhouse gas (GHG) emissions for cultured meat were higher than those of poultry and pork, but emissions were 75% lower compared to beef. Land usage and agricultural input of cultured meat were lower, and so was the eutrophication potential [2]. Cultured meat is seen as a more environmentally friendly and carbon-efficient substitute for conventional meat, with a 78–96% reduction in GHG emissions, an 82–96% decrease in water use, 7–45% lower energy use, and a 99% decrease in land use [29]. These figures are noteworthy, however they are predicated on theories.

Three different beef production systems were compared with cultured meat production on the basis of GHG emissions by Lynch and Pierrehumbert. The results

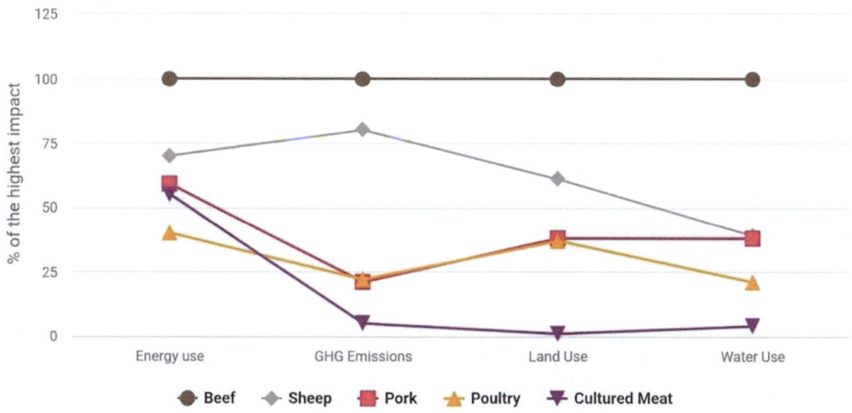

Fig. 11.4 Comparison of resources used for conventional meat production with cultured meat production [27]

favored cultured meat, showing it usually produces less pollution than beef, but one scenario conflicted that it is always climate-friendly as it produces energy and releases carbon dioxide into the environment, which stays in the atmosphere longer than nitrous oxide or methane released during traditional meat production. The study brought up the possibility that, over time, cultured meat could pose an even greater threat to the climate than beef. In any case, the conclusion is rather ambiguous because the outcome in the simulations doesn't happen for hundreds of years [30].

Industries are advancing their approach to cultured meat production by focusing on sustainability using several strategies. Culturing meat in bioreactors reduces water and land use and lowers GHG emissions compared to conventional production. For increasing production and overcoming technological challenges, partnerships among the food industries, life sciences, and academia are important [31].

Regulatory status for the cultured meat industry was first initiated by the European Union (EU) on recognizing these benefits. Singapore and Israel have also shown positive participation, while the United States has been the most active promoter. The United States, The US Department of Agriculture Food Safety and Inspection Service (USDA–FSIS), the Food and Drug Administration (FDA), jointly introduced a regulatory framework for cultured meat. In other countries, such as Australia and China, statutory authorities responsible for developing food standards have suggested including cultured meat as a new food category [32].

11.4 Role of Artificial Intelligence in Artificial Meat Production

An emerging part of the industry and one of the most popular topics all over the world is AI technology. For implementing non-destructive prophecies of the quality of meat, evaluation, and supplying essential technical support for online meat grading, AI technology has become important [8].

The favored techniques used in the food industry include fuzzy logic, which deals with incomplete and indistinct information; machine learning, which solves complicated tasks including constantly changing and complex data; expert systems, used for decision-making; and adaptive neuro-fuzzy inference system (ANFIS), used in complex systems with complicated functional relationships. Due to its capacity for integrating expert knowledge, handling indistinct information, and designing nonlinear food systems, fuzzy logic is one of the most efficient techniques [33].

The growth of AI applications in the food industry over the years can be attributed to their roles in accelerating the food production process, enhancing quality, parameter prediction, food sorting, food safety, and classification. With the growth of AI, food recipes based on personal dietary tastes and restrictions can be created. To mimic the human nose in sensing flavors or odors, an instrument called the Electronic Nose (E-nose) has been developed. The E-nose is used in gas sensing, where it analyzes components or mixtures of gasses and vapors, recognizing both complex and simple odors [33]. However, AI depends on the quality of the data for precise predictions of results. Lack of standardization and quality of data concerns the growing technology. Further, AI implementation has been unable to justify the return on investment (ROI) which leads to small or medium-sized enterprises facing deployment challenges. Also, AI is still emerging to be accepted for its delivery ability or value [34].

11.4.1 Case Studies

AI technology is being used by several companies to enhance artificial meat production processes. To amplify the food value of its cultured meat and to fine-tune its cell proliferation process, BioCraft Pet Nutrition Inc., incorporated in Delaware, USA, is applying AI to small-scale, more focused evaluations in the lab. It finds low-cost components and inputs that are unlikely to cause problems during the regulatory approval procedure. In the US, animal agriculture, attributed to feeding pets, causes an environmental impact of more than 25%. BioCraft's cultivated meat offers nutrients that dogs and cats require without steroids, antibiotics, chemical contaminants, or other hormones, thus there is a lower chance of harboring diseases like Salmonella, E. coli, and other typical fecal-borne germs [35].

Firmenich, founded as a startup in Geneva, Switzerland, is one of the world's largest fragrance and taste manufacturers. It produced the first AI flavor in history—a natural beef flavor—for use in plant-based meat substitutes. The taste was created using a rule-based formula generation methodology that used Firmenich's vast formula databases and held raw material consumption records [36].

In addition, plant-based meat products, similar in taste and structure to meat, are being produced by several companies in Europe and the United States, such as Beyond Burger, Schnitzel, and Impossible Burger, at present. Impossible Burger uses AI to study and select plant-based proteins and other molecules that replicate the taste, texture, and cooking properties of beef. Beyond Meat also leverages AI to refine its formulations and improve the sensory experience of its products. Ingredients like mushrooms, legumes, wheat gluten, rice, flavoring additives, and soy protein are used to process nutrient qualities and the aesthetic of meat [6].

11.5 Cost Analysis

When cultured meat was first introduced to the public, the cost of producing a single cultured beef burger was around $325,000 in 2013 [1]. This high cost was primarily due to the novelty of the technology and the small scale of production.

The primary determinants of the cost of producing cultured meat are:

- Cell Culture Medium: The growth medium used to nourish the cells is one of the most expensive components of cultured meat production. This medium must be rich in nutrients and growth factors, which are often costly. Efforts to reduce costs include developing serum-free and plant-based media [5].
- Bioreactor Scaling: Scaling up bioreactors to industrial levels can significantly reduce costs through economies of scale. Larger bioreactors can produce more meat at a lower relative cost, but they require significant upfront investment [37].

The cost of producing cultured meat has, however, significantly decreased, according to current trends. Producing cultured meat came at a very high initial cost. The first cultured beef hamburger was created, and it cost over $325,000 USD when Dr. Mark Post unveiled it in 2013. According to current projections, the cost of cultivating beef will drop to roughly $5.66 USD per kilogram by 2030 [38], reflecting advancements in technology and production scales.

When Dr. Mark Post unveiled the original hamburger made of cultured beef in 2013, it cost about $325,000 USD to produce. Whereas the average cost of ground beef in the United States is about $5.30 per pound [39]. However, the anticipated cost of manufacturing cultured meat will decrease significantly in the future, potentially reaching as low as $5.66 USD per kilogram by 2030, owing to technological advancements. It is anticipated that cultivated meat products will become competitively priced with conventional meat products, making them accessible to consumers at similar price points [38].

11.5.1 Technological Advancements in Cultured Meat Production

Improved Cell Culturing Techniques: Advances in cell culture methods have significantly improved the efficiency of growing in vitro muscle cells. This includes employing optimized growth media that support faster cell proliferation and differentiation [5].

Scaffold Materials: The development of biodegradable scaffolds that mimic the structure of animal tissues has improved the texture and consistency of cultured meat [6].

Automation and AI: Integration of automation and artificial intelligence in the production process has streamlined operations, reducing labor costs and improving consistency [6].

11.6 Consumer Acceptance

Consumer acceptance of cell-based meat is a complex issue influenced by various factors such as taste, safety, ethics, and sustainability. While there is growing interest in cultured meat as a potential solution to issues related to conventional meat production, several challenges remain regarding its widespread adoption (Fig. 11.5).

Here are some key problems with consumer acceptance of cultured meat production:

Taste and texture: One of the primary concerns among consumers is whether cultured meat will taste and feel like traditional meat. Research has indicated that when it comes to consumers' adoption of substitute meat products, taste and texture play a critical role. Customers may reject cultured meat if it is unable to match the

Fig. 11.5 Challenges associated with acceptance of cultured meat by consumers

sensory experience of traditional meat. In order to enhance the flavor and texture of cultured meat and increase its consumer appeal, research and development are still being conducted [28].

Perception of safety: Customers might be worried about the safety of cultured meat because of its use of cutting-edge technologies and the possibility of contamination during the manufacturing process. Gaining the trust of consumers requires ensuring the safety of cultured meat products through stringent testing and regulatory control [2].

Ethical and cultural considerations: Some consumers may have ethical objections to cultured meat, viewing it as unnatural or questioning the ethics of growing meat in a laboratory setting. Cultural factors also influence consumer attitudes towards food, and taking on cultured meat may vary across different places and communities [40].

Price: Currently, one of the biggest obstacles to the global adoption of cultured meat is the expense of production. Cultured meat production involves expensive technologies and processes, which could result in higher prices in contrast to traditional meat. As of recent estimates, the cost of producing cultured meat has decreased to about $116–$22,423 per kg [38], whereas the average cost of ground beef in the United States is about $5.30 per pound [39]. Making cultured meat affordable for consumers will require advancements in production efficiency and scale, as well as cost reduction strategies [9].

Communication and Marketing: For cultured meat to be accepted, it is essential to address consumer concerns and effectively communicate the benefits of the product. Misinformation or a lack of understanding about the technology behind cultured meat could lead to skepticism or distrust among consumers. Clear and transparent communication regarding the production process, safety, and sustainability benefits is essential for building consumer confidence [41].

11.7 Conclusions

Artificial meat production has shown great promise for transforming the world's food system. It supports the worldwide trend towards more ethical, sustainable, and efficient food sources by providing a productive and sustainable substitute for conventional meat production. Studies show that cultural meat is a carbon-improved and sustainable alternative to conventional meat, leading to reduced energy use, GHG footprint, land use, and water use. As the field continues to evolve, the collaboration between biotechnology and AI has been proven to be pivotal in shaping a future where artificial meat becomes a mainstream and reliable food source. This fact is supported by the usage of AI technologies by various companies in producing artificial meat products. To improve consumer acceptance, technological improvements are made that have proven to reduce costs significantly. However, there are still certain challenges associated with consumer acceptance, like the taste and texture, safety, ethical and cultural considerations, costs, and lack of effective marketing of cultured meat. Further, technical hurdles, such as scaling up production and ensuring consistency in product quality, remain significant. Regulatory frameworks also need to be established and harmonized globally to ensure safety and consumer trust.

Despite the challenges, ongoing research and development efforts make the future prospects of artificial meat production look promising. Cultured meat has the potential to address food security and sustainability challenges on a global scale. As production methods become more efficient and scalable, cultured meat could provide a viable alternative to conventional meat production, particularly in regions facing environmental constraints or resource shortages.

References

1. M.J. Post, Cultured beef: medical technology to produce food. J. Sci. Food Agric. **94**(6), 1039–1041 (2013)
2. C.S. Mattick, A.E. Landis, B.R. Allenby, N.J. Genovese, Anticipatory life cycle analysis of in vitro biomass cultivation for cultured meat production in the United States. Environ. Sci. Technol. **49**(19), 11941–11949 (2015)
3. Smail, Green Econ. J. (2022), https://greeneconomyjournal.com/explainer/the-environmental-impact-of-livestock-farming/. Accessed 20 Jun 2024
4. Kevany, Vox (2023), https://www.vox.com/future-perfect/23738600/un-fao-meat-dairy-livestock-emissions-methane-climate-change. Accessed 20 Jun 2024
5. E.A. Specht, D.R. Welch, E.M. Rees Clayton, C.D. Lagally, Opportunities for applying biomedical production and manufacturing methods to the development of the clean meat industry. Biochem. Eng. J. **132**, 161–168 (2018)
6. G. Zhang, X. Zhao, X. Li, J. Du, J. Xu, Decellularized plant scaffold-based hybrid system for cultured meat production. Food Biosci. **31**, 100438 (2020)
7. N. Stephens, L. Di Silvio, I. Dunsford, M. Ellis, A. Glencross, A. Sexton, Bringing cultured meat to market: technical, socio-political, and regulatory challenges in cellular agriculture. Trends Food Sci. Technol. **78**, 155–166 (2018)
8. Y. Shi, X. Wang, M.S. Borhan, J. Young, D. Newman, E. Berg, X. Sun, A review on meat quality evaluation methods based on non-destructive computer vision and artificial intelligence technologies. Food Sci. Anim. Resour. **41**(4), 563 (2021)
9. C. Bryant, J. Barnett, Consumer acceptance of cultured meat: a systematic review. Meat Sci. **143**, 8–17 (2018)
10. T. Mateti, A. Laha, P. Shenoy, Artificial meat industry: production methodology, challenges, and future. Jom **74**(9), 3428–3444 (2022)
11. A.L. Van Eenennaam, Genetic modification of food animals. Curr. Opin. Biotechnol. **44**, 27–34 (2017)
12. P. Dolma, A. Ali, R.C. Sobti, A. Chauhan, Animal cloning: perspectives for futuristic medicine, in *Advances in Animal Experimentation and Modeling* (Academic Press, 2022), pp. 53–67
13. M.N. Verzijden, Z.W. Culumber, G.G. Rosenthal, Opposite effects of learning cause asymmetric mate preferences in hybridizing species. Behav. Ecol. **23**(5), 1133–1139 (2012)
14. T.U.S. Hasibuan, A. Muslim, Inovasi industri daging buatan dalam perspektif Fiqh Syafi'i. J. Kaji. Islam. Interdisip. **7**(1), 87–108 (2022)
15. A.C. Schnitzler, A. Verma, D.E. Kehoe, D. Jing, J.R. Murrell, K.A. Der, M.S. Rook, Bioprocessing of human mesenchymal stem/stromal cells for therapeutic use: current technologies and challenges. Biochem. Eng. J. **108**, 3–13 (2016)

16. Z.F. Bhat, H. Bhat, V. Pathak, Chapter 79—prospects for in vitro cultured meat—a future harvest, in *Principles of Tissue Engineering*, 4th edn., ed. by R. Lanza, R. Langer, J. Vacanti (Academic Press, 2014), pp.1663–1683
17. J. Junaidi, A. Asmadin, In vitro meat industry: legal analysis of manufacture and consumption in FIQIH perspective. Proceeding Annu. Conf. Islam. Econ. Law **2**(1), 68–83 (2023)
18. M. Javaid, A. Haleem, 4D printing applications in medical field: a brief review. Clin. Epidemiol. Glob. Health **7**(3), 317–321 (2019)
19. https://aleph-farms.com/journals/aleph-farms-and-the-technion-reveal-worlds-first-cultivated-ribeye-steak/#:~:text=Together%20with%20our%20research%20partner,meat%20%E2%80%94%20real%20cow%20cells%2C%20without. Accessed 24 Jun 2024
20. J. Mullins, The stuff of beams. New scientist (2551), 44–47 (2006)
21. D.K. Banerjee, A.K. Das, R. Banerjee, M. Pateiro, P.K. Nanda, Y.P. Gadekar, S. Biswas, D.J. McClements, J.M. Lorenzo, Application of Enoki mushroom (Flammulina Velutipes) stem wastes as functional ingredients in goat meat nuggets. Foods (Basel, Switzerland) **9**(4), 432 (2020)
22. A.K. Das, P.K. Nanda, S. Bandyopadhyay, R. Banerjee, S. Biswas, D.J. McClements, Application of nanoemulsion-based approaches for improving the quality and safety of muscle foods: a comprehensive review. Compr. Rev. Food Sci. Food Saf. **19**(5), 2677–2700 (2020)
23. R. Ben Ayed, M. Hanana, Artificial intelligence to improve the food and agriculture sector. J. Food Qual. **2021**, 1–7 (2021)
24. P. Kumar, A.A. Abubakar, A.K. Verma, P. Umaraw, M. Adewale Ahmed, N. Mehta, A.Q. Sazili, New insights in improving sustainability in meat production: opportunities and challenges. Crit. Rev. Food Sci. Nutr. **63**(33), 11830–11858 (2023)
25. H.L. Tuomisto, The eco-friendly burger: could cultured meat improve the environmental sustainability of meat products? EMBO Rep. **20**(1), e47395 (2019)
26. M.C. Parlasca, M. Qaim, Meat consumption and sustainability. Annu. Rev. Resour. Econ. **14**, 17–41 (2022)
27. H.L. Tuomisto, M.J. Teixeira de Mattos, Environmental impacts of cultured meat production. Environ. Sci. Technol. **45**(14), 6117–6123 (2011)
28. M.J. Post, Cultured meat from stem cells: challenges and prospects. Meat Sci. **92**(3), 297–301 (2012)
29. S. Hubalek, M.J. Post, P. Moutsatsou, Towards resource-efficient and cost-efficient cultured meat. Curr. Opin. Food Sci. **47**, 100885 (2022)
30. J. Lynch, R. Pierrehumbert, Climate impacts of cultured meat and beef cattle. Front. Sustain. Food Syst. **3**, 421491 (2019)
31. Life Science Industry's Role in Cultured Meat Industry Scale Up, Food Engineering (foodengineeringmag.com). Accessed 24 Jul 2024
32. X. Guan, Q. Lei, Q. Yan, X. Li, J. Zhou, G. Du, J. Chen, Trends and ideas in technology, regulation and public acceptance of cultured meat. Futur. Foods **3**, 100032 (2021)
33. N.R. Mavani, J.M. Ali, S. Othman, M.A. Hussain, H. Hashim, N.A. Rahman, Application of artificial intelligence in food industry—a guideline. Food Eng. Rev. **14**(1), 134–175 (2022)
34. https://www.linkedin.com/pulse/ai-food-industry-what-benefits-challenges-applications-binmile?utm_source=perplexity. Accessed 24 Jul 2024
35. BioCraft Begins Using AI in Cultivated Meat Production, Food Engineering (2023), https://www.foodengineeringmag.com/articles/101609-biocraft-begins-using-ai-in-cultivated-meat-production. Accessed 12 Jun 2024
36. Firmenich, The World's First AI-Created Flavor, https://www.firmenich.com/taste-and-beyond/blog-article/worlds-first-ai-created-flavor. Accessed 12 Jun 2024
37. S. Welin, C. Van Der Weele, Minding animals: awareness, emotions, and heart, in *The Oxford Handbook of Animal Studies* (2017)
38. Good seed Ventures (2021), https://goodseedventures.com/cultivated-meat-production-costs-past-present-future-2/. Accessed 31 May 2024
39. U.S. Bureau of Labor Statistics, Average Price: Ground Beef, 100% Beef (Cost per Pound/453.6 Grams) in U.S. City Average [APU0000703112], retrieved from FRED, Federal Reserve Bank of St. Louis (2024), https://fred.stlouisfed.org/series/APU0000703112

40. L.I. Laestadius, M.A. Caldwell, Is the future of meat palatable? Perceptions of in vitro meat as evidenced by online news comments. Public Health Nutr. **18**(13), 2457–2467 (2015)
41. W. Verbeke, A. Marcu, P. Rutsaert, R. Gaspar, B. Seibt, D. Fletcher, J. Barnett, 'Would you eat cultured meat?': Consumers' reactions and attitude formation in Belgium, Portugal and the United Kingdom. Meat Sci. **102**, 49–58 (2015)

Chapter 12
Bio-Plastics: A Sustainable Alternative for a Fossil Fuel-Reliant Industry

Parshav Jain, Pranay Bisht, and Komal Aggarwal

Abstract Plastics due to their durability, lighter weight nature, and versatility have become a necessary part of our lives. Conventional plastics have been derived predominantly from non-renewable fossil fuels, like, crude oil or natural gas. World-wide production of plastic is growing significantly every year; while the amount of plastic that has been recycled effectively is quite low, forcing a negative impact on the environment and sustainability. Larger plastic items decompose into smaller particles over time known as microplastics. These microplastics can enter the food chain, potentially harming living organisms and posing risks to human health. Conventional plastics also play an important role in greenhouse gas emissions, contributing to climate change. Bioplastics like PLA, PHA and PHB are becoming increasingly popular as a sustainable alternative to traditional petroleum-based plastics due to their eco-friendly nature. In addition, the bioplastics were easily recycled and biodegraded completely in the soil within short span of time. In this review, we discuss the recent developments in the field of bio bioplastics and their applications. Initially, we introduce the types of green plastic and their bio sources. Special importance will focus on their various applications in daily life and their biodegradable nature which make bio-plastics a sustainable alternative. The purpose of this study is to investigate the latest progress and the future perspective of biodegradable plastics and their impact on the environment.

Keywords Bioplastics · Polylactic acid (PLA) · Polyhydroxyalkanoates (PHA) · Polyhydroxy butyrate (PHB)

P. Jain · P. Bisht · K. Aggarwal (✉)
School of Engineering and Technology, Vivekananda institute of professional studies–Technical campus, New Delhi, India
e-mail: komal.aggarwal@vips.edu

A. Kumar et al. (eds.), *Selected Proceedings of the 1st International Conference on Advanced Materials for Sustainable Innovation; IC-AMSI 2024; 28–30 August; New Delhi; India*, Springer Proceedings in Energy,
https://doi.org/10.1007/978-981-96-4492-6_12

12.1 Introduction

Plastic materials are all around us. Despite the fact that plastics were first produced on an industrial scale in the 1950s, their low bulk density, great durability, superior processability, excellent mechanical and barrier qualities, and low cost have made them quickly indispensable for both high-end and daily uses [1]. They are now widely used manufacturing materials with applications in many different industries, including packaging, toy manufacturing, plastic cutlery, shopping bags, straws, automotive industry and 3D printed rocket nozzles [2–6].

According to new estimates, the COVID-19 pandemic epidemic is producing over 1.6 million tons of plastic garbage every day worldwide in the form of single-use surgical facemasks, face shields, syringes, latex or nitrile gloves, medical gowns, shoe covers, and sanitizer containers [7].

For example, the COVID-19 pandemic has resulted in the disposal of almost 3.4 billion single-use face masks and face shields globally per day. Despite a minor decline in worldwide plastics production in 2020, demand for the material rose by 40%, and the amount of plastic trash produced in the form of medical waste and personal protective equipment surged by 370% [8].

Plastic is used to make almost everything, including bottles, printers, cell phones, ordinary household packaging, and other items. It also finds use in the manufacturing of pharmaceuticals and cars, among other sectors. Because their structure can be chemically altered to take on a variety of strengths and forms, they are essential synthetic polymers because it enables the fabrication of materials with a larger molecular weight, lower reactivity, and longer durability. Plastics are an essential material since they are dependable and reasonably priced for all, but they have grown to be a major environmental problem because they account for 25% of all landfill volume, which leads to pollution [9].

People are becoming more conscious of the harm that plastics made from petrochemicals due to the environment. Numerous studies have been done to develop environmentally friendly plastic substitutes to control the world's plastic trash. Because of microbial enzymatic activity, bioplastics are an environmentally benign substitute that can be disposed of and break down quickly in the environment [10]. Methane, carbon dioxide, water, biomass, and quickly removed humic detritus are the products of the breakdown of biodegradable polymers [11].

Several rules have been issued by government authorities in recent years to reduce the usage of plastic that is not biodegradable. Several scientific communities have also made an effort to create biodegradable materials, including polylactic acid, polyhydroxyalkanoates, and bioplastic based on starch [12].

Bioplastics, which are plastics that are biodegradable, renewable resource-based, or made of biologically generated polymers like starch or the cellulose found in wood, straw, and cotton, offer a viable substitute [13–16]. A plastic material is considered bioplastic, according to European Bioplastics, if it is either biobased, biodegradable, or possesses both qualities [17]. While bio derived bioplastics enable a significant reduction in the carbon footprint at the resource extraction stage, biodegradable bioplastics offer alternate methods for trash disposal, minimizing the quantity of plastic waste ending up in our environment [18].

12.2 Bio Plastics

Bioplastics are not only biobased but also biodegradable because they can be broken down by microorganisms into CH_4, CO_2, water, and compost under controlled conditions. Examples of these biopolymers include polylactic acid, polybutylene succinate, polyhydroxyalkanoates, starch, cellulose, protein, and lipid-based biopolymers. Bioplastic helps the environment by lowering carbon emissions, using fewer fossil fuels, producing less plastic waste, and solving environmental issues. Biodegradable, recyclable, or compostable, bioplastics are 100% biodegradable [19, 20]. Biodegradability also extends to polybutylene adipate terephthalate and polycaprolactone, which are generated via petrochemical processes. Figure 12.1 illustrates the petrochemical generated and non-biodegradable nature of polyethylene, polypropylene, poly(trimethylene terephthalate), polyvinyl chloride, high-density polypropylene, and low-density polypropylene. While bioresources can be used to manufacture polyethylene terephthalate, polypropylene, polyamide 11, and poly (trimethylene terephthalate), these materials cannot naturally biodegrade [21].

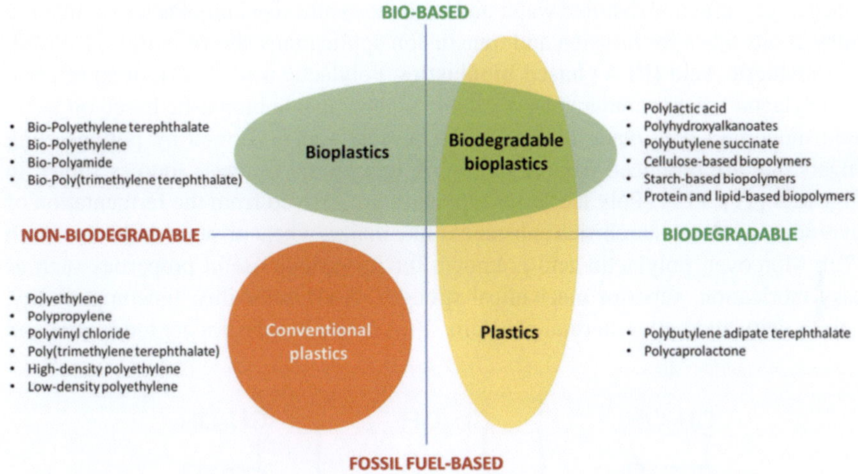

Fig 12.1 Biodegradability of plant-based and fossil fuel-based plastic resins [21]

12.2.1 Types of Bioplastics

Starch based bioplastics. All green plants generate starch, which is a powdery substance with a white hue and no flavor. Granular in nature, it is organic and insoluble in alcohol, cold water, and certain other solvents. As seen in Fig. 12.2, the chemical formula for starch is $(C_6H_{11}O_5)n$ [22]. On earth, starch is a type of carbohydrate that is widely distributed. Starch-based polymers can be derived from any plant that has a greater starch content, however they are primarily generated from sources of wheat, potatoes, rice, and corn. Complex blends of starch and biodegradable substances such as PLA, PBAT, PBS, PCL, PHB, and PHAs are known as starch-based plastics [23]. Products such as plates, cups, and eating utensils are made of polymers based on starch. It holds 80% of the market share for bioplastics [24, 25]. Conventional processing techniques like extrusion, blown film, injection and blow molding, and thermoforming can be used to create starch-based biodegradable polymers. Traditionally, starch-based biodegradable polymers can be produced by injection and blow molding, blown film, extrusion, and thermoforming [26].

Sultan et al. developed a starch-based bioplastics using potato waste a eco-friendly material in the form of peel and pulp as a source of starch. It was found that the prepared biofilm showed high tensile strength and more biodegradable in nature [27].

Gnanendra and their research group prepared and characterized starch based bioplastics using a mixture of banana peel as a main source of starch, corn starch, vinegar, glycerin and distilled water and used the synthesized bioplastics for making intravenous tubes for infusion and transfusion applications shown in Fig. 12.3 [28].

Polylactic Acid (PLA) based bioplastics. Polylactic acid (PLA), often referred as polylactide, is a commercially available biodegradable bioplastic based on lactic acid monomer. It is made of lactide/ lactic acid and is formed by polymerizing sugars that are produced from a variety of agricultural biomass sources shown in Fig. 12.4 [11]. PLA (Poly lactic acid) plastics are derived from the fermentation of byproducts such as starch-rich substances like maize, wheat or sugar and corn starch [29]. Moreover, polylactic acid is known for its various useful properties such as easy fabrication, superior mechanical strength, biodegradability, biocompatibility, non-toxicity and better thermal plasticity. PLA based bioplastics are too fragile and

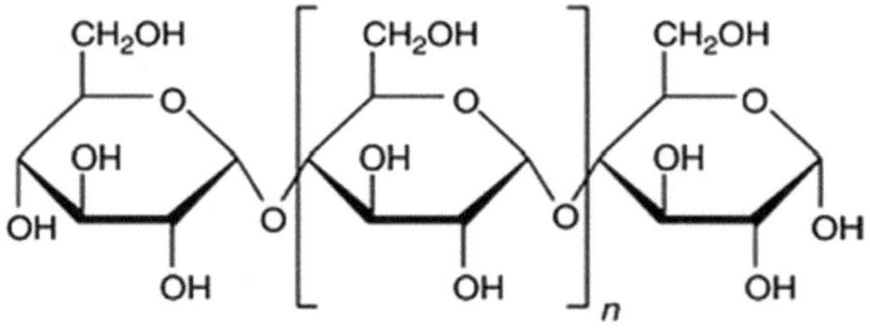

Fig. 12.2 Molecular structure of starch based plastics

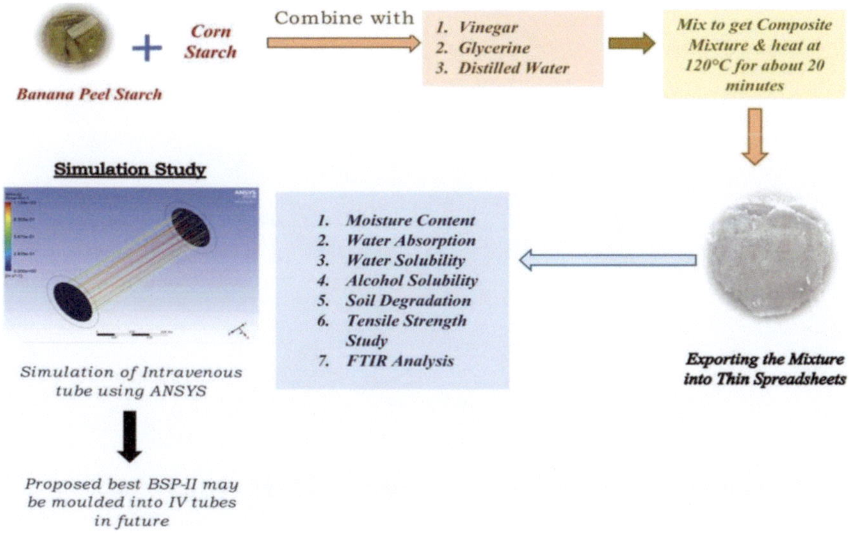

Fig. 12.3 Synthesis of starch based bioplastics using banana peel [28]

cannot be used for various packaging manufacturing processes. Therefore, additives have been used to make PLA more durable. Moreover, PLA is the most biodegradable thermoplastic, typically degrading via hydrolysis. Additionally, polylactic acid does not release toxic intermediates and byproducts upon hydrolysis. The emission rate of greenhouse gas of polylactic acid based polymers has a relatively lower than other synthetic polymers. While one ton of polylactic acid releases nearly 1600 kg CO_2, polyethylene terephthalate, polypropylene and polyamide 11 emit approximately 4140, 2740 and 7150 kg CO_2 per ton, respectively [30]. It is used in food packaging items because it has been certified for direct contact with food and because of its great scent barrier, high transparency, gloss, rigidity, and printability [31]. PLA is frequently blended with other bio-based and/or biodegradable polymers to increase stiffness and strength while lowering costs [32].

Tarsila Rodrigues Arruda et at. developed a polylactic acid based bioplastic sheets by incorporating β-acid rich hop extract (KBAE). It was observed that loss in mass of the prepared bioplastic sheets after burring in the soil for 180 days, shows their biodegradable nature shown in Fig. 12.5 [34].

Polyhydroxyalkanoates (PHA) based bioplastic. The polyhydroxyalkanoates (PHA) class of bio polyesters is produced by bacterial fermentation of lipids, carbohydrates, or chemical agents (methylene or propylene chloride, chloroform) and undergoes complete degradation in the environment in a short span of time [35], as illustrated in Fig. 12.6. Gram-positive and gram-negative bacteria naturally produce polyhydroxyalkanoate, an elastomeric or thermoplastic polyester of (R)-3-hydroxyalkanoic acid monomers, as intracellular energy and carbon storage molecules [36]. Numerous hydroxyalkanoate units have been separated with various –R groups, so the pendant alkyl (–R) group can vary from methyl to tridecyl [37].

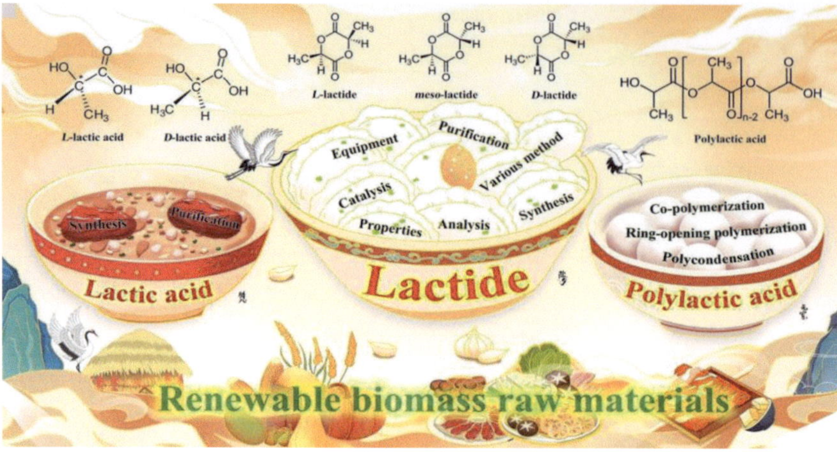

Fig. 12.4 Reactions of the polymerization of lactic acid to obtain PLA [33]

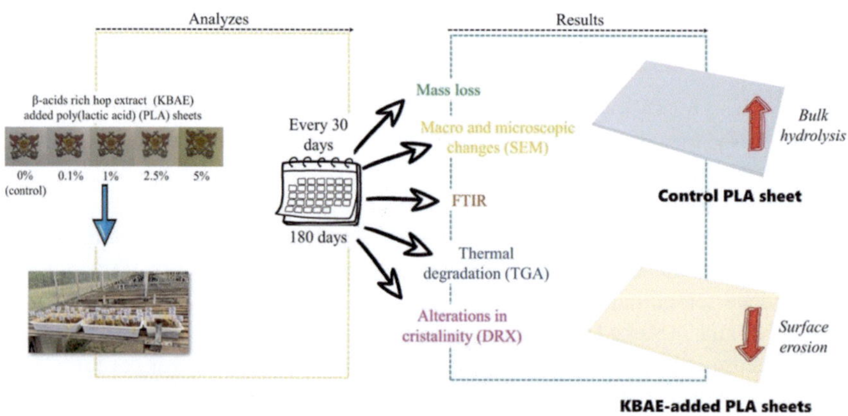

Fig. 12.5 Biodegradation analysis of PLA based bio sheets [34]

Numerous distinct bacterial cultures manufacture PHAs. The most well-researched microorganism is Cupriavidus necator, formerly known as Ralstonia eutropha or Alcaligenes eutrophus [38]. Bacillus species, Alcaligenes species, Pseudomonas species, Aeromonas hydrophilia, Rhodopseudomonas palustris, Escherichia coli, Burkholderia sacchari, and Halomonas boliviensis are a few significant additional microbe strains that have been the subject of current research [39]. There are two phases involved in PHA fermentation. Creating a culture with a high cell density was the first step; increasing the PHA concentration was the second. The conditions for fermentation are set by the needs of the microorganisms. There is less dissolved oxygen tension when modest stirrer speeds and a temperature range of 30–37 °C are

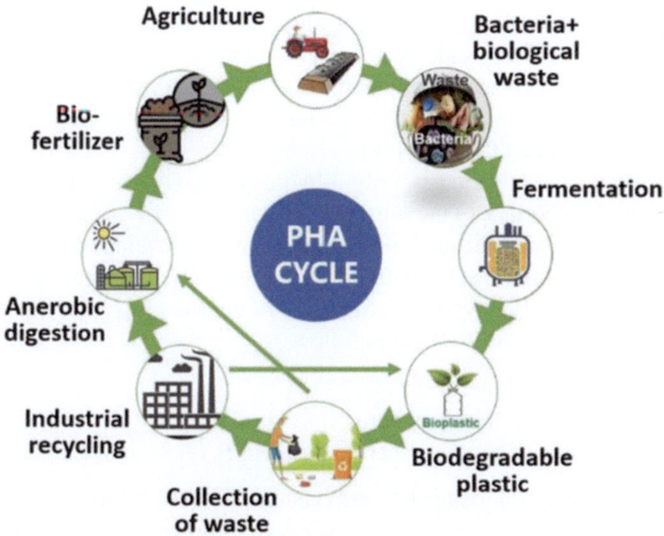

Fig. 12.6 PHA formation and degradation cycle [43]

utilized. Either no monitoring is done, or pH maintenance is controlled by adding a substrate, such as glucose [40].

Because of their piezoelectric qualities, they are also utilized as osteo synthetic materials in blood vessel replacements, bone plates, and surgical sutures to promote the formation of new bone [41]. A wide variety of PHA items, including golf tees, pens, bottles, coating and packaging, and cosmetic containers, are covered by PHA patents [42].

Ting Kang Yong and their research group was prepared a high quality polyhydroxyalkanoates (PHA) based bioplastic by enzymatic hydrolysis using sugarcane bagasse (SCB) as the primary feedstock in the presence of Bacillus cereus. The soil burial test of the prepared polymer showed a high rate of biodegradability about 41.73% in a week [44]

Polyhydroxybutyrate (PHB) based bioplastic. Polyhydroxybutyrate (PHB) are a widely used bioplastic substitute for polyester that is derived from PHA. Microorganisms, such as bacteria, algae, and genetically engineered plants, are the source of PHB. These organisms also directly synthesize the polymer [45]. The actual enzymatic process that produces the polymer inside the cell is intricate. Firstly, β-ketothiolase catalyzes the condensation of two acetyl COA molecules to produce acetoacetyl-CoA. Subsequently, β-hydroxybutyryl-CoA is produced when acetoacetyl-CoA reductase lowers the acetoacetyl-CoA. As seen in Fig. 12.7, the polymerization of β-hydroxybutyryl-CoA to PHB is catalyzed by PHB synthase in its final phase [46]. The PHB is found in the cell's cytoplasm as cysts. Consequently, to extract the PHB, the cell must be killed [47]. The polymer is made by the cell through a complex enzymatic process. In order to create acetoacetyl-CoA, two acetylCoA molecules must first condense, which is catalyzed by ketothiolase. Subsequently, the hydroxybutyryl-CoA that yields acetoacetyl-CoA is lowered by acetoacetyl-CoA reductase. The final step in the polymerization of hydroxybutyryl-CoA to PHB is catalyzed by PHB synthase. PHB cysts are located in the cell's cytoplasm. The cell has to be killed to extract the PHB [48].

In addition to these advantages, poly-3-hydroxybutyrate's high production costs have led to a global shortage of the substance. In comparison to comparable polymers like bio-based polylactic acid and fossil-based polyethylene terephthalate, which sell for about USD $2600 and USD $1000 per ton, respectively, the current global market price of poly-3-hydroxy-butyrate is USD $11,424 per ton [49].

A few characteristics of PHB are as follows:

(1) They have a 100% water resistance.
(2) They decompose entirely in nature.
(3) They are capable of handling thermoplastic [50].

Mohammad et al. developed polyhydroxybutyrate (PHB) from the potato peel as a cheap and abundant carbon source under nutritional stress conditions; using efficient

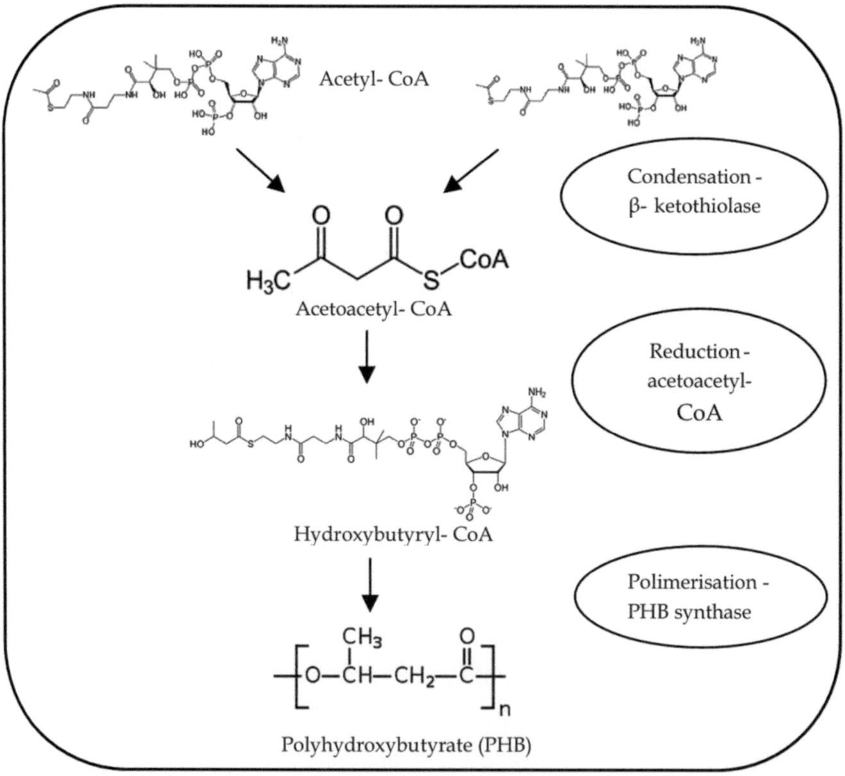

Fig. 12.7 Synthesis of PHB [39]

bacterial strains A Pseudomonas stutzeri PSB1 bacterial strain has been used for the generation of PHB from high-carbon wastes i.e. potato peel. They further developed a way to enhance PHB production through sequential optimization media of PHB chronic assay by screening 11 Sudan Black B and Nile red positive shown in Fig. 12.8. It was observed that the recovered PHB biopolymer from Pseudomonas stutzeri PSB1 is a good alternative to agricultural mulch films and non-biodegradable plastics [51].

12.2.2 Non-Biodegradable Plastics

A substance called ethanol is used to make biobased polyethylene, or bio-polyethylene, which is derived from wheat grain, sugar beet juice, sugar cane juice, and other related crops [52]. Because bio-polyethylene has physio-mechanical characteristics with synthetic polyethylene (PE), it is used as a polymeric material to make bioplastics. These polymers are derived from bio-based resources, but manufacturing processes have altered their chemical makeup, rendering them non-biodegradable [53]. The same process, tools, and equipment are used in the manufacturing of bioplastic products based on renewable or natural feedstock, such as polyhydroxyalkanoates, thermoplastic starches, and polylactic acid (PLA). Since they are effectively a "biosimilar" copy of conventional (fossil-based) plastics, they cannot be composted or biodegraded [54].

12.2.3 Environmental Impact of Bio-Based Plastics

The fossil fuel based plastics emit a significant amount of CO_2 and other toxic emissions from recycling processes. The recycling of bioplastics is a process with low carbon emissions and therefore, bioplastics can decrease greenhouse gas emissions by partially or fully eliminating the usage of fossil fuel derivatives with plant-based fibers and polymers shown in Fig. 12.9 [55].

Some benefits of bioplastics, shown in Fig. 12.10, that derived from renewable resources, on environment are:

- The bioplastic products are produced from renewable resources, and it contributes to reduction of greenhouse gases emission through reduced carbon footprint [56].
- The bio plastics can evade environmental problems like uncontrolled dumping of wastes on land and disposal to sea, and the related emission of toxic substances. However, effective implementations of collection, sorting and recycling practices and public awareness are also required to premium the assets of bio plastics [57].
- Bio plastic from plant resources can manage the plant waste or plant residues in an efficient manner [58].

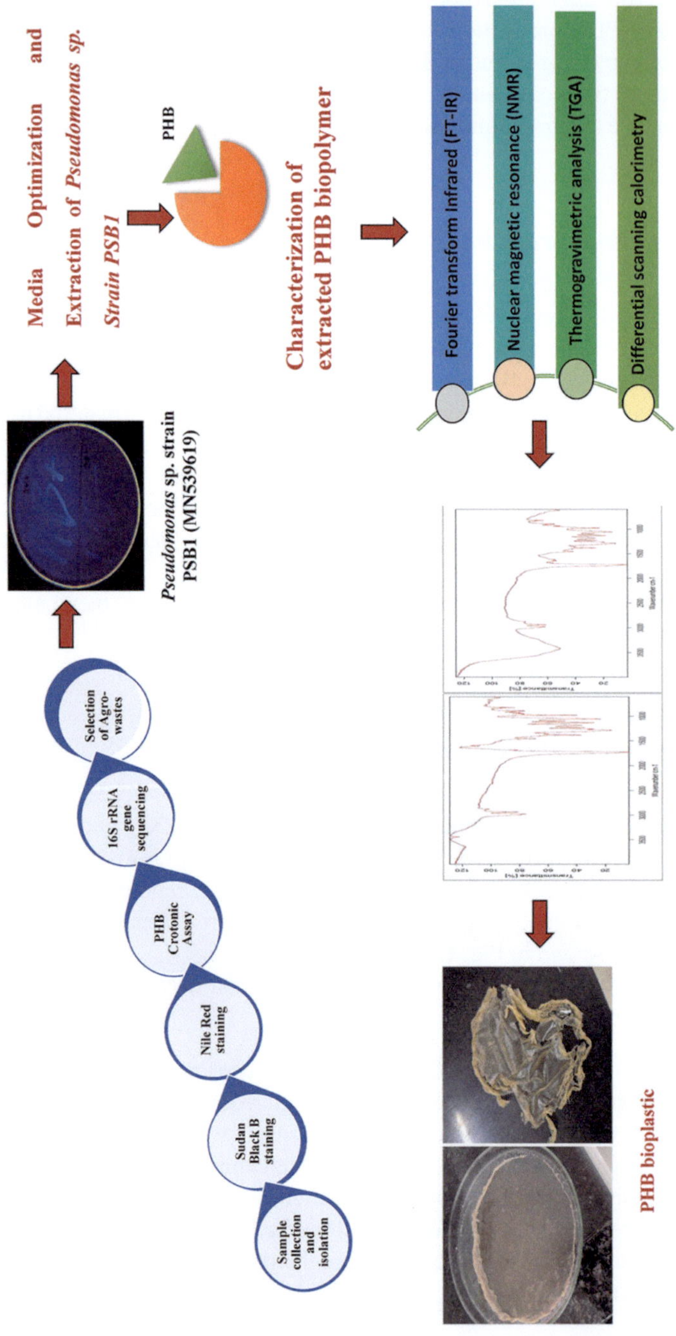

Fig. 12.8 Production of PHB from pseudomonas stutzeri PSB1 strain [51]

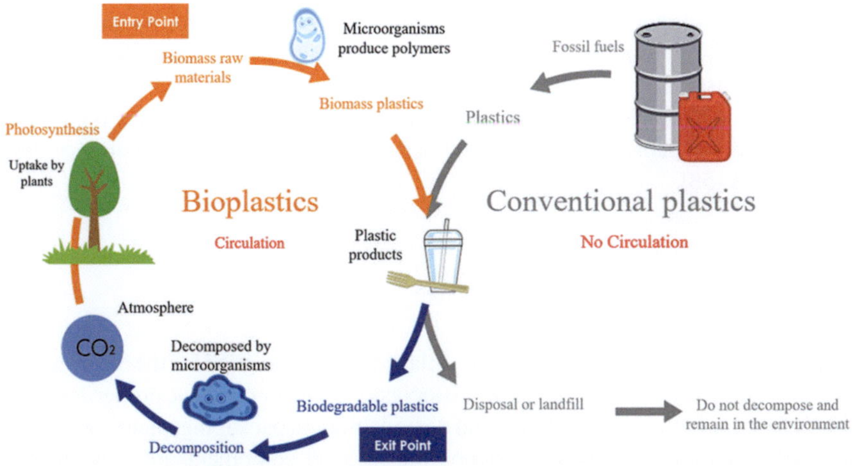

Fig. 12.9 Environmental impacts of fossil fuel-based plastics and plant-based bioplastics [55]

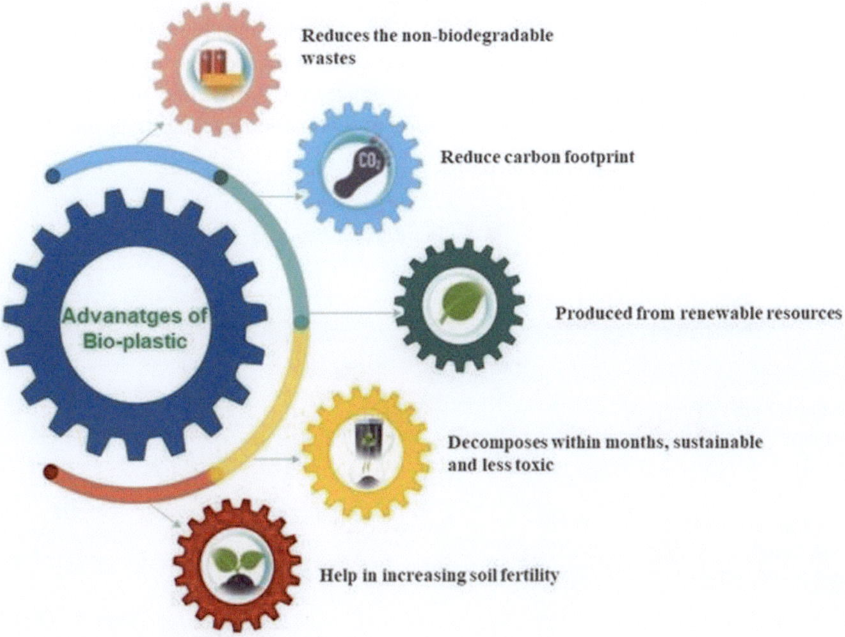

Fig. 12.10 Benefits of bioplastics [58]

- Bioplastics support the principles of a circular economy by providing alternative end-of-life options, such as composting and anaerobic digestion, which can convert waste into valuable resources like compost and biogas [59].
- Some bioplastics offer unique features such as improved moisture resistance, enhanced mechanical properties, and better compatibility with certain recycling processes [58].

12.2.4 Applications of Bioplastics

Bioplastics are a substitute for traditional plastics which are good for the environment and can be used across many industries. Bioplastics have higher value properties, such as resistance to boundaries, softness and degradability in packaging, plant containers, automobile interior components, disposable medical equipment, toys and fabrics shown in Fig. 12.11. These applications contribute to reducing the environmental impact and advancing sustainability in everyday products.

- Packaging: Bioplastics are extensively used in packaging due to their ability to be tailored for specific purposes like food containers, bottles, films, and trays. They offer properties such as barrier resistance, flexibility, and moisture resistance, making them suitable for both food and non-food packaging applications.

Fig. 12.11 Applications of bioplastic [60]

Bioplastics can help reduce the environmental footprint of packaging materials by replacing conventional plastics derived from fossil fuels [61].

- Plant Pots and Seedling Trays: Bioplastics are used to manufacture pots and trays for planting seedlings. These products can be planted directly into the soil, where they degrade naturally, reducing transplant shock and plastic waste [62].
- Interior Components: Bioplastics are utilized for manufacturing interior components such as dashboards, seat cushions, and door panels in automobiles. They contribute to lightweighting and improve fuel efficiency while reducing the environmental impact compared to traditional plastics [61].
- Disposable Medical Devices: Bioplastics are used in the production of disposable medical supplies such as syringes, surgical gloves, and packaging for pharmaceutical products. They offer a safe and sustainable alternative to conventional plastics in healthcare settings [58].
- Toys and Household Items: Bioplastics are used in the manufacturing of toys, utensils, and various household items, providing non-toxic and eco-friendly alternatives to conventional plastics [63].
- Fibers and Fabrics: Bioplastics are used to produce fibers and textiles for clothing, upholstery, and industrial applications. They offer properties such as breathability, moisture resistance, and biodegradability, catering to sustainable fashion trends [64].

12.2.5 The Life Cycle of Bioplastic

The life cycle of bioplastics involves complex interactions between biomass production, bioplastic manufacturing processes, utilization, disposal pathways, and environmental impacts shown in Fig. 12.12. Understanding these stages is essential for maximizing the environmental benefits of bioplastics compared to conventional plastics derived from fossil fuels. Here's a detailed explanation of each stage:

1. Carbon Source (Biomass Production): Bioplastics begin with the cultivation and growth of biomass, such as plants (e.g., corn, sugarcane), algae, or bacteria. These organisms utilize photosynthesis to convert carbon dioxide (CO_2) from the atmosphere into organic compounds, primarily carbohydrates. This biomass serves as the raw material for bioplastic production [65].
2. Biomass Processing: After biomass harvesting, it undergoes processing to extract valuable components such as cellulose, starches, or oils, which are then used as feedstocks for bioplastic production. Processing methods vary depending on the biomass source and desired bioplastic type [66].
3. Bioplastic Production: The extracted biomass components undergo conversion into bioplastic polymers through processes such as fermentation (e.g., for polylactic acid or PLA) or polymerization of monomers derived from biomass (e.g., polyhydroxyalkanoates or PHAs). These processes result in the formation of biodegradable or compostable materials suitable for use in various applications [67].

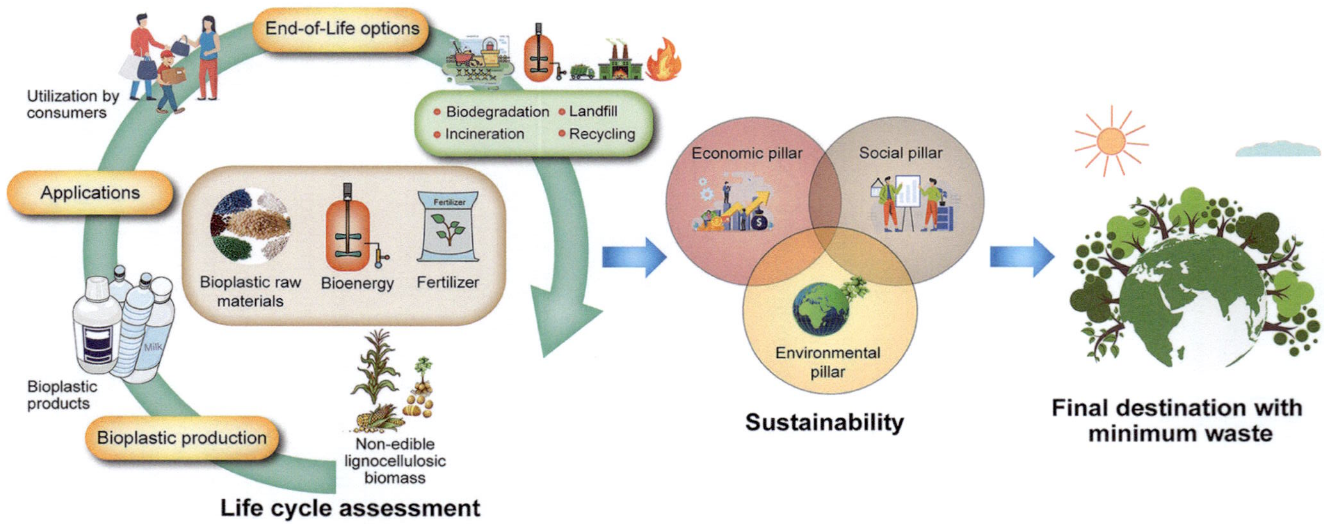

Fig. 12.12 The life cycle of bioplastic [67]

4. Use and Disposal: Bioplastics are utilized in diverse applications including packaging, consumer goods, and medical devices. During their use phase, bioplastics can undergo degradation depending on their composition and environmental conditions they are exposed to [67].
5. End-of-Life Options: The fate of bioplastics at the end of their use phase significantly impacts the carbon cycle:

   (i) Biodegradation: Some bioplastics are designed to biodegrade under specific conditions, breaking down into simpler organic compounds and releasing CO_2 back into the atmosphere or contributing to soil carbon storage [68].
   (ii) Recycling or Upcycling: Bioplastics can potentially be recycled or upcycled, though current infrastructure limitations and material compatibility challenges may restrict widespread adoption [68].

6. Carbon Sequestration and Environmental Impact: Managed sustainably, biomass used for bioplastic production can contribute to carbon sequestration through carbon storage in soils or biomass, offsetting CO_2 emissions associated with traditional plastic production from fossil fuels [69].
7. Life cycle Assessment (LCA): Conducting a lifecycle assessment (LCA) is crucial for evaluating the overall environmental impact of bioplastics, considering factors such as carbon footprint, energy use, and resource efficiency from cradle to grave [70–72].

12.3 Conclusion

- Plastics have become an essential part of our daily requirements, but their non-renewable origin makes plastic a great threat to our environment, as their non-biodegradable nature. Herein, we discussed a comprehensive overview of the recent advances in bioplastics.
- Bioplastic is a key component of today's research needed in the modern world because it reduces pollution caused by plastic waste derived from petrochemicals. Their biodegradable nature will solve many issues like dumping, recycling, and degradation time, making it a sustainable and environmentally benign alternative.
- This paper has addressed its types, classification, renewable sources, advantages, applications in different fields, and the life cycle of bioplastics.
- Additionally, we discussed the recently developed bioplastics originate from starch, polylactic acids (PLA), Polyhydroxyalkanoates (PHA), Polyhydroxybutyrate (PHB) and their applications and biodegradable nature.

Acknowledgements The authors are thankful to VIPS-TC, New Delhi, India for the infrastructural support.

References

1. R. Geyer, J.R. Jambeck, K.L. Law, Production, use, and fate of all plastics ever made. Sci. Adv. **3**, 1700782 (2017). https://doi.org/10.1126/sciadv.1700782
2. J.A. Brydson, Plastics materials: introduction and historical development, in *Plastics Materials*, ed. by J.A. Brydson, 7th edn., (Elsevier, 1999), pp. 1–18. https://doi.org/10.1016/B978-0-323-35824-8.00001-3
3. H. Abedsoltan, Applications of plastics in the automotive industry: current trends and future perspectives. Polym. Eng. Sci. **64**, 929–950 (2024). https://doi.org/10.1002/pen.26604
4. P.N. Tudu, V. Mishra, An eco-friendly alternative to plastic cutlery and food packaging: a case of ecoware in India. Asian J. Manag. Cases (2023). https://doi.org/10.1177/09728201231190189
5. S. Mangaraj, T.K. Goswami, P.V. Mahajan, Applications of plastic films for modified atmosphere packaging of fruits and vegetables: a review. Food Eng. Rev. **1**, 133–158 (2009). https://doi.org/10.1007/s12393-009-9007-3
6. S.A. Pradeep, R.K. Iyer, H. Kazan, S. Pilla, 30–Automotive applications of plastics: past, present, and future, in *Applied Plastics Engineering Handbook*, 2nd edn., (Plastics Design Library, 2017), pp. 651–673. https://doi.org/10.1016/B978-0-323-39040-8.00031-6
7. N.U. Benson, D.E. Bassey, T. Palanisami, COVID pollution: impact of COVID-19 pandemic on global plastic waste footprint. Heliyon **7**, e06343 (2021). https://doi.org/10.1016/j.heliyon.2021.e06343
8. G. Gorrasi, A. Sorrentino, E. Lichtfouse, Back to plastic pollution in COVID times. Environ. Chem. Lett. **19**, 1–4 (2021). https://doi.org/10.1007/s10311-020-01129-z
9. G. Unmar, R. Mohee, Assessing the effect of biodegradable and degradable plastics on the composting of green wastes and compost quality. Biores. Technol. **99**(15), 6738–6744 (2008). https://doi.org/10.1016/j.biortech.2008.01.016
10. Role of energy in the development of agriculture sector in South Asia, in *The South Asian Energy Corridor Toward Sustainability and Security.* (Apple Academic Press, 2024), ISBN: 9781774914786, https://www.appleacademicpress.com/the-south-asian-energy-corridor-toward-sustainability-and-security/9781774914786
11. K. Chauhan, R. Kaur, I. Chauhan, Sustainable bioplastic: A comprehensive review on sources, methods, advantages, and applications of bioplastics. Polym. Plast. Technol. Mater. **63**, 913–938 (2024). https://doi.org/10.1080/25740881.2024.2307369
12. S. Thakur, J. Chaudhary, B. Sharma, A. Verma, S. Tamulevicius, V.K. Thakur, Sustainability of bioplastics: opportunities and challenges. Curr. Opin. Green Sustain. Chem. **13**, 68–75 (2018). https://doi.org/10.1016/j.cogsc.2018.04.013
13. G. Guidotti, M. Soccio, N. Lotti, M. Gazzano, V. Siracusa, A. Munari, Poly (propylene 2,5-thiophenecarboxylate) versus Poly (propylene 2,5-fur-dicarboxylate): two examples of high gas barrier bio-based polyesters. Polymers **10**, 785 (2018). https://doi.org/10.3390/polym10070785
14. D. Garlotta, A literature review of poly (lactic acid). J. Polym. Environ. **9**, 63–84 (2001). https://doi.org/10.1023/A:1020200822435
15. W. Smitthipong, R. Chollakup, M. Nardin, *Bio-based Composites for High-Performance Materials: from Strategy to Industrial Application.* (CRC Press/Taylor & Francis Group, 2015). https://doi.org/10.1201/b17601&type=googlepdf
16. F.M. Lamberti, L.A. Roman-Ramírez, J. Wood, Recycling of bioplastics: routes and benefits. J. Polym. Environ. **28**, 2551–2571 (2020). https://doi.org/10.1007/s10924-020-01795-8
17. M. Van den Oever, K. Molenveld, M. Van der Zee, H. Bos, Bio-based and biodegradable plastics: facts and figures: focus on food packaging in the Netherlands (2017). https://doi.org/10.18174/408350
18. J.G. Rosenboom, R. Langer, G. Traverso, Bioplastics for a circular economy. Nat. Rev. Mater. **7**, 117–137 (2022). https://doi.org/10.1038/s41578-021-00407-8

19. I. Shamsuddin, Bioplastics as a better alternative to petroplastics and their role in national sustainability: a review. Adv. Biosci. Bioeng. **5**, 63(2017). https://doi.org/10.11648/j.abb.20170504.13
20. T. Rasheed, S.H. Vattathurvalappil et al., Recent updates on biodegradability and recyclability of bioplastics–towards a new era in sustainability. Sustain. Mater. Technol. pp. e01051 (2024) https://doi.org/10.1016/j.susmat.2024.e01051
21. S. Nanda, B.R. Patra, R. Patel, J. Bakos, A.K. Dalai, Innovations in applications and prospects of bioplastics and biopolymers: a review. Environ. Chem. Lett. **20**, 379–395 (2022). https://doi.org/10.1007/s10311-021-01334-4
22. J. Jane, Starch properties, modifications, and applications. J. Polym. Sci. Part A Polym. Chem. **32**, 751–757 (1995). https://doi.org/10.1080/10601329508010286
23. X. Yu, L. Chen, Z. Jin, A. Jiao, Research progress of starch-based biodegradable materials: a review. J. Mater. Sci. **56**, 11187–11208 (2021). https://doi.org/10.1007/s10853-021-06063-1
24. R.V. Gadhave, A. Das, P.A. Mahanwar, P.T. Gadekar, Starch based bioplastics: the future of sustainable packaging. Open J. Polym. Chem. **8**, 21–33 (2018). https://doi.org/10.4236/ojpchem.2018.82003
25. P.V. Santos, L. Felipe dos Ouros, Chapter 13—Starch-based packaging and coating polymers for food, in *Starch Industries: Processes and Innovative Products in Food and Non-Food Uses* (Elsevier, 2024), pp. 295–310. https://doi.org/10.1016/B978-0-323-90842-9.00012-1
26. L.V. Siqueira, C.I. La Fuente Arias, B.C. Maniglia, C.C. Tadini, Starch-based biodegradable plastics: methods of production, challenges and future perspectives. Curr. Opin. Food Sci. **38**, 122–130 (2021). https://doi.org/10.1016/j.cofs.2020.10.020
27. A. Sultan, Comparative analysis of physical and mechanical properties of starch-based bioplastic derived from the pulp and peel of potatoes. J. Indian Chem. Soc. **101**, 101301 (2024). https://doi.org/10.1016/j.jics.2024.101301
28. G.V. Kiran et al., Synthesis and characterization of banana peel starch-based bioplastic for intravenous tubes preparation. Mater. Today Commun. **33**, 104464 (2022). https://doi.org/10.1016/j.mtcomm.2022.104464
29. T.A. Swetha, V. Ananthi, A. Bora, N. Sengottuvelan, K. Ponnuchamy, G. Muthusamy, A. Arun, A review on biodegradable polylactic acid (PLA) production from fermentative food waste—its applications and degradation. Int. J. Biol. Macromol. **234**, 123703 (2023). https://doi.org/10.1016/j.ijbiomac.2023.123703
30. M. Karamanlıoğlu, R. Preziosi, G.D. Robson, Abiotic and biotic environmental degradation of the bioplastic polymer poly (lactic acid): a review. Polym. Degrad. Stab. **137**, 122–130 (2017). https://doi.org/10.1016/j.polymdegradstab.2017.01.009
31. T.A. Swetha, A. Bora, K. Mohanrasu, P. Balaji, R. Raja, K. Ponnuchamy, G. Muthusamy, A. Arun, A comprehensive review on polylactic acid (PLA)—synthesis, processing and application in food packaging. Int. J. Biol. Macromol. **234**, 123715 (2023). https://doi.org/10.1016/j.ijbiomac.2023.123715
32. K. Inoue, S. Serizawa, M. Yamashiro, M. Iji, Highly functional bioplastics (PLA compounds) used for electronic products, in *Polytronic 2007—6th International Conference on Polymers and Adhesives in Microelectronics and Photonics* (IEEE, 2007), pp. 73–76. https://doi.org/10.1109/POLYTR.2007.4339141
33. J. Yu, S. Xu, B. Liu, H. Wang, F. Qiao, X. Ren, Q. Wei, PLA bioplastic production: from monomer to the polymer. Eur. Polym. J. **193**, 112076 (2023). https://doi.org/10.1016/j.eurpolymj.2023.112076
34. T.R. Arruda, Effect of the incorporation of β-acid rich hop extract on degradation in soil of polylactic acid (PLA) sheets. React. Funct. Polym. **196**, 105852 (2024). https://doi.org/10.1016/j.reactfunctpolym.2024.105852
35. S. Behera, M. Priyadarshanee, Vandana, S. Das, Polyhydroxyalkanoates, the bioplastics of microbial origin: properties, biochemical synthesis, and their applications. Chemosphere **294**, 133723 (2022). https://doi.org/10.1016/j.chemosphere.2022.133723
36. T. Keshavarz, I. Roy, Polyhydroxyalkanoates: bioplastics with a green agenda. Curr. Opin. Microbiol. **13**, 321–326 (2010). https://doi.org/10.1016/j.mib.2010.02.006

37. G. Mannina, D. Presti, G. Montiel-Jarillo, J. Carrera, M.E. Suárez-Ojeda, Recovery of polyhydroxyalkanoates (PHAs) from wastewater: a review. Bioresour. Technol. **297**, 122478 (2020). https://doi.org/10.1016/j.biortech.2019.122478
38. M. Vaneechoutte, P. Kampfer et al., Wautersia gen. nov., a novel genus accommodating the phylogenetic lineage including Ralstonia eutropha and related species, and proposal of Ralstonia [Pseudomonas] syzygii (Roberts et al. 1990) comb. nov. Int. J. Syst. Evol. Microbiol. **54**, 317–327 (2004). https://doi.org/10.1099/ijs.0.02754-0
39. E.R. Oliveira-Filho, J.G.C. Gomez, M.K. Taciro, L.F. Silva, Burkholderia sacchari (synonym Para Burkholderia sacchari): an industrial and versatile bacterial chassis for sustainable biosynthesis of polyhydroxyalkanoates and other bioproducts. Bioresour. Technol. **337**, 125472 (2021). https://doi.org/10.1016/j.biortech.2021.125472
40. Y.J. Chung, H.J. Cha, J.S. Yeo, Y.J. Yoo, Production of poly(3-hydroxybutyric-co-3-hydroxyvaleric) acid using propionic acid by pH regulation. J. Ferment. Bioeng. **83**, 492–495 (1997). https://doi.org/10.1016/S0922-338X(97)83009-7
41. J.G. Wang, L.R. Bakken, Screening of soil bacteria for poly-β-hydroxybutyric acid production and its role in the survival of starvation. Microb. Ecol. **35**, 94–101 (1998). https://doi.org/10.1007/s002489900063
42. J.C.C. Yeo, J.K. Muiruri, X. Fei, T. Wang, X. Zhang, Y. Xiao, W. Thitsartarn, H. Tanoto, C. He, Z. Li, Innovative biomaterials for food packaging: Unlocking the potential of polyhydroxyalkanoate (PHA) biopolymers. Biomaterials Adv. **2024**, 213929 (2024). https://doi.org/10.1016/j.bioadv.2024.213929
43. Z. Zhang et al., Towards scaling-up implementation of polyhydroxyalkanoate (PHA) production from activated sludge: progress and challenges. J. Clean. Prod. **447**, 141542 (2024). https://doi.org/10.1016/j.jclepro.2024.141542
44. T.K. Yong, M.N. Syahmi Musa, R. Abdulla, E. Derman, J.A. Gansau, M. Rajin, Synthesis and characterization of bioplastics, polyhydroxyalkanoates produced from sugarcane bagasse by using Bacillus cereus. Biocatal. Agric. Biotechnol. **58**, 103181 (2024). https://doi.org/10.1016/j.bcab.2024.103181
45. D. Özçimen, B. İnan, O. Morkoç, A. Efe, A review on algal biopolymers. J. Chem. Eng. Res. Updat. **4**, 7–14 (2017). https://doi.org/10.15377/2409-983X.2017.04.2
46. M. Czemplik, Flax engineering for biomedical application, in *Biomedical Engineering, Trends, Research and Technologies* (2011), https://www.researchgate.net/publication/221909920_Flax_Engineering_for_Biomedical_Application
47. E. Galindo, C. Peña, C. Núñez, D. Segura, G. Espín, Molecular and bioengineering strategies to improve alginate and polyhydroxyalkanoate production by Azotobacter vinelandii. Microb. Cell Fact. **6**, 7 (2007). https://doi.org/10.1186/1475-2859-6-7
48. M. Nieder-Heitmann, K. Haigh, J.F. Görgens, Process design and economic evaluation of integrated, multi-product biorefineries for the co-production of bioenergy, succinic acid, and polyhydroxybutyrate (PHB) from sugarcane bagasse and trash lignocelluloses. Biofuels Bioprod. Biorefin. **13**, 599–617 (2019). https://doi.org/10.1002/bbb.1972
49. Y. Kumar, P. Shukla, P. Singh, Bioplastics: a perfect tool for eco-friendly food packaging: a review. J. Food Prod. Dev. Packag. **1**(1), 1–6 (2014), http://www.jakraya.com/journal/jfpdp
50. E. Bugnicourt, P. Cinelli, A. Lazzeri, A. Alvarez, Polyhydroxyalkanoate (PHA): review of synthesis, characteristics, processing and potential applications in packaging. Express Polym Lett **8**, 791–808 (2014). https://doi.org/10.3144/expresspolymlett.2014.82
51. A. Mohammad, S.F. Rizvi, M. Asif, A. Ahmad, M.B. Alshammari, A. Gupta, M.R. Zaheer, R. Roohi, Polyhydroxybutyrate (PHB) bioplastic characterization from the isolate Pseudomonas stutzeri PSB1 synthesized using potato peel feedstock to combat solid waste management. Biocatal. Agric. Biotechnol. **57**, 103097 (2024). https://doi.org/10.1016/j.bcab.2024.103097
52. S. Kumar, K.S. Thakur, Production and their potential food applications. J. Hill Agric. **8**, 118–129 (2017). https://doi.org/10.5958/2230-7338.2017.00024.6
53. M. Amin, B. Abu-Sharkh, M. Al-Harthi, Effect of starch addition on the properties of low-density polyethylene for developing environmentally degradable plastic bags. J. Chem. Eng. **26**, 38–40 (2012). https://doi.org/10.3329/jce.v26i1.10180

54. A. Demirbas, Biodegradable plastics from renewable resources. Energy Source **29**, 419–424 (2007). https://doi.org/10.1080/009083190965820
55. Science Technology Innovation Japan: practical use of biodegradable plastics: Addressing the challenge of altering plastic circulation in various environments. JST News (2021), https://sj.jst.go.jp/stories/2021/s0927-01j.html
56. P. Prasteen, Y. Thushyanthy, T. Mikunthan, M. Prabhaharan, Bio-plastics–an alternative to petroleum-based plastics. Int. J. Res. Stud. Agric. Sci. **4**(1), 1–7 (2018). https://doi.org/10.20431/2454-6224.0401001
57. *The New Plastics Economy: Rethinking the Future of Plastics*, (World Economic Forum, Ellen MacArthur Foundation, McKinsey & Company, 2016), https://www.ellenmacarthurfoundation.org/the-new-plastics-economy-rethinking-the-future-of-plastics
58. M. Rujnić-Sokele, A. Pilipović, Challenges and opportunities of biodegradable plastics: a mini review. Waste Manag. Res. **35**, 132–140 (2017). https://doi.org/10.1177/0734242X16683272
59. European bioplastics: applications of bioplastics (2020), https://www.european-bioplastics.org/market/applications-sectors/
60. C. Claus, Bioplastics: innovations shaping the future of packaging. Food Infotech (2024), https://www.foodinfotech.com/bioplastics-innovations-shaping-the-future-of-packaging/
61. E. Shlush, M. Davidovich-Pinhas, Bioplastics for food packaging. Trends Food Sci. Technol. **125**, 66–80 (2022). https://doi.org/10.1016/j.tifs.2022.04.026
62. B. Tomadoni, D. Merino, V.A. Alvarez, Biodegradable materials for planting pots, in *Advanced Applications of Biodegradable Green Composites*, Chapter 4 (2020), pp. 85–103. https://doi.org/10.21741/9781644900659-4
63. L. Shen, E. Worrell, M.K. Patel, Life cycle impact assessment of bio-based plastics from sugarcane ethanol. J. Environ. Manage. **90**, 114–127 (2015). https://doi.org/10.1016/j.jclepro.2014.11.071
64. K. Kucharska, E. Słupek, H. Cieśliński, M. Kamiński, Advantageous conditions of saccharification of lignocellulosic biomass for biofuels generation via fermentation processes. Chem. Pap. **74**, 1199–1209 (2020). https://doi.org/10.1007/s11696-019-00960-1
65. V. Streov, T.J. Evans, *Biomass Processing Technologies.* (CRC Press/Taylor & Francis Group, 2015), https://books.google.co.in/books?hl=en&lr=&id=AePMAwAAQBAJ&oi=fnd&pg=PP1&dq=biomass+processing&ots=j3Rc4L6Itm&sig=qk5G3k34-hze5UroTAIrM3nLEBQ&redir_esc=y#v=onepage&q=biomass%20processing&f=false
66. R. Hatti-Kaul, L.J. Nilsson, B. Zhang, N. Rehnberg, S. Lundmark, Designing biobased recyclable polymers for plastics. Plastics **38**, 50–67 (2019). https://doi.org/10.1016/j.tibtech.2019.04.011
67. O. García-Depraect, S. Bordel, R. Lebrero, Inspired by nature: microbial production, degradation and valorization of biodegradable bioplastics for life-cycle-engineered products. Biotechnol. Adv. **50**, 107772 (2021). https://doi.org/10.1016/j.biotechadv.2021.107772
68. G. Atiwesha, A. Mikhael, C.C. Parrish, J. Banoub, T.T. Le, Environmental impact of bioplastic use: a review. Heliyon **7**, e07918 (2021). https://doi.org/10.1016/j.heliyon.2021.e07918
69. S. Dahiya, R. Katakojwala, S. Ramakrishna, S.V. Mohan, Biobased products and life cycle assessment in the context of circular economy and sustainability. Circ. Econ. Sustain. **2**(7), (2020). https://doi.org/10.1007/s42824-020-00007-x
70. G. Bishop, D. Styles, P.N.L. Lens, Environmental performance comparison of bioplastics and petrochemical plastics: a review of life cycle assessment (LCA) methodological decisions. Resour. Conserv. Recycl. **168**, 105451 (2021). https://doi.org/10.1016/j.resconrec.2021.105451
71. S.S. Ali, E.A. Abdelkarim, T. Elsamahy, R. Al-Tohamy, F. Li, M. Kornaros, A. Zuorro, D. Zhu, J. Sun, Bioplastic production in terms of life cycle assessment: a state-of-the-art review. Environ. Sci. Ecotechnol. **15**, 100254 (2023). https://doi.org/10.1016/j.ese.2023.100254
72. D. Park, H. Lee, W. Won, Unveiling the environmental gains of biodegradable plastics in the waste treatment phase: a cradle-to-crave life cycle assessment. Chem. Eng. J. **487**, 150540 (2024). https://doi.org/10.1016/j.cej.2024.150540

Chapter 13
Studies on Durability Properties of Green Concrete by Partial Replacement of Coarse Aggregate with Jhama Bricks

Harideep Kumar Danthala, V. Mallikarjuna Reddy, and C. Vivek Kumar

Abstract Green Concrete which undergoes further improvements in the mix design and placement to ensure an extended lifespan and minimal maintenance surface is referred to as "green concrete." This type of concrete takes into account all aspects related to performance and requires considering all stages of the lifespan of a concrete structure, including production, finishing, and building design. Green concrete can be made by supplementary cementitious materials (SCMs) like Fly Ash, Silica fume and Recycled glass. For this study, used 20% of Fly Ash (FA) in place of cement to make green concrete and this is economical and environmentally friendly. The Jhama Bricks (JB) are over burned bricks, these are fired at high temperature for a long period of time and sometimes the strength of jhama bricks is found to be equivalent or higher than first class bricks. This study focused on the durability properties of green concrete by partial replacement of coarse aggregate with jhama bricks. For this study the concrete mix was prepared as 10% replacement of coarse aggregate with jhama bricks and 20% replacement of cement with fly Ash. Then Hydrochloric acid (HCl) and sulfuric acid (H_2SO_4) are used to test the durability performance of materials in terms of water absorption, sorptivity and resistance to acid attack. Sorptivity and Water absorption tests indicate that the alternative concrete is more susceptible to capillary action and water infiltration than usual concrete, having its low permeability. Concrete samples exhibited minimized weight loss and cracking in the structure after exposure to both HCl and H_2SO_4 solutions, indicating the material's resistance to abrasive environments in this investigation.

H. K. Danthala (✉)
Research Scholar, Structural Engineering, Gokaraju Rangaraju Institute of Engineering and Technology, Hyderabad, India
e-mail: harideepdanthala1818@gmail.com

V. Mallikarjuna Reddy · C. Vivek Kumar
Civil Engineering, Gokaraju Rangaraju Institute of Engineering and Technology, Hyderabad, India

A. Kumar et al. (eds.), *Selected Proceedings of the 1st International Conference on Advanced Materials for Sustainable Innovation; IC-AMSI 2024; 28–30 August; New Delhi; India*, Springer Proceedings in Energy,
https://doi.org/10.1007/978-981-96-4492-6_13

Keywords Jhama bricks · Water absorption · Sorptivity · Acid attack

13.1 Introduction

A major development in sustainable construction methods is the investigation of green concrete, with a particular focus on improving durability by partially substituting jhama bricks for CA [1, 2]. The primary method by which that green concrete differs from ordinary concrete in terms of its environmental impact is by adding SCMs such FA, silica fume, and recycled materials like glass [3, 4]. In this regard, jhama bricks are a viable substitute for CA in concrete mixtures. Jhama bricks are renowned for their excellent strength and durability because of their over-burned production process [5]. Jhama bricks are intended to substitute a part of conventional CA in concrete structures in an effort to enhance not only the structural integrity but also the extended lifespan of these constructions [6]. The study focuses on assessing this green concrete formulation's varied durability features [7, 8]. Important characteristics are sorptivity, or the capacity to take in water through capillary action [9], water absorption and resistance to acid attacks such those caused by HCl and H_2SO_4. To test the resistance of OPC concrete samples (measured $100 \times 100 \times 100$ mm^3) and binders (made of FA and Jhama bricks) to an acidic attack, those samples were immersed in 5% H_2SO_4 acid solutions for 28 days. To keep the pH constant, solutions of acids have been switched out after 14 days. Following 28 days, the collected specimens were placed in the oven to achieve a constant weight, and the detachable particles have been eliminated [10]. The determination of acid resistance can be performed through comparison of the weights of the dry specimens before and right after the acid attack, as the CS of specimens immersed in acidic solution was not able to establish. The sorption experiment illustrates the blockage in water flow caused by capillary suction on concrete specimen surfaces, which is typically influenced by the pore structure and curing time of the concrete [9, 11]. Concrete's sorption capacity is one of its durability attributes that should be as low as feasible to maximise its longevity [12]. The mass of water that gradually flows through the specimen's pores is assessed during sorptivity metrics, which only involve one end of the sample coming into interaction with water [10, 13]. The sorptivity is given by the slope of water that is absorbed per unit surface area of the submerged surface versus the square root of the period. The measurement of water absorption was a vital tool for evaluation used by contractors and architects to assess the water resistance along with durability of concrete helps confirms that the concrete satisfies the required specifications and norms for the particular use and offers beneficial data about the consistency as well as the quality of the concrete mixture [14, 15]. The outcomes of the water absorption test may additionally be utilized for deciding whether specific type of concrete is best for a given project and ensuring that the concrete will be functioning as planned for the duration of its projected lifespan [16, 17].

13.2 Materials and Methodology

The first step of this procedure is to prepare a concrete mix by incorporating 20% of FA as a substitution of cement and substituting 10% of the CA with jhama bricks. For this investigation, a total of 12 cubes with the size of 100 × 100 × 100 mm were cast and after 28 days of curing period 3 cubes were tested for sorptivity, 3 were tested for water absorption and another 6 cubes were tested for acid attack. In that acid attack, 3 cubes are done with HCl and 3 cubes were done with H_2SO_4.

For this investigation, OPC 53 grade cement was operated, available in Miyapur. The relative density of this cement was determined to come 3.155 g/cc. According to IS 2720—part, the specific gravity of cement range is 3.150 g/cc. Fine aggregate for this project was sourced locally, with a specific gravity determined to be 2.7 g/cc. According to IS 2386:1963-part 3, the unit weight of fine aggregate is 2.50 to 3.0. Using CA helps better durable attributes of concrete and other materials, refining their strength and durability. The determined denseness of CA is 2.80 g/cc. According to IS 2386:1963-part 3, the specific weight of fine aggregate is 2.50 to 3.0. The quantity coarse aggregate depends on the percentage of replacement jhama bricks. If 10% jhama bricks are replaced, then 10% of coarse aggregate is decreased by its volume. Jhama bricks are over burnt bricks. These bricks are typically class 4 bricks. Jhama bricks have specific uses in construction due to their unique properties [4, 18]. These bricks are cost-effective and eco-friendly. Fly ash, due to its beneficial properties, is widely utilized in the construction industry, offering both economic and environmental advantages. The quantity of FA is increased to make green concrete. Also used super plasticizer and water along with these materials (Fig. 13.1).

Fig. 13.1 Jhama bricks

13.3 Experimental Investigation

See Table 13.1.

Based on the Indian Code for Mix Design approach in IS 10262:2019 the mix proportioning for concrete grade of M30 was tabulated in Table 13.2. Material quantity for jhama brick depends on the percentage replacement of coarse aggregate. If jhama brick was replaced by 10% weight of CA, then decreased the quantity of CA by 10% (Fig. 13.2).

Table 13.1 Physical properties of cement, FA and CA

Materials	Properties	Value
Cement	Fineness [%]	9
	Grade [MPa]	53
	Specific gravity	3.155
Fine aggregate	Specific gravity	2.7
Coarse aggregate	Specific gravity	2.6

Table 13.2 Materials quantities for kg/m^3

Material	Quantity (kg/m^3)
Cement	307.2
Fly ash	76.8
Fine aggregate	730.2
Coarse Aggregate	1135
Super plasticizer	3
Water	192

Fig. 13.2 **a** Workability of concrete **b** Slump cone test

13.4 Results and Discussion

The test involves a mould shaped like a cone, with a top dia of 10 cm, a bottom dia of 20 cm, and a height of 30 cm. The slump cone has a volume of 0.0068 m^3. The slump value obtained for the conventional concrete was 90 mm. It was noted that the percentage of FA expands, the workability (Slump Value) of green concrete become greater. Similarly, the percentage of jhama bricks increased, the workability (Slump Value) of green concrete decreased (Table 13.3).

For this investigation, 12 concrete cube samples of size 100*100*100 mm with 20% substitution of FA in place of cement and 10% substitution of CA with jhama bricks are cast and cured for 28 days. The following tests are conducted to determine the durability properties of green concrete.

1. Sorptivity of concrete
2. Water Absorption
3. Acid Attack
   (i) Hydrochloric Acid (HCl)
   (ii) Sulfuric Acid (H_2SO_4)

13.4.1 Sorptivity of Concrete

3 samples of size 100*100*100 mm concrete cubes were taken from the curing tank after 28 days of curing and these samples are oven dried at 100° C for 24 h shown in Fig. 13.3. Then the 4 sides of the specimen are covered with tape and bottom side of the cube is open to absorb the water from bottom of the cube. Then the weight of the specimen is recorded. The specimens are then placed in a tray with 5 mm water filled in it. The water should only have absorbed from the bottom side of the specimen. Again, the weight of the specimen was noted and repeats the experiment for 5, 10, 15, 20, 25, 30 min (Fig. 13.4 and Table 13.4).

Here, I = Sorptivity absorption Rate = $\frac{W2-W1}{A*d}$ S = Sorptivity coefficient = $\frac{I}{\sqrt{t}}$

Where, W_1 = Initial Weight of sample, W_2 = Final Weight of sample, d = Density of water, I = Absorption rate, A = Area of c/s and t = Time in minutes.

The average value of sorptivity absorption rate of conventional concrete (C30) was found to be 0.55 mm as well as for green concrete (G30) was found to be 0.293 mm and the average value of sorptivity coefficient is found to be 0.1 mm/min$^{0.5}$ for C30

Table 13.3 Workability (Slump in mm) for different percentages of FA and jhama brick

% replacement	With FA (mm)	With Jhama bricks (mm)
0	90	111
10	105	115
20	98	102
30	86	93

Fig. 13.3 Cubes are oven dried at 100°C for 24 h

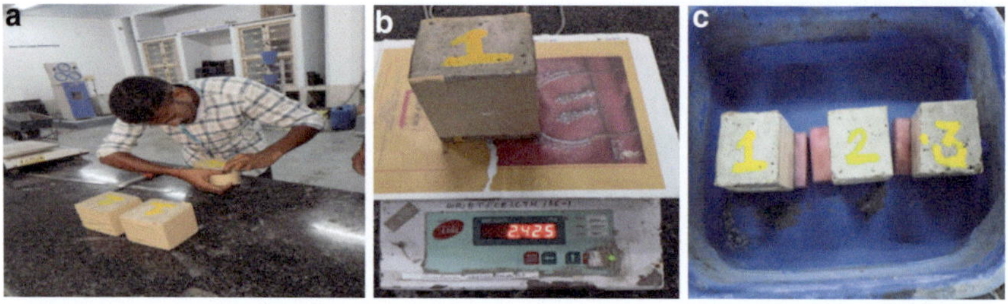

Fig. 13.4 **a** Cubes covered with tape **b** Weight of sample **c** Specimens are placed in a tray

Table 13.4 Sorptivity of concrete results for C30 and G30

Sample	Initial Weight (W1)gm	Weight after immersion in water (W2 gm)						$I=\frac{W2-W1}{A*d}$ (mm)	$S=\frac{I}{\sqrt{t}}$ (mm/ $\sqrt{}$min)
		5 Mins	10 Mins	15 Mins	20 Mins	25 Mins	30 Mins		
C30- M1	2531	2539	2547	2556	2563	2567	2571	0.49	0.089
C30- M2	2392	2405	2413	2422	2431	2440	2448	0.56	0.102
C30- M3	2169	2182	2193	2204	2213	2221	2229	0.60	0.109
G30-M1	2422	2428	2433	2436	2437	2439	2451	0.29	0.052
G30-M2	2361	2365	2368	2370	2376	2379	2394	0.33	0.060
G30-M3	2393	2395	2400	2400	2400	2403	2419	0.26	0.047

as well as 0.053 mm/min$^{0.5}$ for G30. Green concrete often shows improved sorptivity results due to the inclusion of supplementary cementitious materials (SCMs) like fly ash, slag, or silica fume. These materials typically refine the pore structure, reduce pore connectivity, and improve the overall density of the concrete. As a result, green concrete tends to exhibit lower sorptivity values compared to conventional concrete (Fig. 13.5).

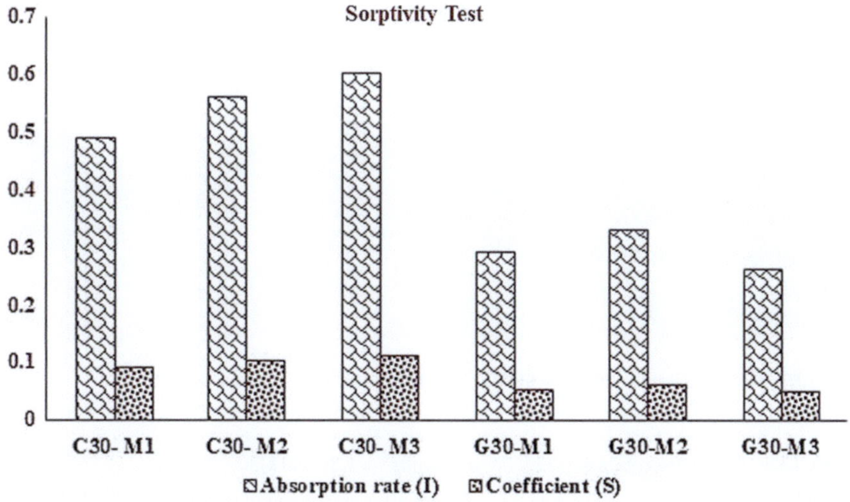

Fig. 13.5 Graphical representation of sorptivity test

13.4.2 Water Absorption of Concrete

For this test, 3 cubes are casted with conventional concrete and 3 cubes are casted with green concrete (i.e., 20% substitution of FA in place of cement as well as 10% substitution of jhama bricks in place of CA) and these cubes are taken from the curing tank after 28 days of curing and note the weight of sample (i.e., initial weight of sample, W_1). This test was performed immediately after the samples are taken from the curing tank and dry the samples in the oven for 24 h at 60°C and then take the samples out from the oven and noted the weight of samples. Then after samples are immersed in a water for 24 h and record the weights of samples for 1, 2, 3, 6, 12 and 24 h (Figs. 13.6 and 13.7 and Table 13.5).

Fig. 13.6 **a** Cubes casted and cured for 28 days **b** Initial weight of sample (W_1) **c** Cubes are oven dried at 100°C for 24 h

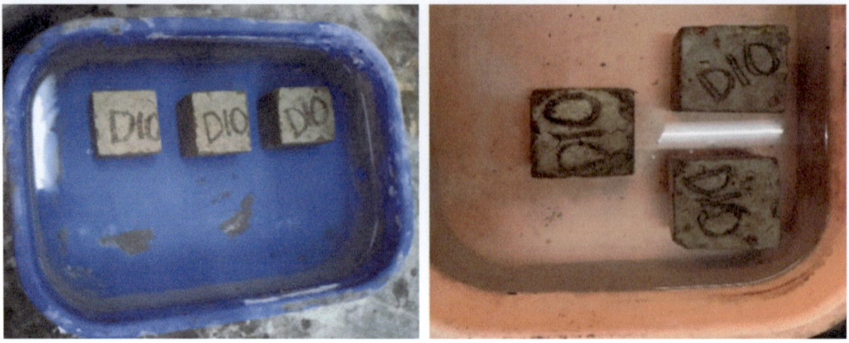

Fig. 13.7 Green concrete and conventional concrete cubes were immersed in water

Table 13.5 Water absorption capacity (W_{AC}) results for C30 and G30

Sample	Initial weight	Dry weight	Weight after immersion in water (gm)				WAC %
			1 Hour	2 Hours	3 Hours	24 Hours	
C30- M1	2563	2420	2452	2464	2470	2491	2.89
C30- M2	2575	2362	2409	2436	2468	2511	2.34
C30- M3	2515	2391	2418	2429	2437	2451	2.61
G30-M1	2425	2304	2342	2358	2371	2390	1.46
G30-M2	2380	2243	2289	2306	2321	2342	1.62
G30-M3	2465	2293	2346	2378	2401	2415	2.07

The average value of Water absorption capacity for conventional concrete (C30) was found to be 2.61% as well as for green concrete (G30) was found to be 1.71%. So, 0.9% of water absorption capacity is more in C30 compared to G30 (Fig. 13.8).

13.4.3 Acid Attack

For this test, 6 cubes are casted with conventional concrete and 6 cubes are casted with green concrete (i.e., 20% substitution of FA in place of cement as well as 10% substitution of jhama bricks in place of CA) and these cubes are taken from the curing tank after 28 days of curing and note the weight of sample (i.e., initial weight of sample, W_1). In that 3 conventional concrete as well as 3 green concrete cubes are reacted with HCl and another 3 conventional concrete and also 3 green concrete cubes are reacted with H_2SO_4. For these cubes 5% of acid is added by its volume of water (Fig. 13.9).

After examined the below Fig. 10a and b, The cubes exhibited changes in size, shape and color (white), with an irregular surface. Most of the cement and FA content leached into the water, leaving visible CA as well as Jhama bricks within the cubes

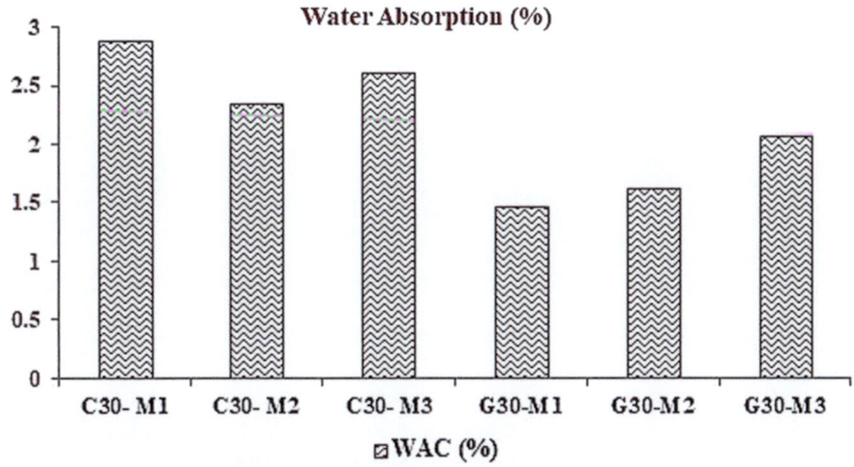

Fig. 13.8 Graphical representation of water absorption test

Fig. 13.9 Cubes immersed in 5% HCl and H_2SO_4 solution

when reacted with HCl solution. Although, the size and shape of the cubes did not change significantly, their color turned reddish color after exposure to H_2SO_4 solution. Furthermore, there was a decrease in both the weight percentage and compressive strength when exposed to HCl and H_2SO_4 solution. Subsequently, these cubes were tested at 15, 30 and 45 days after acid exposure (Tables 13.6 and 13.7).

The ultimate weight loss of conventional concrete (C30) and green concrete (G30) after 45 days of immersion in 5% HCl solution were found to be 9.56%, 7.96% respectively. The weight loss in G30 is 1.60% less than C30 when reacting with HCl solution. In the same way, the weight loss in G30 is 0.93% less than in the case of C30 with 5% H_2SO_4 solution, and the proportions of C30 and G30 after absorption in the solution were found to be 7.27% and 6.34%, respectively. Because G30 generally lose a smaller amount of weight that C30 when it reacts with HCl and H_2SO_4 solution. It is because G30 often includes SCMs which may strengthen its chemical resistance, such as FA, slag, or SF. C30, which primarily consists of

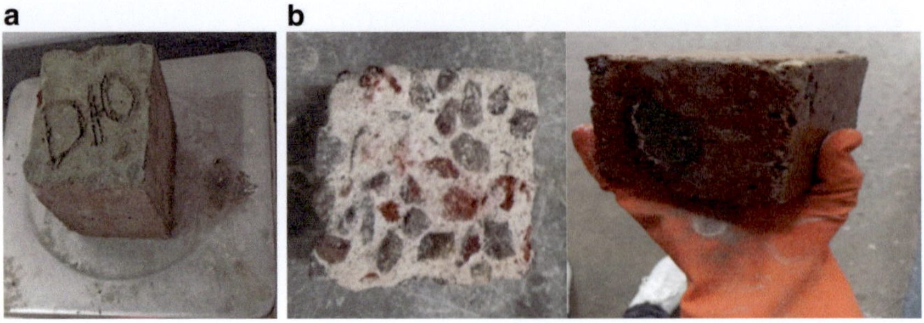

Fig. 13.10 **a** Cube before reacting with acids **b** Cubes after reacted with HCl and H_2SO_4 solution

Table 13.6 Weight loss in percentage for HCl solution

Samples	Immersion period (days)	Initial weight (gm)	Reduction in weight immersed in HCl (gm)	Loss in weight (%)
C30	0	2446	2395	–
G30		2563	2502	–
C30	15	2395	2320	3.23
G30		2575	2500	3.01
C30	30	2462	2335	5.43
G30		2336	2226	4.94
C30	45	2234	2039	9.56
G30		2128	1971	7.96

Table 13.7 Weight loss in percentage for H_2SO_4 solution

Samples	Immersion period (days)	Initial weight (gm)	Reduction in weight immersed in H_2SO_4 (gm)	Loss in weight (%)
C30	0	2446	–	–
G30		2380	–	–
C30	15	2425	2355	2.97
G30		2563	2505	2.31
C30	30	2465	2348	4.98
G30		2521	2453	4.48
C30	45	2329	2171	7.27
G30		2483	2410	6.34

Table 13.8 CS loss in percentage for HCl and H_2SO_4 solution

Samples	Immersion period (days)	Actual strength (MPa)	Strength after immersion in HCl (MPa)	Strength after immersion H_2SO_4 (MPa)	Strength loss in CS (%) HCl	Strength loss in CS (%) H_2SO_4
C30	15	38.5	37.6	37.9	2.39	1.58
G30		38.2	37.5	37.8	1.86	1.05
C30	30	38.5	36.9	37.2	4.33	3.49
G30		38.2	37.2	37.5	2.68	1.86
C30	45	38.5	35.6	36.5	8.14	5.47
G30		38.2	35.9	36.6	6.40	4.37

Portland cement, is often more vulnerable to acid attack and may lose more weight in the presence of HCl and H_2SO_4 solution. At last, when compared with H_2SO_4 solution, the weight loss in % is greater in HCl (Table 13.8).

The ultimate CS loss of conventional concrete (C30) and green concrete (G30) after 45 days of immersion in both 5% HCl and 5% H_2SO_4 solution were found to be 8.14%, 6.40% and 5.47%, 4.65% respectively. The CS losses in G30 is 1.74%, 1.10% less than C30 when reacting with HCl and H_2SO_4 solution respectively. After examined above CS losses, G30 concrete cubes are lost less CS than C30 concrete cubes as well as C30 and G30 concrete cubes are over react with HCl than H_2SO_4 solution after 45 days of immersion.

13.5 Conclusions

Based upon the durability studies conducted in the green concrete with FA and Jhama bricks, the following conclusions were made in this investigation as follows.

1. The Sorptivity absorption rate of conventional concrete (C30) was found to be 0.55 mm as well as for green concrete (G30) was found to be 0.293 mm.
2. Sorptivity coefficient is found to be 0.1 mm/min$^{0.5}$ for C30 as well as 0.053 mm/min$^{0.5}$ for G30. As a result, green concrete tends to exhibit lower sorptivity values compared to conventional concrete.
3. The water absorption capacity of conventional concrete (C30) is 2.61% as well as for green concrete (G30) 1.71%. So, 0.9% of water absorption capacity is more in C30 compared to G30.
4. The ultimate weight loss of conventional concrete (C30) and green concrete (G30) after immersed in 5% HCl solution were found to be 9.56%, 7.96% respectively.
5. The weight loss in G30 is 1.60% less than C30 when reacting with HCl.

6. Accordingly, the weight loss in G30 is 0.93% less than in C30 with 5% H_2SO_4 solution, and the proportions of C30 and G30 after absorption in the solution were found to be 7.27% and 6.34%, respectively. Finally, when compared with H_2SO_4 solution, the weight loss in % is greater in HCl.
7. After examined above CS losses, G30 concrete cubes are lost less CS than C30 concrete cubes as well as C30 and G30 concrete cubes are overreact with HCl than H_2SO_4 solution after 45 days of immersion.

References

1. J. Maheswaran, M. Chellapandian, N. Arunachelam, M. Naga Theera Hari, Thermal and durability characteristics of optimized green concrete developed using slag powder and pond ash. IOP Publishing Ltd, Mater. Res. Express **10** (2023)
2. S. Murkute, S.H. Madan, V.A. Patil, Comparative study of green concrete and conventional concrete on strength and durability properties. Int. Res. J. Adv. Eng. Sci. **3**(4), 229–232 (2018)
3. F. Hatamia, M. Amiri, Experimental study of mechanical properties and durability of green concrete containing slag, waste rubber powder and recycled aggregate with artificial neural network. Clean. Mater. **5** (2022)
4. A. Arun Kumar, A. S. Santhi, G. Mohan Ganesh, Strength and durability of green concrete. Int. J. Struct. Eng. **6**(1) (2015)
5. P. Hiremath, A. Basit Parray, A. Hassan, P. Kankeri, Durability of green concrete composed of fly Ash, GGBS, quarry dust, and RCA after acid exposure. Pract. Period. Struct. Des. Constr. **29**(4)
6. Andreola VM, da Gloria MYR, Pepe M, Toledo Filho RD, A comprehensive experimental study on the physical performance and durability of bamboo bio-concrete. Sustainability **16**(13) (2024)
7. A.N. Swaminathen, C. Vivek Kumar, S. Robert Ravi, S. Debnath, Evaluation of strength and durability assessment for the impact of rice husk ash and Metakaolin at high performance concrete mixes. Mater. Today: Proc. **47**(14), 584–4591 (2021)
8. P. Magudeaswaran, C. Vivek Kumar, K. Vamsi Krishna, A. Nagasaibaba, R. Ravinder, Investigational studies on the impact of supplementary cementitious materials (SCM) for identifying the strength and durability characteristics in self curing concrete. Mater. Today: Proc. (2023)
9. A. Krishnan, S.S. Subramanian, Effect of green gram pod ash (GGPA) as supplementary cementitious material (SCM) in mechanical and durability properties of concrete. Constr. Build. Mater. **411** (2024)
10. C. Vivek Kumar, M. Palanisamy, C. Balakrishna, S. Pooja Sri Reddy, S. Robert Ravi, Evaluation of strength characteristics and identifying the optimum dosage with the impact of partial replacement of recycled fine and coarse aggregate from construction and demolition waste. Mater. Today: Proc. 66(4), 1699–1709 (2022)
11. M.I. Khan, V.V. Ram, V.I. Patel, Durability performance of alkali-activated concrete with pre-treated coarse recycled aggregates for pavements. Sci. Rep. **14**, 13776 (2024)
12. S. Panda, N. Zade, P. Sarkar, R. Davis, Chemical durability evaluation of copper grit aggregate concrete against Alkali-Silica-Reaction, carbonation and chloride penetration. J. Build. Eng. **87** (2024)
13. P Magudeaswaran, C. Vivek Kumar, R. Ravinder, Prediction of strength and durability properties of HPC composites using adaptive neuro-fuzzy inference system, in E3S Web of Conferences, 184 (2020)
14. L. Cheng, H. Jin, J. Liu, F. Xing, A comprehensive assessment of green concrete incorporated with municipal solid waste incineration bottom: experiments and life cycle assessment (LCA). Constr. Build. Mater. **413** (2024)

15. M.F. Javed, M. Khan, M. Fawad, et al., Comparative analysis of various machine learning algorithms to predict strength properties of sustainable green concrete containing waste foundry sand. Sci. Rep. **14** (2024)
16. A. Al-Hamrani, M. Kucukvar, W. Alnahhal, E. Mahdi, NC Onat, Green concrete for a circular economy: a review on sustainability, durability, and structural properties. Materials **14**(2) (2021)
17. M.H. Islam, D.W. Law, C. Gunasekara, M.H.R. Sobuz, M.N. Rahman, M.A. Habib, A.K. Sabbir, Assessing the influence of banana leaf ash as pozzolanic material for the production of green concrete: a mechanical and microstructural evaluation. Materials **17**(3) (2024)
18. M. Hou, Z. Li, V.C. Li, Green and durable engineered cementitious composites (GD-ECC) with recycled PE fiber, desert sand, and carbonation curing: Mixture design, durability performance, and life-cycle analysis. Constr. Build. Mater. **414** (2024)

Chapter 14
Transforming Renewable Energy: Discovering the Development of Piezoelectric Substance

Utsav Singhal, Ansh Chawla, Amit Kumar, and Vanita Bhardwaj

Abstract It has long been known that piezoelectric materials have potential to convert electrical energy into mechanical energy and vice versa, providing chances to find a variety of uses in sensors, Mechanisms and energy harvesting systems. In order to give a thorough overview of the historical development and current achievements in piezoelectric materials—with a focus on their use for sustainable energy harvesting—this study collects insights from a variety of research papers. First, we discuss how this unique property in materials like quartz was first discovered by scientists all those years ago. We then follow their adventure as they discover more advanced materials, such as PZT and barium titanate. There's a catch, though: some of these materials involved lead, which made them bad for the environment. So, researchers started looking for safer alternatives, like bismuth sodium titanate. These days, harvesting energy from vibrations and other processes is of great importance to us when it comes to piezoelectric materials. This is really useful for creating devices that run off of electricity without the need for batteries. Even more flexible materials are being developed that have the ability to detect vibrations and convert them into electricity for use in sensors and phones. This research is historical and modern research to explain how piezoelectric materials allow the development of more environmentally friendly and effective technologies.

Keywords Piezoelectric materials · Energy harvesting · Sustainable technology · Lead-free alternatives · Flexible piezoelectrics

U. Singhal · A. Chawla · A. Kumar
Department of Artificial Intelligence and Machine Learning, Vivekananda Institute of Professional Studies, Delhi, India

V. Bhardwaj (✉)
Department of Applied Science and Humanities, Vivekananda Institute of Professional Studies, Delhi, India
e-mail: vanita.bhardwaj@vips.edu

A. Kumar et al. (eds.), *Selected Proceedings of the 1st International Conference on Advanced Materials for Sustainable Innovation; IC-AMSI 2024; 28–30 August; New Delhi; India*, Springer Proceedings in Energy,
https://doi.org/10.1007/978-981-96-4492-6_14

14.1 Introduction

For over a century, the abilities of piezoelectric materials [1, 2] to transform mechanical energy into electrical energy and vice versa [1, 3, 4] has been fundamental to the evolution of technology. The Curie brothers made the initial discovery of this phenomena in natural minerals such as stone in the late 1800s [5]. Because of their superior piezoelectric qualities, synthetic materials like lead, barium titanate, and zirconate titanate or PZT became industry standards as a result of further study.

However, the hunt for safer substitutes was ignited by the health and environmental dangers linked to lead in PZT. As a result, lead-free piezoelectric materials such as bismuth sodium titanate have emerged, providing similar performance without the risks. These days, creating self-powered gadgets that do not require batteries for everything from sensors to portable electronics depends on the capacity of piezoelectric materials to capture energy from vibrations and other mechanical processes.

The paper provides a thorough analysis of the evolution of piezoelectric materials during history, with a focus on sustainable energy harvesting. We highlight the importance of these materials in developing efficient and environmentally friendly technology by tracking their development from early discoveries to current successes.

14.2 Primary Principle Behind this Phenomenon

The capacity of producing electric charge from materials, under mechanical stress and vice versa is the basis of the amazing phenomenon known as the piezoelectric effect. The original research by Jacques and Pierre Curie in 1880 [6], who found that several crystals, including quartz and tourmaline, generate electrical charges under pressure, is where this effect originated. Simply put, the word "piezoelectric" comes from the Greek terms "Piezo," which means to push, and "piezein," which means to squeeze or press [7]. Basically, the piezoelectric effect is the result of mechanical energy converted in piezoelectric materials into electrical energy and vice versa. An external electric field is produced when these materials undergo mechanical stress, such as compression or tension, which causes the movement of the positive and negative charge centers inside the material. The piezoelectric effect is the name given to this phenomenon. On the other hand, piezoelectric materials experience physical deformation in response to an external electric field; this phenomenon is known as the inverse piezoelectric effect [8].

The piezoelectric effect is very adaptable and applicable in a wide number of disciplines due to its reversible nature. Technically speaking, the piezoelectric effect finds many uses in the real world, such as Microbalances, high voltage, electronic frequency, sound generation and sensing, and ultra-fine focusing of optical assemblies [9]. Additionally, it is the underlying theory of several scientific instrumental methods, like scanning probe microscopes, which allow for atomic-level resolution. Also, the piezoelectric effect has significantly improved common technologies, serving as the source of ignition for lighters and cigarettes and supporting the development of hydrophones, microphones, pressure sensors, and other sensing devices. Moreover, the inverse piezoelectric effect makes it possible to transform electrical energy into mechanical energy, which creates uses for sonar transducers, buzzers, speakers, and ultrasound equipment in medicine.

All things looked at, the piezoelectric effect is an intriguing mixture of mechanical and electrical phenomena that gives an infinite number of chances for creativity and advances in technology in a variety of fields.

14.3 Discovery and Early Studies

In 1880's the phenomenon of piezoelectric effect was invented by the Curie Brothers. Experimentation with ceramics and crystals was a popular area of study for researchers at that time due to the recent discovery of thermoelectric effect. While experimenting with the quartz crystal, Pierre and Jacques Curie found that under the application of pressure, the crystal developed potential difference across its ends. This phenomenon was named as piezoelectric effect. The following year converse piezoelectric effect was discovered. The converse piezoelectric effect demonstrated whenever electric field is applied across piezoelectric materials, they get elongated or compressed depending upon the polarity of the electric field.

The piezoelectric effect gathered significant interest among the researchers, extensive experimentation was done on it but no major breakthrough occurred before World War I. As the world war broke out heavy investments were done I the study of ultrasonic technology. These studies displayed the potential of quartz crystal to be used as resonators to detect ultrasound pulses in the SONAR. Using quartz and Rochelle salt in a certain structure also allowed them to act as transducers by emitting ultrasound pulses of certain frequencies. Another breakthrough occurred in World War II when BaTiO3 was discovered in search of compact high capacitance "condensers" for compact radar systems. BaTiO3 allowed for easy fabrication at very low costs.

14.4 Development in the Domain of Piezoelectric Material

Piezoelectric materials have made big improvements lately to work better, be more self-sufficient, and be used in more ways. These materials make electricity when they feel pressure and are getting noticed in many fields because of their special qualities and the new things they can do.

One big area of change is finding green options to replace the usual piezoelectric materials that use lead. Lead works really well, but it can be bad for the environment and our health. So, scientists are looking more at lead-free options like BNT and

KNN. These materials can do the same job as piezoelectric materials without causing environmental problems, which makes them better for the planet.

Moreover, the study of combinations of organic and inorganic materials, along with flat two-dimensional (2D) materials, has led to new opportunities for using piezoelectric technology. 2D materials such as graphene show excellent electrical properties and strong mechanical qualities, which can be used to make very effective and bendable piezoelectric devices. These materials are promising for future electronic devices, where being flexible and small is important.

In the field of Micro-Electro-Mechanical Systems (MEMS), piezoelectric materials and devices are seeing significant performance enhancements. Methods like adding impurities, mixing different materials, and using nanotechnology have been key. Adding impurities, which changes the material's properties, has been shown to improve the piezoelectric effect and the durability of these materials. Mixing different materials has led to combined effects that further boost performance.

Energy harvesting continues to be a key area of study in piezoelectric research. The aim is to improve how effectively energy is converted and to create systems that can collect energy from various sources like vibrations, temperature changes, and biological movements. This is especially useful for powering devices far from power sources and wearable electronics, reducing the need for conventional batteries. Progress in this field has resulted in the development of very sensitive sensors and flexible, wearable tech that can gather energy from body movements, opening up new possibilities for sustainable energy.

The use of piezoelectric materials is growing and now includes consumer electronics, medical devices, cars, and aerospace. In consumer electronics, piezoelectric parts are used in screens that respond to touch, systems that provide tactile feedback, and microphones. Medical devices also use piezoelectric sensors and actuators in a variety of applications, from ultrasound imaging to precision drug delivery systems. The automotive and aerospace industries utilize piezoelectric materials for vibration control, noise reduction, and energy harvesting systems, enhancing performance and sustainability.

Future studies on piezoelectric materials will likely continue to concentrate on environmentally friendly materials, better energy efficiency, and creating new uses by working together across different fields. New ideas are expected to come from combining material science, nanotechnology, and engineering, resulting in more sophisticated and adaptable piezoelectric systems.

14.5 Experimental Studies

Piezoelectric materials have found a wide range of applications in various industries due to their unique ability to convert mechanical energy into electricity and vice versa. Here are some important useful materials that have been investigated and developed show in Fig. 14.1.

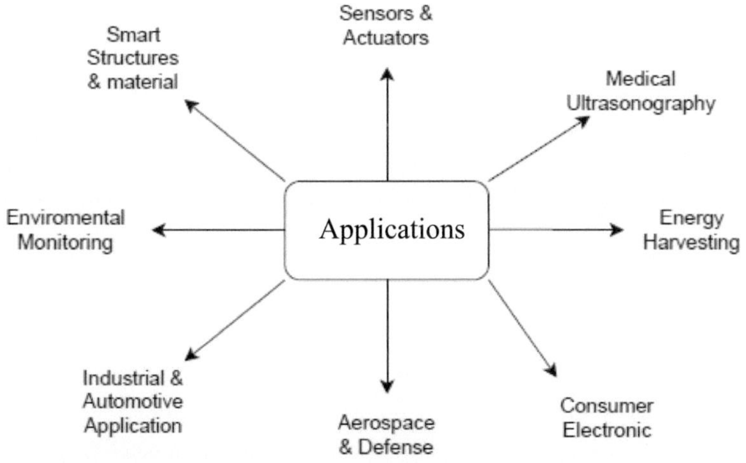

Fig. 14.1 Various applications of piezoelectric materials

14.5.1 Sensors and Actuators

Piezoelectric materials are commonly used in sensors and actuators. In sensors, mechanical stresses are converted into electrical signals, which can be used for strain sensing, bubble detection, and vibration monitoring. In contrast, in actuators, electrical signals can provide mechanical mobility, which is useful in precision positioning systems and industrial automation applications.

14.5.2 Medical Ultrasonography

One of the major applications of piezoelectric materials is medical ultrasonography. Piezoelectric transducers emit and receive ultrasound waves, enabling non-invasive imaging of anatomical structures. This technology is increasingly important in prenatal care, cardiology, and diagnostic imaging.

14.5.3 Energy Harvesting

Piezoelectric materials are increasingly being used for energy harvesting. These materials can convert environmental mechanical energy such as vibration from machinery, walking, or even ocean waves into electricity. This harvested energy can power small electronic devices, sensors, and wireless networks, helping to create self-monitoring systems and reducing reliance on batteries.

14.5.4 Consumer Electronics

In consumer electronics, piezoelectric materials are used in devices. For example, piezoelectric buzzers and speakers are common in alarms, clocks, and telephones. In addition, piezoelectric materials are used in touch-sensitive devices and haptic feedback systems, providing tactile feedback in touchscreens and game controllers.

14.5.5 Aerospace and Defense

The aerospace and defense industries use piezoelectric materials in many important applications. These include vibration monitoring in flight systems, sonar systems for underwater navigation, and precision instrumentation for satellite positioning. Piezoelectric materials contribute to the performance and reliability of these systems in challenging environments.

14.5.6 Industrial and Automotive Applications

In industrial and automotive applications, piezoelectric sensors are used for monitoring and monitoring. Examples include collision monitors in engines, which detect collisions and changes during ignition, and industrial vibration systems that monitor mechanical health to prevent and enable failures proper maintenance system.

14.5.7 Environmental Monitoring

Piezoelectric materials are also used in environmental monitoring. It is used to detect and measure pollutants, seismic activity, and weather. These sensors provide valuable data for environmental assessment, disaster preparedness, and climate change monitoring.

14.5.8 Smart Structures and Materials

Piezoelectric materials are an important source of smart structures and materials. These systems can adapt to changing conditions by sensing and responding to external stimuli.

Applications include self-healing materials, adaptive building systems, and intelligent navigation systems that can monitor and respond to stress or damage. The versatility of piezoelectric materials continues to drive innovation in many areas, demonstrating the tremendous impact on contemporary technology and the potential for future development.

14.6 Conclusion

Since the late nineteenth century, when the Curie brothers discovered piezoelectric materials, a lot has happened. Many technological advances have been made possible by their exceptional capacity to transform mechanical energy into electrical energy and vice versa. From early experiments with quartz to the development of lead-based materials like PZT and the subsequent search for environmentally friendly alternatives, the journey of piezoelectric materials is a testament to human ingenuity and the relentless pursuit of better, safer technologies.

Today, piezoelectric materials are at the forefront of sustainable energy solutions, providing innovative ways to harness energy from everyday activities and environmental sources. The shift towards lead-free materials, such as bismuth sodium titanate, and the exploration of flexible and organic–inorganic hybrids, highlight the commitment to reducing environmental impact while enhancing performance.

The applications of piezoelectric materials are vast and varied, spanning across sensors, medical ultrasonography, energy harvesting, consumer electronics, aerospace, defense, industrial, automotive, environmental monitoring, and smart structures. These materials have not only revolutionized existing technologies but have also opened new avenues for research and development, leading to more efficient, self-sustaining, and intelligent systems.

In conclusion, the continuous advancements in piezoelectric materials promise a future where technology and sustainability go hand in hand. By tapping into the inherent properties of these materials, we can create a world that is not only technologically advanced but also environmentally conscious, ensuring a better future for generations to come.

References

1. A. Aabid, M.A. Raheman, Y.E. Ibrahim, A. Anjum, M. Hrairi, B. Parveez, J. Mohammed Zayan, A systematic review of piezoelectric materials and energy harvesters for industrial applications. Sensors **21**(12), 4145 (2021)
2. M.C. Sekhar, E. Veena, N.S. Kumar, K.C.B. Naidu, A. Mallikarjuna, D.B. Basha, A review on piezoelectric materials and their applications. Cryst. Res. Technol. **58**(2), 2200130 (2023)
3. N. Wu, B. Bao, Q. Wang, Review on engineering structural designs for efficient piezoelectric energy harvesting to obtain high power output. Eng. Struct. **235**, 112068 (2021)
4. N. Sezer, M. Koç, A comprehensive review on the state-of-the-art of piezoelectric energy harvesting. Nano Energy **80**, 105567 (2021)
5. S. Katzir, The discovery of the piezoelectric effect, in *The beginnings of piezoelectricity: a study in mundane physics*. (Springer, Netherlands, Dordrecht, 2006), pp.15–64

6. C. Jacques, C. Pierre, Development, via compression, of electric polarization in hemihedral crystals with inclined faces. Bull. Soc. Minérologique Fr **3**, 90–3 (1880)
7. B. Ajit, Piezoelectric materials (2022). https://doi.org/10.1007/978-3-030-80359-9_2
8. R.W. Greaves, D.R. Lamb, Observation of the inverse piezoelectric effect in polyethylene while under a polarising field. J. Mater. Sci. **6**, 74–79 (1971). https://doi.org/10.1007/BF00550294
9. Z. Xinhua, Piezoelectric ceramic materials: Processing, properties, characterization, and applications. Piezoelectric Mater.: Struct., Prop. Appl. 1–36 (2010)
10. D.M. Damjanovic, Highly anisotropic electromechanical properties in modified lead titanate ceramics, Ph.D. dissertation, Pennsylvania State Univ., University Park, PA (1987)
11. W.A. Smith, The role of piezocomposites in ultrasonic transducers, in *Proceedings, IEEE Ultrasonics Symposium*, 755–766 (1989). R.Y. Ting, Evaluation of new piezoelectric composite materials for hydrophone applications. Ferroelectrics **67**, 143 (1986)
12. T.R. Gururaja, W.A. Schulze, L.E. Cross, R.E. Newnham, Piezoelectric composite materials for ultrasonic transducer applications. Part 11: Evaluation of ultrasonic medical applications. IEEE Trans. Sonics Ultrason. **SU-32**, 499–513 (1985)
13. R.Y. Ting, Evaluation of new piezoelectric composite materials for hydrophone applications. Ferroelectrics **67**, 143 (1986)
14. A.M. Nicholson, Piezophony. US Patent 1,495,429; Filed April 10, 1918, Patented May 10;1924. Jaffe B. Piezoelectric transducers using lead titanate and lead zirconate. US Patent **2**(708), 244 (1955)
15. S. Tuncdemir, S.O. Ural, B. Koc, K. Uchino, Design of translation rotary ultrasonic motor with slanted piezoelectric ceramics. Jpn. J. Appl. Phys. **2011**(50), 027301 (2011)
16. B. Jaffe, H. Jaffe, W.R. Cook, *Piezoelectric ceramics*, 1st edn. (Academic Press, London, 1971)
17. R.E. Newnham, Functional composites for sensors and actuators: smart materials. The Pennsylvania Academy of Science, PA, USA (1998)
18. T.A. Anderson , D.W. Sexton, A vibration energy harvesting sensor platform for increased industrial efficiency, in *Proceedings of Smart Structures and Materials Conference Proceedings SPIE* **6174** 61741Y (2006)
19. A.A. Bent, N.W. Hagood, J.P. Rodgers, Anisotropic actuation with piezoelectric fiber composites. J. Intell. Mater. Syst. Struct. **6**, 338–49 (1995)

Chapter 15
Performance Analysis of Energy Harvesting for Sustainable Satellite-Terrestrial Relay Communication System with Path Selection in Eavesdroppers Presence

Varun Jain, B. V. R. Reddy, and Ashish Payal

Abstract Harvesting energy and improving security in physical layer is now an approach that shows promise, which is securing information not only from eavesdropper without data encryption at upper layer but at the same time improves efficiency of the network. However, it through some new challenges like other person can trace the transmitted confidential message between the sender and final destination via relay. So, the idea to avoid this the transmitted power of signal should be enough large for harvesting energy at the same time it needs to be compact/small enough to prevent eavesdropping. This would be more difficult when relaying is applied. Considering things above discussed our work proposes two protocols/techniques i.e., Random Path Protocol selection and Best Path Protocol selection techniques. These techniques will improve the multipath network security for cooperative satellite and terrestrial communication system even in the presence of eavesdropper. Furthermore, we drive closed form expression and asymptotic expression of the likelihood of an outage for each method in the presence of eavesdropper. Whereas simulation results also plotted with exact and asymptotic results.

Keywords Maximal ratio combiner (MRC) · Outage probability (OP) · Physical layer security (PLS)

V. Jain (✉) · B. V. R. Reddy · A. Payal
USICT, GGSIPU, Dwarka, New Delhi, India
e-mail: varunjain1202@gmail.com

B. V. R. Reddy
e-mail: profbvrreddy@gmail.com

A. Payal
e-mail: ashish@ipu.ac.in

A. Kumar et al. (eds.), *Selected Proceedings of the 1st International Conference on Advanced Materials for Sustainable Innovation; IC-AMSI 2024; 28–30 August; New Delhi; India*, Springer Proceedings in Energy,
https://doi.org/10.1007/978-981-96-4492-6_15

15.1 Introduction

Satellite communication is a two-sided sword when we talk about sustainability. On the other hand side, it offers tremendous benefits that can come up to a more reliable future. On the other hand, the industry needs to address its environmental impact and resource consumption to be truly sustainable [1].

Impacts:

- Enabling Sustainable Development:—Satellites provide crucial data for environmental monitoring, like tracking deforestation, illegal fishing and climate change. This data empowers governments and organizations to make informed decisions for a healthier planet.
- Bridging the Connectivity Gap:—Satellite communication brings internet and communication access to remote areas, facilitating tele-education, healthcare and disaster response. This can improve lives and livelihoods in underserved communities.
- Optimizing Resource Use:—Satellite data helps with precision agriculture, optimizing water usage and fertilizer application. It can also track logistics, optimize transportation routes, and reduce fuel consumption.

By addressing these impactful challenges, the satellite communication industry can continue to play a vital role in a sustainable future for our planet. In our paper, we are considering multi-path and multi-point satellite communication network in presence of eavesdropper where sender and intermediator relay points working as energy harvester from booster to further move that data forward to final destination.

Our main motivation to write this paper is little summarized here:

- Here we proposed path selection techniques such as Random Path Protocol Selection (RPS) and Best Path Protocol Selection (BPS) techniques. In RPS, sender selects some random path to communicate with final destination point. In BPS, as name suggests chooses best path and provide high channel capacity and optimal outage performance of the system [1–3].
- The Radio Freq. technique applied by the sender and relay node to avoid the eavesdropper from mixing of sender data getting over multi-point. Moreover, these specialized transmitters are capable to adjust the transmitting power to decrease channel capacity obtained on eavesdropper links [4].
- Here we are considering both eavesdropper and non-eavesdropper scenario in this our paper [5].
- We drive closed form expression of OP over proposed fading channel [6–8].

15.2 System Model and Performance Evaluation

We suggest model in Fig. 15.1 of proposed protocol techniques in which sender S, communicates with final destination D, in multi-point manner. The M probable paths are between S and D out of which one is selected to S to D communication. In between S to D, K active eavesdroppers are exists to overhear the transmitted message from S to D via R (relay) nodes. To avoid mixing of received data from the eavesdroppers and the MRC combiner, radio freq. technique is applied at each point on the true selected path. Here, we pre-assume that absolutely all devices i.e. S, D, R are energy constrained so for that booster is deployed in the network system.

Let us here we denote R_u relay nodes and we pre-assume that there are no hardware impairments at uth path selected for data transmission. Equation (15.1) and (15.2) are considered as system model equations. Where P_R denotes the transmitted power by transmitter $R_{u,j}$. Equation (15.3) represents the energy harvesting expression for beacon present at the near to relay node. In Eq. (15.4) specifically tells the channel capacity for receiver in the rpesence of eavesdropper with noise denoted with N_0. Where as Eqs. (15.5) and (15.6) explain the OP for both the techniques suggested.

$$Y_{\mathrm{R_{u,j}R_{u,j+1}}} = \sqrt{P_{\mathrm{R_{u,j}}}} h_{\mathrm{R}_{u,j}\mathrm{R_{u,j+1}}} S + \nu_{\mathrm{R_{u,j}R_{u,j+1}}} \tag{15.1}$$

$$Y_{\mathrm{R_{u,j}E_k}} = \sqrt{P_{\mathrm{R_{u,j}}}} h_{\mathrm{R}_{u,j}\mathrm{E_k}} S + \nu_{\mathrm{R_{u,j}E_k}} \tag{15.2}$$

$$EH_{\mathrm{R_{u,j}}} = \eta\alpha\tau_u P_{\gamma B} \tag{15.3}$$

$$C_{\mathrm{R_{u,j}E_k}} = (1-\alpha)\tau_u log_2\left(1 + \frac{P_E}{kP_E + N_0}\right) \tag{15.4}$$

Outage Probability for Random Path calculated as

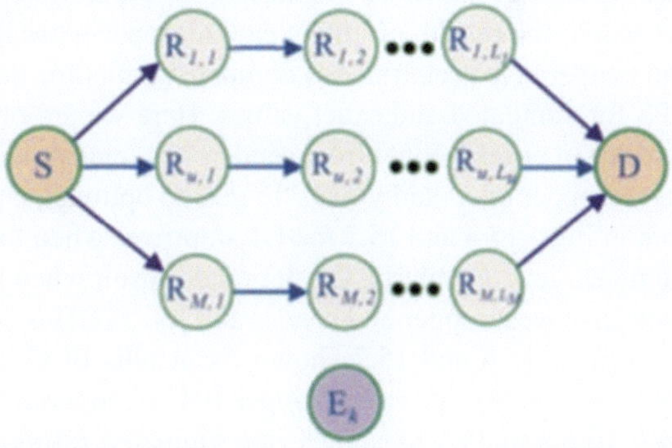

Fig. 15.1 System Model, where Sender, Relay and destination node is shown with eave-droppers node

$$= \frac{1}{M} \sum_{u=1}^{M} P_r\left(C_r^e < C_{\text{th}}\right) \tag{15.5}$$

Outage Probability for Best Path calculated as

$$= P_r\left(C_b^e < C_{\text{th}}\right) \tag{15.6}$$

15.3 Simulation Results

In this part, we confirming our derived findings with provided Monte-Carlo (MC) simulations. The results we are obtained here by using Matlab®R2021a simulation tool. To obtain the OP for the proposed protocols from one end to another end, we perform 10^5 independent trials. We construct Rayleigh channel coefficients for each link that has been set up in the network. In simulation ecosystem, we consider a 2-D plane in which the coordinates axis of the S, R, the destination node and the eavesdropper are (0, 0), $\left(\frac{j}{L_k+1}, 0\right)$ and the eavesdropper $(x_{\text{E}}, y_{\text{E}})$ respectively.

In Figs. 15.2 and 15.3 we show the effect of SNR or Tp transmitted power of booster B (in dB) and the values of OP in case when we take into consideration of eavesdroppers Un-cooperate and cooperate. Where Table 15.1 represents all the parameters we are considering during the simulation and comparison of those results with theoretical results. It has been noticed that when B (in dB) is negative small i.e. approx. -5 dB the OP value reaches 1 and when a value of B is increased the OP value decreases. This means we confirm that increasing the transmitted B (in dB) value can enhance security in physical layer against eavesdropper attacks. In Figs. 15.2 and 15.3 confirm that the OP of BPS technique is lower than that of RPS. As we can see in Figs. 15.2 and 15.3 the solid line represents simulated results and dotted line is representing asymptotic results. We can also observe in Fig. 15.3 some values of simulated results of BPS does not lie on the values of exact results (i.e. solid line) that because of hardware impairment or other system noise. In Figs. 15.4 and 15.5 we shown the results of OP vs eavesdropper moves near or far for Un-cooperative and cooperative scenario. We compare results for both the techniques i.e. RPS and BPS for simulated and exact values. Here we are observing for different values of distance of eavesdropper from sender. As already mentioned in Table 15.1 the axis of eavesdropper on x- and y-axis. To get the optimal result we set the $x_{\text{B}} = 0.35$ as been seen in Figs. 15.4 and 15.5 the OP improves when the eavesdropper moves faraway from data link. Similarly, OP get outdo result when $y_{\text{E}} = 0$ that means the distance between eavesdropper and sender and relays (R) is zero or may be the shortest.

In Figs. 15.6 and 15.7 shown the results of OP vs EHR (or α) as function of harvested energy for eavesdropper is Un-cooperative and cooperative scenario. In Figs. 15.6 and 15.7 as shown our simulated results are lie on exact values of the curve. In this case, EHR is crucial since it influences not only the power received at BPS but also energy sent by the S and R nodes.

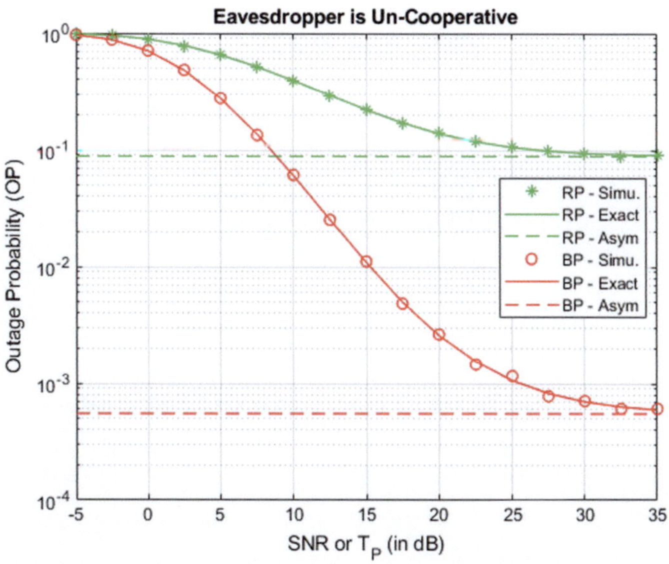

Fig. 15.2 Graph between OP verses SNR (in dB) for RP and BP selection where eavesdropper is Un-cooperative

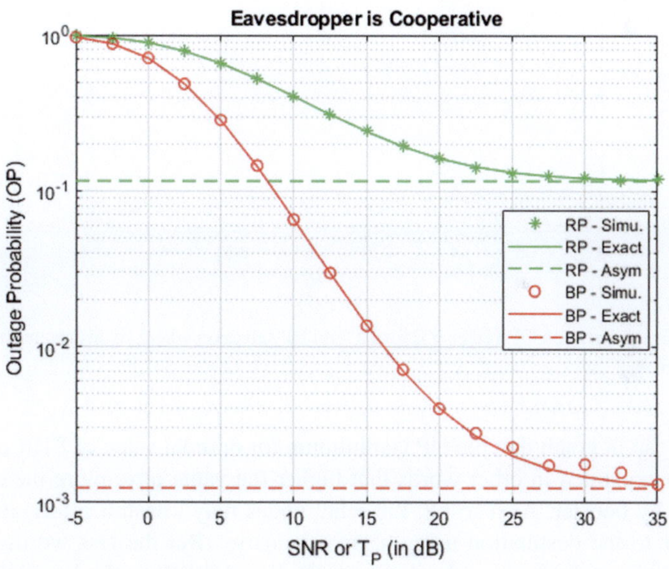

Fig. 15.3 Graph between OP verses SNR (in dB) for RP and BP selection where eavesdropper is cooperative

Table 15.1 Parameters and values we are considering throughout in this paper

Path loss exponent, β	3
No. of relay nodes, L	(2,3,4)
Channel gain of the booster, R	0.5
Hardware impairment, k	0.1
Hardware impairment, K	2
Energy harvesting Ratio, α	0.1
Energy conversion efficiency, η	0.1
Booster position on axis, (xB, yB)	(0.5,1)
Eavesdropper position on axis, (xE, yE)	(0.5,1)

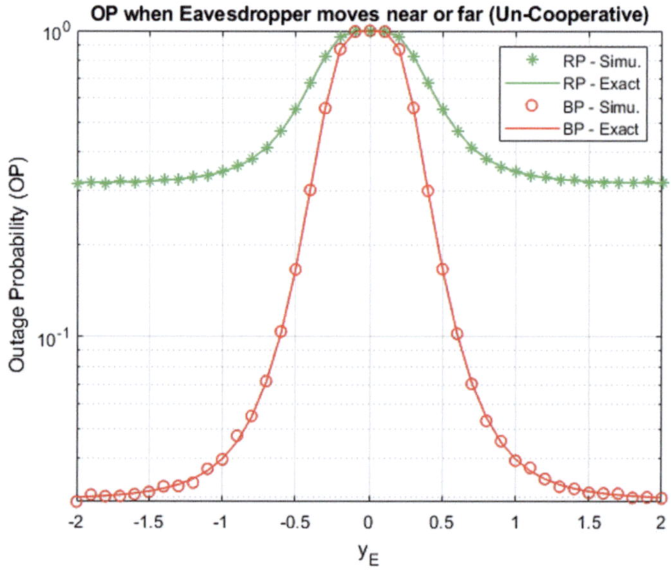

Fig. 15.4 Graph between OP verses YE for RP and BP selection where eavesdropper (moving) is Un-cooperative

As we can in graph that the OP is minimum for optimal value of EHR or α. This thing also explained in other words that higher the value of α, more the energy is harvested by booster. As a result, the relay nodes may transmit information from the sender to the destination using a greater energy. But the less worthwhile the communication time is $(1-\alpha)\tau_u$ between the S and the R, or the R and the R, the greater the value of α. We may therefore get the best OP when α reaches its ideal value.

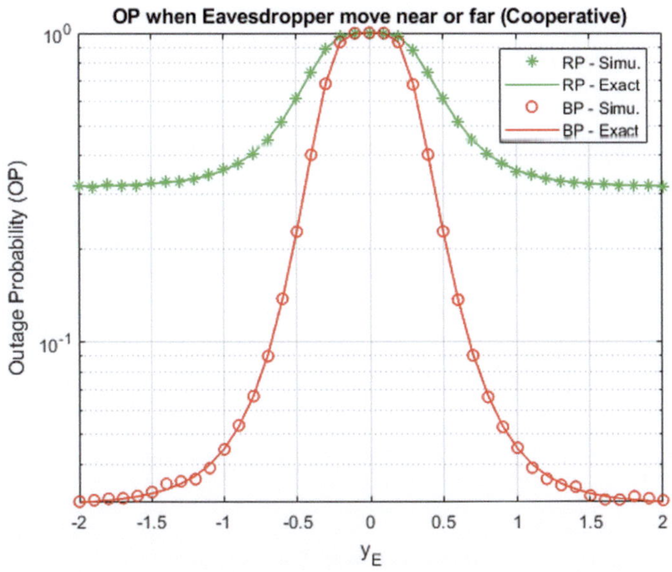

Fig. 15.5 Graph between OP verses YE for RP and BP selection where eavesdropper (moving) is cooperative

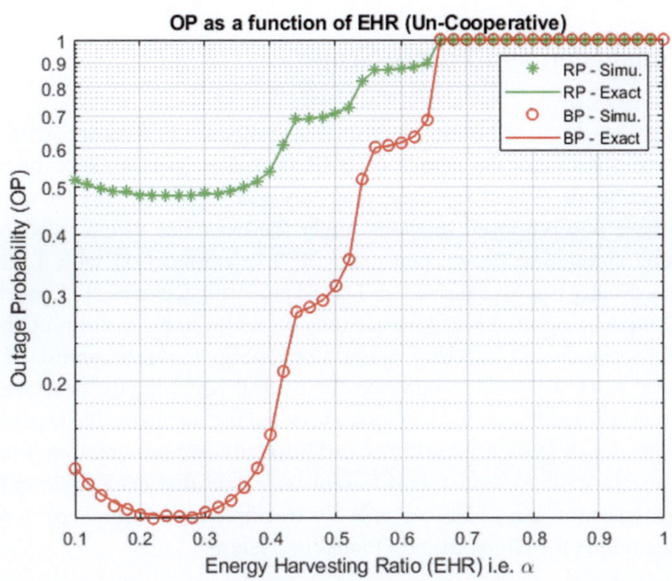

Fig. 15.6 Graph between OP verses EHR for RP and BP selection where eavesdropper is Un-cooperative

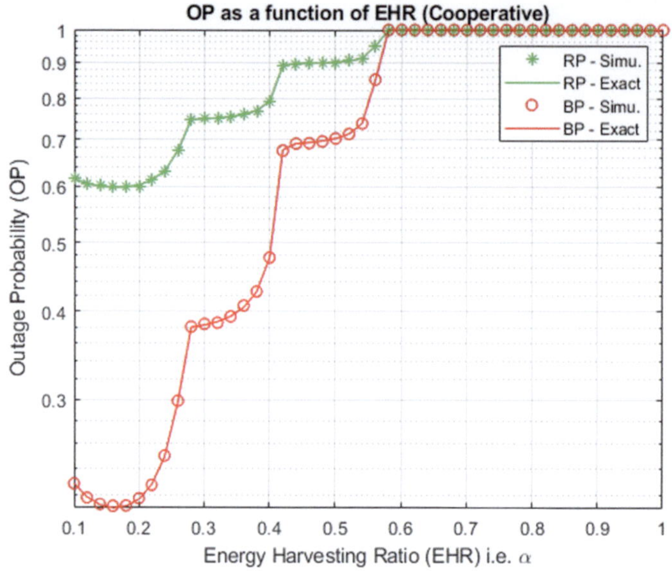

Fig. 15.7 Graph between OP verses EHR for RP and BP selection where eavesdropper is cooperative

15.4 Conclusion and Future Scope

In this paper, we discussed two path selection protocol i.e. RPS and BPS protocol techniques to investigate the results of EHR on outage performance of multipath cooperative network. Where, we drive precisely and asymptotically the outage likelihood of these two proposed methods under the existence of eavesdroppers. The results of the simulation shown that using BPS in conjunction with a multi-relay scheme may greatly improve the secured act of performance of the EH under consideration. Ultimately, optimal performance may be achieved by positioning the energy booster appropriately and selecting an optimum energy harvesting ratio (EHR) or α.

For future work we want to enhance the overall performance described in this manuscript. Additionally, we wish to look into what is causing the basic performance restriction on the suggested algorithms. Additionally, we are creating a new path-selection system method that uses different way data that each node empirically gathers and transmit. And also we will try to implement improved or new path selection technique for High Altitude Platform System.

Acknowledgements I would like to thank Prof. (Dr.) B. V. R. Reddy for their exceptional support. His enthusiasm, knowledge, and experience have been an inspiration and kept me on track. I am also grateful for Prof. (Dr.) Ashish Payal for the insightful suggestions.

Disclosure of Interests The authors have no relevant financial or non financial interests to disclose.

References

1. T.D. Hieu, T.T. Duy, B.S. Kim, Performance enhancement for multihop harvest-to-transmit WSNs with Path-selection methods in presence of eaves-droppers and hardware noises. IEEE Sens.S J. **18**(12), 5173–5186 (2018). https://doi.org/10.1109/JSEN.2018.2829145
2. E. Soleimani-Nasab, S. Coleri, A unified framework for multi-hop wireless relaying with hardware impairments. IEEE Trans. Veh. Technol. **73**(4), 5674–5689 (2024)
3. D. Tung Vo, T. Van Chien, T.N. Nguyen, D.-H. Tran, M. Voznak, B. Seo Kim, L. Thanh Tu, SWIPT-enabled cooperative wireless IoT networks with friendly jammer and eavesdropper: outage and intercept probability analysis. IEEE Access **11**, 86165–86177 (2023)
4. P. Minh Nam, P. Van Ca, P. Viet Tuan, T. Trung Duy, V. Nguyen, Q. Bao, Security versus reliability study for multi-hop cognitive M2M networks with joint impact of interference constraint and hardware noises. Int. Conf. Adv. Technol. Commun. 259–264 (2018)
5. M.K. Arti, M.R. Bhatnagar, Beamforming and combining in hybrid satellite-terrestrial cooperative systems. IEEE Commun. Lett. **18**(3), 483–486 (2014)
6. M.R. Bhatnagar, M.K. Arti, On the closed-form performance analysis of maximal ratio combining in Shadowed-Rician fading LMS channels. IEEE Commun. Lett. **18**(1), 54–57 (2014)
7. M.K. Arti, V. Jain, Relay selection based hybrid satellite-terrestrial communication systems. IET Commun. **11**, 2566–2574 (2017)
8. M.R. Bhatnagar, M.K. Arti, Performance analysis of hybrid satellite-terrestrial FSO cooperative system. IEEE Photon. Technol. Lett. **25**(22), 2197–2200 (2013)

Chapter 16
Solar Energy Harvesting Methods for Sustainable Development

Bharat Kumar Nandmehar, Manasvi, and Navneet Kumar

Abstract As we know that in our day today life, we require energy to perform tasks. As it has been quoted that "Energy can neither be created nor be destroyed but it can change from one form to another" that is the principle of conservation of energy. The term Harvesting means the activity of taking or collecting a useful substance likewise Energy Harvesting means to collect the waste power and bring it back to the system. As per the growing energy demands in the world there is so much of need to harvest the energy for the future generation to survive and maintain sustainability. Sustainability is a core principle driving the adoption of solar energy worldwide. This study explores how solar power contributes to sustainability by reducing environmental impacts, promoting energy independence, and fostering economic growth. Solar energy represents a momentous solution to the world's growing energy demands due to its abundance, renewability, and environmental benefits. This research work highlights some of the ongoing exploration and researches in solar energy technology. Also a comparative analysis has been made on solar energy harvesting method and effecting parameters are compared and discussed which is more focused on the cost effectiveness, eco-friendly, neutral carbon emission which will promote sustainable future of our natural resources.

Keywords Solar energy · Harvesting energy methods · Sustainability · Efficiency · Neutral carbon emission

B. K. Nandmehar (✉) · Manasvi · N. Kumar
School of Engineering and Technology, Vivekananda Institute of Professional Studies-Technical Campus, IndiaDelhi
e-mail: bharat.vips2023@gmail.com

A. Kumar et al. (eds.), *Selected Proceedings of the 1st International Conference on Advanced Materials for Sustainable Innovation; IC-AMSI 2024; 28–30 August; New Delhi; India*, Springer Proceedings in Energy,
https://doi.org/10.1007/978-981-96-4492-6_16

16.1 Introduction

Renewable resources are natural resources that can be regenerated over short periods of time and are thus considered sustainable. Environmental sustainability growth refers to the advancement and expansion of practices and technologies aimed at maintaining and improving the health of the environment while supporting economic and social development. Recent advancement in solar cells, solar water heater and solar thermal energy system and decrease in cost have been a key player for more sustainable energy landscape. Even though it has great potential but it faced several challenges and opportunities Because of weather and climatic conditions the adoption of solar energy by solar cells, solar water heater and solar thermal system is compromised according specific conditions. The research paper is focussed on the different methods of harvesting solar energy, effectiveness and potential for future advancement. Knowing these methods is helpful for converting renewable resources and creating sustainability goals.

16.2 Objective

The objective of this research paper is to explore various solar energy harvesting methods. This study shows a comparative analysis of various parameters such as efficiency and potential advancement of solar energy harvesting technologies like solar cells, solar thermal system and solar water heater. It aims on analysing how different technologies convert solar energy into usable energy and to analyse multiple perspectives. This research work focus on the various technologies and methods taken in consideration in harvesting solar energy, emphasizing on photovoltaics (PV), solar thermal energy, solar water heater. As mentioned above about the energy conservation principle, PV systems utilize semiconductor materials to directly convert light energy into electricity. Solar thermal energy system harnesses the sunlight (solar energy) into thermal energy then that thermal energy is converted into the electrical energy. Solar water heater is a device that receive the heat from sun and it converts the cold water into hot water. Ongoing advancements in battery technology and grid integration strategies are pivotal in enhancing the reliability and stability of solar power within the energy landscape.

16.3 Solar Cells

Solar cell is a type of electrical device that converts Sunlight into electrical energy with the phenomenon of photoelectric effect. Solar cell is also known as photovoltaic cell. It is basically a p–n junction diode whose electrical characteristics varies when exposed to light. Combination of solar results in the formation of module known as

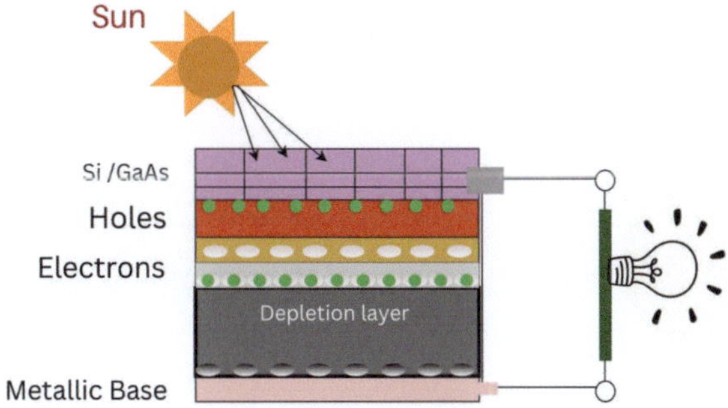

Fig. 16.1 Working of solar cell

solar panel. A single-junction of solar cell can generate a maximum voltage of about 0.5 to 0.6 V. As it is very small value of voltage but when we combine number of solar cell then we can produce large amount of electricity.

Photoelectric effect:- It is the phenomenon in which when the light of suitable frequency fall on the metal surface electrons comes out from the metal surface and those electrons are called photoelectron and by which effect the electron are emitted is known as photoelectric effect.

16.3.1 Construction

Working of solar cell is shown in Fig. 16.1. Semiconductor is covered with protected glass on this metallic strip are made also known as finger metallic strip through which emitted electron flows. In solar cell there are basically two layers the upper layer is made up of N-type material and lower layer is made up of P-type material, On the top of the solar cell transparent glass is placed as a protective cover. Si or GaAs semiconductor material are used for making solar cell. N-type layer is very thin and highly doped it is made very thin because the light that is incident crosses the N-type layer and reaches the junction, the thickness of the N-type layer is about 0.3 μm. P-type layer is thick and lightly doped, the thickness of P-type layer is about 300 μm. Metallic base is attached to the lower portion of solar cell also known as substrate base. With the help of metallic base and metal strip the immediate electron are collected and they behave as electrode.

16.3.2 Working

As electrons majority charge carriers in N-type material and holes are majority charge carriers in P-type material. When they are combined together then electrons from N-type layer crosses the boundary and enters in p-type layer, electrons then fill the holes and at the place of electron, holes are created. As the electrons are negative

Table 16.1 Parameters of a solar cell

S. No.	Solar cell	Parameters
1	Temperature	Residential solar panels are typically evaluated at a standard temperature of 77°F and are designed to operate at optimal efficiency within a temperature range of 59°F to 95°F [1]
2	Efficiency	Solar panel efficiency typically ranges from 15% to 20%, while in certain cases, the efficiency of solar cells can reach up to 42% [2]
3	Cost	On average, a commercial solar system can cost around ₹1,60,000 per kW [3]

charge carrier and the holes are positive charge carrier because of this positive and negative charge carrier a region is created known as depletion region. In depletion region holes and electrons are in neutral state. When sun ray are inn electromagnetic wave in the form of photons incidents on N-type layer of solar cell, since N-type layer is very thin it crosses the N-type layer, and strikes on the depletion layer. When photons strike on this layer, semiconductor material emits electrons that remains free to move. When the electron is emitted, it moves towards the positive charge carrier and holes moves towards the negative charge carrier. When upper finger metallic strip and lower metallic base that is behaving as an electrode are short through wire a close loop is created, by the help of positive and negative electrode. Emitted electron from the upper metallic strip flows in metallic base via wire and as we know rate of flow of electron is known as current, and as a load if we connect a bulb at that point the bulb will glow. Light intensity is directly proportional to the Output current (Table 16.1).

16.4 Solar Thermal Energy System

Solar thermal power plant is a type of electricity generation plant. By the help of solar energy steam is generated in solar power plant then the generated steam rotates or moves the turbine to produce mechanical energy and through that mechanical energy electricity is generated. All solar thermal power systems use solar energy collectors, which consist of two main parts:

(1) Reflectors
(2) Receiver

reflectors (mirrors) that collect and direct sunlight onto a receiver.

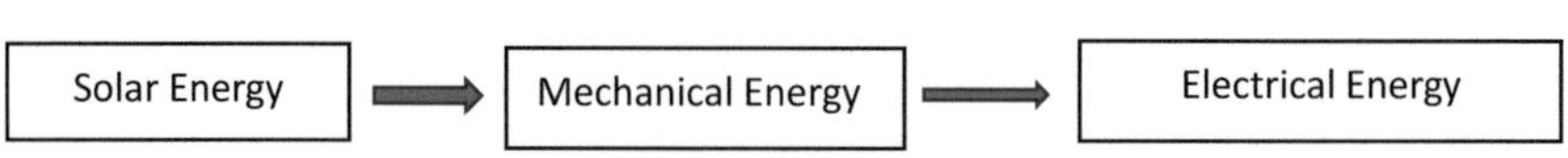

Solar thermal power plant is based on the renewable source of energy so that we can generate electricity in huge quantity.

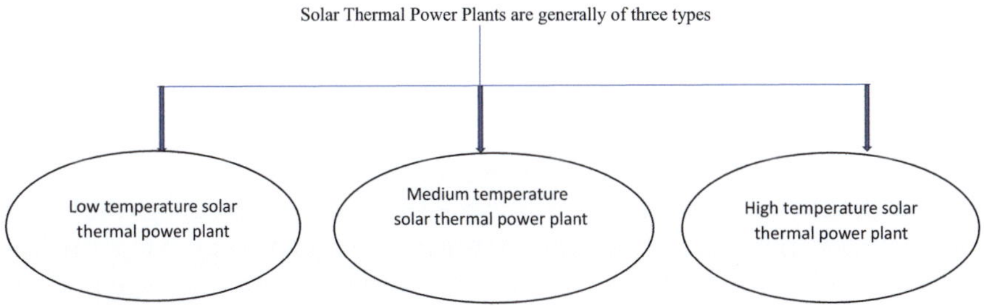

16.4.1 Low Temperature Solar Thermal Power Plant

16.4.1.1 Construction

A schematic of a low-temperature system utilizing flat-plate collectors and its operational details are illustrated in Fig. 16.2. In this thermal power plant, we have a flat plate collector that consist with black absorber plate, its main function is to receive heat from solar radiation. A hollow tube is connected with flat plate collector in which the water is present (or filled) that is circulated with the help of pump. We have butane boiler in this boiler the butane is in the form of fluid (liquid), when the heat is exchanged in the butane boiler then the fluid butane is converted into steam (vapour).

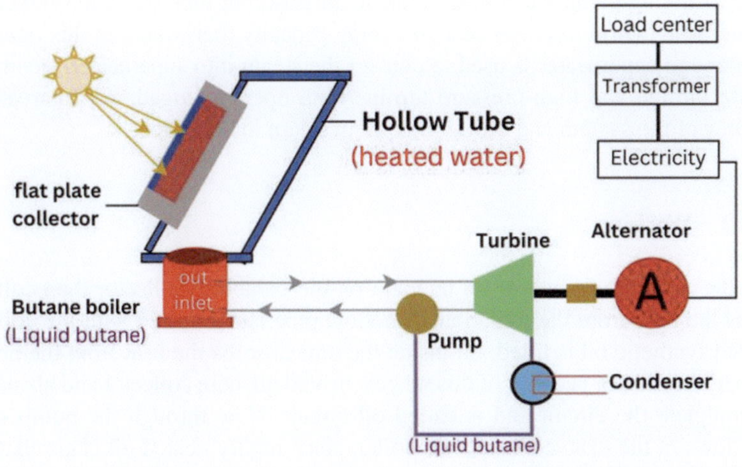

Fig. 16.2 Design and operation of a low-temperature system employing flat-plate collectors

16.4.1.2 Working

When the solar radiation is incident on the flat plate collector then the black absorber plate captures the heat from the radiation since the hollow tube is connected with flat plate collector in which the water is present so because of the captured heat the water temperature also increases, and when the heated water circulates through the butane boiler then heat transferred in liquid butane happens and temperature of liquid butane increases and it evaporated and converted into steam and when the high pressure vapour strikes the turbine blade then the turbine starts to rotate and mechanical energy is produced since the alternator shaft is coupled with turbine shaft so alternator coverts the mechanical energy into electrical energy and generated electricity voltage is stepped and transferred it to the load centre through the transmission line when the high pressure butane strikes the turbine blade then the pressure of butane is reduced then the low pressure vapour gets transferred in condenser where vapour is converted into liquid butane and with the help of pump liquid butane flows into the butane boiler and repeats the cycle.

16.4.2 Medium Temperature Solar Thermal Power Plant

16.4.2.1 Construction

Schematic diagram of a typical plant is shown in Fig. 16.3. In this method of thermal power plant, we can heat the water up to 400 °C. In this type of solar power plant we use focussing parabolic collector, inside the absorber tube we have synthetic oil filled because the oil has high heat absorbing capacity than water, in this operation preheater and superheater is used to convert the steam into superheated steam, low pressure turbine and high-pressure turbine both operates together to increase the efficiency of the system and utilise maximum steam in operation.

16.4.2.2 Working

When the radiation from the sun incident on the parabolic collector then collector absorbs the heat from the radiation. A hollow pipe is connected with the collector in which synthetic oil is filled, oil inside the tube absorbs the heat from the heating chamber of collector because of this oil gets heated up, here collector and absorption tube complete the circuit and is called oil circuit. Now through the pump stored water flows in the heat exchanger or boiler, since highly heated oil is circulated in the absorption tube so water absorbs the heat from the absorption tube which is filled with heated oil (transfer of heated oil high temperature to water) as a result water is converted into a steam due to heat inside the absorption tube, and that steam directly strikes on the high pressure turbine blades then after the steam goes to the preheater where the temperature present due to oil in tube reheat the steam and the

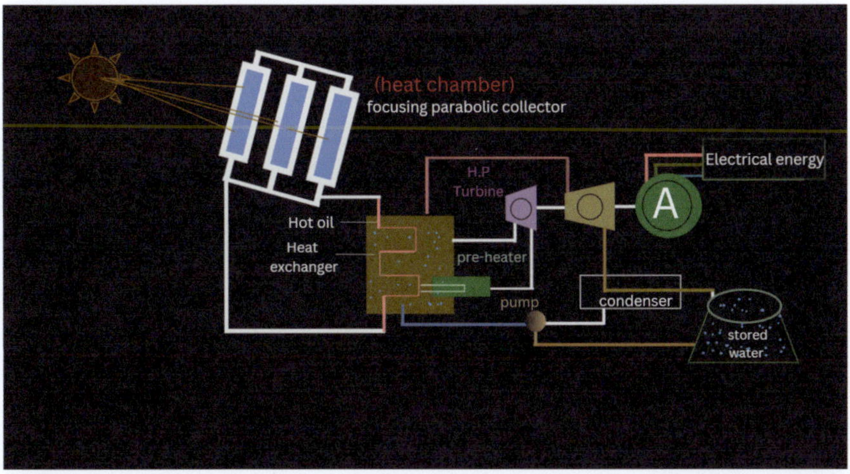

Fig. 16.3 Design and operation of a medium-temperature system with concentrating parabolic collectors

pressure of steam again increases and that steam transfer to the low pressure turbine by the nozzle thus both the high pressure turbine and low pressure turbine rotates together by same shaft also an alternator is connected with the same shaft of turbine so alternator shaft also rotates when mechanical input is provided to the alternator by prime mover or turbine then alternator converts the mechanical to electrical energy, and after striking the steam on low turbine, it goes to the condenser for colling where the steam is converted into the water and that water through pump send to the heat exchanger or boiler and again heat exchanger converts the water into steam this same process repeats itself and continues to produce electrical energy by solar radiation.

16.4.3 High Temperature Solar Thermal Power Plant

16.4.3.1 Construction

With the help of this solar power plant method the temperature can be obtained from about 500 °C–1500 °C. In this type of power plant, we use either parabolic dish type collector or central receiver. In the parabolic dish we use mirror and solar tracking system by which very high intensity solar rays incident on common point, because of this fluid get high heated and converts water into steam and that steam rotates the prime mover. Central receiver is a transparent or hollow pipe coated with black material to receive heat, in this system sun tracking mirrors is used to reflect the solar radiation on receiver also called heliostat, when the solar radiation incident on these mirrors then it gets reflected back and gets concentrated on receiver that is used to convert fluid into steam for generation of electrical energy.

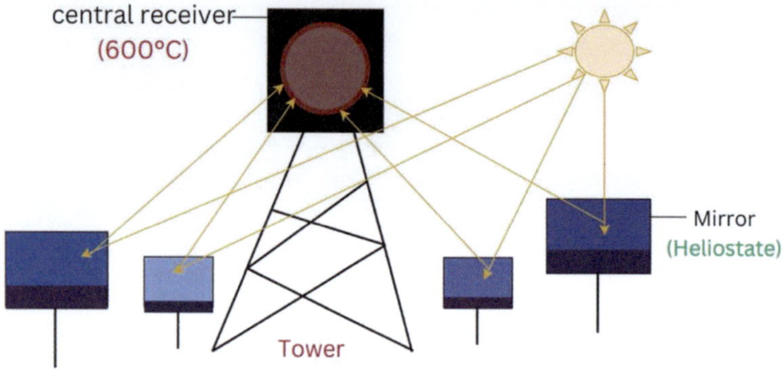

Fig. 16.4 Working of solar power tower

16.4.3.2 Working

Solar power tower and its working is shown in Fig. 16.4. A large number of flat mirrors are used in solar tower system that move to follow the sun. These mirrors, called heliostats, reflect and focus sunlight onto a receiver located on the top tower [4]. Many of these mirrors are arranged on the ground in a paraboloid shape, in between the boiler also known as absorber is kept on a high support. The sun's radiation falls on all mirrors and is reflected by it. These mirrors are arranged in such a way that the radiation is reflected and collected on the absorber, as water passes through the absorber it is converted into high temperature steam this passes to the steam turbine so that the heat is converted into mechanical energy. The alternator connected to the turbine converts mechanical energy into electrical energy. Steam from the turbine is taken to the condenser at a low pressure than the atmosphere. Cold water is circulated from the tube of the condenser, this water takes the heat of the exhaust steam and the steam condenses down, the water is sent to the boiler by a condensate extraction pump. The hot water comes out from condenser tube is sent into the cooling tower where it cools, the water which is cooled get circulated back from the tube to the condenser.

16.5 Parabolic Dish Type Collector

Parabolic dish type collector and working is shown in Fig. 16.5. This type of solar collector is parabolic in shape that's why it is known as parabolic dish collector. This parabolic structure is called parabolic dish reflector main purpose of this reflector is when the solar radiation is incident on this reflector it reflects the solar radiation and concentrate those solar radiation at a common point. At the front portion of the reflector, we place mirror or aluminium foil with the help of this we can reflect 90% of the solar radiation. Some of the parabolic collectors are solar tracking mechanism and

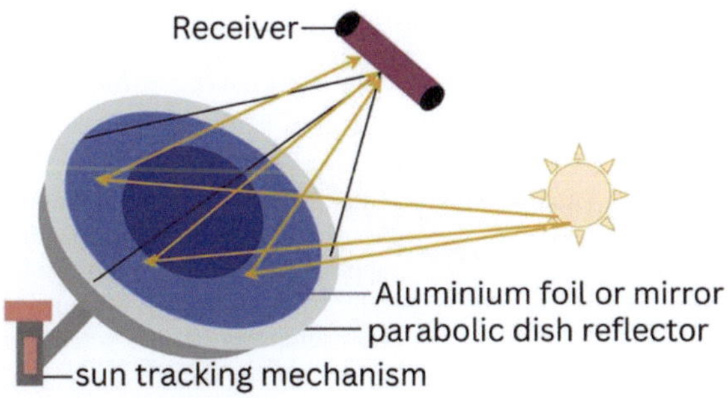

Fig. 16.5 Working of parabolic dish type collector

non-solar tracking mechanism with normal stand. This is tracking mechanism that turns the reflector in direction of sun. Receiver is a transparent hollow pipe coated with heat absorbing material the receiver absorbs heat from direct and reflected radiation. when the solar radiation is incident on the reflector they reflect and concentrate on the receiver due to the parabolic shape of reflector it reflects the overall radiation on receiver, receiver absorbs the heat from solar radiation and temperature rises in the receiver if the water is filled in this receiver then water converts into steam by receiving the temperature and that steam utilize for electricity generation using turbine and alternator, such solar collector is used in Stirling engines to convert thermal energy into electrical energy (Table 16.2).

16.6 Solar Water Heater

A diagram of a Solar Water heater and working is shown in Fig. 16.6. It is a system that receive the heat from sun and it converts the cold water into hot water. This technology operates on renewable energy, as we know that the solar radiation is free of cost as well as it is available in unlimited quantity. Solar water heater is also called as domestic hot water system, temperature of water can be increased till 60 °C to 80 °C by the help of solar water heater. For the domestic purpose the capacity of solar water heater is about 100–300 L.

16.6.1 Construction

The solar water heater consists with different main parts like solar collector generally we use tube solar collector here the tubes are parallel placed, it is made up of transparent glass and black coating placed at inner portion main purpose of this black coating is to receive more heat from solar radiation. Storage tank is placed on the top of device by which remaining tubes are connected. It consists with a special metal of thick layer to provide insulation to maintain the heated water temperature, tubes

Table 16.2 Classification of solar thermal energy system and various parameters

S. No.	Solar thermal energy system parameters	Low temperature	Medium temperature	High temperature
1	Temperature	Uppermost temperatures is approximately 100 °C, Low-temperature cycles operate on it [5]	Uppermost temperatures is approx. 400 °C, Medium-temperature cycles operate on it [5]	At temperature exceeding 400 °C, High-temperature cycles operate on it [5]
2	Efficiency	For the cycle illustrated in Fig. 16.1, Rankine cycle efficiency is 7 to 8% when the temperature differential is just 55 °C. Given that the collector system operates at an efficiency of approximately 25%, the overall system efficiency is consequently limited to about 2% [5]	The synthetic oil heated at collector to 400 °C with an efficiency of approximately 0.7 for direct beam radiation. This heated oil is used to generate superheated high-pressure steam, which operates a Rankine cycle with an efficiency of 38% [5]	Due to fluctuations throughout the day, the average conversion efficiency typically falls short of the maximum efficiencies. As a result, Net yearly Conversion efficiency of solar energy to electricity in pilot power tower systems. range from 7 to 20% [6] Stirling engines are commonly preferred as the prime movers in these systems. Developments have achieved efficiencies of up to 30%, with power outputs ranging from 8 to 50 kW [5]
3	Cost	Typically, the installed cost is approximately ₹300,000 per kW for systems with 6 to 8 h of daily operation, with the collectors being the primary cost component [5]	A cost of $2,900 per kW is reported recently for 80 MW Plant. Currently, the cost generated is approximately $0.08 per kWh [5]	In India, the cost for CSP systems can range from approximately $3,000 to $5,000 per installed kW

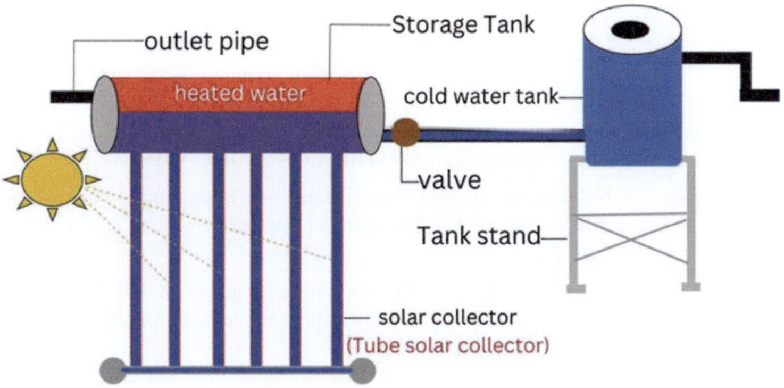

Fig. 16.6 Working of solar water heater

and storage tank are bend at some angle with the help of stand so that more solar radiation are incident. Cold water tank is placed at more height than tube it is done so the water gets properly circulated and there is a tank stand on which the cold-water tank is placed. Cold water tank is connected with storage heater tank through pipe and a valve remain connected to control the flow of water this pipe is also known as inlet pipe. At another end we have outlet pipe that remain connected with storage tank heater through which heated water flows out.

16.6.2 Working

Solar water heater works on the principle of thermosiphon system. Cold water remains filled in the tank that circulate in a storage tank of heater by pipe. Storage tank of solar heater is fully filled by cold water coming from cold water tank. When sun ray incident on the tubes the black material present in the tube absorbs the heat from the solar radiation and because of this heat the water present in the tubes starts to heat and when water gets heated then it flows at upper portion due to lower density and molecules of cold water circulate downwards, when water gets heated and move to the upper portion then cold water comes in tubes from lower portion and extra water flows from main tank through pipe and circulates in the tubes. So here two layers forms that is hot water and cold water and we receive the heated water from outlet pipe for different application.

There are two types of solar water heater:

(1) Flat plate
(2) Evacuated tube collector

Flat plate solar water heater:

Table 16.3 Various parameters of a solar water heater

S. No.	Solar water heater	Parameters
1	Temperature	On a sunny day the temperature raises to maximum of 80 °C while re-circulating the water in the absorber panel in the collector [7]
2	Efficiency	Solar thermal conversion efficiency is approximately 70%, whereas direct solar electrical conversion systems achieve an efficiency of only 17% [8]
3	Cost	On average, the cost of solar panels ranges from INR 37,500 to INR 42,000 per kW, depending on the wattage and type [9]

Flat plate solar water heater has black absorber plate that receives the heat from solar radiation and transfer the heat into liquid.

Evacuated tube collector type solar heater:

In this type of solar water heater double layer borosilicate glass is used and at inner section of tube is coated with black material that receives the heat from solar radiation and transfer the heat into liquid (Table 16.3).

16.7 Conclusion

This research work shows that use of solar energy technology is beneficial, both in how well it works and in terms of cost. The study is more in favour of harvesting solar energy and converting solar energy into useful form, methods have been described to make renewable source cost effective and efficient. On comparing all the three methods Solar cells have the best efficiency of 42% in overall cases. Medium-temperature solar systems utilizing line-focusing paraoolic collectors have achieved a degree of commercialization. The central receiver system has undergone extensive testing at the pilot scale, demonstrating its viability. Additionally, the solar chimney concept has been validated through successful experimental implementations. Also it is very important to set up the panels in the best possible way for a specific location to get the maximum output. From a financial standpoint, the quicker return on investment. Environmentally, solar energy greatly cuts down on carbon emissions, which is a big win for our planet's sustainability goals. As solar technology keeps on improving, with better panel efficiency and lower costs, we expect even more people to choose solar energy. Investment and development in solar energy technologies and research on improving panel designs, exploring new materials, and looking at long-term performance to enhance solar energy. To explore the possibility of more applications from the interpretation of the obtained results, for future work, the authors aim to improve these technologies in term of efficiency and cost effectiveness and to make this process more tireless and effective.

References

1. https://www.bostonsolar.us
2. https://www.greenmatch.co.uk/blog/2014/11/how-efficient-are-solar-panels
3. https://gserenewables.com/blog/solar-panel-installation-cost-india/#:~:text=The%20cost%20of%20commercial%20solar,%E2%82%B920%2C00%2C00%2C000.
4. https://www.eia.gov/energyexplained/solar/solar-thermal-power-plants.php
5. https://www.ias.ac.in/article/fulltext/jcsc/109/06/0521-0531
6. https://en.wikipedia.org/wiki/Concentrated_solar_power#:~:text=Due%20to%20variation%20in%20sun,demonstration%2Dscale%20Stirling%20dish%20systems.
7. https://eerem.delhi.gov.in/eerem/solar-thermal-application
8. S. Sadhishkumar, T. Balusamy, Performance improvement in solar water heating systems—a review. Renew. Sustain. Energy Rev. **37**, 191–198 (2014)
9. https://bluebirdsolar.com/blogs/all/solar-panel-installation-cost-in-india#:~:text=On%20average%2C%20the%20cost%20of,on%20the%20wattage%20and%20type.

Chapter 17
Integrated Cleaner and Sun Tracker for Solar Panels

Siddharth Jha, Sarang, Tejaswi, Toukeer Khan, Monika Bhardwaj, and Pratul Arvind

Abstract This paper presents the development of an innovative integrated solar tracker system with panel cleaning wipers designed to improve the day to day efficiency of the solar panel. The solar tracker constantly adjusts the orientation of solar panels from dawn till dusk to maximize the sunlight exposure on the panel. Meanwhile the cleaning system ensures that the panel remains clean from any dust or debris. Through the integration of sensors, actuators, and control algorithms, the system ensures maximum energy capture by maintaining the most optimized panel alignment with the sun throughout the day. Experimental validation demonstrates the efficacy of the integrated system in improving solar generation efficiency compared to conventional fixed-panel installations. This novel approach offers a practical solution for maximizing solar panel performance, contributing to the advancement of sustainable energy technologies and addressing the challenge of reduction in efficiency due to the dust depositions on the solar installations.

Keywords Solar tracker · Efficient solar panel cleaning · Dust removal · Wiper cleaning mechanism

17.1 Introduction

The energy consumption in the world is soaring high, and in the upcoming decades, it is projected to increase further. Today, about 80% of the world's energy is met from non-renewable energy sources such as fossil fuels, which leads to depletion of the natural resources at fastest ever rate [1]. This leads us nowhere else than an inevitable energy crisis. Furthermore, the process of extracting and burning of these

S. Jha (✉) · Sarang · Tejaswi · T. Khan · M. Bhardwaj
Maharaja Agrasen Institute of Technology, Delhi, India
e-mail: jhasiddharth145@gmail.com

P. Arvind
Dr. Akhilesh Das Gupta Institute of Professional Studies, Delhi, India

A. Kumar et al. (eds.), *Selected Proceedings of the 1st International Conference on Advanced Materials for Sustainable Innovation; IC-AMSI 2024; 28–30 August; New Delhi; India*, Springer Proceedings in Energy,
https://doi.org/10.1007/978-981-96-4492-6_17

fossil fuels emit greenhouse gases in the atmosphere, which leads to many disastrous climate changes like global warming [2]. Hence, it is becoming extremely critical to conserve energy and utilize other sources of energy, preferably renewables, at maximum possible efficiency.

Estimates suggest that more than half of the known oil reserves of the world are possessed by the Middle Eastern countries which can destabilise the economies around the world affecting the geopolitics. As per an estimate, the current system for meeting energy needs can serve us a generation or two [3]. To avert such an extreme energy crisis and mitigate the effects of climate change, it becomes extremely important to ride past from the conventional non-renewable resources and gradually shift towards meeting our energy demands from renewable sources such as solar power [4].

However, the photovoltaic panels are currently constrained by an efficiency ceiling of approximately 15%, with losses attributed to sun tracking, wiring, shading, and soiling [5]. Soiling, particularly prevalent in tropical regions prone to sandstorms and heavy rains, significantly affects PV performance. Regular cleaning mitigates soiling losses by ensuring optimal sunlight exposure and subsequent electricity generation. Soiling loss results from the piling up of dirt, dust, and debris on PV panels, which reduces its light-absorption capacity [6]. Consequently, energy conversion efficiency is also affected. In fact, as calculated in [7], it was concluded that in desert areas, the accumulation of dust reduced the efficiency of solar modules by more than 60%.

The integration of cleaning wipers on solar panels, along with a solar tracker mechanism, provides a novel approach to improve the power output and efficiency of solar panels. The document presented here discusses the developing and implementation of such a system. The proposed method seeks to enhance the day to day performance of solar panels by following the motion of the sun and cleaning the dust that accumulates on the panels over time. The primary objective is to improve the efficiency of solar panels, which is a crucial aspect in the quest for proper utilization of renewable sources of energy. Since the panel efficiency decreases significantly due to deposition of dust, the power generated gets reduced to a significant extent [8]. Dust becomes a barrier between sunlight and solar cells, preventing the panels from functioning optimally. Various methods have been proposed to address this issue, including the use of solar trackers, cleaning systems, cooling techniques, and anti-reflecting coatings [9].

The solar tracker mechanism is designed to continuously track sunlight throughout the day, maximizing solar energy production. This system ensures that the plane of the solar panel is perpendicular to the direction of sunlight, thereby increasing its efficiency [10].

The cleaning system is an essential component of the project, as it presents a solution for the removal of dust on solar panel. The cleaning process is automated, using wiper, which is placed at a position such that the panel can be cleaned in an effective manner. This primary goal of this system is that the panels are regularly cleaned, preventing the power loss due to dust which can be as high as 5–10% of the output power in dry and arid areas [11] (Fig. 17.1).

Fig. 17.1 Dust accumulation on solar panel installed at a rooftop

For the efficient and cost-effective operation of solar plants, especially in dusty areas, it is imperative to clean the solar panels once or twice daily. Manual cleaning under these conditions would be otherwise very labour consuming and impractical [12]. Hence, the development of an automatic panel cleaning system becomes a necessity, offering a solution to mitigate power losses due to soiling. Different types of cleaning techniques used in various climates across the globe are listed in Table 17.1.

Table 17.1 Cleaning methods in various climatic regions

Climate	Cleaning techniques
Dry	Sprinkler [13], Electrostatic biasing [14], Autonomous cleaning
Desert	Surface vibrating [15], and Aerodynamic Streamlining
Cold, Moist or Humid	Anti-reflective coatings [16], Autonomous or Robotic cleaning [17], and Sprinkler
Arid, Hot, or Sunny	Sprinkler, Autonomous or Robotic cleaning and Electrostatic biasing
Shaded or Cloudy	Anti-reflective coating, Sprinkler, Robotic cleaning, Aerodynamic Streamlining
Snowy	Stowing or Inverting [18], Anti-reflective coatings

17.2 Materials and Methods

The power ratings of the major components used in the model is listed in Table 17.2. The purpose and a brief information about the components is discussed below:

1. **Transformer 220/12 V**: Reduces mains voltage (220 V) to a safer level (12 V) for working of the DC motor and other electronic components.
2. **LM358 Comparator IC**: Compares the voltage signal from the two LDRs and gives outputs a high or low signal based on the comparison. The data is then sent to the Arduino to turn on the DC motor for alignment of the panel.
3. **Arduino NANO**: The microcontroller board, which is the brain of the project. It reads sensor data, control motors, and implement the tracking and cleaning logic.
4. **DC Stepper Motor**: Provides precise angular control, being used for the solar panel tracking mechanism.
5. **Ultrasonic sensor**: Detects the amount of dust on the panel; sends the signal to Arduino for further processing.
6. **Servo Motor:** Provides rotational power, used for driving the wiper mechanism.
7. **LDR (Light Dependent Resistor):** Detects light intensity, used to track the Sun's position.
8. **H-bridge with BC547 & BC557:** Controls the direction and speed of the DC Stepper motor.

The solar tracker system is designed for the solar panel to follow sun rays, thereby maximizing energy generation. The system is controlled by the Arduino NANO microcontroller, which receives input from various sensors and controls the movement of the solar panels.

To track the sun's position, the system employs two LDRs (Light Dependent Resistors) which continuously send signals to the comparator IC. The comparator IC then passes the information to the Arduino, which sends pulsed signals to the driver module. The driver module adjusts the orientation of the panels by rotating the stepper motor in small steps until there is no difference in the signal received from the LDRs.

Table 17.2 Power ratings of major components

Name of the components	Specifications and ratings
Solar panel	20 Watts; 12 V
Arduino microcontroller	5 V DC; 0.5W
Light dependent resistor	3-5 V DC
Ultrasonic sensor	4 pins; 5 V; 40 kHz
Comparator IC	5 V
Servo motor	5 V
DC motor	12 V DC

Fig. 17.2 Demonstrative model of the proposed system

The system utilizes a combination of electronic components to facilitate the movement of the solar panels. These components include the LM358 comparator IC, an H-bridge with BC547 & BC557 transistors, and a DC motor. The LM358 comparator IC compares the input signals from the two LDRs mounted on the panel, and provides signal to the Arduino for controlling the direction and speed of the DC motor. The H-bridge circuit, along with the PC817 IC driver module, enables bi-directional control of the DC motor, allowing precise adjustment of the solar panel orientation to track the sun's movement throughout the day (Fig. 17.2).

In addition to solar tracking, the system also incorporates a cleaning mechanism to maintain optimal efficiency. The ultrasonic sensor detects any dust or debris which may be present on the panel surface and triggers a servo motor connected to the wiper cleaning mechanism by sending signal to the Arduino, which interprets it and activates the servo motor to move the cleaning wipers across the surface of the solar panels, effectively removing any dust or debris accumulated.

The servo motor is controlled by the Arduino which actuates the cleaning wipers to sweep across the entire surface area of the solar panels. The movement of the servo motor is synchronized with feedback from the ultrasonic sensor to optimize the cleaning process and minimize energy consumption.

17.3 Discussion and Results

Following were the objectives of this project:

- To follow the position of sun with time and align the panel perpendicularly to the sun rays.
- To detect the presence of dirt on the solar panel using the ultrasonic sensor.
- To clean the panel by wiping out the dust using the wiper mechanism.

To track the position of Sun and align the solar panel perpendicularly to the sun rays, a solar tracker mechanism is set up. The system utilizes sensors, a microcontroller (Arduino NANO), and a mechanical arm to adjust orientation of the panel. The microcontroller calculates the amount of rotation needed for the panel to align perpendicularly to the sun and then sends signals to the driver module for precise adjustments.

Secondly, to sense dust on the solar panel, an ultrasonic sensor is employed. The ultrasonic waves are emitted by the trigger pin and are received by the echo pin when the waves gets reflected by the dust particles. When a certain amount of dust accumulates, the sensor sends a signal to the Arduino indicating the need for cleaning.

The servo motor controls the movement of the cleaning wipers across the panel's surface, effectively removing accumulated dust and debris. The cleaning process is automated and ensures that the solar panel remains clean for optimal efficiency (Table 17.3).

The following table (Table 17.4) shows the data from the Solar panel when it is operated without any tracker or cleaning mechanism. The data points of voltage generated, current, generated power output, and efficiency of the panel per square meter is tabulated here.

The graph below (Fig. 17.3), shows that the efficiency of the panel without a cleaning system gradually decreases over time. The efficiency which was 17% on 01–08-2023 decreases constantly over time due to the accumulation of dust or debris on the surface of the panel when a cleaning mechanism is not present.

The second case, when we employ the solar tracker mechanism and the wiper cleaning system, the difference in the power generated by the panel becomes evident.

Table 17.3 Comparison between existing and proposed method

Existing methods	Proposed method
Existing systems generally don't have integrated system for both tracking and cleaning	The proposed method integrates tracker and cleaner to increase efficiency
The systems in cited papers use sophisticated and bulky robotic systems which is heavy and require high maintenance	The systems in cited papers use sophisticated and bulky robotic systems which is heavy and require high maintenance
Most of the existing cleaning systems installed these days are too expensive to sustain	The proposed system is made using easily available components. Hence, it is economical

Table 17.4 Efficiency of panel without cleaning system

S. No	DATE	Voltage (V)	Current (Ampere)	Power (Watts)	Efficiency per m^2
1	01-08-23	12.00	1.25	15.08	17.24
2	02-08-23	11.98	1.25	15.06	17.22
3	03-08-23	11.99	1.24	14.88	17.01
4	04-08-23	12.00	1.22	14.64	16.74
5	05-08-23	12.00	1.20	14.41	16.47
6	06-08-23	11.98	1.21	14.55	16.63
7	07-08-23	11.97	1.20	14.39	16.45
8	08-08-23	11.99	1.17	14.07	16.08
9	09-08-23	12.00	1.16	14.02	16.03
10	10-08-23	11.95	1.16	13.65	15.48
11	11-08-23	11.99	1.09	13.07	14.94
12	12-08-23	12.00	1.09	13.08	14.95
13	13-08-23	11.98	1.07	12.87	14.71
14	14-08-23	11.99	1.06	12.83	14.67
15	15-08-23	12.00	1.05	12.69	14.51

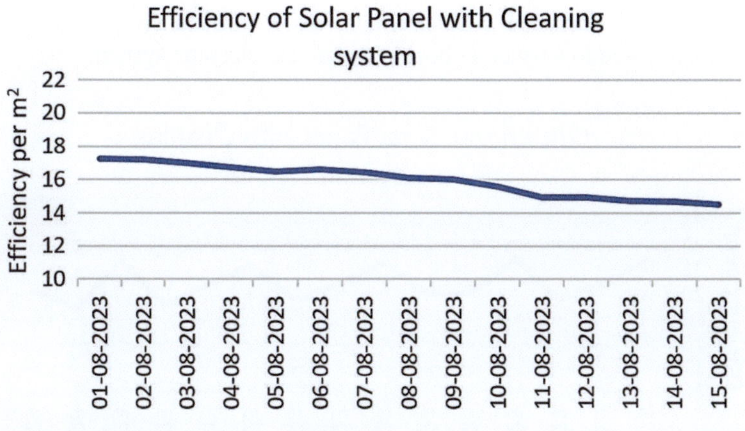

Fig. 17.3 Efficiency of solar panel without cleaning system

The data to support this claim is tabulated below (Table 17.5), where it can be seen that the power generated and the efficiency of the panel, both increases considerably.

From the graph in Fig. 17.4, it may be observed that the solar panel's efficiency increases if the panel is cleaned. However, mild fluctuations can be seen in the graph. These fluctuations are attributed to our method of demonstrating the cleaning system which when turned on, can further increase the efficiency of the panel. This is evident from the graph, which shows that every alternate day, when the cleaning system was turned on, the efficiency values increased. This indicates that the cleaning system

Table 17.5 Efficiency of solar panel with cleaning system

S. No	Date	Voltage (V)	Current (Ampere)	Power (Watts)	Efficiency per m^2
1	01-09-23	11.98	1.56	18.75	21.13
2	02-09-23	11.99	1.49	17.93	20.24
3	03-09-23	11.98	1.44	17.39	19.78
4	04-09-23	12.00	1.56	18.82	21.51
5	05-09-23	11.95	1.46	17.52	20.03
6	06-09-23	11.98	1.46	17.57	20.08
7	07-09-23	11.97	1.55	18.65	21.32
8	08-09-23	11.98	1.49	17.93	20.33
9	09-09-23	12.00	1.44	17.28	19.75
10	10-09-23	11.95	1.55	18.58	21.24
11	11-09-23	11.99	1.48	17.84	20.39
12	12-09-23	12.00	1.43	17.16	19.62
13	13-09-23	11.98	1.56	18.73	21.40
14	14-09-23	11.99	1.44	17.28	19.46
15	15-09-23	11.95	1.47	17.69	20.15

removes dust from the panel's surface, allowing it to maintain a higher level of efficiency compared to a panel without a tracker and cleaning system.

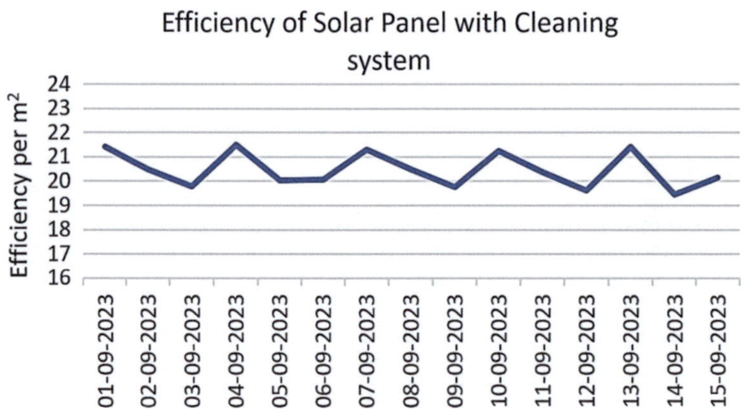

Fig. 17.4 Efficiency of solar panel with cleaning system

17.4 Conclusion

This paper discusses integrated solar tracker and cleaning system using wiper mechanism. The prime focus of this system is to increase the day to day efficiency of the solar panel so that maximum possible electricity can be generated. The tracker keeps track of the position of the sun and alignment of the panel throughout the day to align the panel perpendicular to the sun to maximize the amount of sunlight falling on the panel and the cleaning mechanism by wiper removes the dust or debris accumulated on the panel.

The construction of this system is kept simple and inexpensive. The use of dry wiper-based cleaning ensures no wastage of water. It is a very suitable cleaning method for a country like India where water is getting scarcer by each passing day.

The two graphs (shown in Figs. 17.3 and 17.4) are clearly indicative of two things: first, the efficiency has increased from an average of 15% without the cleaner and tracker to 20% when the proposed method is installed. Secondly, the fluctuations in the graph equipped with the proposed system indicate that the cleaning system can further increase the efficiency of the panel by removing any dust on the panel.

In future, the solar tracker system can be integrated with a number of cleaning methods depending upon the geography and climate of the area. Initially, this system was designed for domestic use. But given the simplicity and cost-effectiveness of the system, this design can be scaled up for implementation in a large solar park where solar energy generation can become even more sustainable.

References

1. M. Höök, X. Tang, Depletion of fossil fuels and anthropogenic climate change—a review. Energy Policy **52**, 797–809 (2013)
2. B.R. Singh, O. Singh O, Trends of fossil fuel reserves and climate change in the 21st century, in *Fossil Fuel and the Environment*, ed by K. Shahriar (InTech, 2012), pp. 167–192
3. M. Hoel, S. Kverndokk, Depletion of fossil fuels and the impacts of global warmin. Resour. Energy Econ. **18**(2), 115–136 (1996)
4. S. Virtanen, Self-cleaning technologies on solar modules and their effect on performance. JAMK Univ. Appl. Sci. 198–874 (2017)
5. S. Ghosh, J.N. Roy, C. Chakraborty, A model to determine soiling, shading and thermal losses from PV yield data. Clean Energy **6**(2), 372–391 (2022)
6. A. Kalra, S. Srivastava, S.P. Singh, S.G. Mohite, Automatic rain sensing wiper. IJCRT (2020)
7. A. Hussain, A. batra, R. Pachauri, An experimental study on effect of dust on power loss in solar photovoltaic module, Springer Open (2017)
8. N. Najmi, A. Rachid, A review on solar panel cleaning systems and techniques. Energies **16**(24), 7960 (2023)
9. R. Nazar, Improvement of efficiency of solar panel using different methods. IJEEE **07**(01) (2015)
10. M. Soni, D. Chauhan, Analysis of solar tracker system to increase the efficiency of solar panel: a study. IRE J **1**(9), ISSN: 2456–8880 IRE 1700235 (2018)
11. R. Innico, Karthish, A. Durai, M. Poovalingam, Solar panel cleaning using wiper. EPRA Int. J. Multidiscip. Res. (IJMR)-Peer Rev. J. **9**(2) (2023)

12. S. Alghamdi, A.S. Bahaj, L.S. Blunden, Y. Wu, Dust removal from solar PV modules by automated cleaning systems. Energies **12**, 1–21 (2019)
13. N. Sugiartha et al., Preliminary design and test of a water spray solar panel cleaning system. J. Physics: Conf. Series **1450**, 012108 (2020)
14. H. Kawamoto, Electrostatic cleaning equipment for dust removal from soiled solar panels. J. Electrostat. **98**, 11–16 (2019)
15. R. Shenouda, Abd-Elhady, A. R. Ali, H. A. Kandil, Influence of vibration time on dust accumulation on PV panels that operate light posts. Energy Rep. **8**, 309–318 (2022)
16. J. J. John, et al. (eds.), *Novel PV module cleaning system using ambient moisture and self-cleaning coating*, in Photovoltaic Specialists Conference (PVSC), 39th IEEE, (2013)
17. B. Parrott, P.C. Zanini, A. Shehri, K. Kotsovos, I. Gereige, Automated, robotic dry-cleaning of solar panels in Thuwal, Saudi Arabia using a silicone rubber brush. Sol. Energy **171**, 526–533 (2018)
18. S. Mondal, A.K. Mondal, A. Sharma, V. Devalla, An overview of cleaning and prevention processes for enhancing efficiency of solar photovoltaic panels. Curr. Sci. **115**(6), 1065–1077 (2018)

Chapter 18
A Review on Reliability and Maintenance of Photovoltaic Thermal Systems

Ravi Kumar, Deepali Atheaya, and Anil Kumar

Abstract Power generation through photovoltaic thermal (PVT) systems is an effective technology to solve the global energy scarcity problem. Therefore, reliability and maintenance of PVT systems is of utmost importance as it ensures the proper functioning of these systems. The various benefits of using photovoltaic thermal systems includes generation of electrical power, thermal energy, increased system efficiency and reduction in CO_2 emissions. This work aims to review the reliability and maintenance analysis of works done by various researchers. It has been observed that the evaluation of reliability and maintenance of renewable energy systems is a significant means to achieve environment Sustainability. The reliability of solar photo voltaic thermal (PVT) systems are calculated on various environmental conditions like high temperature, UV radiation, outdoor exposure etc. And different methods like fault tree analysis, finite element method and probabilistic technique are utilized for the computation of system reliability.

Keywords PVT system · Reliability · Thermal energy · Efficiency · Sustainability · Durability

---

R. Kumar · D. Atheaya (✉)
Department of Mechanical Engineering, Bennett University, Plot Nos 8, 11, TechZone, Greater Noida, Uttar Pradesh 201310, India
e-mail: deepali.atheaya@bennett.edu.in

A. Kumar
Department of Mechanical Engineering, Delhi Technological University, Delhi 110042, India

Clean Energy: Nodal Centre of Excellence in Energy Transition (NCEET), Delhi Technological University, Delhi 110042, India

A. Kumar et al. (eds.), *Selected Proceedings of the 1st International Conference on Advanced Materials for Sustainable Innovation; IC-AMSI 2024; 28–30 August; New Delhi; India*, Springer Proceedings in Energy,
https://doi.org/10.1007/978-981-96-4492-6_18

18.1 Introduction

Recent rapid population expansion, which has increased the price of non-renewable energy sources, has led to a major increase in the consumer appetite for solar energy. With an expanding population and improvements in nations' economic standing with regard to income growth, the use of primary sources are running out quickly, including fossil fuels and other non-renewable resources. In addition to negatively impacting the ecosystem, excessive use of primary resources has caused severe weather variations, including melting glaciers, global warming, and ozone layer degradation [1]. Consequently, it is believed that employing photovoltaic (PV) cells to capture solar energy can assist in addressing environmental challenges and climatic circumstances, particularly in areas with high levels of solar radiation. PV technology is an environmentally friendly and widely accessible answer to the issues related to the usage of fossil fuels [2]. Solar photovoltaic systems convert sunlight which falls into the system into electrical energy with efficiency levels between 5 and 25 percent, indicating that a major portion of sunlight is also transformed into heat energy. Approximately 80% of the residual solar energy is either radiated onto the surrounding area or absorbed by photovoltaic cells [3]. As lot of solar energy is wasted, therefore it is necessary to utilize the solar energy. The two main types of solar energy utilization are solar heat energy, which transforms solar radiation into thermal energy, and photovoltaic energy, which transforms solar radiation into electrical energy. PV-thermal collectors are modules that extract heat in a variety of ways and are utilized in various thermal collectors. For both residential and commercial heating, a liquid or gas is heated in a thermal collector and then let to flow close to a circuit via heat transfer [4]. PVT stands for photovoltaic thermal systems, which use absorber plate-integrated PV cells to concurrently generate electrical and thermal energy. PVT technology raises the electrical conversion efficiency of photovoltaic modules, which increases their energy efficiency since heat produced in the solar panels is recovered as low-temperature heat radiation. Figure 18.1 gives the idea of how photovoltaic thermal systems (PVT) works in the presence of solar energy in which solar energy is converted into heat and electrical energy by the solar PVT system. Throughout the years, a great deal of theoretical and experimental research was conducted, but the majority of the studies described the use of air and water as the two primary collectors for extracting heat from PV modules [5]. There are questions about what defines failure or inadequate performance for the majority of PVT equipment. Notably, maintaining safety as one ages is essential to safeguarding both people and property. PVT modules that can operate in the field for at least 25 years are required by the solar sector, as is a trustworthy method for assessing their viability. These are essential for important safety issues as well as for product development, warranty considerations, risk to financial stakeholders, and overall economic sustainability. Therefore, reliability study of PVT systems needs to be done to achieve above concerns. This review paper presented the research on PVT system reliability, which has a major impact on a PVT system's entire lifespan.

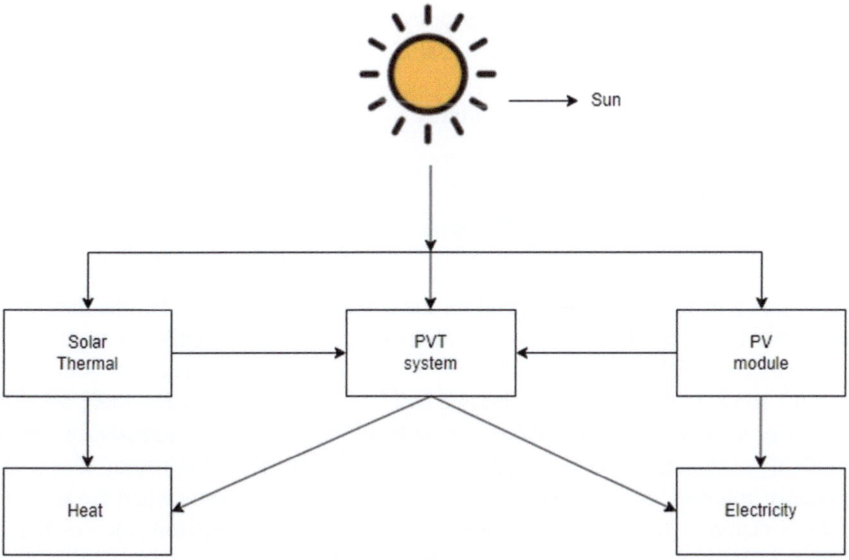

Fig. 18.1 Idea of photo voltaic thermal (PVT) system

18.2 Reliability

The probability that a component or system will perform a required function for a given period of time when used under stated operating conditions" is the definition of reliability in engineering. In practically all branches of engineering, measurement science, and qualitative research, the word "reliability" is used extensively. It denotes "repeatability" or "consistency" in measurement science. When the same measurement is made repeatedly on the same part feature using the same equipment and operator, it is said to be reliable. Part failure data is used to calculate part dependability through statistical techniques. System reliability is calculated using probability approaches based on the knowledge of system structure and part reliability. The exponential, Weibull and gamma distributions are among the often employed distributions. Equations pertaining to an exponential reliability distribution are as follows [6]:

$$\mathrm{R(t)} = \left(e^{\lambda t}\right) \tag{18.1}$$

$$\lambda = \left(\frac{1}{MTBF}\right) \tag{18.2}$$

$$\lambda_s = \lambda_1 + \lambda_2 + \lambda_3 + \lambda_4 \tag{18.3}$$

where, R(t) = Reliability at any given time t

λ = Failure rate in failures per hour or cycle.

λ_s = System failure rate, and individual λ's in Eq. (18.3) are failure rates of parts in series.

MTBF = Mean Time Between Failures.

18.2.1 Reliability of PVT System

Reliability considerations are crucial for the advancement of PVT, but the literature has given them very little attention. Furthermore, there are currently no established standards for PVT reliability testing. While most pertinent features are addressed by the testing standards for solar thermal collectors and PV laminates, there are still several particular concerns that are unique to PVT that are not addressed by these standards. The stagnation temperature in a traditional thermal collector system is probably going to happen a few times a year during certain hours when the storage is fully charged. In these situations, the heat is not removed from the collector, leading to elevated module temperatures. In a traditional thermal collector, stagnation temperatures typically range from 180 to 200 °C. A heated stagnant collector may receive the introduction of cool collecting fluid upon activation of the PVT-system pump. Large temperature variations are thus produced across the PV laminate. It ought to be impervious to damage or breaking due to the accompanying heat stresses. Enough electrical insulation between the PV and the metal absorber is necessary to prevent safety hazards [7]. Regarding further reliability-related concerns, like hot spots, thermal behaviour of collectors, UV radiation, etc. Still, these are critical concerns for PVT's continued development [8]. Durability refers to PV modules' slow power loss that stays within the warranty's restrictions, whereas reliability refers to significant power degradation that happens after the guarantee has expired and is considered a hard failure. If a photovoltaic module exhibits a high likelihood of efficiently performing its intended function for a period of 25–30 years, given the specific operating conditions to which it is exposed, it is deemed reliable [9]. Several works are found in the literature on reliability of PVT systems. Some of them are discussed below.

Föste et al. [10] found efficiency of the prototype flat plate collector was enhanced by the implementation of gas-filled low-e glazing. A reliable collector that is capable of enduring high temperatures, UV radiation, and outdoor exposure was developed. Prototype measurements for efficiency and reliability assessment are conducted. Test methodologies are devised for evaluating the temperature and UV-resistance of glazing materials. Gas filling levels are measured to assess the gas tightness of sealing materials. The stability in the face of UV radiation and thermo-mechanical stresses is investigated. High-transmittance low-emissivity coatings were formulated for mass production. The reliability of the glazing of the collector was effectively assessed under extreme heat and ultraviolet radiation. Following a year of outdoor exposure, the prototype collectors demonstrated a reduction in efficiency of 1.2%.

The recommended operational temperature range for the collector is between 80 °C and 110 °C.

Maghzian et al. [11] developed a solar heating system for an educational building in Tehran entails a focus on ensuring reliability. The assessment of reliability is conducted through the utilization of RAM Commander software to evaluate the system's performance. To enhance reliability, a parallel auxiliary system is incorporated, which consists of collectors and a boiler. The system's reliability is guaranteed through the utilization of a parallel boiler system, resulting in an impressive reliability rate of 99.9922%.

Stember [12] studied on the reliability assessment of solar photovoltaic systems in the absence of comprehensive data is presented. The paper delves into methodologies for evaluating the reliability and availability of both existing and upcoming solar PV systems. Failure modes and effects analyses (FMEA) are employed to conduct subsystem analysis. Fault tree analysis is utilized for predicting system failures and calculating probabilities. Techniques related to reliability and availability are applied in the design of photovoltaic systems. Early analysis of reliability contributes to the enhancement of system design and provides guidance for maintenance. The incorporation of redundancy is instrumental in improving the reliability of solar photovoltaic systems.

Park et al. [13] carried out research investigation to assess the reliability of Solar PV/T systems in light of various environmental conditions. The reliability experiments revealed a notable impact of hydraulic pressure on the lifespan of the system. The analysis of reliability yielded a shape parameter value of 5.6658. Reliability evaluation encompasses examinations of performance, environmental impact, safety considerations, and longevity assessments. The hydraulic pressure examination is iteratively carried out on the PV/T framework.

Zini et al. [14] found the reliability of large-scale grid-connected photovoltaic systems is evaluated through the application of fault tree analysis and probabilistic methodology. The methodology is based on FTA and utilizes an exponential distribution to model fault probability density functions, with a focus on understanding the impact of individual components on system reliability. The study highlights how crucial it is to take repairable failures and component deterioration into account in subsequent research in order to evaluate energy and financial losses and enhance PV efficiency in real-world scenarios. Numerous investigations have looked into the dependability of various PV system components, such inverters and battery voltage regulators, employing various techniques and field research to comprehend failure reasons and maintenance needs.

Rohouma et al. [15] paper examines the reliability of PV systems by analyzing various configurations for their performance. Analysis of AC bus versus DC bus system configurations is undertaken in this study. The impact of inverter configuration on the performance of PV systems is quantitatively measured. An assessment of the reliability of various system configurations is also conducted. Module integrated inverters provide increased reliability and decreased cable losses. The influence of shading is minimized when utilizing module integrated inverters. The enhanced modularity facilitates seamless incorporation of additional panels.

Tshemese et al. [16] examination focuses on the technologies, reliability, and long-term performance of photovoltaic systems. It underscores the importance of enhancing methodologies for forecasting photovoltaic system performance. Accelerated lifetime testing is utilized to forecast the performance of photovoltaic (PV) modules. Analysis of performance and degradation of PV modules is conducted. Photovoltaic technology plays a crucial role in mitigating the impacts of global warming. Current assessments of the reliability of photovoltaic modules lack adequacy in forecasting long-term performance. Enhanced approaches are essential to effectively replicate genuine operating conditions of the system.

Al-Habahbeh et al. [17] in the paper reliability of the solar concentrator receiver was evaluated through the utilization of Finite Element Method (FEM) and probabilistic techniques. The reliability of Concentrated Solar Power (CSP) receiver is assessed through Finite Element Method (FEM) and Latin Hypercube (LHC) sampling techniques. Probabilistic approaches are employed to forecast the Probability Density Function (PDF) of the operational lifespan, taking into account the effects of transient thermal stresses. The tube absorber demonstrates resilience to elevated temperatures.

18.3 Conclusion

Despite the recent and increasing interest in the subject of hybrid Photovoltaic-Thermal systems, no specific approach for PV-T collector characterization has yet been defined. There are tests that must be performed in order to determine reliability, such as mechanical loads and long-term exposure. To account for the PV-T requirements, certain tests, such as thermal cycling and insulating stability against voltage, must be improved.

Photo voltaic thermal system life prediction is a developing field that frequently needs methods and data that are not readily available. With the exception of real-time 25-year testing, no test program can 100% predict that a module will function correctly in an environment for more than 25 years, but an enhanced methodology that more closely resembles the real environment while maintaining a reasonable acceleration can undoubtedly be a useful tool to address PVT system reliability.

References

1. R. Muhammad Qamar, M. Nadarajah, C. Ekanayake, On recent advances in PV output power forecast. Sol. Energy **136**, 125–144 (2016). https://doi.org/10.1016/j.solener.2016.06.073
2. A. Deepali, et al., Analytical characteristic equation for partially covered photovoltaic thermal (PVT) compound parabolic concentrator (CPC). Sol. Energy **111**, 176–185 (2015). https://doi.org/10.1016/j.solener.2014.10.025
3. M.C. Browne, B. Norton, S.J. McCormack, Heat retention of a photovoltaic/thermal collector with PCM. Sol. Energy **133**, 533–548 (2016). https://doi.org/10.1016/j.solener.2016.04.024

4. S.S. Joshi, A.S. Dhoble, Photovoltaic-thermal systems (PVT): technology review and future trends. Renew. Sustain. Energy Rev. **92**, 848–882 (2018). https://doi.org/10.1016/j.rser.2018.04.067
5. S. Ashish, D. Atheaya, A. Kumar, Study of hybrid photovoltaic thermal systems, in *IOP Conference Series: Materials Science and Engineering*, vol. 748. No. 1. IOP Publishing (2020)
6. R.V. Narang, Introducing reliability and maintainability in engineering and technology, in *2012 ASEE Annual Conference & Exposition* (2012)
7. H.A. Zondag, Flat-plate PV-thermal collectors and systems: a review. Renew. Sustain. Energy Rev. **12**(4), 891–959 (2008). https://doi.org/10.1016/j.rser.2005.12.012
8. G. Abdülkerim, Reliability and ecological aspects of photovoltaic modules. IntechOpen (2020)
9. P. Rong, J. Kuitche, G. Tamizhmani, Degradation analysis of solar photovoltaic modules: Influence of environmental factor, in *2011 Proceedings-Annual Reliability and Maintainability Symposium*. IEEE (2011). https://doi.org/10.1109/RAMS.2011.5754514
10. S. Föste, et al. Performance and reliability of a high efficiency flat plate collector–final results on prototypes. Energy Procedia **48**, 48–57 (2014)
11. A. Maghzian, et al., Reliability assessment and design of a solar heating system for an educational building in Tehran. People 24772.20727: 0
12. L.H. Stember, Reliability considerations in the design of solar photovoltaic power systems. Solar Cells **3**(3), 269–285 (1981). https://doi.org/10.1016/0379-6787(81)90008-9
13. T.-K. Park, et al., A study on the reliability assessment of solar photovoltaic and thermal collector system. New & Renew. Energy **16**(4), 49–64 (2020). https://doi.org/10.7849/ksnre.2020.0021
14. G. Zini, C. Mangeant, J. Merten, Reliability of large-scale grid-connected photovoltaic systems. Renew. Energy **36**(9), 2334–2340 (2011). https://doi.org/10.1016/j.renene.2011.01.036
15. W.M. Rohouma, I.M. Molokhia, A.H. Esuri, Comparative study of different PV modules configuration reliability. Desalination **209**(1–3), 122–128 (2007). https://doi.org/10.1016/j.desal.2007.04.020
16. Z. Tshemese, et al., Reliability study of solar photovoltaic systems for long-term use, in *Electrode Materials for Energy Storage and Conversion*. CRC Press, 309–320 (2021)
17. O.M. Al-Habahbeh, R.A. Alomosh, D.K. Aldebie, Reliability simulation of solar concentrator receiver. Int. J. Sustain. Energ. **35**(8), 793–801 (2016). https://doi.org/10.1080/14786451.2014.950963

Chapter 19
Energy Saving Enhanced Brain Tumor Classification Accuracy in Intelligent Systems Using Deep Convolutional Neural Network

Abhimanu Singh and Smita Jain

Abstract Brain tumors are among the most fatal diseases. If brain tumors are diagnosed in the early stage, then an effective treatment procedure can be planned otherwise the chances of survival are very few. Manual diagnosis of brain tumor is not possible as the location of brain tumor is within the skull. Thus, for the diagnosis purpose, an accurate and non-invasive method is needed. For correct diagnosis of a brain tumor, a machine with high accuracy is desired. Computerized diagnosis systems are proving to be the most efficient in detection and classification tasks. Advancements in deep learning have revolutionized the medical imaging related research. Deep learning methods have outperformed the other traditional methods. Continuous efforts are being made by the researchers to enhance the detection accuracy. This work proposes an architecture for enhanced brain tumor detection accuracy. Ours proposed architecture has an accuracy of 98.85% on the test dataset, which is better than the other methods stated in the literature survey. Our proposed architecture has a very low complexity as compared with other architectures which saves the time as well as the computational efforts, and hence saves energy.

A. Singh (✉)
Department of Applied Mathematics, Bhagwan Parshuram Institute of Technology, Rohini, Delhi, India
e-mail: asingh19669@gmail.com

Research Scholar, Department of Mathematics, JECRC University, Vidhani, Jaipur, Rajasthan, India

S. Jain
Department of Mathematics, JECRC University, Vidhani, Jaipur, Rajasthan, India

A. Kumar et al. (eds.), *Selected Proceedings of the 1st International Conference on Advanced Materials for Sustainable Innovation; IC-AMSI 2024; 28–30 August; New Delhi; India*, Springer Proceedings in Energy,
https://doi.org/10.1007/978-981-96-4492-6_19

19.1 Introduction

Brain tumors pose significant challenges to healthcare professionals due to their complex nature and life-threatening consequences. Timely accurate diagnosis of a brain tumor facilitates effective treatment.

19.1.1 Background

MRI modality, is most effective modality for detection and classification of brain tumors. However, the interpretation of these images often requires extensive expertise and can be subjective, leading to variability in diagnosis process. Automated, non-invasive detection methods using deep learning algorithms such as Convolutional Neural Networks (CNNs) are greatly assisting in the process of disease diagnosis.

19.1.2 Significance of the Problem

Despite the advancement in medical imaging and machine learning, accurate classification of brain tumors still remains a challenging task. The diversity of tumor types, variations in size and location, and the presence of noise in imaging data contribute to the complexity of the problem. Inaccurate diagnosis or delayed diagnosis can lead to serious consequences for the survival of patients. It emphasizes the need for robust and reliable classification methods.

19.1.3 Existing Methods

Traditional machine learning techniques, such as feature engineering combined with classifiers, have been widely employed for brain tumor classification. However, these methods often rely on handcrafted features. These methods are not capable of recognizing complex patterns present in medical images. On the other hand, deep learning methods, particularly CNNs, have shown remarkable success in capturing complex features from raw imaging data, leading to superior classification performance.

19.1.4 Purpose of the Study

The aim of this work is to investigate an energy saving (low complexity) novel architecture of deep CNNs for enhanced brain tumor classification accuracy using MRI images.

19.2 Convolutional Neural Network

Convolutional neural network is a special kind of artificial neural networks. In the operation of convolutional, a filter convolves over the input image and extracts important features from the input image. A convolutional neural network consists of three types of layers. Convolution layer assisted by a maxpooling layer extract complex features directly from an input image. Important features are retained and the less important features are filtered out. It helps in reducing the computational efforts required. Third layer, a fully connected layer, performs the task of classification of the input image on the basis of features extracted by the first two layers. In medical image classification, the performance of convolutional neural networks are outstanding.

19.2.1 Convolutional Operation

Mathematical equation for the convolution operation between an input tensor X and a kernel tensor K is:

$$Y[i, j] = (X * K)[i, j] = \sum_{m=0}^{M-1} \sum_{n=0}^{N-1} \sum_{p=0}^{P-1} \sum_{q=0}^{Q-1} X[m + p, n + q].K[M - 1 - p, N - 1 - q] \quad (19.1)$$

where:

- $Y\left[i, j\right]$ is the response at position (i, j) in the output feature map.
- X is input tensor.
- K is kernel tensor (also referred to as the filter).
- M and N are the spatial dimensions of the input tensor X.
- P and Q are the spatial dimensions of the kernel tensor K.

19.2.2 ReLU (Rectified Linear Unit)

It is the activation function which is commonly used in neural networks. Mathematically it is defined as

$$f(x) = \max(0,\ x) \tag{19.2}$$

It returns x, if x is more than zero, and returns zero otherwise. Relu activation function is simple yet effective. It allows the model to learn complex patterns or feature maps in an image by introducing non-linearity.

19.2.3 Sigmoid Function

A sigmoid function performs binary classification tasks. The sigmoid function is defined as:

$$\text{Sigmoid}(z) = \frac{1}{1 + e^{-z}} \tag{19.3}$$

The range of sigmoid function lies between 0 and 1. The sigmoid function has an "S" shaped curve, which is why it's also known as the sigmoid curve.

19.2.4 Confusion Matrix

A confusion matrix is a square matrix which compares the actual values (results) with the predicted values (results). Contents of confusion matrix are used to evaluate the performance metrics of a machine learning model. Order of a confusion matrix is N $\times$ N, if the number of target classes is N. In case of binary classification, the number of target classes is 2, and therefore the corresponding confusion matrix will be of order 2 $\times$ 2. The target classes are Positive and Negative.

True Positive (TP): It is the case where in actual a sample is positive and it is predicted as positive by the Machine Learning model.

False Positive (FP): It is the case where in actual a sample is negative and it is predicted as positive by the Machine Learning model.

False Negative (FN): It is the case where a sample in actual is positive and it is predicted as negative by the Machine Learning model.

True Negative (TN): It is the case where a sample in actual is negative and it is predicted as negative by the Machine Learning model.

19.2.4.1 Accuracy

Accuracy is defined as.

$$\text{Accuracy} = \frac{\text{Number of correct predictions made}}{\text{Total number of predictions made}} = \frac{\text{TP} + \text{TN}}{\text{TP} + \text{FP} + \text{FN} + \text{TN}}$$

It gives a better measure of performance when there is equal number of samples belonging to each target class; otherwise, it gives a false sense of higher accuracy.

19.3 Material and Methods

A. Akter, N. Nosheen, et al., in their research study employed segmentation approach in their deep learning architecture for brain tumor classification and obtained highest accuracy of 97.7% [1]. Ramakrishnan, Sridevi, et al., in the work achieved an accuracy of 96.2% [2]. Deshpande, Estrela et al., in their research study achieved 98.14% accuracy with their architecture [3]. Gupta, Gaurav, et al., in their research for classification of brain tumor and achieved accuracy of 90% [4]. Hari Mohan Rai and Kalyan Chatterjee in their research paper used U-Net for classification of brain M R images into normal and abnormal classes. They performed the classification of their data using Le-Net and VGG16 and their proposed model (U-Net) and reported 98% accuracy [5]. Md. N. Islam, Md. S. Azam, et al., in their research paper obtained an accuracy of 98.82% [6]. Al-Jammas, Al-Sabawi, et al., in their research paper achieved accuracy of 97.4% [7]. Khairandish, M. Sharma, et al. in their research study obtained an overall accuracy of 98.4959% [8]. Patil, and Kirange, in their research paper accuracy to 97.77% [9]. Shanthi, Saradha, et al., in their research paper reported 97.5% accuracy [10]. Rahman, and Md. Islam, in their research paper reported accuracy of 97.33%, 97.60%, and 98.12% on three datasets [11]. Zar Nawab Khan Swati and, Qinghua Zhao, et al. in their research work achieved an average accuracy of 94.82% [12]. Dataset used in this study is publicly available at Kaggle database [13].

19.3.1 Proposed Methodology

In the proposed network architecture, we applied five convolutional layers each with 'relu' activation function, 'same' padding, maxpooling. Dropout is 0.30, Dense layer is 1 with unit 1, and sigmoid activation function. Optimizer used is Adam (epsilon = 0.0001). Loss is binary_crossentropy. Callback used is earlystopping with patience = 20, min_delta = 0.0001. Batch size is 64 and epochs employed = 30. Computation is performed on Google Colab and the code is written in Keras library of TensorFlow framework.

19.3.2 Dataset Used

Kaggle's Dataset Brain_Tumor_Detection_MRI. Training data consists of 2023 MR images of size (128, 128, 3) labelled as Tumor or No Tumor. Test data consists of 868 MR images of size (128, 128, 3) labelled as Tumor or No Tumor.

19.4 Results

The maximum validation accuracy of our model is 98.85%. The accuracy shown by our model is better than other models. Figure 19.1 presents training and validation accuries. Figure 19.2 presents loss of the proposed model on training and validation data. Table 19.1 presents the accuracies of the models stated in the literature review and our proposed model.

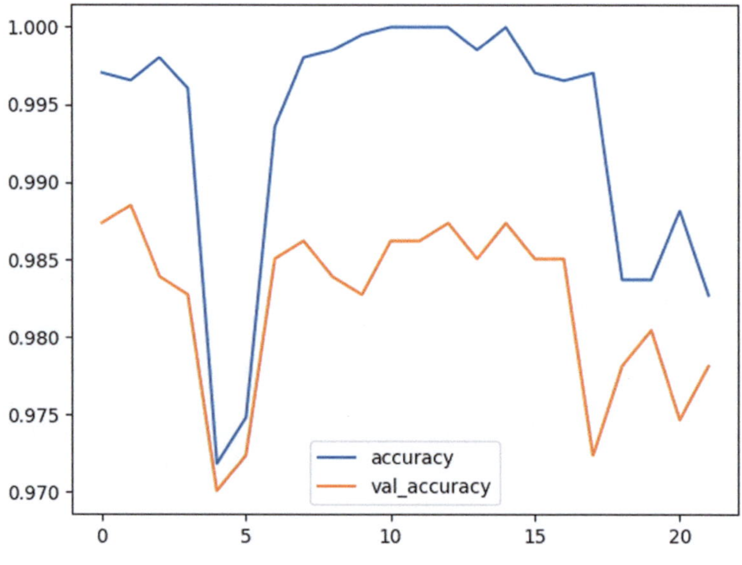

Fig. 19.1 Accuracy versus epochs

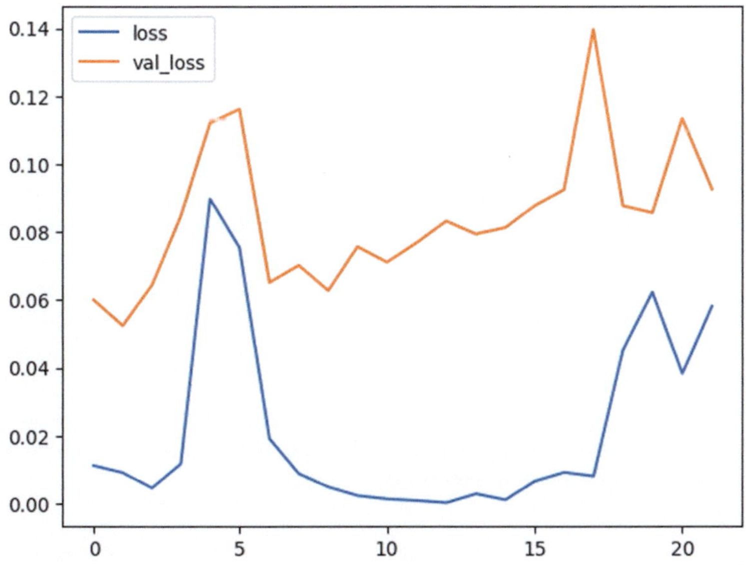

Fig. 19.2 Loss versus epochs

Table 19.1 Comparative accuracy of models

S. No.	Ref	Year	Method	Accuracy (%)
1	A. Akter, N. Nosheen, et al	2024	CNN + U-Net	97.70
2	A. B. Ramakrishnan, M. Sridevi, et al	2024	Hybrid CNN	96.20
3	A. Deshpande, V. V. Estrela et al	2021	DCT-CNN-ResNet50	98.14
4	B. B. Gupta, A. Gaurav, et al	2024	Deep CNN	90.00
5	Hari Mohan Rai and Kalyan Chatterjee	2020	Novel Lu-Net	98.00
6	Md. N. Islam, Md. S. Azam, et al	2024	CNN + Ensemble	97.40
7	M. H. Al-Jammas, E. A. Al-Sabawi, et al	2024	Deep learning	98.82
8	M. O. Khairandish, M. Sharma, et al	2022	CNN-SVM	98.50
9	S. Patil, and D. Kirange	2023	Ensemble DL	97.77
10	S. Shanthi, S. Saradha, et al	2022	Hybrid DNN	97.50
11	T. Rahman, and Md. S. Islam	2023	Parallel Deep CNN	98.12
12	Zar Nawab Khan, Qinghua Zhao, et al	2019	Transfer learning	94.82
13	Our proposed model		Deep CNN	98.85

19.5 Conclusion

Objective of this study was to develop an energy saving deep CNN architecture with enhanced accuracy of classification of brain tumors. Our proposed model achieved a high classification accuracy with very low complexity (saving energy), on the validation data, outperforming existing methods and demonstrating significant potential for clinical application. The implications of these findings are substantial for the medical community, as the deployment of such a model could enhance diagnostic accuracy, aiding radiologists in accurate identification of brain tumors, leading to more effective treatment plan and improved patient outcomes. However, this study is not without limitations. The dataset, while comprehensive, may not encompass all the variability present in real-world scenarios. Additionally, the performance of model on rare tumor types and its ability to generalize across different MRI machines and settings were not thoroughly evaluated. Addressing these limitations will be crucial for the practical application of the proposed model.

References

1. A. Akter, N. Nosheen et al., Robust clinical applicable CNN and U-Net based algorithm for MRI classification and segmentation for brain tumor, expert systems with applications. Elsevier Ltd **238**, 122347 (2024). https://doi.org/10.1016/j.eswa.2023.122347
2. A.B. Ramakrishnan, M. Sridevi, et al., Optimizing brain tumor classification with hybrid CNN architecture: Balancing accuracy and efficiency through one API optimization. Inform. Med. Unlocked, Elsevier Ltd., 44101436 (2024). https://doi.org/10.1016/j.imu.2023.101436
3. A. Deshpande, V.V. Estrela et al., The DCT-CNN-ResNet50 architecture to classify brain tumors with super-resolution, convolutional neural network, and the ResNet50. Neurosci. Inform. , Elsevier, Masson, SAS (2021). https://doi.org/10.1016/j.neuri.2021.100013
4. B.B. Gupta, A. Gaurav, et al., Deep CNN based brain tumor detection in intelligent systems. Int. J. Intell. Netw., Elsevier B.V. **5**, 30–37 (2024). https://doi.org/10.1016/j.ijin.2023.12.001
5. H.M. Rai, K. Chatterjee, Detection of brain abnormality by a novel Lu-Net deep neural CNN model from MR images. Mach. Learn. Appl., Elsevier Ltd. (2020). https://doi.org/10.1016/j.mlwa.2020.100004
6. Md.N. Islam, Md.S. Azam et al., An improved deep learning-based hybrid model with ensemble techniques for brain tumor detection from MRI image. Inform. Med. Unlocked, Elsevier Ltd. **47**, 101483 (2024). https://doi.org/10.1016/j.imu.2024.101483
7. M.H. Al-Jammas, E.A. Al-Sabawi et al., Brain tumors recognition based on deep learning, e-Prime—advances in electrical engineering. Electron. Energy, Elsevier Ltd. (2024). https://doi.org/10.1016/j.prime.2024.100500
8. M.O. Khairandish, M. Sharma, et al., A Hybrid CNN-SVM threshold segmentation approach for tumor detection and classification of MRI brain images, IRBM, Elsevier, Masson, France 290–299 (2022). https://doi.org/10.1016/j.irbm.2021.06.003
9. S. Patil, D. Kirange, Ensemble of Deep Learning Models for Brain Tumor Detection. Int. Conf. Mach. Learn. Data Eng., Procedia Comput. Sci. Elsevier B.V. **218**, 2468–2479 (2023). https://doi.org/10.1016/j.procs.2023.01.222
10. S. Shanthi, S. Saradha, et al., An efficient automatic brain tumor classification using optimized hybrid deep neural network. Int. J. Intell. Netw., Elsevier B.V. **2**(2022), 188–196 (2022). https://doi.org/10.1016/j.ijin.2022.11.003

11. T. Rahman, Md. S. Islam, MRI brain tumor detection and classification using parallel deep convolutional neural networks. Meas.: Sens., Elsevier Ltd. **26** (2023). https://doi.org/10.1016/j.measen.2023.100694
12. Z.N.K. Swati, Q. Zhao, et al., Brain tumor classification for MR images using transfer learning and fine-tuning. Comput. Med. Imaging Graph., Elsevier Ltd., 34–46 (2019). https://doi.org/10.1016/j.compmedimag.2019.05.001
13. [Data] A. Panigrahi, Brain_Tumor_Detection_MRI, Kaggle (2021). https://www.kaggle.com/abhranta

Chapter 20
Simulation and Modelling of Nano Grid System Using Solar PV and Hydro Power Plant

Vandana Sondhiya and Kaustubh Dwivedi

Abstract An effective grid integrated nanogrid model can be able to enhance the use of distributed energy resources. In the presented nano-grid model the idea includes photovoltaic (PV) and micro hydro power (MHP) systems with grid interconnected as distributed energy resources for power plants. The modelling results are examined using HOMER and MATLAB software. With three different options—50 kW, 100 kW, and 150 kW for MHP capacity—varying with the dynamic loads and water supplies, the HOMER imitate a nano-grid model. The nano-grid model with MHP design demonstrated lower energy costs as the grid reached peak load (150 KW), responsible CO_2 emission depletion, and the highest percentage of renewable energy. According to the simulation results, the model nano-grid with the largest MHP capacity (150 kW) generated the least amount of energy at the lowest energy prices, the greatest reduction in carbon emissions is observed, and the best possible use of the electric output from renewable energy power plants.

Graphical Abstract

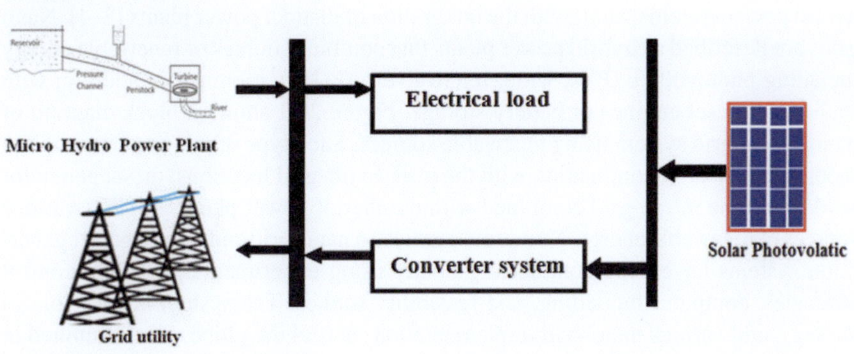

V. Sondhiya (✉) · K. Dwivedi
Department of Electrical and Electronics Engineering, University Institute of Technology(UIT), Bhopal, India
e-mail: vandanasondhiya95@gmail.com

A. Kumar et al. (eds.), *Selected Proceedings of the 1st International Conference on Advanced Materials for Sustainable Innovation; IC-AMSI 2024; 28–30 August; New Delhi; India*, Springer Proceedings in Energy,
https://doi.org/10.1007/978-981-96-4492-6_20

Keywords Nano grid · Photovoltaic · Micro hydro power · MATLAB. HOMER · Carbon emission · Distributed energy resources

20.1 Introduction

Conventional Power plants depends on fossil fuels like coal, natural gas, and oil. With the globally increase in industrialization and transportation oil and natural gas will eventually run out (except coal). Additionally, Carbon overload is caused with increase in burning of fossil fuel to meet the continuously increasing demand, cut down trees, industrialization, burning of forests and globalization. Innovations in renewable energy sources might pave the way for clean, environmentally friendly technology. Coal, oil, gas, and are considered to be the prime energy sources for conventional power plants. The increase in carbon emissions brought on by the application of fossil fuel powered power plants would be impacted by this situation. To reduce carbon emissions produced from fossil fuels, usage of renewable energy sources are need to be increased. The remote renewable energy power plants demands to boost electricity supply due to uneconomical and reliable system. The application of renewable energy fluctuates according on the weather conditions. In contrast, the rate of generated energy for plants that rely only on sources of renewable energy is superior to the cost of electricity sold on the local grid. Such type of condition arrived due to government subsidising the retail price of electricity in the grid system. Additionally, rural areas have lesser load factors for electrical loads [1], therefore employing a stand-alone power system. An effective strategy to maintain the dependability of the energy supply from renewable energy sources is to use hybrid power systems, alongwith the integration of distinct power plants [2–4]. Nano grids are described as hybrid power plants that combine sources of renewable energy including photovoltaic (PV), wind, micro-hydro (MHP), biomass, in addition with traditional diesel engines or battery storage. Figure 20.1 shows a block diagram of nano hybrid grid system using renewable sources. Such type of power plants can run independently or in conjunction with the grid. In off-grid locations, diesel generator works with the micro-grid combined with number of power plants employs various sources of renewable energy. Numerous studies on nano-grid and hybrid power generation systems have been conducted globally using experimental research, control strategies, computer modelling, and feasibility studies. The system layout, control strategy, and various nano-grid implementations across the globe were examined in the previous research [2]. The studies [3–8] looked at the utility of nano-grids that combined micro hydro turbines with photovoltaic energy sources. Previous works have also covered the feasibility and optimization of HOMER based nano-grid/hybrid model, appropriate for remote rustic areas [8–11]. The building industry heavily relies on renewable energy sources and energy-saving measures to reduce conventional energy usage. Based on energy utilization and dynamic loads analysis of load pattern studies has been conducted for residential and commercial buildings. Solar PV and battery systems are recalculated to determine savings in investment if buildings share their production and storage [12]. The popularity of solar PV-based DC Nano-grids in residential and commercial buildings is driven by energy efficiency and environmental friendliness. However, traditional solar photovoltaic systems (SPV) are unreliable due to external factors and module mismatch losses. A bidirectional modular PV battery system (BMPBS) is proposed to address these issues and enhance the reliability of SPV systems. The system uses non-isolated buck and boost converters and a battery storage system for efficient power extraction [13]. A unique adaptive control technique based on the ANOVA Kernel Kalman Filter (AKKF) is used in a grid-integrated solar energy conversion system (SECS) to provide precise harmonic support. The AKKF is hybridized to increase estimation accuracy, and the system functions as a Distribution Static Compensator (DSTATCOM) when solar power is unavailable. Experiments validate the control approach's performance under adverse grid conditions, solar insolation variations, and load imbalance conditions [14]. A steepest descent Laplacian regression (SDLR) based adaptive control technique for optimal operation of a three-phase single-stage grid-tied solar photovoltaic system is proposed. The technique aims to mitigate harmonics and improve grid currents, with an additional feature of DSTATCOM for reactive power support during night or zero power generation [15]. The perturb & observe (P&O) algorithm is widely used for maximum power point tracking in solar photovoltaic systems. However, it faces issues with varying irradiations and oscillations around the maximum power point. A circle center-line concept-based P&O (CCCP&O) algorithm is introduced, combining the concept of a circle and its center with the P&O algorithm, reducing iterations and settling time. The algorithm's performance is demonstrated in simulation and experimental testing [16]. A novel DFOGI-based control algorithm for grid-integrated solar photovoltaic systems is introduced, using the HPO algorithm for GMPPT tracking. The algorithm provides rapid and accurate global maximum power peak searching, ensuring steady-state and dynamic performance even in rapid solar irradiance changes. The goal is to transfer solar power to the grid at unity power factor, maintaining power quality during abnormal conditions. Experimental results meet the objectives and meet IEEE-519 standard parameters [17]. A model predictive-based control for a solar PV system integrated into the grid, aiming to optimize power transfer and management is studied. The controller uses a modified-dual second-order generalized-integrator to estimate power requirements based on system parameters. The controller's performance is validated through simulation and hardware implementation, adhering to the IEEE-519 standard, demonstrating remarkable performance [18]. This study reveals a simplified nano-grid model that enhances the utilisation of solar photovoltaic renewable energy with micro hydro power plant. The application of non conventional energy sources causes environmental problem therefore to reduce the global warming issues the use of renewable energy sources are adopted for power generation and connected to grid supply system to enhance clean and green energy technologies for nano grid systems.

In this work, following are objectives are taken into consideration:

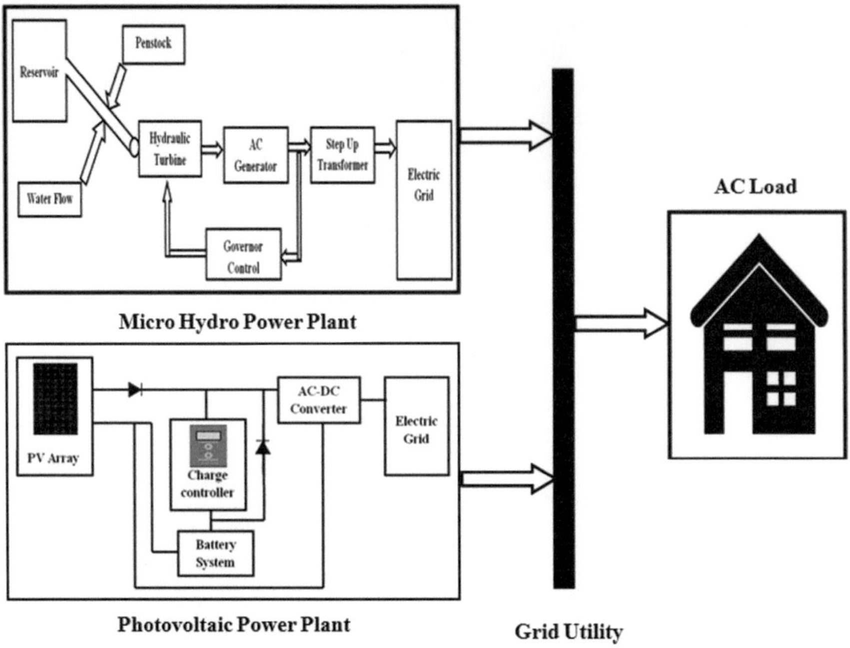

Fig. 20.1 Block diagram of Nano-Hybrid grid system using renewable sources

- The opportunity of developing an elementary design of nano-grid model that enhances the application of specific renewable energy for on-grid locations is explored.
- The HOMER software is employed to optimise the scale and different types of energy conversion technology.
- The costing of energy and the reduction in carbon emissions are significant variables in the meantime.
- In addition, overall performance the nano-grid model is examined using Simulink-MATLAB software.

20.2 Methodology

SIMULINK-MATLAB mode is being designed to study the behaviour of nanogrid system for dynamic load. The simulink model diagram of proposed system is revealed in Fig. 20.2. The model is framed by using MHP, PV power generation, Grid and load block.

The river stream flow was suggested as a source of power for a micro hydro power plant (MHP). The flow of stream was lowest during the month of April (2141 L/s. 1251 L/s, or 1.28 m^3/s) can be utilised to generate power when the remaining flow of 1000 L/s is taken into account. Equation [19] can be used to determine the nominal

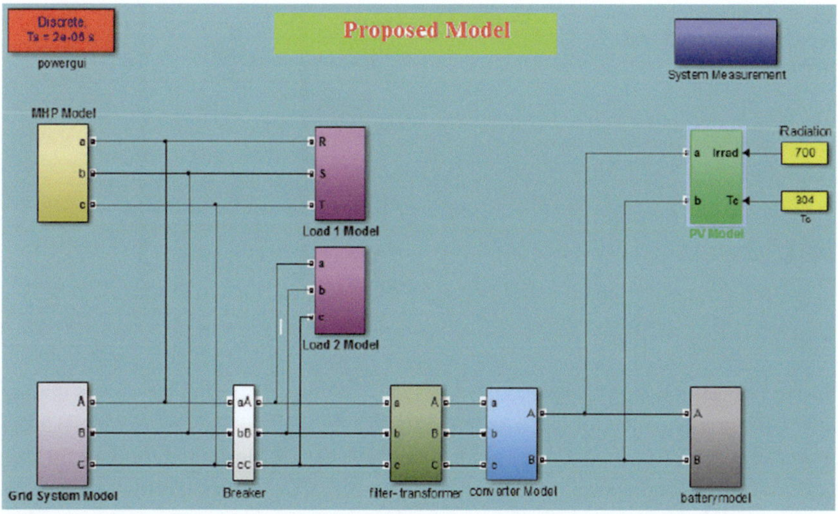

Fig. 20.2 Simulink model of proposed work

power of MHP.

$$P_{MH} = g\ \eta.h.Q_d \tag{20.1}$$

where, g is the acceleration caused by gravity [m/s^2], P_{MH} is the power output of the micro hydro power plant in kW, design flow is Q_d [m^3/s] and gross head h in meter, η is MHP efficiency. The design flow alternatives for the micro hydro power plant and simulated as 0.8, 0.7, and 0.5 m^3/s while taking the loads and water resources accessibility into account. Meanwhile, the average gross head for all phases was 25 m [12–16]. The replacement cost is used to determine the cost of replacing energy equipment. Using the following equation, HOMER calculates the PV power generation's power output:

$$P_{pV} = K_{PV} G_{PV} I_S / I_R \tag{20.2}$$

where, I_s is the solar radiation (kW/m^2), K_{PV} is the rated capacity of PV (kW), the derating factor of photovoltaic cell is G_{PV}, and I_R is the standard amount radiation in 1 kW/m^2. Photovoltaic modules are connected in different connection (i.e. series and parallel connection). The cost of purchasing solar panels, mounting panels, control systems, wiring, and installation are included in the capital cost (USD 2000) per kW [8]. The overall age of a solar module is estimated to be 25 years based on the specifications [17, 18, 20]. To enhance the performance of nano grid, battery system is necessary to respond to transient disturbances and fluctuations in solar irradiation.

The number of batteries with rated power is used to match the power and voltage ranges of the photovoltaic generator. Also, converter may be used for electrical power

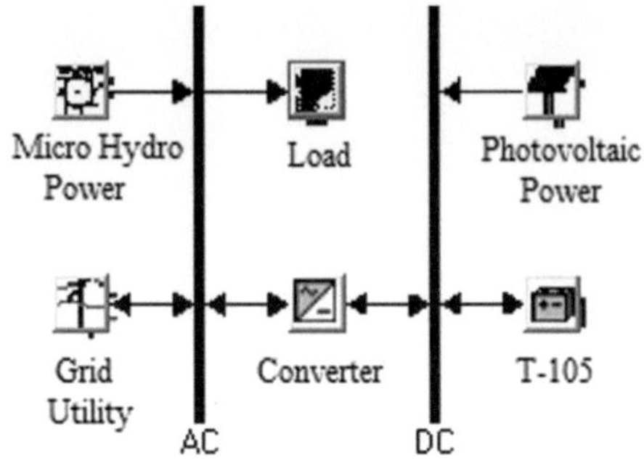

Fig. 20.3 Homer software based model of nano grid system

flow in two directions (bidirectional) [21–24]. Therefore, depending on the flow of power, the power converter can function as both a rectifier and an inverter. A 10.5 kW converter is used for steam power plants that serve as electric source for the grid emit gases during normal operation. The carbon dioxide (CO_2) emissions have dangerous impact on global warming [25]. As a result, the current analysis exclusively considers CO_2 emissions. The grid's emission coefficients are calculated by considering two major factors i.e. emission coefficient and the electrical power system [26, 27]. It is necessary to model the performance parameters of the nano-grid using Simulink-MATLAB for validation process. Figure 20.3 displays Homer software based model of nano grid system. A Photovoltaic power, MHP, the grid and the load are used to build such model.

The converter, transformer, filter, and PV module build the PV power generation block. Such model puts phase II configuration into practice. PV modules were connected in both series and parallel. An induction machine supplied by the MATLAB software was employed as MHP block. Simulation of nanogrid model is being analysed for the various dynamic loads to optimize economic benefits.

20.3 Result and Discussion

The results of the simulation grid performance using Simulink MATLAB are revealed in Fig. 20.4a shows the micro hydro power has been providing electricity to grid and load system before t = 0. At t = 0 photovoltaic power coupled to the grid. This graph illustrates photovoltaic and micro hydro grid system shared power to grid and load (Fig. 20.4b). The active power is increased upto the value of 4123W at time t = 0.32, while no change is considered in reactive power during this period. And, at t = 0.4 s and the active reduces from 4123 to 3984 W. On the other hand, the grid will boost power for the micro-grid to prepare for the rising load (Fig. 20.4c), photovoltaic and micro hydro power keep power at standard level. In this situation, the grid, MHP and photovoltaic systems will contribute to powering the load. At t = 0.7 s, it is observed that real and reactive power increased up-to 3852W and 4798VAr respectively (MHP). Since the supply of micro hydro power and photovoltaic can

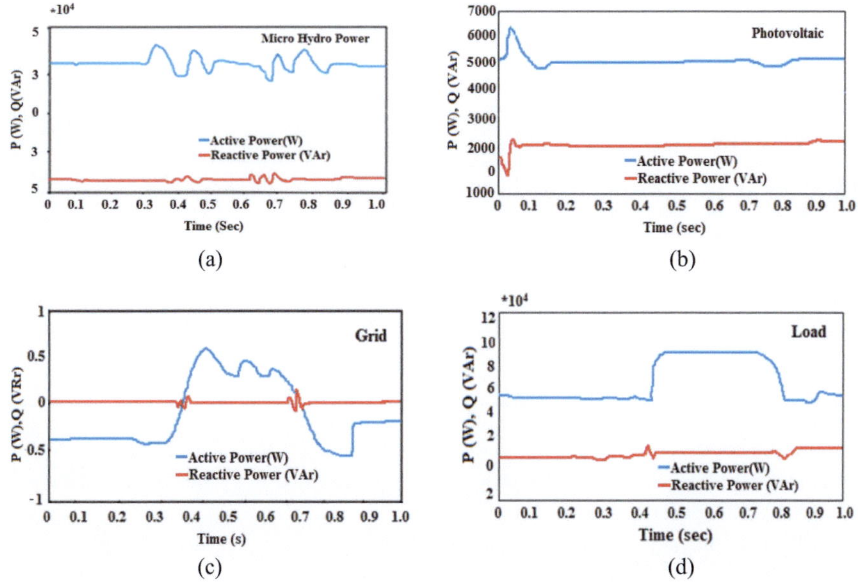

Fig. 20.4 Power and reactive power profile of **a** micro hydro power, **b** Photovoltaic system **c** Grid **d** Load using MATLAB simulation

satisfy various requirements of load side (Fig. 20.4d), the grid system will be divided from the nano-load grid's system. In a nutshell, the planned micro-electricity grid's supplied by renewable energy sources were optimised, and the grid system's power was adjusted with the demands of the load.

Figure 20.5a shows optimization results of nano grid using considering 50 KW MHP utilising the HOMER whereas Fig. 20.5b, c reveals optimization results of nano grid using considering 100 and 150KW MHP respectively. It has been observed that nano-grid model with 150 KW MHP reduces CO_2 emission, and share largest load as it approached the peak load capacity. Additionally, a minimum capacity of 5 kW for PV systems was usually recommended.

In order to create the suggested nano-grid model, the configuration incorporating all nearby renewable energy sources (PV and MHP) was selected for each phase (Table 20.1). The suggested nano-system grid's performance needed to be improved, that necessitated the battery system. Phase I revealed more expensive in terms of initial capital expenses. As opposed to phase III, that has a higher energy cost compared to the power grid's tariff, phases I and II have lower energy prices. The phase I and phase II have negative operating costs as a result of the sale of electrical power generated by surplus power to grid-system. The system's operational expenses will be covered by the earnings from the sale of excess power. Furthermore, the cost of MHP for phases I and II was still less expensive than the grid electricity rate. The measure of carbon emissions produced by nanogrid system as well as number of carbon emissions reductions on the environment were connected

Sensitivity Results | Optimization Results

Sensitivity variables

Load (kWh/d) 1552 | Hydro Capital ($) 440,0 | Hydro Replacement ($) 111,0 | Hydro O&M ($/yr) 21,00

Double click on a system below for simulation results.

PV (kW)	Hydro (kW)	T-105	Conv. (kW)	Grid (kW)	Initial Capital	Operating Cost ($/yr)	Total NPC	COE ($/kWh)	Ren. Frac.
	50			100	$ 450,000	-29,915	$ 259,301	0.030	1.00
	50	25	7.0	100	$ 457,625	-28,856	$ 244,397	0.030	1.00
7.5	50			100	$ 450,000	-29,497	$ 267,393	0.030	1.00
7.5	50			100					1.00
				100	$ 0	44,061	$ 302,831	0.082	0.00
		25	7.0	100	$ 7,625	45,471	$ 317,735	0.086	0.00
				100	$ 25,125	44,985	$ 329,305	0.086	0.00
		25	7.0	100	$ 20,125	45,121	$ 337,641	0.086	0.00

(a)

Sensitivity Results | Optimization Results

Sensitivity variables

Load (kWh/d) 1,425 | Hydro Capital ($) 323,0(| Hydro Replacement ($) 83,31(| Hydro O&M ($/yr) 13,05(

Double click on a system below for simulation results.

PV (kW)	Hydro (kW)	T-105	Conv. (kW)	Grid (kW)	Initial Capital	Operating Cost ($/yr)	Total NPC	COE ($/kWh)	Ren. Frac.
	100			200	$ 320,000	-4,066	$ 202,903	0.052	0.99
				200	$ 0	44,060	$ 302,801	0.082	0.00
	100	20	7.5	200	$ 302,605	-3,600	$ 307,808	0.055	0.99
5	100		7.5	200	$ 340,125	-4,101	$ 316,002	0.056	0.99
		20	7.5	200	$ 7,625	45,101	$ 310,735	0.086	0.00
5	100	20	7.5	200	$340,625	-3400	$ 329,305	0.054	0.99
5			7.5	200	$ 20,125	44,985	$ 324,345	0.090	0.00
5		20	7.5	200	$ 25,125	45,471	$ 337,641	0.092	0.00

(b)

Sensitivity Results | Optimization Results

Sensitivity variables

_oad (kWh/d) 1,455 | Hydro Capital ($) 455,0(| Hydro Replacement ($) 115,5(| Hydro O&M ($/yr) 22,50(

Double click on a system below for simulation results.

PV (kW)	Hydro (kW)	T-105	Conv. (kW)	Grid (kW)	Initial Capital	Operating Cost ($/yr)	Total NPC	COE ($/kWh)	Ren. Frac.
	150			200	$ 430,000	-28,715	$ 254,397	0.030	1.00
	150	25	7.0	200	$ 454,623	-26,856	$ 249,300	0.030	1.00
5	150		7.0	200	$ 430,133	-27,497	$ 277,373	0.032	1.00
5	150	25	7.0	200	$ 434,123	-28,477	$ 257,354	0.030	1.00
				200	$ 0	44,061	$ 302,831	0.082	0.00
		25	7.0	200	$ 25,125	45,478	$ 334,601	0.092	0.00
5			7.0	200	$ 20,125	44,885	$ 329,300	0.092	0.00
5		25	7.0	200	$ 25,125	45,478	$ 334,601	0.092	0.00

(c)

Fig. 20.5 **a** Optimization results of nano grid using considering 50 KW MHP **b** Optimization results of nano grid using considering 100 KW MHP **c** Optimization results of nano grid using considering 150 KW MHP

Table 20.1 Phases of nano grid system

Phases	Photovoltaic power (KW)	MHP (KW)	Grid power (KW)	Photovoltaic (Reduction in CO_2 emission) (Kg/Year)	MHP (Reduction in CO_2 emission) (Kg/Year)	Grid (Reduction in CO_2 emission) (Kg/Year)	Carbon emission (Kg/ Year)
Phase I	5	150	200	6,956	1,110,744	10	765,321
Phase II	5	150	200	6,956	798,385	10,938	498,986
Phase III	5	150	200	6,956	399,230	150,529	301.332

with the percentage of renewable sources. The CO_2 emissions created by phases I and II for the nano-grid are negative, refers to a nano-grid that lowers grid system CO_2 emissions as a result of supplying the grid with electrical energy from renewable energy sources. The measure of renewable energy produced overall as well as the carbon emission coefficient of the economy together define the overall amount of CO_2 reduction on the environment. In phase I, renewable energy sources (MHP and PV) met the whole demand for electricity. The nano-grid system sells roughly 56% of the energy, generate to the grid in sellback the price that exceeds the MHP cost.

The nano-energy grid's cost will drastically fall as a result in such circumstances. Additionally, photovoltaic and MHP power generation cost is greater than the rate of grid. Energy prices for such choice will be higher due to the requirement than the tariff of grid electricity.

20.4 Conclusion

This article concluded that in order to maximise the use of local renewable energy for on-grid areas, the potential for developing a simple nano-grid model is used. Three different design alternatives for the MHP model such as 150 kW, 100 kW, and 50 kW are taken into consideration depends upon the load profile and the accessibility of water supply. The nano-grid model with MHP design generated economic energy at minimum cost, accountable drop in carbon emission, and maximum share of renewable energy as it approached the peak load capacity (150 kW). Adversely, the requirement demanded high preliminary capital expense. While a model of nano-grid using MHP design approaches load capacity (50 kW) requires cheap initial capital expenditures, in addition, high energy costs and limited CO_2 emission reduction. However, The proposed nanogrid model proven to be benefits with the increase in the load simultaneously optimum utilization of the resources. Considering the rise in electrical load and the decline in solar unit cost, the development of photovoltaic units on the proposed nano-grid is quite promising. MATLAB-SIMULIK simulation was used to monitor the nano-grid performance. According to the simulation's findings,

Hydro power power plants produced the most efficient amount of electricity. To optimise the utilisation of locally available renewable energy, the study findings must be maintained through the development of a nano-grid control strategy.

References

1. D.A. Torrey, A.M. Al-Zamel, Single-phase active power filters for multiple nonlinear loads. IEEE Trans. Power Electron. **10**(3), 263–272 (1995)
2. D. Venkatramanan, V. John, Integrated higher-order pulse-width modulation filter-transformer structure for single-phase static compensator. IET Power Electron. **6**(1), 67–77 (2013)
3. T.-F. Wu, K.-H. Sun, C.-L. Kuo, C.-H. Chang, Predictive current controlled 5-KW single-phase bidirectional inverter with wide inductance variation for DC microgrid applications. IEEE Trans. Power Electron. **25**(12), 3076–3084 (2010)
4. Z. Xinxin, X. Lan, W. Zilong, L. Yun, W. Chuyang, Control strategy without phase-locked loop based on coordinate transformation for three phase AC/DC converter. IET Power Electron. **8**(9), 1701–1709 (2015)
5. S. Yang, Q. Lei, F.Z. Peng, Z. Qian, A robust control scheme for grid-connected voltage-source inverters. IEEE Trans. Ind. Electron. **58**(1), 202–212 (2011)
6. D. Souvik, S. Sanjibkumar, P. Sanjibkumar, Single-phase inverter control techniques for interfacing renewable energy sources with microgrid-Part I: parallel-connected inverter topology with active and reactive power flow along with grid current shaping. IEEE Trans. Power Electron. **26**(3), 717–731 (2011)
7. V.B. Foroutan, M.H. Moradi, M. Abedini, Optimal operation of autonomous microgrid including wind turbines. Renew Energy **99**, 315–324 (2016)
8. K. Sureshkumar, V. Ponnusamy, Hybrid renewable energy systems for power flow management in smart grid using an efficient hybrid technique. Trans. Inst. Meas. Control. **42**(11), 2068–2087 (2020)
9. V. Gajula and R. Rajathy, An agile optimization algorithm for vitality management along with fusion of sustainable renewable resources in microgrid. Energy Sources, Part A: Recov Util. Environ. Eff. 42 (13)1580–1598 (2020)
10. Chen, J., Zhou, Z., Karunakaran, V., Zhao, S., An efficient technique-based distributed energy management for hybrid MG system: a hybrid QOCSOS-RF technique. Wind. Energy **23**(3), 575–592 (2020)
11. R. Lingamuthu, R. Mariappan, Power flow control of grid connected hybrid renewable energy system using hybrid controller with pumped storage. Int. J. Hydrogen Energy **44**(7), 3790–3802 (2019)
12. K. Meenakshi, A.E. Wilson, A. Abraham, J. Thomas, R. Chacko, Economic feasibility investigation of integrated Nano-Grid with solar PV for community deployment based on load pattern survey. Mater. Today: Proc. **58**, 293–298 (2022)
13. N.J. Bodele, P.S. Kulkarni, Modular battery-integrated bidirectional single-stage DC–DC converter for solar PV based DC Nano-grid application. Sol. Energy **259**, 1–14 (2023)
14. N. Kumar, B. Singh, B.K. Panigrahi, ANOVA Kernel Kalman filter for multi-objective grid integrated solar photovoltaic-distribution static compensator. IEEE Trans. Circuits Syst. I Regul. Pap. **66**(11), 4256–4264 (2019)
15. B. Singh, N. Kumar, B.K. Panigrahi, Steepest descent Laplacian regression based neural network approach for optimal operation of grid supportive solar PV generation. IEEE Trans. Circuits Syst. II Express Briefs **68**(6), 1947–1951 (2021)
16. P. Kumar, H.K. Channi, R. Kumar, C. Prakash, A. Bhowmik, et al., Efficient integration of photo voltaic and hydro energy technologies for sustainable power generation in rural areas: a case study. Mater. Sci. Energy Technol. **7**, 297–308 (2024)

17. N. Kumar, B. Singh, P., Bijaya, Grid Synchronization framework for partially shaded solar PV Based Microgrid using Intelligent control strategy. IET Gener., Transm. & Distrib. (2018)
18. V. Saxena, N. Kumar, B. Singh, B.K. Panigrahi, An MPC based algorithm for a multipurpose grid integrated solar PV system with enhanced power quality and PCC voltage assist. IEEE Trans. Energy Convers. **36**(2), 1469–1478 (2021)
19. M. Singh, V. Khadkikar, A. Chandra, R.K. Varma, Grid interconnection of renewable energy sources at the distribution level with power-quality improvement features. IEEE Trans. Power Delivery **26**(1), 307–315 (2011)
20. M. Ansarian, S.M. Sadeghzadeh, M. Fotuhi-Firuzabad, Optimum generation dispatching of distributed resources in smart grids. Int. Trans. Electr. Energy Syst. **25**(7), 1297–1318 (2015)
21. M. Durairasan, D. Balasubramanian, An efficient control strategy for optimal power flow management from a renewable energy source to a generalized three-phase microgrid system: A hybrid squirrel search algorithm with whale optimization algorithm approach. Trans. Inst. Meas. Control. **42**(11), 1960–1976 (2020)
22. K. Venkatesan, U. Govindarajan, Optimal power flow control of hybrid renewable energy system with energy storage: a WOANN strategy. J. Renew. Sustain. Energy **11**(1), 015501 (2019)
23. X. Wu, W. Cao, D. Wang, M. Ding, A multi-objective optimization dispatch method for microgrid energy management considering the power loss of converters. Energies **12**(11), 2160 (2019)
24. J. Garcia-Guarin, D. Rodriguez, D. Alvarez, S. Rivera, C. Cortes, A. Guzman, A. Bretas, J.R. Aguero, N. Bretas, Smart microgrids operation considering a variable neighborhood search: the differential evolutionary particle swarm optimization algorithm. Energies **12**(16), 3149 (2019)
25. H. Qiu, W. Gu, P. Li, X. Zhang, H. Long, Z. Wu, CRSO approach for microgrid power dispatching. IET Gener. Transm. Distribution **13**(11), 2208–2215 (2019)
26. K. Meenakshi, A.E. Wilson, A. Abraham, J. Thomas, R. Chacko, Economic feasibility investigation of integrated Nano-Grid with solar PV for community deployment based on load pattern survey. Mater. Today: Proc. **58**(Part 1), 293–298 (2022)
27. P. Malik, M. Awasthi, S. Sinha, A techno-economic investigation of grid integrated hybrid renewable energy systems. Sustain. Energy Technol. Assess. **51**, 101976 (2022)

Chapter 21
An Investigation of Mismatch Between Energy Yields of Building Integrated Photovoltaic Systems

Mansi Singh and Anil Kumar Rai

Abstract Over the years, building integrated photovoltaic system has been considered as a practical and climate conscious approach for building design, development and comfortable living. This paper investigates the mismatch in energy yields of PV system that is connected to grid situated over a building in Northern India for viability purposes. The system under consideration is of 330 kWp capacity; it consists of a large number of subarrays and experimental measurements were carried out of two subarrays (units) of identically rated components to explore the disparity between rated energy output at Standard Test Condition (STC) and the energy actually produced by the system as well as variations in electrical energy output between the units. It is found that the monthly mean energy differences between the subarrays is 75 kWh whereas the minimum and maximum energy differences are 12 kWh and 180 kWh respectively. These results show that there is notable decrease in energy output due to mismatch in system subarrays. The root cause mechanism causing the mismatch between the electrical energy output of two units is also discussed.

Keywords Solar photovoltaic system · Subarray · STC · Energy yield

21.1 Introduction

Present day buildings are an intense consumer of energy. The technological options for practical and climate consensus approach for buildings include thermal as well as electrical energy efficiency, use of renewable energy resource etc. Building integrated solar photovoltaics have obtained steadily increased attention as a means to supply/

M. Singh (✉) · A. K. Rai
Ajay Kumar Garg Engineering College, Ghaziabad, Uttar Pradesh, India
e-mail: mansi2221002m@akgec.ac.in

A. K. Rai
e-mail: raianil@akgec.ac.in

A. Kumar et al. (eds.), *Selected Proceedings of the 1st International Conference on Advanced Materials for Sustainable Innovation; IC-AMSI 2024; 28–30 August; New Delhi; India*, Springer Proceedings in Energy,
https://doi.org/10.1007/978-981-96-4492-6_21

save energy in the buildings. Roof-top building integrated solar photovoltaic system is the simplest version of BIPV. There is currently increasing hope that solar PV power systems of this kind may provide the world's severe power deficit with a solution.

Various organizations have initiated the installation of rooftop solar power systems to bridge the difference between energy demand and supply in recent years. Rooftop photovoltaic (PV) systems avoid costs associated with land acquisition, site development, foundations, and structural support required for the setup of PV array. The effectiveness and efficiency of these solar power systems depend on location, orientation and tilt angle of the array.

Location- The quantity of sunlight received has an impact on the amount of electricity produced.

Orientation- Two crucial elements to optimize energy output are orientation of building and how the SPV arrays are installed on the top of the building. Northern Hemisphere buildings can achieve optimal solar energy capture by facing south. This orientation maximizes sunlight exposure for the panels, leading to the most efficient electricity generation.

Tilt Angle- In the field condition, solar PV modules/array are tilted to obtain higher incident radiation of solar on the surface of the array to increase the energy yield. Mounting of modules/array can be done in several ways: Rooted tilt to maximize the yearly mean energy yield and according to the energy needed, seasonally adjusted tilt in summer, equinoxes and winter.

Around the globe, several studies have been carried out on Photovoltaic System with grid connection on the roof. These studies [1–6] show that the yield of energy supplied to grid is lesser than the estimated value. The effectiveness analysis of a 20 kWp roof-top grid-interactive SPV system installed on a library building at the Indian Institute of Science, Bengaluru, India, has been conducted using capacity utilization factor and performance ratio as grading systems [7]. In the view of the above, to investigate the difference between electrical energy output we have considered identical rated of two units of 20kWp each in Northern India.

21.2 Methodology

The system under consideration is mounted on the top of a building in Northern India. The longitude is 77.50^0 E and latitude of the site is 28.67^0N. The solar irradiance (in kWh/m^2) and temperature (in ^0C) were collected during April 2017 to March 2018. The maximum monthly mean temperature of 29.9^0C observed in the month May while the maximum monthly mean radiation of 5.62 kWh/m^2 noted in the month of April.

A 40 kWp, two units each of 20 kWp as shown in Fig. 21.1, Solar photovoltaic power plant is considered. Each unit having 68 opaque Modules (17 opaque modules in series and 4 such series connected modules in shunt) and each module is made by ALPEX SOLAR India of capacity 300 Wp $\pm$ 3% having 72 (12 solar cells in connected in series and 6 such strings shunt connected) solar cells made of Mult crystalline-silicon. The modules fixed at an angle of 30^0 facing south are free from shading effects. Every fifteen days, the PV modules' surface is washed with water to avoid accumulation of dust over modules. The schematic of installed systems is depicted in Fig. 21.1.

The details of these opaque PV modules are given in Table 21.1. To fed the power into grid, a three-phase inverter manufactured by Delta Electronics Pvt. Ltd. has been used to convert the DC into AC power and the type of the inverter is RPI M20A. The efficiency (rated) of the inverter is 98.5%.

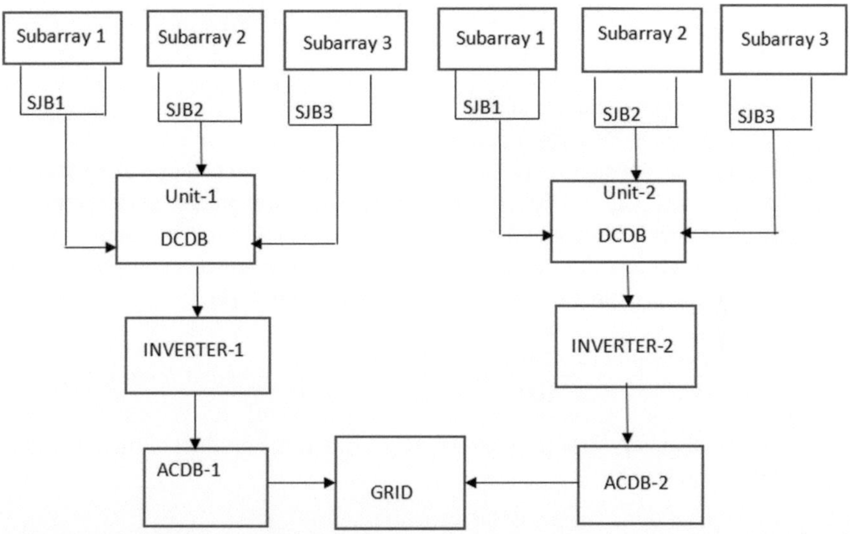

Fig. 21.1 Schematic layout of installed SPV system

Table 21.1 Module details

Solar PV modules (Multi-crystalline-Si)	12 × 6
P_{max}	300Wp ±3%
Open circuit voltage, Voc	44.64 V
Short circuit current, Isc	8.75 A
Current at maximum power	8.35 A
Voltage at maximum power	35.93 V
Fill-factor	0.76
Module area	1.64 m^2

21.3 Results and Discussion

The generated energy from PV module in kWh which is the multiplication rating of module at STC (Wp) and Equivalent Hours of Full Sunlight (EHFS). The module power output also depends on cell operating temperature.

When the solar cell operating temperature rises above STC (25 °C), it has a negative impact on its performance. The Nominal Cell Operating Temperature (NOCT) is determined to be 45^0 C for the climate of India. The NOCT can be related with ambient temperature and junction temperature of the cell [8].

$$T_{junc} = NOCT + T_{ambient} - 20 \; or, T_{junc} = T_{ambient} + 25 \tag{21.1}$$

where;

T_{junc}= junction temperature of solar cell (°C).

$T_{ambient}$ = Ambient air temperature (°C).

The PV module's energy output decreases with increasing cell operating temperature above the STC (25 °C). To account the effect of temperature, EHS is replaced by another parameter known as Unit Array Output (UAO) and is examine in the system design to account for the temperature effect and ensure an appropriate safety margin. The UAO is expressed in Wh/Wp/day and given as [9].

$$UAO = [1 - \frac{percentage\ temperature\ coeffient}{100} \times (Max.Ambient\ Temp.)] \times EHFS$$

The electrical energy generated by the SPV array is the product of module rating at STC (Wp) and UAO.

21.4 Incident Solar Radiation Calculation on Tilted Surface

The incident solar irradiation on horizontal surface at the site under examination is collected. The monthly daily mean global radiation on inclined plane is calculated for twelve months, which is given by using the expression [9]

$$C\,(t) = G\,(t) \times [G\,(\beta)](t) \; in\; kWh/m^2/day \tag{21.2}$$

where G(t) = monthly daily mean global radiation on horizontal plane in kwh/m^2/ day for twelve months, multiplying factor, G (β) is used to multiply global radiation on a horizontal plane in order to produce radiation on a south facing surface kept at tilt angle of β is given by:

***

$$G(\beta) = R(1 - r) + \frac{r}{2}(1 + cos\beta) + \frac{A}{2}(1 - cos\beta) \tag{21.3}$$

and R is given by

***

$$R = [\frac{\cos(\varnothing - \beta)cos\delta sin\omega_{st} + \omega_{st}\sin(\varnothing - \omega_{st})sin\delta}{cos\varnothing cos\delta sin\omega + \omega sin\varnothing sin\delta}]$$

ω_{st}, ω, are sunset hour angles on tilted and horizontal surface respectively. δ and ϕ are declination angle and latitude of the site respectively. On a horizontal plane, r is the ratio of the diffuse to the global component of incident solar radiation. A is the portion of sunlight that is reflected off the earth.

Calculate the cumulative annual global radiation in kWh/m^2/day for different tilt angle β (0 to 75^0) for twelve months.

$$C_c = \sum_{t=1}^{12} C(t)\left[Total\ number\ of\ days\ of\ corresponding\ month\ M\ (t)\right]$$

Calculate monthly daily mean of EHFS for twelve months.

21.5 Calculation of Yearly Daily Mean Value of Unit Array Output

Calculate monthly daily mean value of Unit Array Output for different tilt angle β_{tilt} (0 to 75^0) for twelve months.

$$U(t) = [1 - \frac{\%Temp.\ Coeff.}{100} \times (Max.Ambient\ Temp.\ in\ a\ month)] \times EHFS(t)$$

Calculate cumulative annual Unit Array Output in kwh/m^2

$$U_c = \sum_{t=1}^{12} U(t)\left[Total\ number\ of\ days\ of\ corresponding\ month\ M\ (t)\right]$$

Optimum array tilt angle ($\beta_{opt.}$) Will be selected for the maximum value of cumulative annual Unit Array Output (U_{cmax}.).

Calculate yearly monthly mean value of Unit Array Output

$$U_L = \frac{U_{cmax}}{12} \tag{21.4}$$

The ambient temperature, irradiance and unit array out at the site under consideration are shown in Table 21.2.

Table 21.2 Calculated unit array output

Month	Ambient temp (°C)	Irradiance level (kwh/m^2)	Unit array output (Wh/Wp/day)
Jan	14.76	3.3	3.069
Feb	18.79	4.21	3.85
March	24.07	4.59	4.09
Apr	28.97	5.62	4.88
May	29.9	4.84	4.18
June	29	4.94	4.29
July	26.86	4.36	3.83
Aug	26.64	4.32	3.80
Sept	26.48	4.57	4.02
Oct	26.06	4.69	4.14
Nov	19.89	3.05	2.77
Dec	17.01	3.28	3.02

The energy yield for every month in the year-round cycle for both subarrays are collected and shown in Table 21.3 and energy yield of both the units vs rated capacity is shown in Fig. 21.2. It has been observed that there is significant difference in energy yield from both the units. The energy yield is maximum and minimum in the month of April and January respectively. The total energy yield and energy yield imbalance between two subarrays are depicted in Figs. 21.3 and 21.4 respectively.

21.6 Causes for Unit-Wise Variations in Energy Yield

Some potential reasons for the variation in electrical energy output are as follows:

Module characteristics tolerances set by the manufacturer- In the field conditions, series/parallel connections are made of PV modules to match the voltage and current rating of the load. These PV modules display deviations in their current–voltage traits, despite stringent controls and automation, owing to manufacturer tolerances. Electrical mismatch arises when connecting these mass-produced solar cells, which possess distinct current–voltage properties, either in series or parallel configurations. This inconsistency in module characteristics within a series string constrains the Isc to a value at the lower spectrum of the distribution's lower spectrum. As a result, in real-world scenarios, the combined highest power output often less than the summation of maximum power output of individual module's maximum power output. It has been shown that there is power loss resulting from mismatched cell characteristics within a network [10]. The authors have also demo that this power loss tends to escalate from 2 to 12% from the aging of solar cell.

Table 21.3 Energy yield of both units

Month	Rated energy yield (kwh) of 20kWp unit	Unit-1 energy yield (kWh)	Unit-2 energy yield (kWh)	Differences between both units (kWh)	Total energy yield (kWh)
January	1902.78	1619.43	1479.93	145.64	3099.2
February	2233	2196.33	2111.8	84.71	4308.13
March	2535.8	2133.88	2108.65	42.49	4242.52
April	2988	2985.86	2986.49	16.49	5972.35
May	2653.6	2633.98	2637.07	11.79	5271.05
June	2574	2136.94	2116.41	55.67	4253.35
July	2374.6	2143.25	2154.29	14.9	4297.54
August	2356	1853.62	1845.6	75.44	3699.22
September	2412	1903.58	1938.25	70.45	3841.83
October	2566.8	2211.82	2157.37	77.03	4369.19
November	1662	1633.8	1508.9	124.9	3142.71
December	1872.4	1763.79	1583.71	180.08	3347.49
Total annual	28,130.98	25,216.28	24,628.47	899.59	49,844.58
Annual average	2344.25	2101.36	2052.37	74.97	4153.72

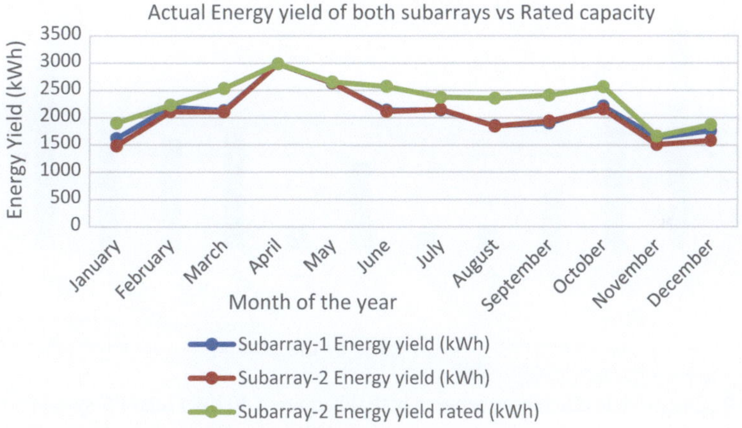

Fig. 21.2 Energy yield subarrays vs rated capacity

Irregular deposition of dust and other materials over PV system due to weather condition- The effectiveness of solar panels can be reduced by dirt, dust, bird droppings, or other material building up on their surface, which prevents sunlight from reaching the photovoltaic cells. Sometimes cleaning is not smooth. Due to accumulation of rice husk, a reduction in power output of 3.88 W of PV module of rating

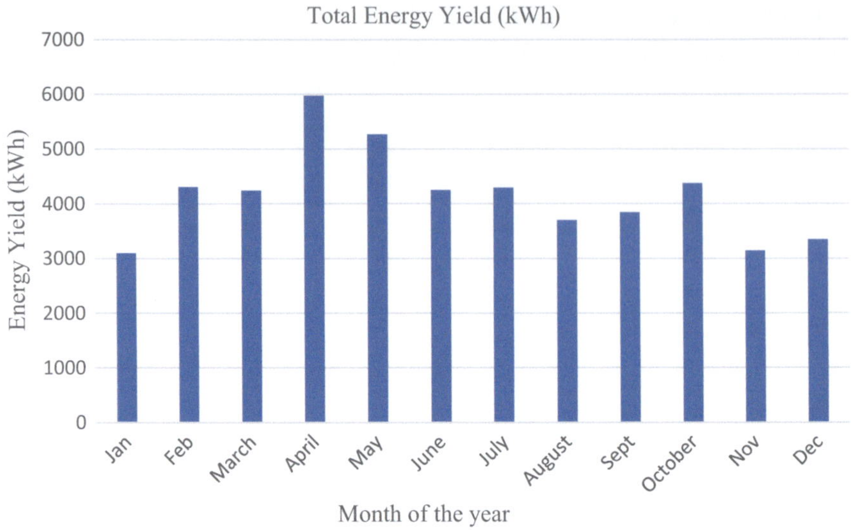

Fig. 21.3 Total energy yield from 40 kWp Solar PV System

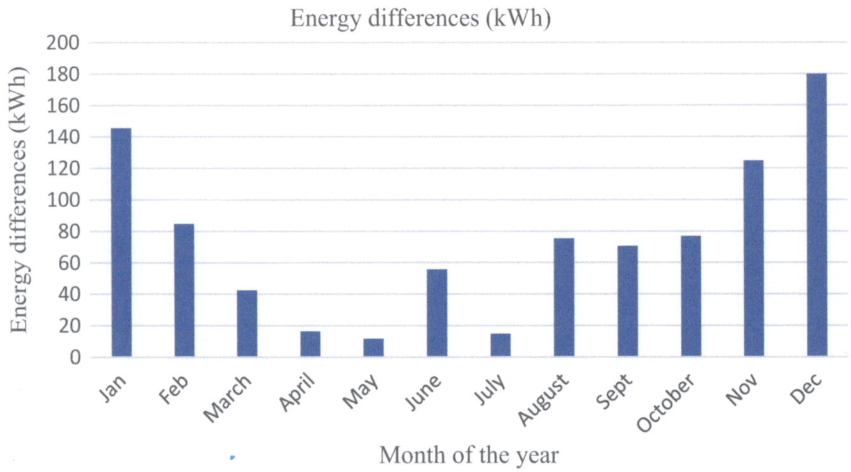

Fig. 21.4 Energy yield differences between both subarrays of 40 kWp solar PV system

60 Wp polycrystalline PV module has been reported [11]. Several authors have also reported about reduction in energy generation due to accumulation dust over PV modules [12, 13].

Inverter Issues- Inverters serve a crucial role in a solar PV system by transforming the produced DC electricity by the PV modules into usable AC electricity. Any faults or malfunctions in inverters can lead to diminished energy production or even the complete shutdown of the system.

Variation in grid voltage and grid outage- Variations in grid voltage levels have the potential to affect the inverter's operation and reduce the amount of electricity it generates and influence of SPV systems that is connected to grid on stability of voltage of the 53-Bus distribution grid has been investigated [14]. In addition to this, in some situations, the grid may not be able to handle excess energy produced by the solar system because of voltage restrictions or grid outages, which would require reducing solar output.

21.7 Conclusions

In the present work the investigation of mismatch between energy output of two units of identical rated 20kWp each has been carried out. The average energy difference between the two subarrays is 75 kWh with minimum and maximum energy difference 12 kWh and 180 kWh respectively. These results indicate that the system subarray has a substantial mismatch. The root cause mechanism causing the discrepancy in electrical energy output between two identical rated subarray is also discussed. It has been found that loss of power due to variation in module characteristics tends to escalate from 2 to 12% if variation is more than 3%.

References

1. J.D. Mondol, Y. Yohanis, M. Smyth, B. Norton, Long term performance analysis of a grid connected photovoltaic system in Northern Ireland Energy Convers. Manage **47**(18), 2925–2947 (2006)
2. V. Sharma, S.S. Chandel. Performance analysis of a 190 kWp grid interactive solar photovoltaic power plant in India Energy **55**, 476–485 (2013)
3. K.S.S. Pundir, N. Varshney, G.K. Singh, Comparative study of performance of grid connected solar photovoltaic power system in IIT Roorkee campus, in *Paper of International Conference on Innovative Trends in Science*. (Engineering and Management held at New Delhi, India, 2016), pp.422–431
4. S. Goel, R. Sharma, Analysis of measured and simulated performance of a grid-connected PV system in Eastern India. Environ. Dev. Sustain. **23**(1), 451–476 (2021). https://doi.org/10.1007/s10668-020-00591-7
5. R. Arora, R. Arora, S.N. Sridhara, Performance assessment of 186 kWp grid interactive solar photovoltaic plant in Northern India. Int. J. Ambient Energy **43**(1), 128–141 (2022). https://doi.org/10.1080/01430750.2019.1630312
6. M.K. Gaur, A. Shrivastava, R.K. Pandit, Performance analysis and sustainability assessment of a building integrated solar PV system. Int. J. Ambient. Energy **45**(1) (2024). https://doi.org/10.1080/01430750.2023.2283762
7. M.S. Vasisht, J. Srinivasan, S.K. Ramasesha, Performance of solar photovoltaic installations: Effect of seasonal variations. Sol. Energy **131**, 39–44 (2016)
8. F.C. Treble, *Generating electricity from the sun* (Pergamon Press, New York, 1991)
9. T. Bhattacharya, Terrestrial solar photovoltaic, Narosa Publishing House (1998)
10. N.D. Kaushika, A.K. Rai, An investigation of mismatch losses in solar photovoltaic cell networks. Energy **32**(5), 755–759 (2007)

11. H. Athar, A. Batra, R. Pachauri, An experimental study on effect of dust on power loss in solar photovoltaic module. Renew.S: Wind., Water, Sol. **4**, 9 (2017). https://doi.org/10.1186/s40807-017-0043-y
12. B. Hammad, M. Al-Abed, A. Al-Ghandoor, A. Al-Sardeah, A. Al-Bashir, Modeling and analysis of dust and temperature effects on photovoltaic systems' performance and optimal cleaning frequency: Jordan case study. Renew. Sustain. Energy Rev. **82**, 2218–2234 (2018). https://doi.org/10.1016/j.rser.2017.08.070
13. A. Al-Badi, Performance assessment of 20.4 kW eco-house grid connected PV plant in Oman. Int. J. Sustain. Eng. **13**(3), 230–241 (2020). https://doi.org/10.1080/19397038.2019.1658824
14. A.S. Saidi, Impact of grid-tied photovoltaic systems on voltage stability of Tunisian distribution networks using dynamic reactive power control. Ain Shams Eng. J. **13**, 101537 (2022). https://doi.org/10.1016/j.asej.2021.06.023

Chapter 22
Hybrid PVT-TEG Collector Coupled with VAR System: A Comparative Study

Abhishek Tiwari and Shruti Aggarwal

Abstract The refrigeration application of renewable energy technologies is essential for improving energy efficiency and sustainability. This study explores the potential of hybrid PVT systems coupled with TEGs to achieve these goals. The hypothesis is that a semi-transparent or glass-to-glass PV module in a PVT-TEG system would outperform an opaque PV module in terms of exergy output and refrigeration efficiency. MATLAB-based simulations were used to develop thermal models and evaluate the performance of two systems: one featuring an opaque PV module (case 1) and the other with a semi-transparent PV module (case 2), under clear-day conditions in Srinagar during May. The results indicate that the semi-transparent system's daily overall exergy output was 27.74 Wh higher than that of the opaque system, and the COP of the VAR system was also higher in case 2 by 37.17% (peak value). These findings suggest that utilizing a glass-to-glass PV module in a PVT-TEG system enhances exergy output and refrigeration efficiency, making it a more effective solution for critical applications like vaccine refrigeration and industrial hot water supply in remote areas.

Keywords Solar panel · Thermoelectric module · VAR system · COP

22.1 Introduction

In this communication, a MATLAB simulation-based performance comparison between case 1 and case 2 as described in the abstract section has been conducted across key parameters namely overall exergy, overall exergy efficiency and COP. The performance analysis of both the systems as well as performance comparison between both the systems have been done for clear day in Srinagar city for the month of May. Wu et al. [1] proposed a novel concentrated PV system attached with

A. Tiwari (✉) · S. Aggarwal
USBAS, GGSIPU, Delhi, India
e-mail: abhishek.usbas@ipu.ac.in

A. Kumar et al. (eds.), *Selected Proceedings of the 1st International Conference on Advanced Materials for Sustainable Innovation; IC-AMSI 2024; 28–30 August; New Delhi; India*, Springer Proceedings in Energy,
https://doi.org/10.1007/978-981-96-4492-6_22

thermoelectric modules featuring a self-cooling mechanism, triggering once PV cell temperature exceeded the threshold temperature at a concentration ratio of 25. Nazri et al. [2] carried out both theoretical and experimental investigations on a hybrid PVT-TE system. Khalili et al. [3] investigated a PV panel integrated with a TEG, heat transfer tube, and a confined jet impingement heat sink. Singh et al. [4] investigated a PVT-TEG based solar dryer and compared it with open sun drying. Tiwari et al. [5] evaluated the performance of a CPC-PVT-TEG collector integrated with a VAR system across various types of PV panels. Athaeya et al. [6] conducted an exergy study of PVT-CPC at constant collection temperature mode. Tiwari et al. [7, 8] conducted their research work on organic Rankine cycle through solar power and N solar collectors connected in series. While previous research has explored various configurations of hybrid PVT systems integrated with TEGs, this study uniquely contributes by directly comparing the performance of semi-transparent versus opaque PV modules in a PVT-TEG system coupled with a VAR system. This novel approach provides valuable insights into optimizing hybrid energy systems. The findings reveal that the semi-transparent PV module-based system outperforms its opaque counterpart, offering significant advantages in exergy output and refrigeration efficiency. These results have important implications for the development of sustainable energy solutions, particularly for critical applications like vaccine refrigeration in remote areas and industrial hot water supply. This study not only advances the field but also supports global efforts in enhancing energy sustainability and resilience, especially in resource-limited environments.

22.2 System Description

A PVT-TEG collector + VAR system has been presented in Fig. 22.1. The design of the hybrid collector and VAR system is consistent both scenarios, with the only variation being in the type of PV module utilised. The detailed functioning of the hybrid collector (for both scenarios) and the VAR system has been previously described by Tiwari et al. [9].

22.3 Modelling

22.3.1 Case 1

The expression for $T_{SC,1}$ for an opaque PV module is given as [9, 10]:

$$T_{SC,1} = (J_1 + J_2 + J_3 + J_4 - J_5 - J_6)/(U_{L-op} - J_7) \tag{22.1}$$

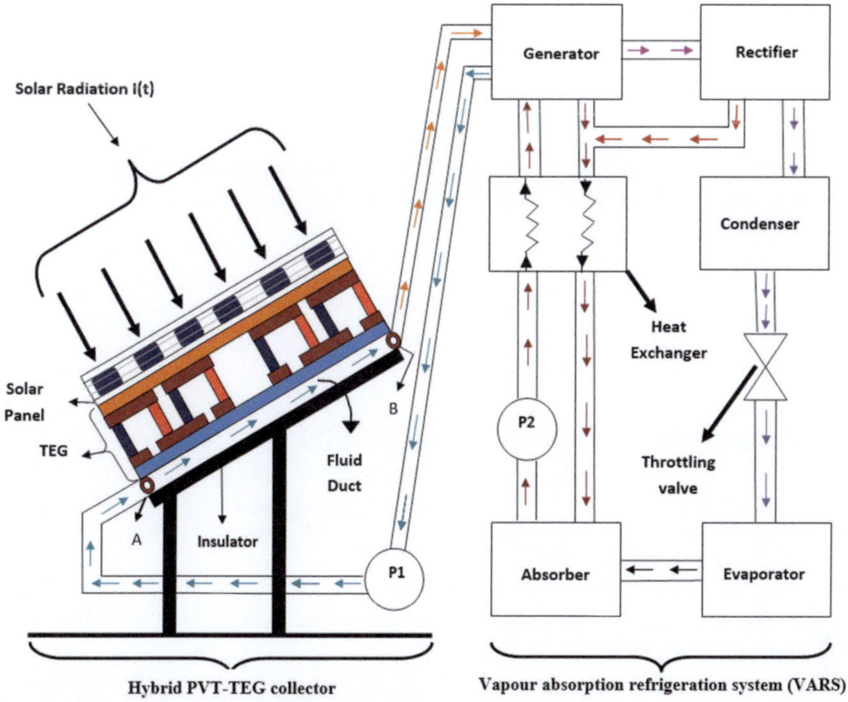

Fig. 22.1 PVT-TEG collector coupled with VAR system

The expression for J_1, J_2, J_3, J_4, J_5, J_6 and J_7 has been provided in appendix. The mathematical expression for PV cell efficiency ($\eta_{SC,1}$) is given as [9, 10]:

$$\eta_{sc,1} = \eta_o[1 - \beta_o(T_{SC,1} - T_o)] \quad (22.2)$$

The expression for the electrical efficiency ($\eta_{PV,1}$) and electrical energy output ($E_{PV,1}$) of the PV module is given as [9, 10]:

$$\eta_{PV,1} = \eta_{SC,1}\tau_g\beta_{PV} \quad (22.3)$$

$$E_{PV,1} = \eta_{PV,1}I(t)A_{PV} \quad (22.4)$$

Following the energy balance equations for TEG [9, 10] heat absorbed by the TEG, $\dot{Q}_H$ is given as:

$$\dot{Q}_H = U_{BH}A_{PV}(T_{SC,1} - T_H) \quad (22.5)$$

The electrical output of TEG, $E_{TEG,1}$ and its efficiency, $\eta_{TEG,1}$ are given below [9, 10]:-

$$E_{TEG,1} = [(s_{teg} - \mu_{teg})(\Delta T_{teg})]^2/4R_{teg} = \dot{Q}_H - \dot{Q}_C \quad (22.6)$$

$$\eta_{TEG,1} = E_{TEG,1}/\dot{Q}_H \quad (22.7)$$

$\dot{Q}_C$ is the amount of heat transferred from the cold side to the flowing fluid. The electrical output of the hybrid collector $E_{HC,1}$ and its electrical efficiency, $\eta_{HC,1}$ is given below [9, 10]:

$$E_{HC,1} = E_{PV,1} + E_{TEG,1} \quad (22.8)$$

$$\eta_{HC,1} = E_{HC,1}/I(t)A_{PV} \quad (22.9)$$

The output fluid temperature ($T_{FO,1}$) is given as [9]:

$$T_{FO,1} = \frac{v(t)}{a}\left(1 - e^{-aL_{PV}}\right) + T_a e^{-aL_{PV}} \quad (22.10)$$

where $a = ((F'U_C b)/\dot{m}_F c_F) + (bU_B/\dot{m}_F c_F))$ and $v(t) = ((F'U_C bT_C)/\dot{m}_F c_F) + (bT_a U_B/\dot{m}_F c_F)$. The thermal efficiency, thermal exergy, overall exergy and overall exergy efficiency are given as [6]:

$$\eta_{TH,1} = \dot{m}_F c_F (T_{FO} - T_a)/I(t)A_{PV} \quad (22.11)$$

$$E\dot{_{XC,1}} = \dot{m}_F c_F (T_{FO} - T_a) - \dot{m}_F c_F (T_a) ln(\frac{T_{FO}}{T_a}) \quad (22.12)$$

$$E\dot{_{EX,1}} = E_{HC,1} + E\dot{_{XC,1}} \quad (22.13)$$

$$\eta_{EX,1} = \eta_{HC,1} + E_{EE} \quad (22.14)$$

The expression for E_{EE} has been given in the appendix section. The expression for the COP of the VAR system is given as [9]:

$$COP = \frac{T_E(T_G - T_a)}{T_G(T_a - T_E)} \quad (22.15)$$

where, generator temperature, $T_G = T_{FO,1}$ and evaporator temperature, $T_E = 278$ K.

22.3.2 *(Case 2)*

The expression for glass-to-glass PV module $T_{SC,2}$ is given as:

$$T_{SC,2} = (J_8 - J_9 - J_{10} + J_{11} + J_{12})/(U_{L-st} - J_{13}) \tag{22.16}$$

The expression for J_8, J_9, J_{10}, J_{11}, J_{12} and J_{13} has been shown in the appendix. The expression for PV cell efficiency ($\eta_{SC,2}$) is given as [9, 10]:-

$$\eta_{SC,2} = \eta_o[1 - \beta_o(T_{SC,2} - T_o)] \tag{22.17}$$

The expression for the electrical efficiency ($\eta_{PV,2}$) and electrical energy output ($E_{PV,2}$) of the PV module are given below [9, 10]:

$$\eta_{PV,2} = \eta_{SC,2}\tau_g\beta_{PV} \tag{22.18}$$

$$E_{PV,2} = \eta_{PV,2}I(t)A_{PV} \tag{22.19}$$

Following the energy balance equations for TEG [9, 10], the heat absorbed by the TEG, $\dot{Q}_H$ is given as:

$$\dot{Q}_H = U_{bh}A_{PV}(T_{SC,2} - T_h) + \tau_g{}^2\alpha_{TEG}(1 - \beta_{PV})A_{PV}I(t) \tag{22.20}$$

The expressions for the rest of the relevant parameters for case 2 are the same as that of case 1 with subscript 1 being replaced by subscript 2.

22.4 Results

The value of the design parameters has been sourced from Tiwari et al. [9, 10]. The graph of E_{HC} and η_{HC} vs time for both cases has been illustrated in Fig. 22.2. It is seen in Fig. 22.2 that for both cases, the E_{HC} value increases and attains a maxima at around noon as I(t) increases in this duration which increases electrical energy generation. After noon, the E_{HC} value starts decreasing as I(t) decreases in this duration. It is also observed in Fig. 22.2 that η_{HC} for both the cases decreases and attains a minima between 12 noon and 1 pm as in this duration $T_{SC,OP}$ and $T_{SC,ST}$ increases which decreases the efficiency of the opaque and semi-transparent PV module. After 1 pm, η_{HC} increases as in this duration $T_{SC,OP}$ and $T_{SC,ST}$ decreases which increases η_{HC}. Based on simulation results, it is also noted that case 2 has attained higher values of E_{HC} and η_{HC} throughout the observation period than case 1. The upper and lower limit value of E_{HC} obtained was 97.18 Wh (case 2) and 39.7 Wh (case 1) respectively. The upper and lower limit value of η_{HC} obtained was 11.82% (case 2) and 9.61% (case 1) respectively. The graph of η_{TH} vs time for both the cases has been shown in Fig. 22.3. It is observed in Fig. 22.3 that for both the cases, η_{TH} increases and attains a maxima at around 1 pm as in this duration I(t) increases. After 1 pm, η_{TH} decreases as I(t) decreases in this duration. On the basis of simulation results, it is also noted that case 2 has attained a higher value of η_{TH} than case 1 throughout the period (8 am - 5 pm). The upper and lower limit value of η_{TH} obtained was 17.84% (case 2) and 12.24% (case 1) respectively.

The graph of E_{XC} vs time for both cases has been shown in Fig. 22.4. It is noted in Fig. 22.4 that for both cases, E_{XC} value increases and attains a maxima at around noon as in this duration I(t) increases. After noon, E_{XC} decreases as in this duration

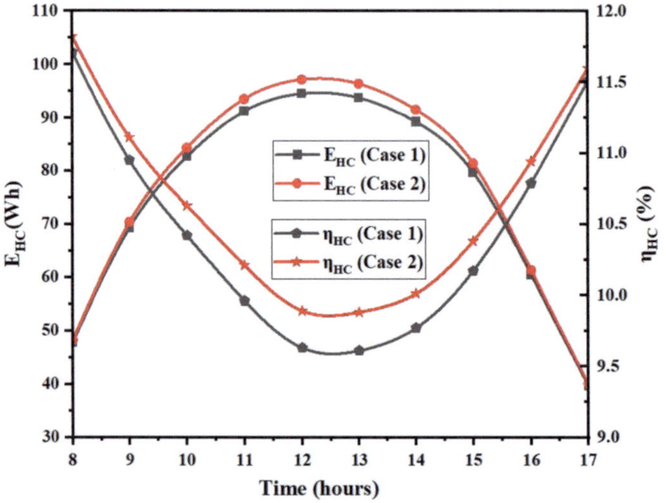

Fig. 22.2 E_{HC} and η_{HC} versus time

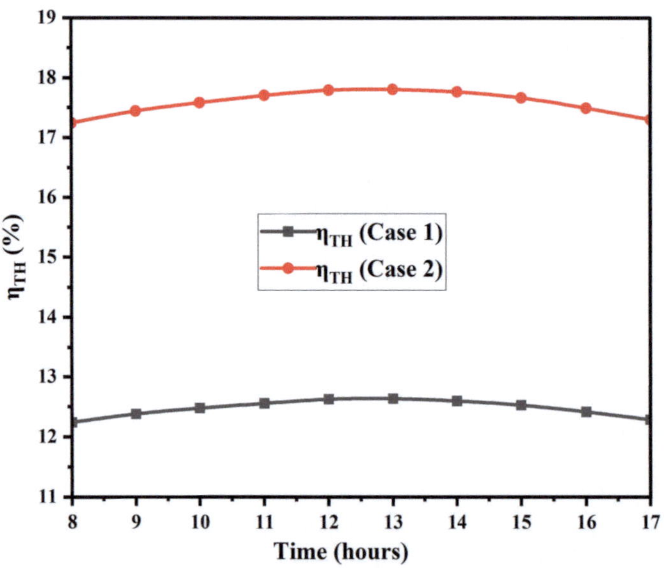

Fig. 22.3 η_{TH} versus Time

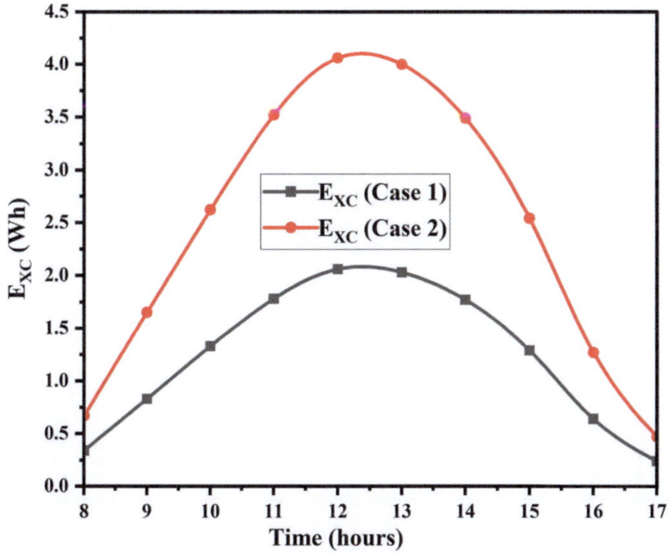

Fig. 22.4 E_{XC} versus time

I(t) decreases. On the basis of simulation results, it is also noted that case 2 has attained a higher value of E_{XC} than case 1. The upper and lower limit value of E_{XC} obtained was 4.06 Wh (case 2) and 0.24 Wh (case 1) respectively.

The graph of E_{EX} and η_{EX} vs time for both cases has been illustrated in Fig. 22.5. It is seen in Fig. 22.5 that for both cases, E_{EX} increases and attains a maxima at around noon as I(t) increases in this duration which increases electrical and thermal energy generation. After noon, the E_{EX} value starts decreasing as I(t) decreases in this duration. It is also observed in Fig. 22.5 that η_{EX} for both cases decreases and attains a minima between 12 noon and 1 pm. After 1 pm, η_{EX} increases. On the basis of simulation results, it is also noted that case 2 has attained higher values of E_{EX} and η_{EX} throughout the observation period than case 1. The upper and lower limit value of E_{EX} obtained was 101.24 Wh (case 2) and 39.94 Wh (case 1) respectively.

The upper and lower limit value of η_{EX} obtained was 11.99% (case 2) and 9.83% (case 1) respectively. The graph of T_{FO} vs time for both cases has been shown in Fig. 22.6.

Based on simulation results, it is noted that case 2 has attained higher values of T_{FO} throughout the observation period than case 1. The upper and lower limit value of T_{FO} obtained was 305.44 Kelvin (case 2) and 294.98 Kelvin (case 1) respectively. The graph of COP vs time for both cases has been shown in Fig. 22.7. On the basis of simulation results, it is noted that case 2 has attained higher values of COP throughout the observation period than case 1.

The upper and lower limit values of COP obtained were 1.07 (case 2) and 0.16 (case 1) respectively. The peak value of COP obtained in case 2 exceeded the peak value of COP obtained in case 1 by 37.17%.

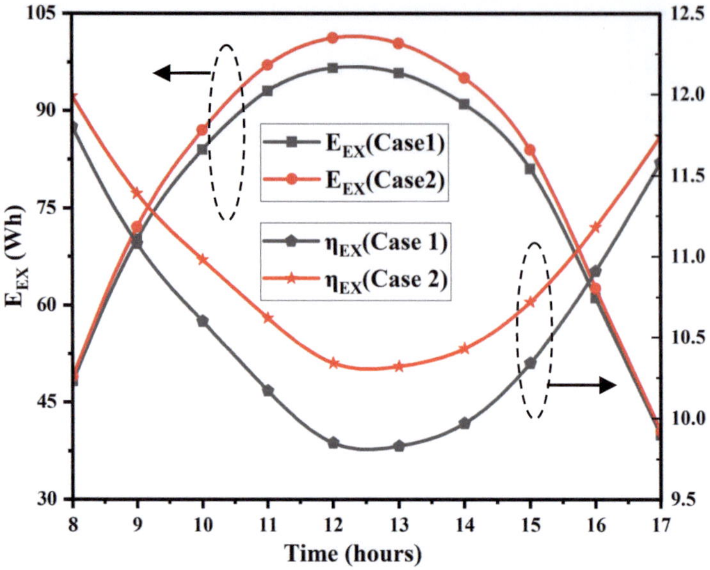

Fig. 22.5 E_{EX} and η_{EX} versus time

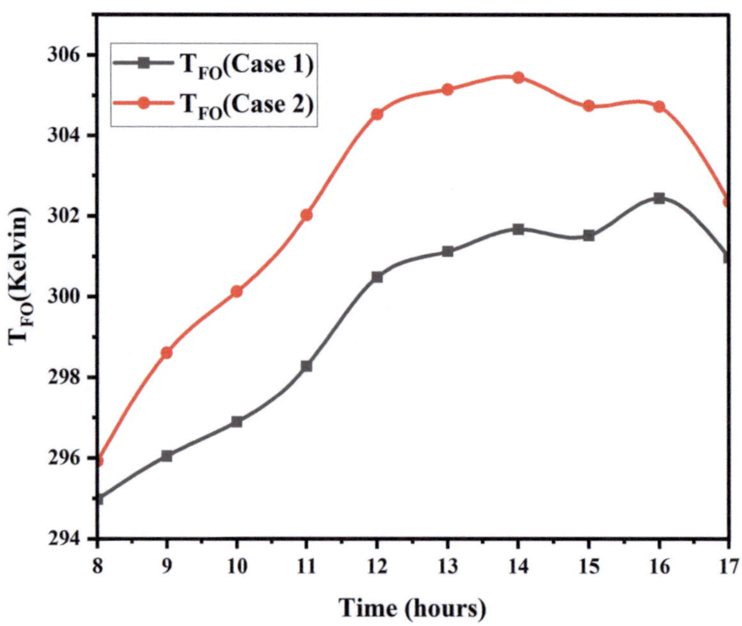

Fig. 22.6 T_{FO} versus time

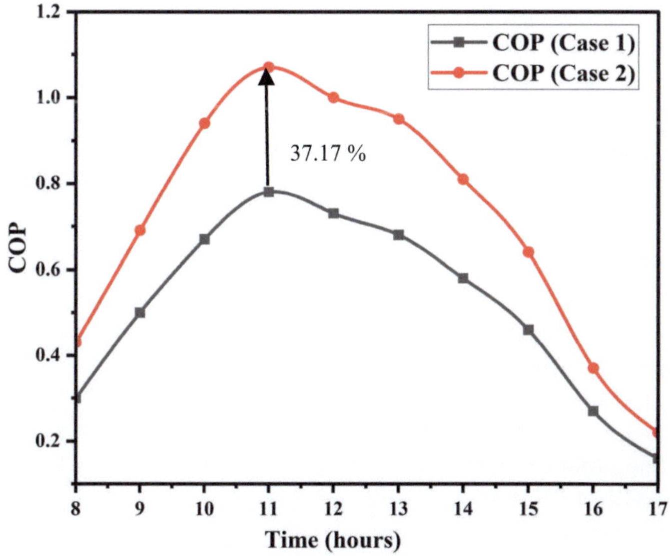

Fig. 22.7 COP versus time

22.5 Conclusions

This study introduces a novel comparison between semi-transparent and opaque PV modules in hybrid PVT-TEG systems. The conclusions drawn are as follows:

1. The electrical efficiency (η_{HC}) and electrical energy output (E_{HC}) for case 2 are higher than case 1 throughout the observation period. This highlights the innovative use of semi-transparent PV modules in PVT-TEG systems, demonstrating their superior capability in harnessing solar energy.
2. The thermal efficiency (η_{TH}) and thermal energy output (E_{XC}) for case 2 are consistently higher than in case 1, underscoring the importance of integrating semi-transparent PV modules to enhance thermal performance. This improvement is critical for applications where both electrical and thermal outputs are vital
3. The overall exergy efficiency (η_{EX}) and overall exergy output (E_{EX}) for case 2 are higher than case 1 throughout the observation period, indicating the innovative advantage of semi-transparent PV modules in optimizing exergy performance. This finding is significant for advancing the design of hybrid systems that aim to maximize energy utilization.
4. The COP value of the VAR system is higher for case 2 than case 1 by 37.17% (peak value). This result emphasizes the importance of semi-transparent PV modules in enhancing the efficiency of refrigeration systems, which is crucial for applications like vaccine refrigeration and industrial hot water supply in remote areas.

The above findings have significant implications for the development of sustainable energy solutions, particularly in critical applications where both electrical and thermal outputs are essential. This research paves the way for broader adoption of advanced PVT-TEG systems in areas with limited resources, contributing to global energy sustainability and access to essential services.

Nomenclature

T_H, T_C	Hot and cold side TEG temperature
L_{PV}	Length of PV module
m_F	Mass flow rate
c_F	Specific heat of water
T_S	Sun surface temperature
b	Width of PV panel
F	Penalty factor
U_C	Thermal conductivity coefficient from cold side to fluid
α_{SC}	Absorptivity of solar cell
α_t	Absorptivity of Tedlar
L_i	Thickness of insulator
h_O	Thermal conductivity coefficient from PV module to ambient
K_g	Glass thermal conductivity
v	Wind velocity
K_{cer}	Ceramic thermal conductivity
$U_{TA\text{-}OP}$	Thermal conductivity coefficient from opaque PV module top to ambient
$U_{BH\text{-}ST}$	Thermal conductivity coefficient from glass-to-glass PV module top to TEG
η_O	Electrical efficiency at STC
β_O	Temperature coefficient of solar cell
T_O	Reference Temperature
τ_g	Glass Transmittivity
β_{PV}	PV panel packing factor
A_{PV}	Area of PV module
U_B	Thermal conductivity coefficient from insulator to ambient
α_{TEG}	Absorptivity of TEG
K_i	Thermal conductivity of insulator
h_i	Heat transfer coefficient
L_g	Thickness of glass
K_t	Thermal conductivity of Tedlar
L_{cer}	Ceramic thickness
L_{Cu}	Thickness of copper
K_{Cu}	Thermal conductivity of copper
$U_{BH\text{-}OP}$	Thermal conductivity coefficient from opaque PV module to TEG

$U_{TA\text{-}ST}$ Thermal conductivity coefficient from glass-to-glass PV module top to ambient

Appendix

$$J_1 = \alpha_{SC}\tau_g\beta_{PV}I(t),\ J_2 = \alpha_t\tau_g(1 - \beta_{PV})I(t),\ J_3 = U_{TA-OP}T_a, J_4 = U_{BH-OP}T_h,$$
$$J_5 = \eta_o\tau_g\beta_{PV}I(t), J_6 = \eta_o\tau_g\beta_{PV}I(t)\beta_o T_o,\ J_7 = \eta_o\tau_g\beta_{PV}I(t)\beta_o,$$
$$J_8 = \alpha_c\tau_g\beta_{PV}I(t),\ J_9 = \tau_g\beta_{PV}I(t)\eta_o, J_{10} = \tau_g\beta_{PV}I(t)\,\eta_o\beta_o T_o,$$
$$J_{11} = U_{BH-ST}T_h,\ J_{12} = U_{TA-ST}T_a,\ J_{13} = \tau_g\beta_{PV}I(t)\eta_o\beta_o.$$

$$E_{EE} = \frac{\dot{E}_{XC}}{A_{AM}\dot{E}_{XI}},\ \text{where } \dot{E}_{XI} = I(t)\left(1 - \frac{4}{3}\left(\frac{T_a}{T_S}\right) + \left(\frac{T_a}{T_S}\right)^4\right)$$

References

1. Z. Wu, G. Xie, F. Gao, W. Chen, Q. Zheng, Y. Liu, Experimental study of a self-cooling concentrated photovoltaic (CPV) system using thermoelectric modules. Energy Convers. Manage. **299**, 117858 (2023)
2. N.S. Nazri, A. Fudholi, E. Solomin, M. Arifin, M.H. Yazdi, T. Suyono, E.R. Prindana, M. Mustapha, M.H. Hamsan, A.H. Hussain, M.F.S. Khaidzir, M.I. Zaini, N.N. Rosli, M. Mohammad, K. Sopian, Analytical and experimental study of hubrid photovoltaic-thermal-thermoelectric systems in sustainable energy generation. Case Stud. Therm. Eng. **51**, 103522 (2023)
3. Z. Khalili, M. Sheikholeslami, Analyzing the effect of confined jet impingement on efficiency of photovoltaic thermal solar unit equipped with thermoelectric generator in existence of hybrid nanofluid. J. Clean. Prod. **406**, 137063 (2023)
4. A.P. Singh, A. Gupta, A. Biswas, B. Das, Experimental study of a novel photovoltaic-thermal-thermoelectric generator-based solar dryer for grapes drying. Int. J. Green Energy **21**(5), 1161–1173 (2023)
5. A. Tiwari, S. Aggarwal, Thermal modelling, performance analysis and exergy study of a concentrated semi-transparent photovoltaic thermal-thermoelectric generator (CSPVT-TEG) hybrid power generation system. Int. J. Sustain. Energ. **40**(10), 947–976 (2021)
6. D. Athaeya, A. Tiwari, G.N. Tiwari, Exergy analysis of photovoltaic thermal (PVT) compound parabolic concentrator (CPC) for constant collection temperature mode. Sol. Energy **135**, 222–231 (2016)
7. D. Tiwari, A.F. Sherwani, D. Athaeya, Thermodynamic analysis of solar power organic Rankine cycle based on experimental data. Int. J. Ambient Energy **42**(2), 179–186 (2021)
8. D. Tiwari, D. Athaeya, A. Kumar, N. Kumar, Analysis of organic Rankine cycle powered by N-number of solar collectors in series. Energy Sources Part A: Recover., Util., Environ. Ef-Fects **44**(3), 6678–6697 (2022)
9. A. Tiwari, S. Anand, S. Aggarwal, Comparative study of compound parabolic concentrator - photovoltaic thermal—thermoelectric generator (CPC-PVT-TEG) collector integrated with vapour absorption refrigeration (VAR) system. Energy Sources Part A: Recover., Util., Environ. Ef-Fects **45**, 10277–103003 (2023)

10. A. Tiwari, S. Aggarwal, A comparative study of concentrated photovoltaic thermal-thermoelectric generator hybrid systems, in *Innovations in Electrical and Electronic Engineering. ICEEE 2022. Lecture Notes in Electrical Engineering*, vol 894, ed. by S. Mekhilef, R.N. Shaw, P. Siano (Springer, Singapore 2022), pp. 469–481

Chapter 23
A Sustainable Enhanced MPPT Approach with Whale Optimization Algorithm (WOA) Tuned PID Control for Solar Water Pumping

Vishal Srivastava and Anuj Rawat

Abstract This study introduces a solar-powered water pumping system driven by a BLDC motor, focusing on the development of a Proportional-Integral-Derivative controller for Maximum Power Point Tracking in photovoltaic systems under dynamic solar intensity and temperature conditions. The research establishes a control scheme for a BLDC-based solar water pump and employs MATLAB simulations to compare various techniques for extracting energy from the PV system to power the motor. Notably, the PID controller gains are optimized using a Whale Optimization Algorithm. The performance evaluation highlights the superior efficiency of the WOA-tuned PID controller compared to the Perturb and Observe method. The comprehensive investigation suggests that the proposed controller represents a promising MPPT approach with high tracking efficiency.

Keywords Proportional integral derivative (PID) · Maximum power point tracking (MPPT) · Whale optimization algorithm (WOA) · Brushless DC motor (BLDC) · Photovoltaic (PV) · Perturb and observe (P&O)

23.1 Introduction

The primary applications of SPV-based water pumping systems include irrigation in agriculture, as well as domestic and industrial water supply [1]. Although sophisticated water pumping methods have been documented in the literature, numerous

V. Srivastava (✉)
School of Engineering and Technology, Vivekananda Institute of Professional Studies - Technical Campus, Delhi 110034, India
e-mail: vishal.srivastava@vips.edu

A. Rawat
Shri Govindram Sekseria Institute of Technology and Science, Indore, India

A. Kumar et al. (eds.), *Selected Proceedings of the 1st International Conference on Advanced Materials for Sustainable Innovation; IC-AMSI 2024; 28–30 August; New Delhi; India*, Springer Proceedings in Energy,
https://doi.org/10.1007/978-981-96-4492-6_23

complex challenges have prompted researchers to explore increasingly innovative techniques, intending to develop a simple, reliable, cost-effective, and efficient system [2]. A more efficient motor-pumping system reduces the size and, consequently, the costs of the required SPV array [2]. Traditionally, water pumps have been powered by DC motors, which exhibit poor efficiency and high maintenance costs due to their switching accessories and brushes. Alternatively, induction motors have been utilized to power water pumps, overcoming the drawbacks associated with DC motors. Induction motors generally exhibit high strength, efficiency, affordability, availability, and low maintenance costs [3]. However, this motor type is not employed in the current study

While other motor types are used in solar-powered water pumping, they present drawbacks such as complex control requirements and overheating risks at low voltages. This study focuses on a feedback-based algorithm for Maximum Power Point Tracking, recognized in the literature for its effectiveness under rapidly changing irradiance conditions. A DC-DC converter [4], coupled with the BLDC, maximizes both PV array's power output and the motor-pump speed. However, traditional controlled DC bus voltage systems operating at high frequencies can lead to increased costs, especially for larger systems. To address these limitations, this research proposes a novel solar water pumping system driven by a BLDC motor [5]. The system features an MPPT control scheme based on a Proportional-Integral-Derivative controller [6], with gains optimally tuned using the Whale Optimization Algorithm. This WOA-based PID controller aims to enhance the tracking speed, accuracy, and stability of the MPPT system, ultimately improving overall efficiency.

23.2 Proposed System Configuration

This research investigates a solar water pumping system utilizing a BLDC motor. The system operates by first capturing solar energy through the SPV array (Fig. 23.1).

MPPT algorithm generates switching pulses for the converter to operate at the maximum power point of the array [7]. VSI then converts DC power to three-phase AC to drive BLDC motor [8]. VSI switching sequence is determined by an electronic switching method, utilizing Hall Effect signals from the BLDC motor's encoder to generate six pulses for precise control [9]. The design process involves selecting a BLDC motor and pump combination that meets the desired water pumping requirements. Subsequently, the appropriate power rating for the PV array is determined to ensure sufficient energy supply under various weather conditions [10]. The system's performance is evaluated through detailed simulations in MATLAB software [11]. Results indicate that the proposed BLDC-driven system achieves satisfactory performance, particularly in maintaining the PV array's operation at its maximum power point [9]. This system demonstrates higher energy conversion efficiency compared to induction motor-based systems, especially under light loads. This advantage stems from the BLDC motor's permanent magnet excitation, which eliminates the excitation losses inherent in induction motors [10].

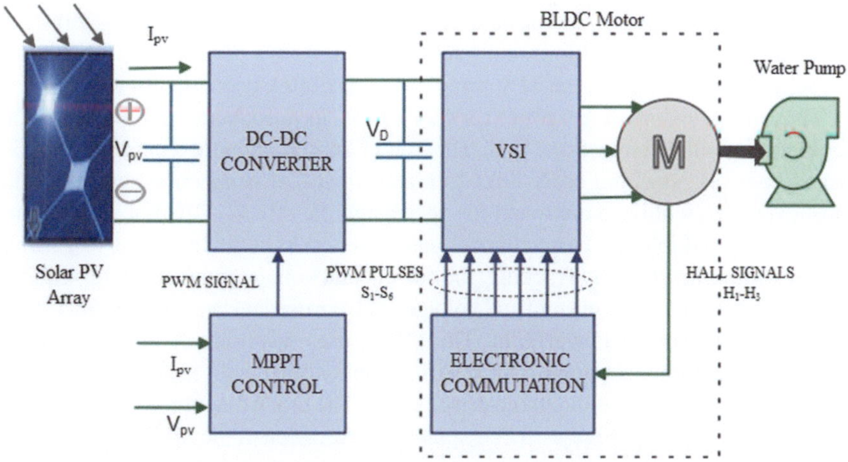

Fig. 23.1 Schematic representation of a PV-powered BLDC water pumping system

23.2.1 System Design and Specifications

The specifications and design of the BLDC motor and solar photovoltaic system are crucial factors that influence the desired operation of the water pump (Fig. 23.2).

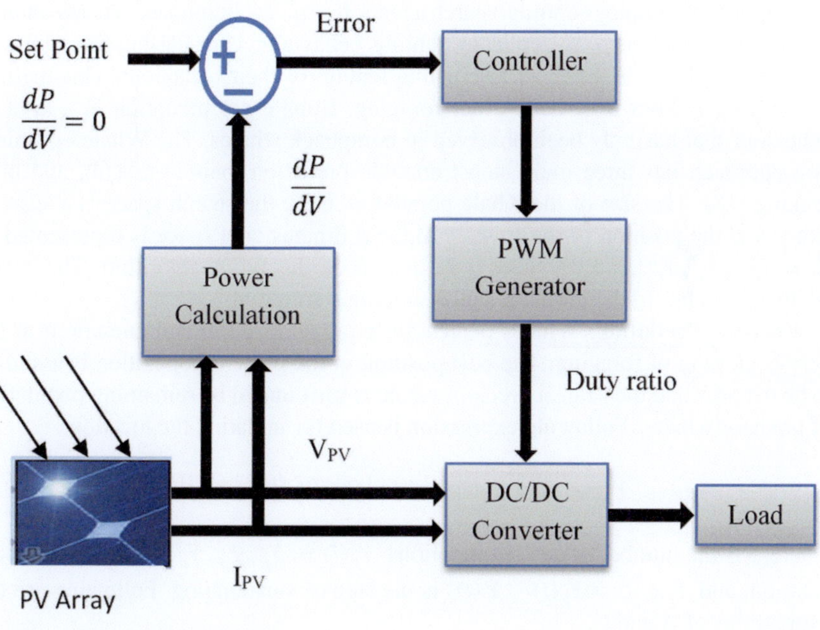

Fig. 23.2 MPPT control structure

The PV array, BLDC motor, and DC link capacitor are selected to withstand undesirable weather conditions and maintain the system's performance. Regarding the solar PV array design, the SPV array has a maximum power of 2.4 kW.

These modules are used to create a solar PV array in various configurations, such as series and parallel connections [12]. The Total Cross Tied connection scheme [13], which involves attaching cross-links to a series-parallel connection system, is the most extensively utilized approach for decreasing the effects of partial shading and losses compared to other basic connections, as reported in the literature. Regarding the Maximum PowerPoint Tracking, the current and voltage samples, along with the preceding samples of PV current and voltage [14], are used to determine the inflection point at the Maximum PowerPoint. The MPPT uses this point of inflection as the controller's reference set-point. This criterion applies universally and is unaffected by environmental conditions or characteristics [15]. The block diagram of the feedback-based MPPT, shown in Figure 23.3, displays the execution of the MPPT scheme. The MPP is continuously tracked and compared to the reference zero by computing the slope [16]. The operation of the feedback based MPPT method involves the use of a controller [6]. The controller accepts the slope-based error signal as input and generates PWM reference voltage as the output [17].

23.3 Whale Optimization Algorithm (WOA)

This algorithms employs optimal search agents, to emulate humpback whales' bubble net assault mechanism, to replicate hunting behaviour. Unusual hunting skills of Humpback whales is the most fascinating feature of their behaviour. This method of foraging is known as bubble net foraging. Bubble net predation is a kind of behaviour that has only been observed in humpback whales. The Whale optimization approach has three main steps: encircle predation, bubble assault, and prey seeking [18]. The size of the whale population is N, the search space is d dimensions, and the position of the t whale in the d dimensional space is represented as $Y_t = (Y_t^1, Y_t^1, Y_t^1, \ldots Y_t^D)$, $t = 1, 2 \ldots \ldots N$ in the WOA algorithm. The prey's position matches to the problem's global optimal solution.

Encircle Predation: Whales detects their prey's location and encircle it at the very beginning of foraging. The best position in the present population is assumed to be the prey and the nearest predator whale is surrounded by remaining population of predator whales. Following expression is used for updating the location:

$$Y(i+1) = Y_p(i) - A \times \left| C \times Y_p(i) - Y(i) \right| \tag{23.1}$$

where, i is the number of current iterations, $Y_p(i) = Y_p^1, Y_p^2, Y_p^3, \ldots Y_p^D$ is the local solution, and $A \times \left| C \times Y_p(i) - Y(i) \right|$ is the step of surrounding. Following are the expressions of A and C.

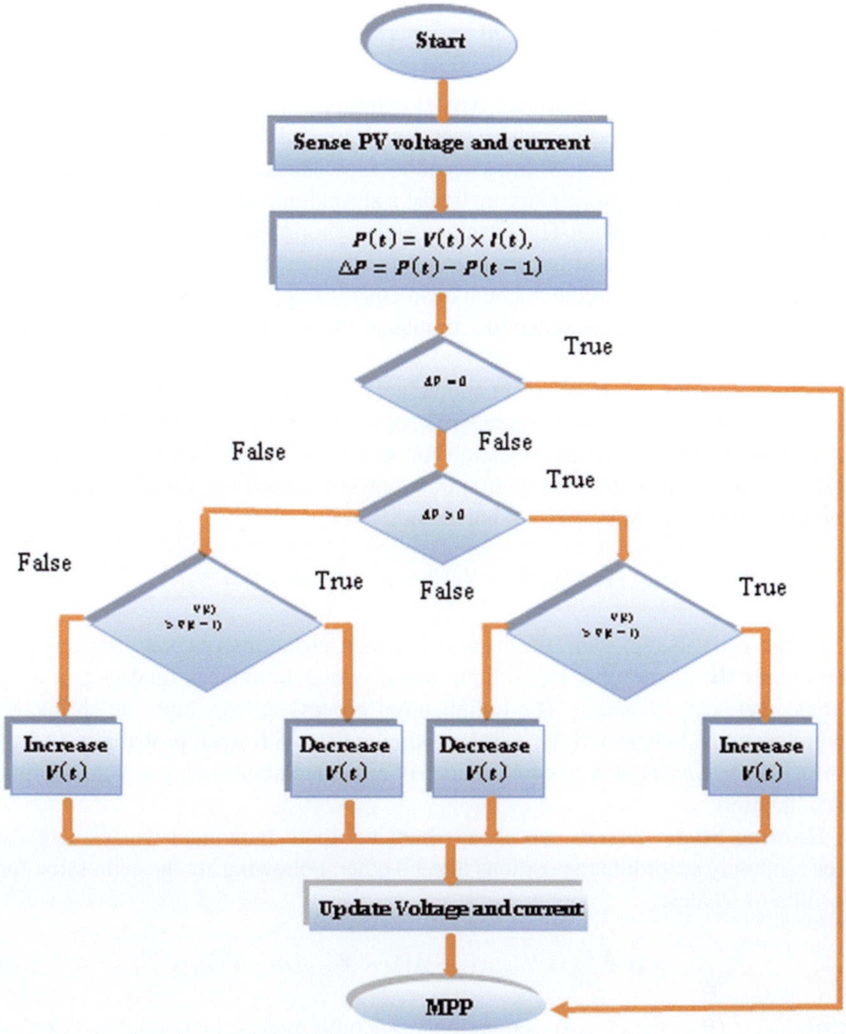

Fig. 23.3 Flow chart of P & O algorithm

$$A = 2m \times rand1 - m \tag{23.2}$$

$$C = 2 \times rand2 \tag{23.3}$$

In this context, and represent random numbers within the range of 0 to 1. The variable serves as a convergence factor, decreasing linearly from 2 to 0 as the number of iterations increases. This convergence factor can be calculated using the following formula:

$$m = 2 - \frac{2i}{i_{max}} \quad (23.4)$$

Bubble Attack: The Humpback whales exhibit a fascinating foraging technique known as bubble-net feeding, where they spiral upward through the water column, encircling and trapping prey within a net of bubbles. This behavior can be mathematically modeled using two primary methods: a shrinking and encircling mechanism, and a spiral position update. The shrinking and encircling mechanism, similar to the hunting strategies observed in certain insects and mites, is simulated by reducing a convergence factor within the mathematical equations governing the whale's movement (Eqs. 6.3–6.5). This reduction simulates the shrinking encirclement of the prey.

Before initiating the spiral ascent, individual whales determine their distance from the targeted prey. The spiral movement pattern is then mathematically modeled using the following equation [insert equation for spiral walk mode here]. This equation describes the whale's position as it spirals upward toward the surface, effectively trapping the prey within the rising bubble net.

$$Y(i+1) = \left|Y_p(i) - Y(i)\right| \times e^{lc} \times cos(2\pi l) + Y_p(i) \quad (23.5)$$

In this scenario, represents the distance between a whale and its prey. The constant determines the shape of a logarithmic spiral, while introduces randomness as a number between –1 and 1. The optimization process utilizes both the shrinkage encirclement technique and the spiral position update with equal probability (0.5 or 50%). This means there is an equal chance of either technique being selected during each iteration.

Hunting Stage: Whales can always hunt for food. Individual whales actually seek randomly according to positions of each other. Following are the expression for position of whales:

$$Y(i+1) = Y_{rand}(i) - A|C \times Y_{rand}(i) - Y(i)| \quad (23.6)$$

where, $Y_{rand}(i)$ is the current population's randomly picked individual position of whales. Objective function in this optimized algorithm is to minimized integral sqaure error (ISE).

23.3.1 Brushless DC Motor Pump Speed Control

The rotational speed of the brushless DC motor is directly influenced by the amount of power generated by the solar photovoltaic system. Environmental factors, such as irradiance and temperature, play a crucial role in the PV array's power output, ultimately affecting the BLDC motor's speed. The motor's speed is regulated by adjusting the input voltage in response to these atmospheric changes. Furthermore, variations in the Hall effect sensor signal frequency from the BLDC motor's encoder impact the fundamental frequency generated for the voltage source inverter. This, in turn, leads to changes in the switching frequency due to the modulation techniques used to control the pulse width of the switching signals [19].

23.4 Result and Discussion

This research investigates the performance of a proposed solar water pumping system under various dynamic conditions using MATLAB/Simulink simulations. The results, presented in Figures 23.4, 23.5, 23.6 and 23.7, highlight the system's behavior under a solar irradiance of 1000 W/m^2.

23.4.1 Performances of SWPS at 1000 W/m^2

Figures 23.4 and 23.5 present a comparative analysis of the proposed PID control technique and the conventional Perturb and Observe method. These figures illustrate the transient and steady-state responses. The simulation results clearly demonstrate that the PID controller-based Maximum Power Point Tracking system outperforms the P&O method. This superior performance is manifested in the system's ability to

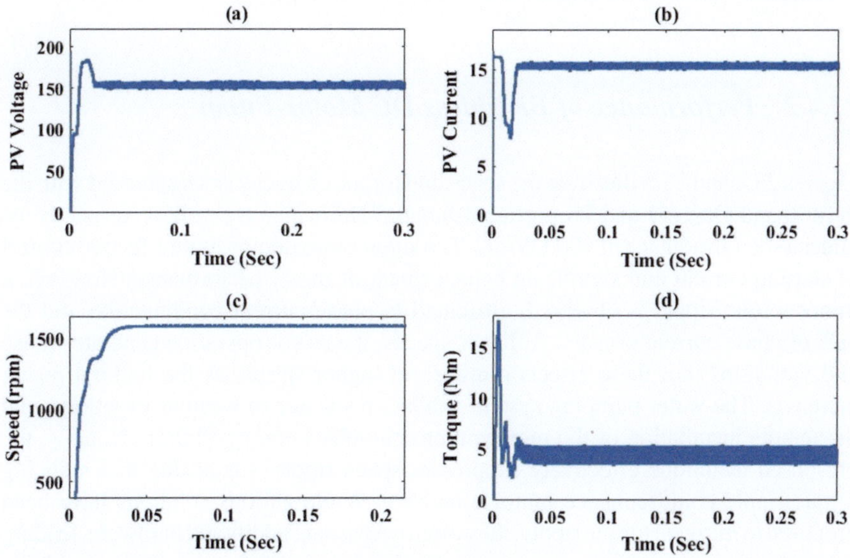

Fig. 23.4 Performance evaluation of a BLDC motor-pump and PV array employing perturb and observe method under 1000 W/m^2 irradiance

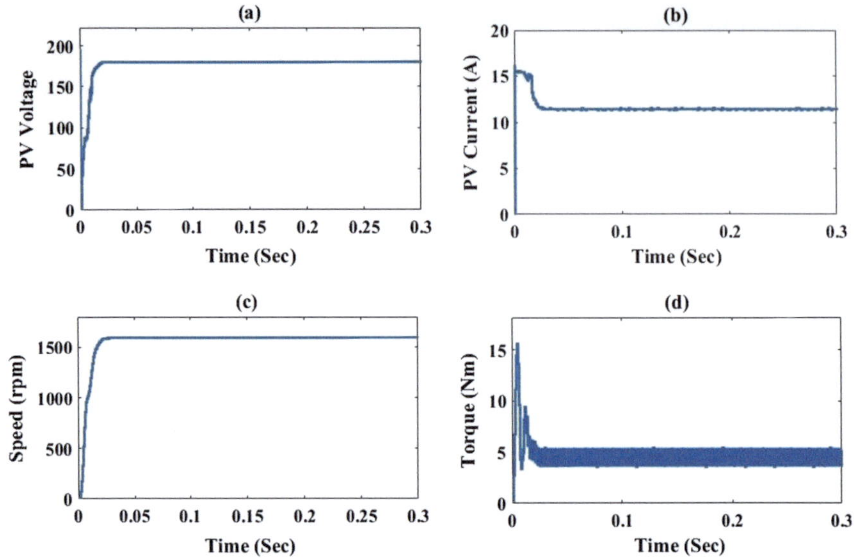

Fig. 23.5 Performance analysis utilizing PID control under 1000 W/m^2 irradiance

rapidly reach the maximum power point and maintain it consistently, as evidenced by the PV voltage and current profiles. These findings strongly suggest that the proposed system effectively and accurately tracks the MPP. Additionally, the system is designed with features that ensure safe motor starting

23.4.2 Performance of Brushless DC Motor-Pump

Figures 23.4 and 23.5 illustrate the speed and torque characteristics achieved with the Perturb and Observe and Proportional-Integral-Derivative controllers, respectively, under a solar irradiance of 1000 W/m2. The simulations demonstrate effective control of starting current and significant enhancement of motor performance. However, a minor torque ripple is observed, attributed to phase current commutation and the lack of phase current sensors. At lower speeds, the motor operation generates noise and vibrations, but these effects diminish at higher speeds as the motor's power increases. The water pumping system exhibits resilience to weather variations, and the remote installation of the motor-pump minimizes noise pollution. Notably, the pro-posed technique effectively suppresses speed ripple, suggesting that reducing torque ripple could enhance motor efficiency. While various solutions have been proposed to mitigate torque ripple, they often necessitate additional hardware, leading to increased system complexity and cost. Furthermore, at higher speeds, torque ripple becomes challenging to control, and even the most effective ripple reduction methods may not necessarily improve efficiency. The solar photovoltaic water pump system is designed for water pump operation at speeds exceeding 1100 rpm.

23.4.3 Performances at 800 W/m²

Figures 23.6 and 23.7 illustrate the robust performance of the proposed water pumping system, even under conditions of a 20% reduction in solar irradiance. The following sections will delve into the transient and steady-state behavior of the system.

Figures 23.6 and 23.7 demonstrate the significant performance improvement of the solar PV array at 800 W/m2, attributed to the accurate tracking of the Maximum Power Point. Specifically, Figure 23.6 illustrates the speed response over time using the Perturb and Observe controller, while Figure 23.7 presents the corresponding response using the Proportional-Integral-Derivative controller. Similarly, Figures 23.6 and 23.7 depict the torque response over time for the P&O and PID controllers, respectively. These results are all obtained under a solar irradiance level of 800 W/m2

As a result of the motor's rated torque, the pump operates at nearly full speed. Optimal PV array power determines the motor speed. Additionally, a sluggish startup is observed under these conditions. The performance achieved by the PID controller is superior to that of the P&O controller.

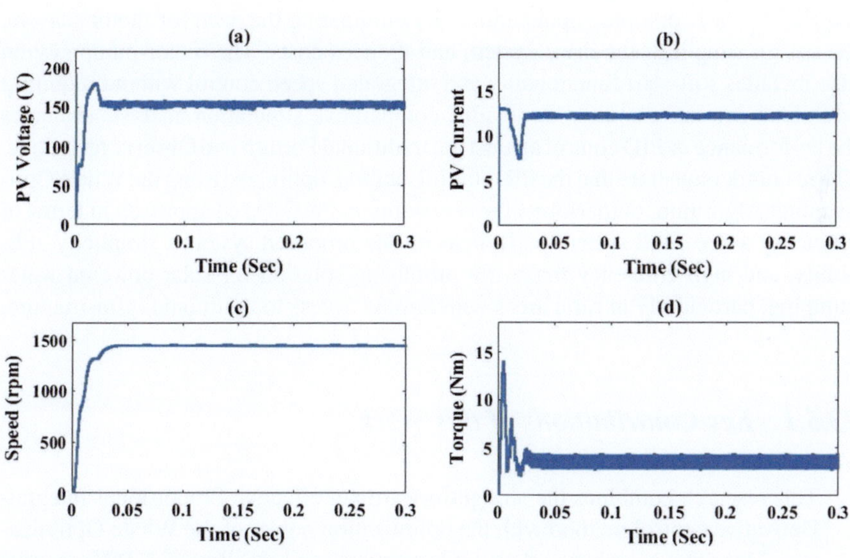

Fig. 23.6 Performance evaluation of a BLDC motor-pump and PV array employing the perturb and observe method under 800 W/m^2 irradiance

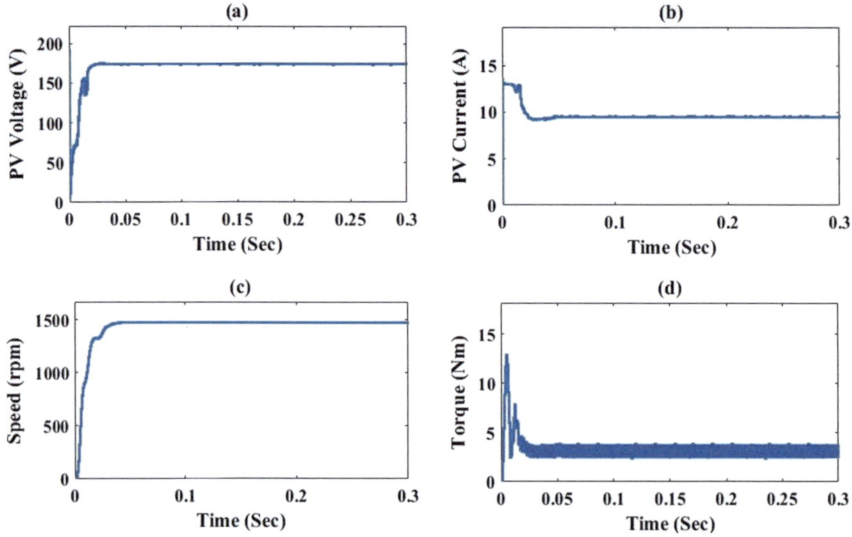

Fig. 23.7 Performance analysis utilizing PID control under 800 W/m^2 irradiance

23.5 Conclusion

This research presents a comprehensive MATLAB simulation model of a single-stage, solar-powered water pumping system utilizing a brushless DC motor. The system incorporates a high-efficiency DC-DC converter specifically designed for PV-fed BLDC water pumping applications. By eliminating the need for motor sensors, the design simplifies the drive system and reduces costs. The motor-pump system also includes soft-start functionality and integrated speed control without requiring additional external circuitry. A detailed comparative simulation analysis evaluates the performance of PID control against the traditional Perturb and Observe technique. The results demonstrate that the PID control method, optimized using the Whale Optimization Algorithm, outperforms the conventional P&O-based approach in terms of efficiency and overall system performance. The proposed system's simplicity, reliability, and high efficiency make it a promising solution for solar-powered water pumping, particularly in rural areas with limited access to traditional infrastructure.

23.5.1 Key Contributions of this Work

1. This research combines the straightforward yet effective Proportional-Integral-Derivative control method with the optimization power of the Whale Optimization Algorithm to enhance the speed, accuracy, and stability of MPPT systems beyond conventional approaches. By optimizing the MPPT process, the system

aims to maximize energy extraction from the photovoltaic array, improving overall efficiency and water pumping performance, especially under fluctuating solar irradiance.
2. The developed model allows for a comprehensive analysis of system behavior under various operating conditions, providing insights into the effectiveness of the proposed Whale Optimization Algorithm-tuned PID controller. This contributes to the advancement of sustainable and efficient water pumping solutions, particularly for off-grid applications. By effectively utilizing solar energy, the system presents an environmentally friendly alternative to traditional fossil fuel-powered pumps.
3. The research findings have practical implications for designing and implementing efficient solar water pumping systems. The proposed Whale Optimization Algorithm-tuned PID controller can be easily implemented using affordable microcontrollers, making it suitable for real-world applications

References

1. Krishnakant, P.S. Kulkarni, Design and simulation of solar photovoltaic water pumping system
2. S. Kumar, S. Kumar, Solar PV powered water pumping system using dc motor drive: a critical review
3. Abdelrahman and M.Z. Youssef, A cost effective magnetic/electronic design for the water pump application drive: analysis, design, and experimentation
4. R. Kumar, B. Singh, Grid interactive solar PV-based water pumping using BLDC motor drive
5. H. He, W. Liu, L. Bao-hua, R. Ma, A novel soft switching inverter for BLDCM EV propulsion system
6. D. Yang, Tuning of PID parameters based on particle swarm optimization
7. M.A. Jusoh, M.F.N. Tajuddin, S.M. Ayob, M.A. Roslan, Maximum power point tracking charge controller for standalone PV system
8. R. Kumar, B. Singh, Solar PV array fed cuk converter-VSI controlled BLDC motor drive for water pumping
9. L. Xu, R. Cheng, J. Yang, A New MPPT Technique for fast and efficient tracking under fast varying solar irradiation and load resistance
10. N. Loganathan, C. Naveen, K. Nivashini, B. Nivin, S. Pavurnaa, Solar photo voltaic based water pumping using BLDC motor
11. D.U. Nandini, J.J. Daniel, Modelling of BLDC motor driven solar water pump with INC-MPPT in MATLAB
12. M. Benzaouia, L. Bouselham, B. Hajji, A. Migan-Dubois, A. Benslimane, M.E. Ouariachi, Design and performance analysis of a photovoltaic water pumping system based on DC-DC boost converter and BLDC motor
13. S.S. Kumar, C. Bibin, K. Akash, K. Aravindan, M. Kishore and G. Magesh, Solar powered water pumping systems for irrigation: a comprehensive review on developments and prospects towards a green energy approach
14. R.K. Megalingam, V.V. Gedela, Solar powered automated water pumping system for eco-friendly irrigation
15. M. Santhosh, R.R.S. Raj, R. Sivakumar, P. Deepa, D. Shivaani, Fractional-order controller design for electrical systems—a study
16. S. Miqoi, A. E. Ougli, B. Tidhaf, Design of an adaptive sliding mode controller for efficiency improvement of the MPPT for PV water pumping

17. P.K. Singh, B. Singh, V. Bist, Brushless DC motor drive with power factor regulation using Landsman converter. IET Power Electron. **9**(5), 900–910 (2016)
18. Mirjalili, A. Lewis, The whale optimization algorithm. Adv. Eng. Softw. **95**, 51–67 (2016)
19. R. Kumar, B. Singh, Solar PV powered BLDC motor drive for water pumping using Cuk converter. IET Electric Power Appl. **11**(2), 222–232 (2017)

Chapter 24
Enhancing Evaporation and Condensation in Solar Desalination Systems Through Hybridization: A Mini-Review

Rahul Patel, Rahul Dev, Ajaya Bharti, Dhananjay Singh, and Anil Kumar

Abstract With the global population surpassing 8.1 billion people, there is a growing strain on freshwater resources. The world is currently facing the challenge of addressing the escalating demand for fresh water, a fundamental necessity crucial for human survival and various other essential activities. Solar desalination addresses the growing demand for freshwater sustainably. By utilizing solar energy, desalination systems convert seawater into potable water. The study delves into various methods and technologies aimed at optimizing the performance of these systems. Specifically, it explores the incorporation of high-frequency ultrasonic wave atomizers in single-slope solar stills, demonstrating an effective evaporation technique resulting in a significant 56% increase in production. Additionally, the effectiveness of external condensers in enhancing the condensation process is discussed. The integration of solar stills with pin fins absorbers and integrated condensers shows a promising 32% increase in production. This review aims to provide valuable insights into strategies for efficiency and output improvement of solar desalination systems through the synergistic combination of different technologies.

R. Patel · R. Dev (✉)
Department of Mechanical Engineering, MNNIT Allahabad, Prayagraj, Uttar Pradesh, India
e-mail: rahuldsurya@mnnit.ac.in

R. Patel
e-mail: rahul.2020rme09@mnnit.ac.in; rahulbestkumar95@gmail.com

A. Bharti
Department of Applied Mechanics, MNNIT Allahabad, Prayagraj, Uttar Pradesh, India
e-mail: abharti@mnnit.ac.in

D. Singh
Department of Chemical Engineering, IET Lucknow, Lucknow, Uttar Pradesh, India
e-mail: dhananjay.singh@ietlucknow.ac.in

A. Kumar
Deptartment of Mechanical Engineering, Delhi Technological University, Delhi, India
e-mail: anilkumar76@dtu.ac.in

A. Kumar et al. (eds.), *Selected Proceedings of the 1st International Conference on Advanced Materials for Sustainable Innovation; IC-AMSI 2024; 28–30 August; New Delhi; India*, Springer Proceedings in Energy,
https://doi.org/10.1007/978-981-96-4492-6_24

Keywords Solar desalination system · Hybridization · Evaporation enhancer · Condensation enhancer

24.1 Introduction

Population growth and the limited availability of freshwater have indeed raised concerns about the supply of clean drinking water and its impact on public health. Researchers have been working on water purification technologies for over a century, but there is still a need for effective policies and methods to ensure clean water for everyone. Many water purification technologies have been developed, and some of them rely on electricity, which can have environmental consequences. Solar distillation is considered an immaculate and eco-friendly method for drinking water treatment. The desalination works on the principle of evaporation and condensation and the use of solar energy works as a heating source [1]. Numerous desalination methods are available for the production of clean and fresh water. Thermal desalination systems, known for their straightforward designs and cost-effective installation and operation, encompass various approaches such as solar stills [2], absorber solar still [3], conical solar still [4], nanofluid [5], hemispherical solar still [6] concept of micro and nano-particles in still [7], modified solar stills [8–10], Triple slope solar still integrated with PV [11], High frequency wave atomizer [12], black dye in the basin of still to enhance the performance [13], humidification and dehumidification effect on the passive solar still [14], as well as MSF desalination [15], Multi-Effect Distillation system [16], and vapor compression process [17]. Membrane-type desalination systems. Front, options like reverse osmosis (RO) [18], organic Rankine cycle based on solar applications [19], and membrane-used desalination and still [20], along with innovative hybrid systems [21], hybrid Reverse osmosis methods [22] are gaining attention in the still for enhanced desalination processes. Hybrid type solar stills shows the effective improvement in performance as production rate and efficiencies.

Hybrid solar desalination systems

Hybrid solar desalination, involving the integration of two or more solar-driven techniques, presents a promising system for more efficient solutions characterized by reduced energy consumption and lower desalinated water costs compared to individual methods [23]. Several prior reviews (Table 24.1) have discussed the historical progress in maximizing heat absorption and enhancing system design in solar-powered desalination systems.

Hybrid solar distillation systems can be broken down into several key sections or components such as solar radiation collection section (by using advanced absorbing materials like nano partials or pin fin), heating section, Evaporation section, condensation section thermal storage and water preheating sections.

In the present study the review has been selected for the section of evaporation and condensation enhancement under the hybridization.

Table 24.1 Literature survey

Authors	Key findings
Rahimi et al. [24]	Enhancing evaporation rates and productivity in HDH systems can be achieved by elevating the inlet air and water temperatures while optimizing airflow rates. Water heated HDH systems generally demonstrate a higher Gain Output Ratio when compared to HDH systems that rely on air or dual heating sources
Lawal and Qasem [25]	Studied the performance of HDH system driven by different power producing sources as power plants, VCRS, geothermal energy, solar thermal and electric systems emerge as preferred choices for hybridization in off-grid regions. These sources offer promising alternatives, considering factors such as water production under low cost and sustainability
Faegh et al. [26]	A thorough investigation into contemporary research on H–D technologies and their integrational effect in desalination technologies
Kasaeian et al. [27]	Solar desalination employing HDH techniques proves to be an efficient solution for remote areas with moderate freshwater needs, demanding minimal operational considerations
Santosh et al. [28]	Shown the enhancements in system performance are attributed to factors as the heat transfer rate, forced convection, a multi pass techniques. The use of nano heat transfer fluid, and the incorporation of PCM. The HDH process proves to be appealing for small-scale units, especially when utilizing cost-effective heat sources
Moumouh et al. [29]	Reviewed several technologies currently under investigation for desalinating saline water through the process of humidification-dehumidification (HD) of air with a specific focus on leveraging solar thermal energy. This examination places particular emphasis on diverse types of solar water heaters and various designs of solar air heaters
Pansal et al. [30]	Studied that by interpolation of Photovoltaic in the solar stills, The overall thermal efficiency (η) is 25.0% higher than that of conventional counterparts
Singh et al. [31]	Solar stills incorporating nano-fluids exhibit superior performance and increased potable water production in comparison to traditional models
Fang et al. [32]	A thorough examination covered planning criteria for desalination, schemes to hold, and economic viability analysis is conducted for a wastewater desalination project in China
Sharshir et al. [33]	Utilizing nanotechnology results in a 30% increase in productivity and a 38% improvement in efficiency
Arunkumar et al. [34]	Exploration of diverse water collection systems, encompassing solar stills, atmospheric water distillers, bio-sand filters, and steam generation utilizing absorber materials (solar radiation)
Srithar and Rajaseenivasan [35]	Found that in the HDH desalination, the bubble column humidifier stands out as superior when compared to packed bed and other humidifier types. Additionally, it has been observed that biomass interpolated desalination systems are particularly well-suited for regions with weak intensity of solar radiation and abundant source of available biomass

(continued)

Table 24.1 (continued)

Authors	Key findings
Khan et al. [36]	A notable decrease in water production costs is observed with hybrid by PV and wind tunnel integrated RO systems, suggesting promising prospects for this technology
Anand et al. [37]	An examination of the integration of PV/T systems with diverse desalination technologies and stills
Dixit et al. [38]	Exploration of MXenes, a category of two-dimensional materials, for applications in water treatment and energy-efficient desalination

24.1.1 Hybridization to enhance evaporation of the still

An evaporation enhancer refers to a device designed to augment the dehumidification process by elevating humidity levels, such as a high-frequency ultrasonic waves atomizer. In another study, an experiment is conducted experiments on a hybrid desalination system involving a tubular solar still. They introduced a vibrator to induce a forced vibrational effect, affecting the surface tension behavior of saline water and boundary layer dynamics. The modified system produced distillate water at a rate of 4.2 L/m^2, demonstrating a 34% improvement over the conventional setup. The estimated cost was reported as 0.031 USD/L.m^2 [12]. A solar desalination system using a humidification-dehumidification (HDH) process with an external collector (evacuated tube type) and an ultrasonic wave atomizer for spraying of hot water inside the basin. The use of ultrasonic techniques significantly enhanced humidification efficiency, rapidly achieving 100% relative humidity. The results showed a 14.6% increase in system productivity, with a maximum yield of 45.0 Liter in a day [39]. Furthermore, an experiment on a solar desalination system employing a fog technique in conjunction with HDH and a solar driven electric heater. The system, powered by fog nozzles with a diameter of 1.0 mm and a fog pump pressure of 40 bar, achieved a maximum efficiency of 39%. The output and distillate cost for the unit were reported as 5.8 l/m^2 per day and 0.035 $/l, respectively [40]. Experimental research to enhance the output of a solar still by incorporating wind turbine. The study compared the performance of a conventional SS with that of a fan-assisted SS at varying water heights and rotational speeds. Results demonstrated a 17% increase in system production when utilizing a water fan in the conventional SS, with the maximum thermal efficiency obtained of the fan-assisted SS reaching approximately 40.0% [41]. Enhancement of SS performance using a vertical rotating wick. System performance was evaluated across different seasons, revealing a system yield of 5.0 L/m2 and 7.2 L/m^2 in the winter and summer respectively. The solar still, with vertical rotating wick (VRW), enhanced productivity by 14.7% and 51.1% in in summer and winter respectively [42]. Experimented with a double slope SS combined with different wicks. Tested the system in throughout year and revealed maximum production rates of 9 L/day for black cotton. With thermal efficiency of 23.0% [43]. The use of an ultrasonic waves (high-frequency) atomizer as a humidifier in a solar

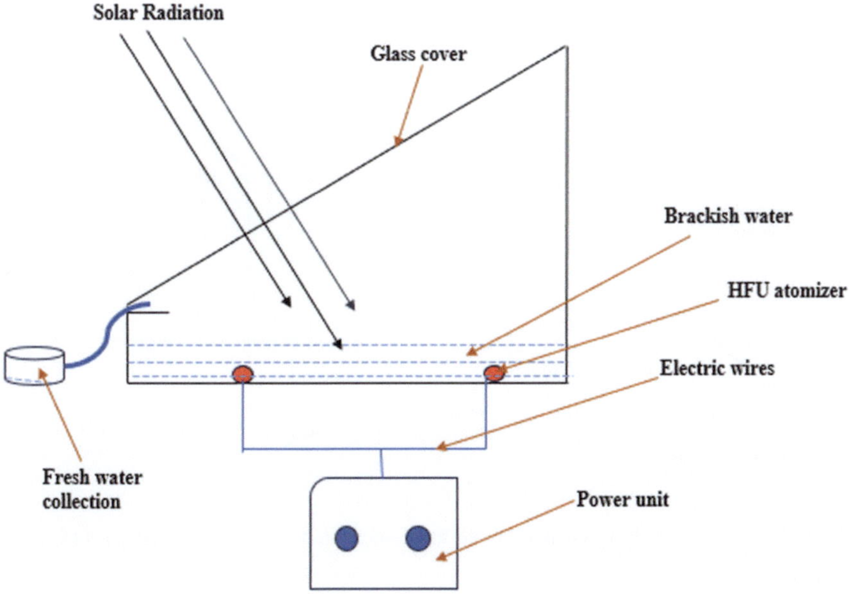

Fig. 24.1 Desalination system with a high-frequency ultrasonic waves atomizer [12]

desalination system (SDS). Their experimental setup as shown in Fig. 24.1, integrated the atomizer with a solar still (SS) and investigated the impact of factors like atomizer number, water depth, and sample water quality on system yield. The results indicated a maximum productivity of 4.41 l/m^2 and a thermal efficiency of 55.75% [12].

Figure 24.2 represents the experimental setup of HDH system. The investigation was conducted on SS using the humidification and dehumidification method, incorporating a hot air stream and heat pump. The study explored the impact of factors such as the total number of high-frequency ultrasound atomizers, water depth, and heated air flow rate. The findings revealed that by increasing the number of atomizers with decreasing water depth resulted higher production. Optimal conditions were identified at six atomizers and 1.0 cm water depth with hot air flow rate of 0.011 kg/sec, leading to a substantial increase in daily freshwater productivity. A daily production of 7.72 L, energy and exergy efficiency of 33.84% and 1.43% respectively, and distillate cost of 0.0112 US$/L [44].

24.1.2 Hybridization to Enhance Condensation of the Still

As in the soar distillation process condensation takes place when the basin water vapor molecules are mixed with still air and get higher kinetic energy which becomes the reason to rise the motion of the mixture and finally collision with all the boundaries after that it gated energy down due to lose in latent heat and condensation takes place.

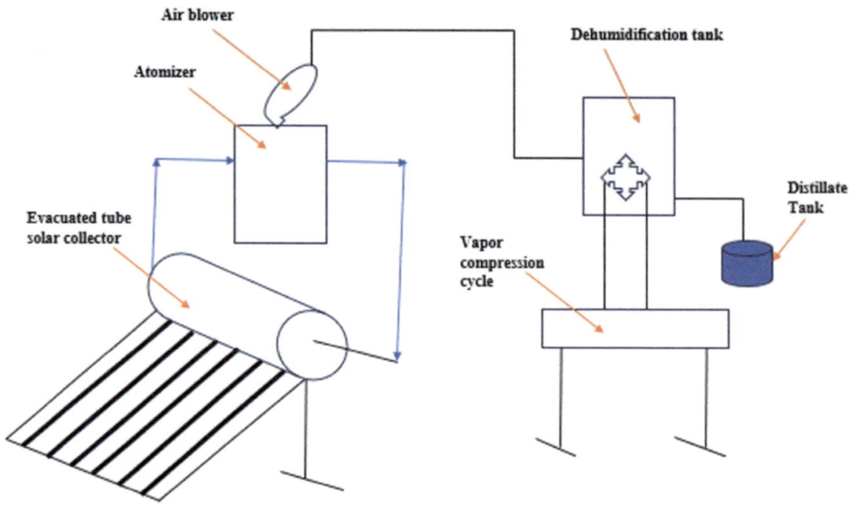

Fig. 24.2 Solar desalination system using the humidification-dehumidification (HDH) [44]

Condensation enhancement involves employing a changed technique to increase the rate of condensed water production. In a separate study, Conducted experiments on a membrane distillation which hybrid with solar systems. The results demonstrated a thermal efficiency of 49.01%, a GOR (Gain Output Ratio) of 0.49, and a daily production of 33.55 L, the pilot plant comprised a two-stage H–D desalination system, and the analysis revealed satisfactory desalination water yield in the second stage operation [45]. Experimental research on hybrid desalination, employing a dehumidification by well-cooling chamber/plant [46]. In the realm of portable solar stills, a system integrated with thermoelectric heating and cooling and heat pipe technologies [47, 48]. Solar stills sensitivity to atm temperature and global radiation intensity, with higher distillate production observed at lower thermoelectric surface temperatures. The maximum efficiency of 7.0% [49]. A modification has been done in solar still by incorporating of pin fins in the absorber surface and integration of external condenser as shown in Fig. 24.3, that underwent experimental observations against different configurations. The modifications demonstrated a substantial 41.95% water production and a 12.9% Instantaneous efficiency increase. The use of an external condenser with pin fins absorber outperformed other setups, resulting by 32.18% in water production enhancement. This study underscores the effectiveness of the external condenser in enhancing solar still efficiency [50].

Figure 24.4 shows a solar still and PV, incorporating a condenser coated with silver/nanofluid. The system utilized thermoelectric elements with dual-acting heating and cooling mechanisms powered by photovoltaic energy. The study revealed a 26.7% increase in system productivity with the use of an external condenser [51].

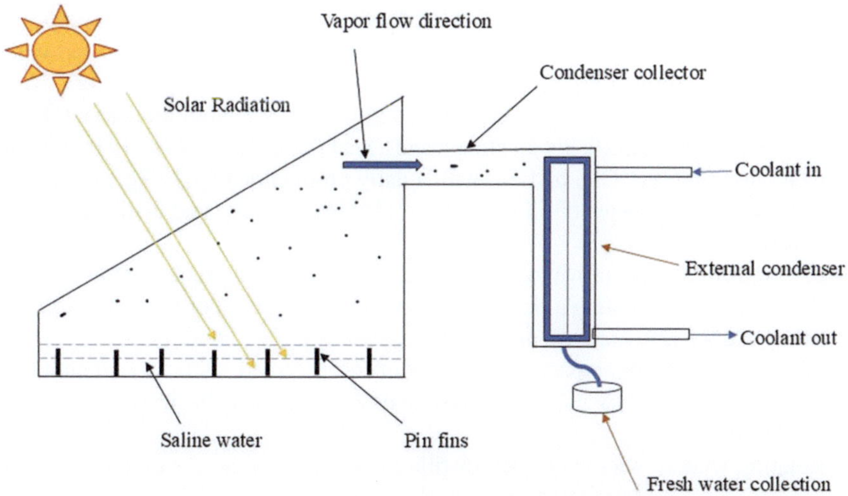

Fig. 24.3 Modified solar still incorporating a pin fins absorber and integrated condenser [50]

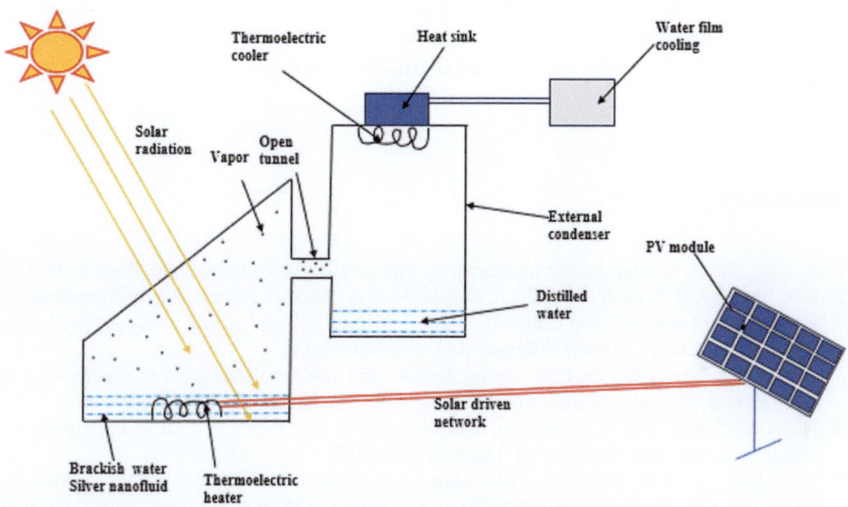

Fig. 24.4 Solar still with condenser coated and silver/nanofluid [51]

24.2 Conclusion

In this study, a comprehensive review has been conducted, and analysis has been carried out focusing on the enhancement of evaporation and condensation in solar-driven distillation systems. The following conclusions have been drawn:

- The introduction of a high-frequency ultrasonic waves atomizer in a conventional solar still demonstrates an effective evaporation method, can be enhanced more than 50% in production.
- Solar desalination system using the humidification-dehumidification (HDH) method, incorporating a hot air stream and a heat pump demonstrated a maximum gain output ratio (GOR) of 1.54, with a daily production of 7.72 kg, and energy and exergy efficiencies of 33.84% and 1.43% respectively, and distillate cost of 0.0112 US$/L.
- External condensers prove to be effective in enhancing the condensation process. Solar stills incorporating a pin fins absorber and integrated condenser exhibit can be more than 32% increase in production.
- Solar still and PV, incorporating a condenser coated with silver/nanofluid. Revealed approx. 25% increase in system productivity with the use of an external condenser.
- Rotating wick or wick type solar distillation system are the cheapest method for enhancing the evaporation rate.
- Nano fluid, PCMs are the advanced type material which enhances the radiation absorption rate of the solar stills.
- The overall conclusion of the study is that the hybridization of solar distillation unit in evaporation and condensation processes shows as an effective method in enhancement of production rate and thermal efficiency.

References

1. R. Dev, Tiwari, Drinking water treatment, Springer Netherlands Amsterdam, Book (2011)
2. S.W. Sharshir, G. Peng, N. Yang, M.A. Eltawil, M.K.A. Ali, A.E. Kabeel, A hybrid desalination system using humidification-dehumidification and solar stills integrated with evacuated solar water heater. Energy Convers. Manage. **124**, 287–296 (2016)
3. R. Dev, S.A. Abdul-Wahab, G.N. Tiwari, Performance study of the inverted absorber solar still with water depth and total dissolved solid. Appl. Energy **88**(1), 252–264 (2011)
4. H.E. Gad, S.S. El-Din, A.A. Hussien, K. Ramzy, Analysis of a conical solar still performance: an experimental study. Sol. Energy **122**, 900–909(2015)
5. A.W. Kandeal, N.M. El-Shafai, M.R. Abdo, A.K. Thakur, I.M. El-Mehasseb, I. Maher, M. Rashad, A.E. Kabeel, N. Yang, S.W. Sharshir, Improved thermo-economic performance of solar desalination via copper chips, nanofluid, and nano-based phase change material. Sol. Energy **224**, 1313–1325 (2021)
6. T. Arunkumar, R. Jayaprakash, D. Denkenberger, A. Ahsan, M.S. Okundamiya, S. Kumar, H. Tanaka, H.S. Aybar, An experimental study on a hemispherical solar still. Desalination **286**, 342–348 (2012)
7. G. Peng, S.W. Sharshir, Y. Wang, M. An, D. Ma, J. Zang, A.E. Kabeel, N. Yang, Potential and challenges of improving solar still by micro/nano-particles and porous materials—a review. J. Clean. Prod. **311**, 127432 (2021)
8. P. Pal, S.K. Patel, A. Bharti, A. Narayan, R. Dev, D. Singh, Energy, exergy, energy matrices, exergoeconomic and enviroeconomic assessment of modified solar stills. Sustainable Energy Technol. Assess. **47**, 101514 (2021)

9. P. Pal, R. Dev, Performance study of modified basin-type single-slope solar distiller. Euro-Mediterranean J. Environ. Integr. **3** (2018)
10. P. Pal, R. Dev, Experimental study on modified double slope solar still and modified basin type double slope multiwick solar still. Int. J. Civ. Environ. Eng. **10**(1), 70–75 (2016)
11. R. Patel, A. Kumar, A. Kumar, R. Dev, V. Narayanan, Water quality enhancement using PV-integrated triple slope solar still. NanoWorld J **9**(S1), 336-S340 (2023)
12. E.M.S. El-Said, G.B. Abdelaziz, Experimental investigation and economic assessment of a solar still performance using high-frequency ultrasound waves atomizer. J. Clean. Prod. **256**, 120609 (2020)
13. B.A. Akash, M.S. Mohsen, O. Osta, Y. Elayan, Experimental evaluation of a single-basin solar still using different absorbing materials. Renew Energy **14**(1–4), 307–310 (1998)
14. A.E. Kabeel, E.M. El-Said, A hybrid solar desalination system of air humidification, dehumidification and water flashing evaporation: Part II. Experimental investigation. Desalination **341**, 50–60 (2014)
15. S. Kabiri, M.K. Manesh, M. Yazdi, M. Amidpour, New procedure for optimal solar repowering of thermal power plants and integration with MSF desalination based on environmental friendliness and economic benefit. Energy Convers. Manage. **240**, 114247 (2021)
16. S. Aly, H. Manzoor, S. Simson, A. Abotaleb, J. Lawler, A.N. Mabrouk, Pilot testing of a novel multi effect distillation (MED) technology for seawater desalination. Desalination **519**, 115221 (2021)
17. M.A. Farahat, H.E. Fath, I.I. El-Sharkawy, S. Ookawara, M. Ahmed, Energy/exergy analysis of solar driven mechanical vapor compression desalination system with nano-filtration pretreatment. Desalination **509**, 115078 (2021)
18. K.M. Kotb, M.R. Elkadeem, A. Khalil, S.M. Imam, M.A. Hamada, S.W. Sharshir, A. Dán, A fuzzy decision-making model for optimal design of solar, wind, diesel-based RO desalination integrating flow-battery and pumped-hydro storage: case study in Baltim Egypt. Energy Convers. Manag. **235**, 113962 (2021)
19. M. Sharma, R. Dev, Review and preliminary analysis of organic rankine cycle based on turbine inlet temperature (2018)
20. D.L.M. Mendez, C. Castel, C. Lemaitre, E. Favre, Membrane distillation (MD) processes for water desalination applications. Can dense selfstanding membranes compete with microporous hydrophobic materials. Chem. Eng. Sci. **188**, 84–96 (2018)
21. S.W. Sharshir, G. Peng, N. Yang, M.O.A. El-Samadony, A.E. Kabeel, A continuous desalination system using humidification–dehumidification and a solar still with an evacuated solar water heater. Appl. Therm. Eng. **104**, 734–742 (2016)
22. M. Abdelgaied, A.E. Kabeel, A.W. Kandeal, H.F. Abosheiasha, S.M. Shalaby, M.H. Hamed, N. Yang, S.W. Sharshir, Performance assessment of solar PV-driven hybrid HDH-RO desalination system integrated with energy recovery units and solar collectors: theoretical approach. Energy Convers. Manage. **239**, 114215 (2021)
23. G.B. Abdelaziz, E.M. El-Said, A.G. Bedair, S.W. Sharshir, A.E. Kabeel, A.M. Elsaid, Experimental study of activated carbon as a porous absorber in solar desalination with environmental, exergy, and economic analysis. Process. Saf. Environ. Prot. **147**, 1052–1065 (2021)
24. Z. Rahimi-Ahar, M.S. Hatamipour, L.R. Ahar, Air humidification-dehumidification process for desalination: a review. Prog. Energy Combust. Sci. **80**, 100850 (2020)
25. D.U. Lawal, N.A. Qasem, Humidification-dehumidification desalination systems driven by thermal-based renewable and low-grade energy sources: a critical review. Renew. Sustain. Energy Rev. **125**, 109817 (2020)
26. M. Faegh, P. Behnam, M.B. Shafii, A review on recent advances in humidification-dehumidification (HDH) desalination systems integrated with refrigeration, power and desalination technologies. Energy Convers. Manag. **196**, 1002–1036 (2019)
27. A. Kasaeian, F. Rajaee, W.M. Yan, Osmotic desalination by solar energy: a critical review. Renew. Energy **134**, 1473–1490 (2019)
28. R. Santosh, T. Arunkumar, R. Velraj, G. Kumaresan, Technological advancements in solar energy driven humidification-dehumidification desalination systems—a review. J. Clean. Prod. **207**, 826–845 (2019)

29. J. Moumouh, M. Tahiri, M. Salouhi, Solar thermal energy combined with humidification–dehumidification process for desalination brackish water: technical review. Int. J. Hydrog. Energy **39**(27), 15232–15237 (2014)
30. K. Pansal, B. Ramani, K. Kumar Sadasivuni, H. Panchal, M. Manokar, R. Sathyamurthy, M. Suresh, M. Israr, Use of solar photovoltaic with active solar still to improve distillate output: a review. Groundw. Sustain. Dev. **10**, 100341 (2020)
31. A.K. Singh, D.B. Singh, V.K. Dwivedi, Tiwari, G.N. and Gupta, A.: Water purification using solar still with/without nano-fluid: a review. Mater. Today: Proc. **21**, 1700–1706 (2020)
32. S. Fang, W. Tu, L. Mu, Z. Sun, Q. Hu, Y. Yang, Saline alkali water desalination project in Southern Xinjiang of China: a review of desalination planning, desalination schemes and economic analysis. Renew. Sustain. Energy Rev. **113**, 109268 (2019)
33. S.W. Sharshir, Y.M. Ellakany, A.M. Algazzar, A.H. Elsheikh, M.R. Elkadeem, E.M. Edreis, A.S. Waly, R. Sathyamurthy, H. Panchal, M.S. Elashry, A mini review of techniques used to improve the tubular solar still performance for solar water desalination. Process. Saf. Environ. Prot. **124**, 204–212 (2019)
34. T. Arunkumar, K. Raj, D.D.W. Rufuss, D. Denkenberger, G. Tingting, L. Xuan, R. Velraj, A review of efficient high productivity solar stills. Renew. Sustain. Energy Rev. **101**, 197–220 (2019)
35. K. Srithar, T. Rajaseenivasan, Recent fresh water augmentation techniques in solar still and HDH desalination–a review. Renew. Sustain. Energy Rev. **82**, 629–644 (2018)
36. M.A. Khan, S. Rehman, F.A. Al-Sulaiman, A hybrid renewable energy system as a potential energy source for water desalination using reverse osmosis: a review. Renew. Sustain. Energy Rev. **97**, 456–477 (2018)
37. B. Anand, R. Shankar, S. Murugavelh, W. Rivera, K.M. Prasad, R. Nagarajan, A review on solar photovoltaic thermal integrated desalination technologies. Renew. Sustain. Energy Rev. **141**, 110787 (2021)
38. F. Dixit, K. Zimmermann, R. Dutta, N.J. Prakash, B. Barbeau, M. Mohseni, B. Kandasubramanian, Application of MXenes for water treatment and energy-efficient desalination: a review. J. Hazard. Mater. **423**, 127050 (2022)
39. A.I. Shehata, A.E. Kabeel, M.M.K. Dawood, M.M.A. Elazm, A.M. Abdalla, A. Mehanna, Achievement of humidification and dehumidification desalination system by utilizing a hot water sprayer and ultrasound waves techniques. Energy Convers. Manage. **201**, 112142 (2019)
40. A.E.A. El-Maaty, M.M. Awad, G.I. Sultan, A.M. Hamed, Solar powered fog desalination system. Desalination **472**, 114130 (2019)
41. Z.M. Omara, A.S. Abdullah, T. Dakrory, Improving the productivity of solar still by using water fan and wind turbine. Sol. Energy **147**, 181–188 (2017)
42. Z. Haddad, A. Chaker, A. Rahmani, Improving the basin type solar still performances using a vertical rotating wick. Desalination **418**, 71–78 (2017)
43. P. Pal, P. Yadav, R. Dev, D. Singh, Performance analysis of modified basin type double slope multi–wick solar still. Desalination **422**, 68–82 (2017)
44. D. Emad, M.A. El-Hameed, A.A. El-Fergany, Optimal techno-economic design of hybrid PV/wind system comprising battery energy storage: case study for a remote area. Energy Convers. Manage. **249**, 114847 (2021)
45. A.E Kabeel, M. Abdulaziz, E.M. El-Said, Solar-based atmospheric water generator utilisation of a fresh water recovery: a numerical study. Int. J. Ambient. Energy **37**(1), 68–75 (2016)
46. C. Chiranjeevi, T. Srinivas, Experimental and simulation studies on two stage humidification–dehumidification desalination and cooling plant. Desalination **376**, 9–16 (2015)
47. N. Rahbar, J.A. Esfahani, Experimental study of a novel portable solar still by utilizing the heatpipe and thermoelectric module. Desalination **284**, 55–61 (2012)
48. J.A. Esfahani, M. Rahbar, M. Lavvaf, Utilization of thermoelectric cooling in a portable active solar still—an experimental study on winter days. Desalination **269**(1–3), 198–205 (2011)
49. K. Rahbar, A. Riasi, Performance enhancement and optimization of solar chimney power plant integrated with transparent photovoltaic cells and desalination method. Sustain. Cities Soc. **46**, 101441 (2019)

50. K. Rabhi, R. Nciri, F. Nasri, C. Ali, H.B. Bacha, Experimental performance analysis of a modified single-basin single-slope solar still with pin fins absorber and condenser. Desalination **416**, 86–93 (2017)
51. S.M. Parsa, A. Rahbar, M.H. Koleini, S. Aberoumand, M. Afrand, M. Amidpour, A renewable energy-driven thermoelectric-utilized solar still with external condenser loaded by silver/nanofluid for simultaneously water disinfection and desalination. Desalination **480**, 114354 (2020)

Chapter 25
Mathematical Modeling for Exergy Analysis of a Sea Water Reverse Osmosis Desalination System

Gaurav Aggarwal and Surendra Singh Kachhwaha

Abstract In the present work, a mathematical model for exergy analysis of a two-pass sea water reverse osmosis (SWRO) desalination system has been developed using Engineering Equation Solver (EES) software and the developed model has been validated using data set reported in the literature. The minimum separation work has been found out as 154.35 kW and second law efficiency were found out as 18.3%. Total exergy destruction was 686.94 kW out of which major exergy destruction of 291.30 kW occurs in SWRO brine disposal, 210.32 kW in pumps, followed by 109.80 kW and 59.76 kW in SWRO and BWRO membranes, respectively. The results for exergy analysis match very accurately with the published results. It is suggested that the measures like use of energy recovery device for SWRO brine disposal, high-performance pumps having variable frequency drive features and high-performance membranes have potential for minimizing the exergy destruction.

Keywords Exergy analysis · Reverse osmosis · Desalination

G. Aggarwal (✉) · S. S. Kachhwaha
Department of Mechanical Engineering, School of Technology, Pandit Deendayal Energy University, Raisan, Gandhinagar, India
e-mail: gaurav.dst@gmail.com

S. S. Kachhwaha
e-mail: Surendra.Singh@sot.pdpu.ac.in

G. Aggarwal
Department of Science and Technology, New Delhi, India

A. Kumar et al. (eds.), *Selected Proceedings of the 1st International Conference on Advanced Materials for Sustainable Innovation; IC-AMSI 2024; 28–30 August; New Delhi; India*, Springer Proceedings in Energy,
https://doi.org/10.1007/978-981-96-4492-6_25

25.1 Introduction

In order to sustain quality life for human beings, one need to take care of changing environmental conditions, depleting natural resources and ever-increasing demand for more resources for sustenance due to increasing population, rapid industrialization enhanced human comforts. Besides energy requirements, clean water and clean air have become core challenges today for researchers across the globe. If we take note of clean water, it has become one of the most fundamental necessities not only for human beings but also for animals, agricultural and industrial requirements. According to United Nations Sustainable Development Goals Report 2023, 2.2 billion people still lacked safely managed drinking water in 2022. Goal six, clean water and sanitation targets ensuring availability and sustainable management of water and sanitation for all and to meet 2030 targets, pace of progress will have to accelerate six times for drinking water, five times for sanitation and three times for hygiene [1]. Desalination is a promising intervention for achieving this target. According to Technology Vision 2035 Document [2], clean air and potable water is one of the twelve sectors chosen for sharper focus; everyone should be able to access safe drinking water and to fulfill it, affordable desalination technology is one of such interventions which requires targeted research and need to be moved from lab to field. Desalination is a useful technology to meet the growing requirement of water by converting available water (sea water, river water, spring water, ground water) into useful water. Among the different desalination technologies, reverse osmosis (RO) desalination is having the highest share of 68.7% (Fig. 25.1).

A desalination system works primarily on separation of saline water from the source (ground water, spring water, river water, sea water etc.) into two components, i.e., product water with low salinity for applications like drinking water, water for agricultural applications, water for industrial applications. Since this separation process is not a natural process, one need to spend some energy from outside for this purpose, as per second law of thermodynamics. In reverse osmosis process, this energy in high-pressure pumping is required to overcome osmotic pressure. The quantum of required energy is a function of various parameters like, salinity,

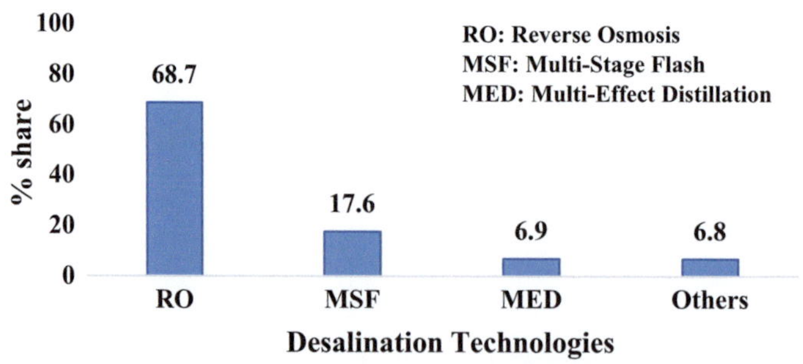

Fig. 25.1 Share of RO in different desalination technologies [3]

temperature and pressure of incoming saline water for the incoming side and also the salinity of product water for the outgoing side. The amount of energy input not only affects the cost of production of fresh water but it also affects the amount of greenhouse gases released to the environment. In view of these very important twin factors coupled with other environmental factors like disposal of waste brine, it is desirable to minimize the requirement of energy input for desalination, particularly the reverse osmosis desalination, which is most widely used among different desalination techniques and which is also energy-intensive one. Hence, reduction in consumption of energy for production of fresh water from RO desalination is a very important research area for the desalination industry. The first RO desalination system used cellulose acetate membranes in 1959 and first commercial RO desalination plant came up in California, USA in 1965 [4]. Since then, continuous efforts are being made to minimize specific energy consumption (SEC). For sea water RO desalination, SEC has come down from 8.5 kWh/m^3 in 1990 to about 3 kWh/m^3 in 2009 [5]. For groundwater source, SEC is 0.48 kWh/m^3 [6].

Exergy analysis uses conservation of mass and energy principles for analysis, design and improvement of thermodynamic systems. For a desalination system also, it helps in identifying the extent and locations of thermodynamic losses and suggest possible improvements in the system [7]. Cerci carried out exergy analysis of a 7250 m^3/day capacity brackish water RO desalination plant operating in California considering saline water as a mixture of pure water and salt, reporting second law efficiency of 4.3% [8]. Vicente et al. also conducted exergy analysis of SWRO desalination plant of capacity, 20,700 m^3/day, operating in Spain [9]. Isam H. Aljundi conducted second law analysis of a RO plant of 552.7 m^3/day capacity, operating in a thermal power station in Jordan, using brackish water, reporting second law efficiency of 4.1% [10]. R. Kempton et al. did thermodynamic analysis of RO desalination systems operating in Western Australia and found that for RO desalination system, specific power consumption was 3.29 kWh/m^3 and greenhouse gases emissions was 2.91 kg CO_2 eq/m^3 using black coal [11]. Mostafa et al., developed correlations for thermo-physical properties of seawater and calculated exergy values of sea water for desalination systems [12, 13]. Nuri et al. performed exergy analysis of SWRO desalination system using sea water as source [14]. Usama A.S. Ezzeghni et al. did exergy analysis of a SWRO desalination unit of 10,000 m^3/day capacity, operating in Libya [15]. M. Sarai Atab et al. performed second law analysis of a brackish water RO desalination plant of 24,000 m^3/day capacity, operating in Iraq [16]. Alanood A. Alsarayreh et al. carried out exergy analysis of an industrial RO desalination system of capacity 1200 m^3/day using brackish water [17]. Yigit Kupcu et al. carried out exergy analysis of purified water plant operating in a pharmaceutical industry [18]. From literature review, it has been observed that researchers across the globe have carried out exergy analysis of sea water and brackish water reverse osmosis desalination plants located in different parts of the world, prominently, in Middle East region, Persian Gulf region, Saudi Arabia, Jordan, Iraq, Libya, USA, Australia, Spain.

25.2 Methodology

The objective of this work is to carry out mathematical modeling for exergy analysis of a sea water reverse osmosis (RO) desalination system. The operating parameters are, mass flow rate, temperature, pressure and salinity of water streams at different state points of RO desalination plant.

25.2.1 *Systems Description and Plant Operation*

In this work, a two-pass sea water RO desalination system used by Nuri et al. [14] has been considered. The process flow diagram of the RO desalination system is shown in Fig. 25.2.

As indicated in Fig. 25.2, the RO desalination system uses two-pass RO membranes, sea water RO (SWRO) membrane and brackish water RO (BWRO) membrane. The sea water at salinity of 36,000 ppm is drawn by pump, P1 at atmospheric pressure and then its pressure increases to 1.5 bar. Thereafter, sea water enters sea water feed pump, P2, which increases its pressure to 65 bar, to overcome the osmotic pressure of SWRO membrane. Here, incoming water is separated into two streams, one having treated water of 486 ppm which goes to BWRO feed pump, P3, where its pressure is increased to 18 bar. The other stream is brine with salinity of 65,000 ppm is rejected to environment. The water coming from SWRO feed pump at 18 bar pressure and 486 ppm enters BWRO membrane where it is separated into two streams, one of brine having salinity of 3189 ppm while the other stream is of product water having salinity of 10 ppm, which is taken out for the intended use. The data set used in the present work is given in Table 25.1.

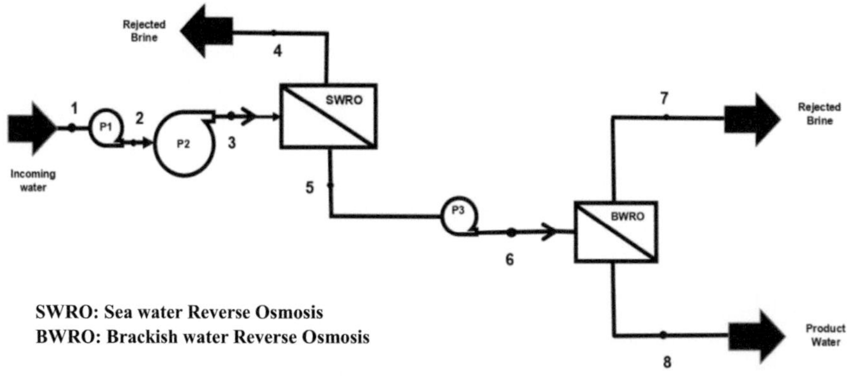

Fig. 25.2 Process flow diagram for SWRO desalination system

Table 25.1 Data set used in the present work [14]

State point	m	T	p	S
1	90.8	25.0	101.3	36.00
2	90.8	25.0	150.0	36.00
3	90.8	25.7	6500.0	36.00
4	50.0	25.7	6200.0	65.10
5	40.9	25.7	200.0	0.49
6	40.9	25.9	1800.0	0.49
7	6.13	25.9	1700.0	3.19
8	34.7	25.9	100.0	0.01

25.2.2 Exergy Analysis

Exergy is the maximum obtainable work when a system is moved from its state to environmental. For a desalination system also, it helps in identifying the extent and locations of thermodynamic losses and suggest possible improvements in the system [7]. Total exergy of any flow stream is given by, E_T,

$$E_T = E_{ph} + E_{ch} + E_p + E_k \tag{25.1}$$

Since RO desalination system is at rest, the terms for total potential and kinetic exergies are neglected. Thus, Eq. (25.1) is reduced to

$$E_T = E_{ph} + E_{ch} \tag{25.2}$$

Specific exergy is given by,

$$e_T = \frac{E_T}{m} \tag{25.3}$$

The specific exergy is further expressed as,

$$e_T = e_{ph} + e_{ch} \tag{25.4}$$

In the RO desalination process, there are three streams. The stream entering the system is sea water while the streams leaving the system are brine and product water. The physical and chemical exergy of these three streams is calculated by correlations developed by Mostafa et al. [12]. Specific physical exergy (e_{ph}) of water is given by the relation,

$$e_{ph} = (h - h_0) - T_0(s - s_0) \tag{25.5}$$

The parameters at environmental dead state in the present work are, $p_0 = 101.3$ kPa, $s_0 = 0.036$ kg/kg and $T_o = 25^0$ C.

For pure water and sea water, density is given by [12],

$$\rho_w = 9.999 \times 10^2 + 2.034 \times 10^{-2}T - 6.162 \times 10^{-3}T^2 + 2.261 \times 10^{-5}T^3 - 4.657 \times 10^{-8}T^4 \quad (25.6)$$

$$\rho_{sw} = \rho_w + S\left(a_1 + a_2T + a_3T^2 + a_4T^3 + a_5ST^2\right) \quad (25.7)$$

Validity for ρ_{sw}, $0 < T < 180^0$ C; $0 < S < 0.16$ kg/kg, Accuracy is $\pm\, 0.1\%$

For pure water and sea water, enthalpy is given by [12],

$$h_w = 141.355 + 4202.07 \times T - 0.535 \times T^2 + 0.004 \times T^3 \quad (25.8)$$

$$h_{sw} = h_w - S[b_1 + b_2S + b_3S^2 + b_4S^3 + b_5T + b_6T^2 + b_7T^3 + b_8ST + b_9S^2T + b_{10}ST^2] \quad (25.9)$$

The constants used in the Eqs. (25.7) and (25.9) are given in Table 25.2.

Validity for h_{sw} and h_w; $10 \leq T \leq 120\ ^0$C; $0 \leq S \leq 0.12$ kg/kg; Accuracy is $\pm$ 0.5%

The effect of pressure on the enthalpy is taken from the relation,

$$h_{sw}(T,\ p,\ w_s) = h_{sw}(T,\ p_0,\ w_s) + 1/(p - p_o) \quad (25.10)$$

For pure water and sea water, entropy is given by [12],

$$s_w = 0.1543 + 15.383 \times T - 2.996 \times 10^{-2} \times T^2 + 8.193 \times 10^{-5} \times T^3 - 1.370 \times 10^{-7} \times T^4 \quad (25.11)$$

$$s_{sw} = s_w - S(c_1 + c_2S + c_3S^2 + c_4S^3 + c_5T + c_6T^2 + c_7T^3 + c_8ST + c_9S^2T + c_{10}ST^2) \quad (25.12)$$

Validity for s_{sw} and s_w; $10 \leq t \leq 120^0$ C; $0 \leq S \leq 0.12$ kg/kg Accuracy is $\pm\, 0.5\%$

Table 25.2 Constants used for density and specific enthalpy [12]

$a_1 = 8.020 \times 10^2$	$b_1 = -2.348 \times 10^4$	$b_6 = -4.417 \times 10^1$
$a_2 = -2.001$	$b_2 = 3.152 \times 10^5$	$b_7 = 2.139 \times 10^{-1}$
$a_3 = 1.667 \times 10^{-2}$	$b_3 = 2.803 \times 10^6$	$b_8 = -1.991 \times 10^4$
$a_4 = -3.060 \times 10^{-5}$	$b_4 = -1.446 \times 10^7$	$b_9 = 2.778 \times 10^4$
$a_5 = -1.613 \times 10^{-5}$	$b_5 = 7.826 \times 10^3$	$b_{10} = 9.728 \times 10^1$

Specific chemical exergy (e_{ch}) of sea water is given by the relation [12],

$$e_{ch} = \mathrm{m_f}(\mu - \mu_0) \quad (25.13)$$

Chemical potential of water in saline water is given by,

$$\mu_w = \frac{\partial G_{sw}}{\partial m_w} = g_{sw} - S\frac{\partial g_{sw}}{\partial S} \quad (25.14)$$

Chemical potential of salts in saline water is given by,

$$\mu_s = \frac{\partial G_{sw}}{\partial m_s} = g_{sw} + (1 - S)\frac{\partial g_{sw}}{\partial S} \quad (25.15)$$

where

$$g_{sw} = h_{sw} - (T + 273.15)s_{sw} \quad (25.16)$$

$$\frac{\partial g_{sw}}{\partial S} = \frac{\partial h_{sw}}{\partial S} - (T + 273.15)\frac{\partial s}{\partial S} \quad (25.17)$$

$$-\frac{\partial h_{sw}}{\partial S} = b_1 + 2b_2S + 3b_3S^2 + 4b_4S^3 + b_5T + b_6T^2 + b_7T^3 + 2b_8ST + 3b_9S^2T + 2b_{10}ST^2 \quad (25.18)$$

$$-\frac{\partial s_{sw}}{\partial S} = c_1 + 2c_2S + 3c_3S^2 + 4c_4S^3 + c_5T + c_6T^2 + c_7T^3 + 2c_8ST + 3c_9S^2T + 2c_{10}ST^2 \quad (25.19)$$

The constants used in the Eqs. (25.12), (25.18) and (25.19) are given in Tables 25.2 and 25.3.

Table 25.3 Constants used for specific entropy and chemical potentials [12]

$c_1 = -4.231 \times 10^2$	$c_6 = -1.443 \times 10^{-1}$
$c_2 = 1.463 \times 10^4$	$c_7 = 5.879 \times 10^{-4}$
$c_3 = -9.880 \times 10^4$	$c_8 = -6.111 \times 10^1$
$c_4 = 3.095 \times 10^5$	$c_9 = 8.041 \times 10^1$
$c_5 = 2.562 \times 10^1$	$c_{10} = 3.035 \times 10^{-1}$

25.3 Model Validation

First, the values of sea water thermodynamic properties, specific volume, specific enthalpy, specific entropy, chemical potentials of water, chemical potentials of salt and flow exergy were calculated using the correlations developed by Sharqawy et al. [12] for different temperatures, ranging from 10–90°C at reference conditions and the calculated values for these properties match perfectly [13]. Thus, the algorithm is validated.

25.4 Results and Discussion

After validation of algorithm, a mathematical model was developed in EES [19] using the layout of RO desalination plant [14] and correlations developed by Sharqawy et al. [12]. Using the developed model, values for exergy flow rate were found at all state points of RO system and results obtained are given in Table 25.4 and the same are compared with published results [14].

The last three streams, 4z, 7z, and 8z mentioned in Table 25.4 represent exergy of rejected brine from SWRO, exergy of rejected brine from BWRO and exergy of product water when they reach environment state. This forms the basis to calculate minimum separation work (W_{min}), which is given by the following equation.

$$W_{min} = X_{4z} + X_{7z} + X_{8z} = 154.35 \text{ kW}$$

Table 25.4 Exergy flow rates at different state points in RO system

State point	m	T	p	S	X		% age deviation
					[14]	[Present work]	
1	90.8	25.0	101.3	36.00	0	0	0
2	90.8	25.0	150.0	36.00	4.32	4.32	0
3	90.8	25.7	6500.0	36.00	567.00	567.41	0.07
4	50.0	25.7	6200.0	65.10	334.00	334.30	0.09
5	40.9	25.7	200.0	0.49	115.00	123.31	7.22
6	40.9	25.9	1800.0	0.49	181.00	186.87	3.24
7	6.13	25.9	1700.0	3.19	23.70	21.55	9.07
8	34.7	25.9	100.0	0.01	97.80	105.56	7.93
4z	50.0	25.0	101.3	65.10	42.90	43.00	0.23
7z	6.13	25.0	101.3	3.19	13.80	13.81	0.07
8z	34.7	25.0	101.3	0.01	97.70	97.54	0.16

Table 25.5 Distribution of exergy destruction across major components of RO desalination system

Component	X_D
Sea water brine disposal	291.30
Pumps	210.32
SWRO membrane	109.80
BWRO membrane	59.76
Brackish water brine disposal	7.74
Product water disposal	8.02
Total	686.94

The values for exergy inputs and other useful parameters has been found out as under.

Exergy input for sea water pump, P1 $= X_2–X_1 = 4.32$ kW

Exergy input for SWRO feed pump, P2 $= X_3–X_2 = 563.09$ kW

Exergy input for BWRO feed pump, P3 $= X_6–X_5 = 63.56$ kW

Exergy input in pumps $= X_{in,pumps} = (1/0.75) \times [(X_2–X_1) + (X_3–X_2) + (X_6–X_5)] = 841.29$ kW

Here, pumps efficiency is taken as 75%.

Second law efficiency is the ratio of minimum separation work to the total exergy input.

$\eta = \text{Wmin/Einput} = 154.35/841.29 = 18.3\%$

Total exergy destruction $= X_D = X_{in}–X_{out}$

Exergy destruction in pumps $= X_{D,pumps} = (1 - 0.75) \times X_{in,pumps} = 210.32$ kW

Exergy destruction in SWRO membrane $= X_{D,SWRO} = X_3 - X_4 - X_5 = 109.80$ kW

Exergy destruction in BWRO membrane $= X_{D,BWRO} = X_6–X_7–X_8 = 59.76$ kW

Exergy of sea water brine disposal $= X_{D,RSWD} = X_4 - X_{4z} = 291.30$ kW

Exergy of rejected brackish water brine disposal $= X_{D,RBWD} = X_7–X_{7z} = 7.74$ kW

Exergy of product water disposal $= X_{D,PWD} = X_8 - X_{8z} = 8.02$ kW.

The distribution of exergy destruction across major components of the RO desalination system is given in Table 25.5.

25.5 Conclusions

It is understood that second law efficiency of RO desalination plant is 18.3% which is low and lot of potential exists for improving the same. As major exergy destruction takes place in SWRO brine disposal, pumps and SWRO membrane, it is suggested that energy recovery device may be used for SWRO brine, pumps with variable frequency drive be used and high-performance membrane using low pressure be used for minimizing the exergy destruction.

Nomenclature

e	Specific exergy, kJ/kg
E	Total exergy, kJ
h	Enthalpy, kJ/kg
m	Mass flow rate, kg/s
p	Pressure, kPa
r	Recovery ratio
S	Salinity, g/kg
s	Entropy, kJ/kg K
T	Temperature, ^{0}C, K
v	Volume
W	Work, kJ/kg
X	Exergy rate, kW

Subscripts

brine	Outgoing brine
ch	Chemical
D	Destruction
ph	Physical
p	Potential
k	Kinetic
sw	Sea water
w	Pure water
s	Salt
T	Total
product	Outgoing product water
o	Environment dead state

Symbol

ρ	Density, kg/m^3

References

1. https://www.un.org/sustainabledevelopment/water-and-sanitation
2. Technology vision 2035 document, Technology information, Forecasting and Assessment Council (TIFAC), Department of science and technology, Government of India, New Delhi (2015)
3. D. Curto et al., A review of the water desalination technologies. Appl. Sci. **11**(2), 670 (2021). https://doi.org/10.3390/app11020670
4. Y. Cerci, et al., Improving the thermodynamic and economic efficiencies of desalination plants: minimum work required for desalination and case studies of four working plants, Report (2003)
5. H. Nassrullah, S.F. Anis, R. Hashaikeh, N. Hilal, Energy for desalination: a state-of-the-art review. Desalination **491**, 114569 (2020)
6. J. Feria-Díaz et al., Commercial thermal technologies for desalination of water from renewable energies: a state-of-the-art review. Processes **9**(2), 262 (2021). https://doi.org/10.3390/pr9020262
7. Exergy, Book exergy: energy, environment and sustainable development, Ibrahim Dincer. Elsevier Science Publishing Co Inc., 3rd edn. (2020)
8. Y. Cerci, Exergy analysis of a reverse osmosis desalination plant in California. Desalination 257–266 (2002)
9. R.T. Vicente, Exergy analysis of a seawater reverse osmosis plant. Desalination 197–207 (2005)
10. I.H. Aljundi, Second-law analysis of a seawater plant in Jordan. Desalination 207–215 (2009)
11. R. Kempton, D. Maccioni, S.M. Mrayed, G. Leslie Thermodynamic efficiencies and GHG emissions of alternative desalination processes. Water Sci. Technol.: Water Supply **10**(3), 416–427 (2010). https://doi.org/10.2166/ws.2010.085.
12. M.H. Sharqawy et al., Thermo-physical properties of seawater: a review of existing correlations and data. Desalination Water Treat. 354–380 (2010)
13. M.H. Sharqawy et al., On exergy calculations of seawater with applications in desalination systems. Int. J. Therm. Sci. **50**(2), 187–196 (2011)
14. N.M. Eshoul et al., Exergy analysis of a two-pass reverse osmosis (RO) desalination unit with and without an energy recovery turbine (ERT) and pressure exchanger (PX). Energies **8**, 6910–6925 (2015). https://doi.org/10.3390/en8076910
15. U.A.S. Ezzeghni et al., Exergy analysis of a 10,000 m^3/day Tajoura SWRO desalination plant, in *ICCPGE 2016*, Al-Mergib University, Alkhoms, Libya (2008)
16. M. Sarai Atab et al., Exergy analysis of reverse osmosis for potable water and land irrigation. World academy of science, Eng. Technol. Int. J. Chem. Mol. Eng. **13**(3) (2019)
17. A.A. Alsarayreh et al., Thermodynamic limitations and exergy analysis of brackish water reverse osmosis desalination process. Membranes **12**, 11 (2022). https://doi.org/10.3390/membranes12010011
18. Y. Kupcu, M. Ince, M. Bayramoglu, Exergy analysis of purified water plant in a pharmaceutical industry. Int. J. Exergy **40**(1) (2023). https://doi.org/10.1504/IJEX.2023.10053571
19. Engineering equation solver © 1992–2018 S.A. Klein, Professional V10.561 (2018-12-20]. www.fChart.com

Chapter 26
Enviro-Economic Assessment of Some Designs of PCM Incorporated Solar Stills

Aseem Dubey and Akhilesh Arora

Abstract Conventional solar stills are suffered from low productivity, but attractive due to their simple construction and less investment. Besides various design modifications, the incorporation of energy storage materials (PCMs) has shown considerable impact on increasing the solar still's night productivity and overall performance. In this study, to find the comparative viability of the solar still, enviro-economic analysis of some designs added with PCMs has been carried out using the uniform approach of interest rate, maintenance cost, salvage value, and life span of the system. In the passive designs, pyramid-shaped v-corrugated PCM-loaded solar still yielded 94% higher than solar without PCM. Considering the twenty-year life span, 44 tCO_2 will be mitigated with distillate production cost of 0.0084 $/kg and a payback period of 1.9 years. The study revealed the feasibility of using PCM-added solar stills in remote locations, enhancing the productivity in the range of 8–94% and lowering the payback period in the range of 1.1–5.6 years, respectively, at 0.05 $/kg selling rate of water in the market.

Keywords Solar still · Phase change material · CO_2 mitigation · Production cost · Payback period

26.1 Introduction

The most vital things for sustainable development are water and energy, and globally, 3rd world countries are facing an acute shortage of both. About 1/3rd of the worldwide population has a shortage of safe potable water, especially with the lack of energy resources, and the situation will be more critical in the future [1]. The use of fossil fuel as an energy source emits greenhouse gases, and India produces 1742 million metric tons of carbon dioxide, accounting for 5.72% of total emissions globally [2]. Various

A. Dubey (✉) · A. Arora
Department of Mechanical Engineering, Delhi Technological University, Delhi 110042, India
e-mail: aseemdubey009@gmail.com

A. Kumar et al. (eds.), *Selected Proceedings of the 1st International Conference on Advanced Materials for Sustainable Innovation; IC-AMSI 2024; 28–30 August; New Delhi; India*, Springer Proceedings in Energy,
https://doi.org/10.1007/978-981-96-4492-6_26

solar-powered techniques such as solar still, humidification-dehumidification, multi-stage flashing, reverse osmosis, and hybrid units have been reported to distill water in an eco-friendly manner [3]. Solar still plays a vital role in solving freshwater scarcity in remote areas as well as in reducing the emission of greenhouse gases to meet the target of sustainable development goal. Solar stills are broadly classified as passive and active. However, low daily productivity of 2.0–4.0 kg/m^2 from passive design is a significant obstacle to commercialize it. Fath et al. [4] reported justifiable cost of fresh water to be $0.03/liter from solar still. To improve the productivity and efficiency of solar still, different techniques, including modified designs or adding external/ internal add-ons, have been carried out in the past [3].

Among the internal add-ons, using PCMs enhance the productivity of solar still. PCMs store thermal energy during available period of radiation and discharge the same during low/off sunshine hours to heat the basin water, thus bridging the gap in unavailability of solar energy and increasing the productivity from solar still by reducing the losses to the environment from basin [5]. Due to the benefits associated with PCMs, much research has been carried out to improve the performance of solar stills by incorporating PCMs [6]. Paraffin wax (PW) has been the most commonly used PCM due to its favorable melting point and high latent heat of fusion, but it is associated with its low thermal conductivity. To study the impact of PW on solar still, Kabeel and Abdelgaied [7] experimented with single-slope (SS) solar still made from GI sheet of 0.0015 m thickness, with 0.072 m^2 area of basin, and insulated with fiberglass of 0.005 m thickness. The condensing glass cover (0.003 m thick) was positioned at an angle of 30 deg. The daily yields from PCM loaded solar still was found to be 7.54 kg/m^2, which was higher than solar still without PCM (4.51 kg/m^2). The economic analysis inferred the decrease in production cost of distillate using the PCM system (0.03 $/kg) compared to system with no PCM (0.032 $/kg), accounting for 340 days, 10 years life span, 30% annual maintenance. Sonker et al. [8] reported that cost reduction of distillate from Rs. 5 to Rs. 3/kg of distillate with a payback time of 4.3 years by incorporating nano-enhanced PCM in novel frugal solar still, accepting the application of solar stills in a resource constrained duopoly market.

To study the effect of PCM with v-corrugated basin liner, Shalaby et al. [9] inserted copper tubes inside the PCM material to take the effect of expansion during melting and tested the still at 0.025 m and 0.035 m water depths and 18 kg PW. A decrease in the daytime yield and a considerable enhancement in the night-time yield were noticed. The PCM loaded still yielded 3.76 kg, higher by 12% than the still without PCM (3.35 kg). Life cycle cost analysis accounting for 300 productive days in a year, a 30% annual maintenance charge, and 10 years life period of the system reveled a CPL as 0.08369 $/kg. Vigneswaran et al. [10] studied the effect of PW-filled GI tubes submerged in basin water on solar stills fabricated using galvanized iron (GBSS), while the basin/cover of the other were made of acrylic (ABSS). It was reported that using acrylic basins reduced the charging time and also delayed the discharge time, enhancing the yield to 4.36 L/m^2/day. The economic analysis was carried out considering 10 years life span, 8% interest rate, 20% salvage value on initial investment, 10% annual maintenance, and accounting 365 days. They inferred CPL as

Rs. 1.1 and 1.23 with PW incorporated GIBSS and ABSS solar stills, respectively. Elashmawy [11] tested tubular solar still incorporated with PCM tubes and integrated with a parabolic solar concentrator (PTC). The CPL of distillate was found as 0.01485 \$/kg, accounting for 292 days in a year, a 6% interest rate, 15% annual maintenance cost, and zero salvage value at the end of 15 year lifespan. Further, an enviro-economic analysis carried out on solar still coupled with a PCM-filled pulsating heat pipe (PHP) and a thermosyphon heat pipe (THP) [12]. They reported the lowest production cost of 0.0456 \$/kg using THP/PW solar still, accounting for 20 years life span, 20% interest rate, 10% annual maintenance, 20% salvage value, and 365 operational days in a year. The CO_2 emission reduction obtained using conventional, THP/PW system, and PHP/PW system was about 14.56 tons, 17.93 tons, and 19.53 tons, respectively.

From the literature survey, it has been noticed that many researchers have carried out the economic analysis of PCM added solar stills using multifarious approaches such as life span of the system, interest rate, annual productive day, annual maintenance, etc. Therefore, comparing the CPL of potable water and payback time of solar stills incorporated with PCMs, requires consideration to check the viability of each solar still under similar variable parameters/approach. Environmental considerations play a crucial role in CO_2 mitigation and carbon credit earned from each configuration of solar stills that not only enhance potable water productivity but also contribute to environmental protection. Thus, it is necessary to assess the benefits of modifications from the economic viewpoint. In this paper, nine different configurations of solar stills added with PCMs and condensing cover slopes equal to the latitude of the place have been considered for investigating the enviro.-economic viability based on 300 clear days in a year [13] under similar variable parameters and using the initial capital investment of each solar still mentioned in the respective publication. Brief details of solar stills under consideration are given in Table 26.1.

The novelty of the present paper exists as similar comparative study hasn't been found in the literature on PCM-loaded solar stills under uniform parameters such as uniform interest rates, maintenance costs, and salvage value. The prime objectives of the manuscript are;

- To carry out a comparative enviro-economic assessment of PCM added solar stills along with their conventional designs, accounting 300 productive days/year.
- To predict the CO_2 mitigation and effect of carbon credit on distillate production cost.

26.2 Enviro-Economic Analysis

The schematic of various configuratios of SS added with PCMs and under consideration are illustrated in Fig. 26.1. The enviro.-economic performance has been carried out to ensure the benefits of the modifications from the environmental and economic viewpoints compared to system with no PCM. Following Kumar and Tiwari [17],

Table 26.1 Different configurations of PCM incorporated solar stills

Ref	Denomination	Solar still type
[7]	[Ia]	SS solar still of GI sheet, with insulation of fiber glass
	[Ib]	Incorporated with PW beneath basin liner (17.5 kg)
[9]	[IIa]	Single slope v-corrugated basin SS of GI sheet, insulated by foam
	[IIb]	V-corrugated SS with PW beneath basin liner (18 kg)
[14]	[IIIa]	Single slope SS of GI sheet and acrylic body, insulated by Armaflex
	[IIIb]	Incorporated with PW beneath the basin (17 kg)
	[IIIc]	Incorporated with PW (17 kg) and pin fins
[10]	[IVa]	Single slope SS of GI sheet, insulated by aluminum foil and plywood
	[IVb]	Solar still made of acrylic sheet incorporating PW filled 21 GI tubes (6 kg) into basin water (ABSS)
	[IVc]	PW filled 21 GI tubes (6 kg) into basin water (GBSS)
[8]	[Va]	SS solar still of stainless steel, insulated by glass wool
	[Vb]	Solar still with PW filled 6 copper cylinders placed in basin (3 kg PCM)
[15]	[VIa]	A pyramid shaped v-corrugated basin of GI sheet, insulated by fiber glass/wooden box
	[VIb]	A pyramid solar still with PW beneath basin (44 kg)
[11]	[VIIa]	Tubular solar still (TSS) coupled with parabolic concentrator
	[VIIb]	TSS-SCTC incorporated with 12 PW filled aluminum tubes (1.035 kg) in basin
[16]	[VIIIa]	Single slope SS of GI sheet, insulated by plywood/wooden box
	[VIIIb]	Still added with 20 cylindrical capsules filled with PW (13 kg) in basin
	[VIIIc]	Active single slope SS with 20 cylindrical filled PW capsules (13 kg) in basin water and coupled with FPC
[12]	[IXa]	A single slope SS of GI sheet, insulated by polyurethane foam
	[IXb]	SS connected with pulsating heat pipe (PHP)
	[IXc]	SS connected PHP filled with PW (2.7 kg)
	[IXd]	SS connected with thermosyphon heat pipe (THP)
	[IXe]	SS connected with THP filled with PW (2.7 kg)

CO_2 emission/mitigation and carbon credit earned by solar still during fabrication, production of distillate and carbon credit earned can be expressed as;

26.2.1 CO_2 Emission

$$CO_2 \text{ emission over life time (ton)} = \frac{E_{in} \times 1.58}{1000} \tag{26.1}$$

where E_{in} is embodied energy (kWh) of respective solar still.

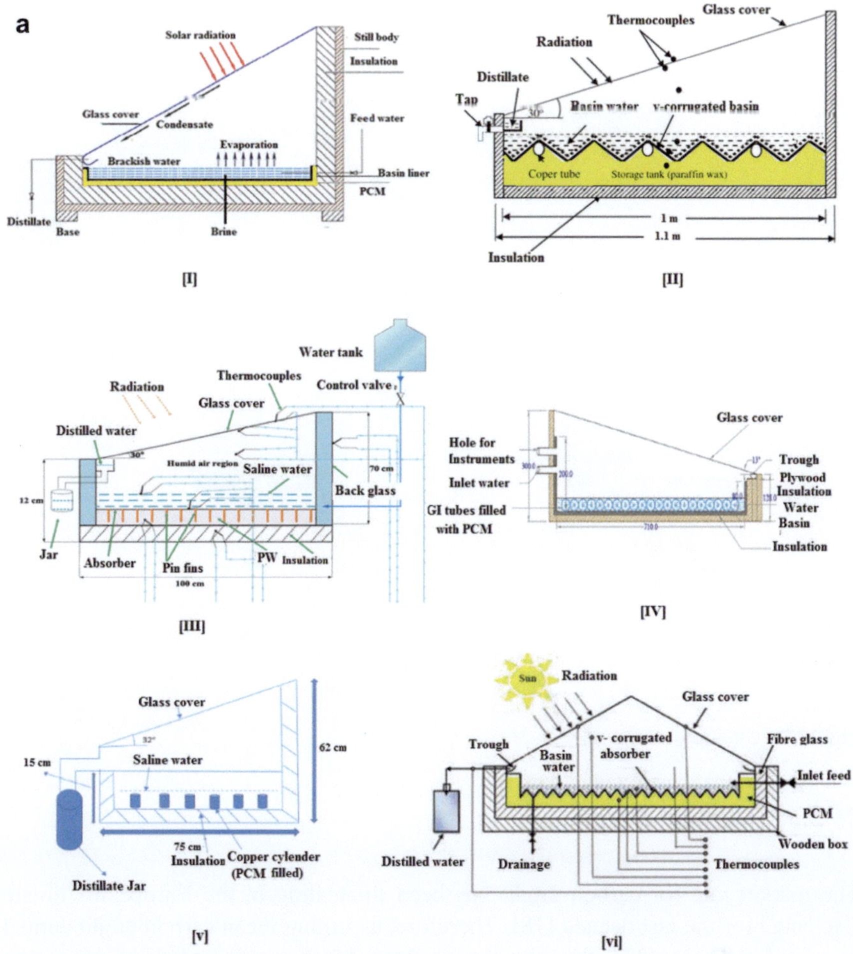

Fig. 26.1 **a** Schematic of passive solar stills incorporated with PCMs. **b** Schematic of active solar stills incorporated with PCMs

26.2.2 CO_2 Mitigation

$$CO_2 \text{ mitigation over life time} = \frac{E_{out} \times n \times 1.58}{1000} \quad (26.2)$$

$$\text{Net } CO_2 \text{ reduction over life time} = \frac{(E_{out} \times n - E_{in})}{1000} \times 1.58 \quad (26.3)$$

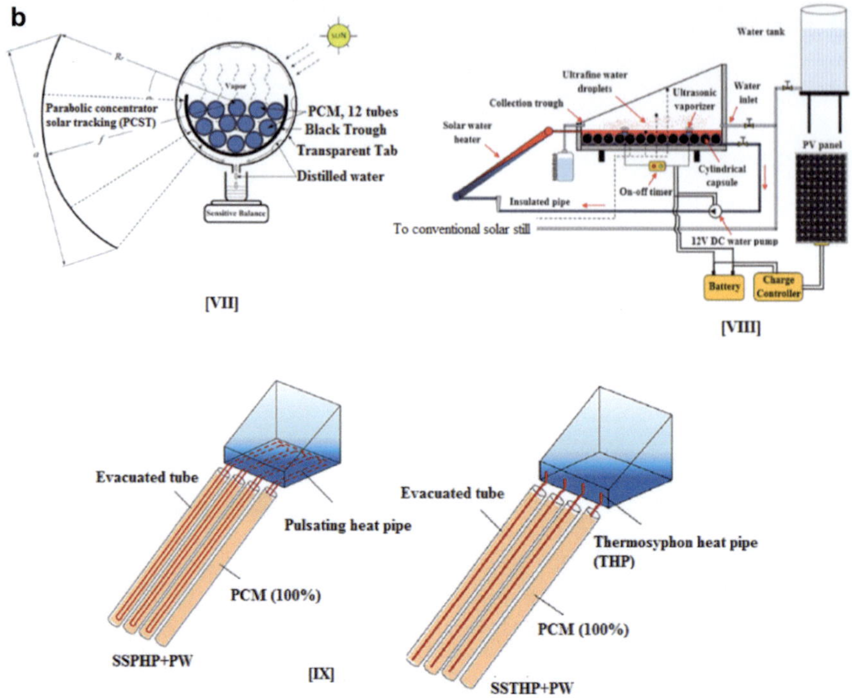

Fig. 26.1 (continued)

26.2.3 Carbon Credit

The market rate for carbon credit has been fluctuating in the European Climate Exchange for the last decade [18]. Therefore, assuming mean carbon credit earned as ~ 10 $/tCO_2$ in 2024, the annual cash flow ($/kg) due to carbon credit can be expressed as;

$$ACC = \frac{(E_{out} \times n - E_{in}) \times 1.58 \times 10}{1000 \times n \times M_y} \quad (26.4)$$

26.2.4 Annualized Cost

The annualized cost for a initial capita;l cost of system can be expressed as [17];

$$\mathrm{UA} = \mathrm{P_s} \times \mathrm{F_{CR,\,i,\,n}} + \left(\mathrm{P_s} \times \mathrm{F_{CR,\,i,\,n}}\right) \times \mathrm{M_s} - \mathrm{S_s} \times \mathrm{F_{SR,\,i,\,n}} \quad (26.5)$$

The capital recovery ($F_{CR,i,n}$) and sinking fund ($F_{SR,i,n}$) factors can be expressed as;

$F_{CR,\,i,\,n} = \frac{i(1+i)^n}{(1+i)^n-1}$ and $F_{SR,\,i,\,n} = \frac{i}{(1+i)^n-1}$.

Accounting carbon credit, the net annualized cost (UA_{net}) can be expressed as;

$$UA_{net} = UA - annual\ credit\ earned \tag{26.6}$$

26.2.5 Production Cost (CPL)

The cost per liter (CPL) of water, based on yearly yield (M_Y) is estimated as;

$$\text{CPL} = \frac{UA_{net}}{M_Y} \tag{26.7}$$

26.3 Results and Discussion

The annual yield obtained from the solar stills under consideration has been considered, accounting 300 clear days and based on the maximum yield reported. The enhancement of yield using PCM incorporated solar still, compared to their own conventional design, is shown in Fig. 26.2. Among them, the maximum enhanced yield by 94.12% has been observed using pyramid-shaped v-corrugated basin solar still, possible due to the higher condensing surface area. Incorporating cylindrical filled PW capsules in basin water enhances the yield by 30.81% (VIIIb), but lower than placing the PW beneath the basin (85.71%, Ib) for single slope solar still. Moreover, integrating solar still with FPC (VIIIC) enhanced the yield further to ~243% higher than conventional solar still without PCM (VIIa).

To check the viability of each solar still, the economic analysis of each system has been performed using eqns. (26.5–26.7) based on initial investment reported by the researchers for the respective designs of solar stills. Cost per liter (CPL) is a major factor in deciding the feasibility of fresh water to market acceptability. Therefore, CPL of potable water has been estimated using the uniform approach with respect to interest rate (5–10%), annual maintenance (10%), salvage value (10%), system life (10–20 years), and Fig. 26.3 shows the CPL estimated. The best production cost obtained using a SS solar still with PCM beneath the basin and found as 0.0072 $/L and 0.0108 $/L at 5% and 10% interest rate for 20 years life period of the system, due to lower investment cost. Though FPC integrated PW added solar still yields significantly higher but, CPL has been estimated as 0.0174 $/L and 0.0262 $/L, respectively, under aforesaid conditions. Moreover, the incorporation of PW with some designs of solar stills [IIb, IIIb, IIc, IVb, IVc, VIb, VIIIb, VIIIc, IXb-e] cost

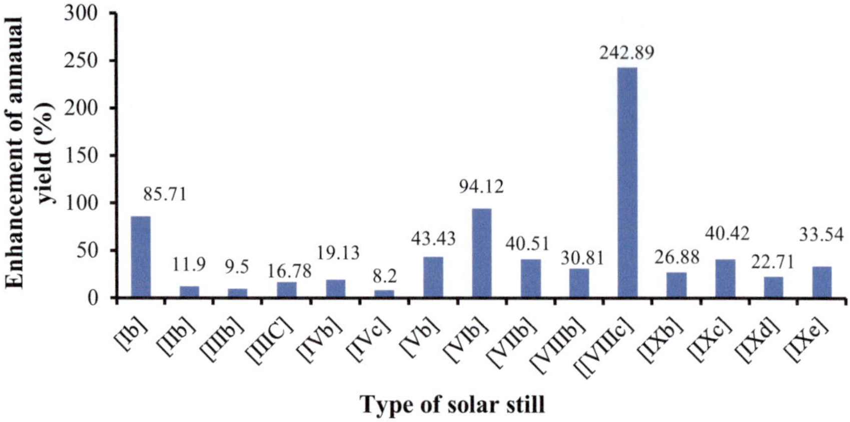

Fig. 26.2 Comparative increase in annual yield using PCM loaded solar stills

marginally higher than their respective conventional designs due to higher initial investment. However, higher yield may justify the application of these solar stills. The CPL increases with an increase in interest rate due to increase in annualized costs.

Embodied energy required to assemble the solar stills has been estimated with and without PCM, and corresponding tCO_2 emission was obtained [13, 17]. The embodied energy per kg of PW produced has been considered as 198 kWh [12]. The estimated lifetime CO_2 emission/mitigation is shown in Fig. 26.4 and obtained in the range of 0.011 tons (single slope without PCM) to 14.39 tons (Pyramid-shaped v-corrugated with PCM) due to 44 kg use of PCM. Therefore, net CO_2 mitigation evaluated for 20 years life is estimated in the range of 7.4 tons (VIIIb) to ~37 tons (Ib

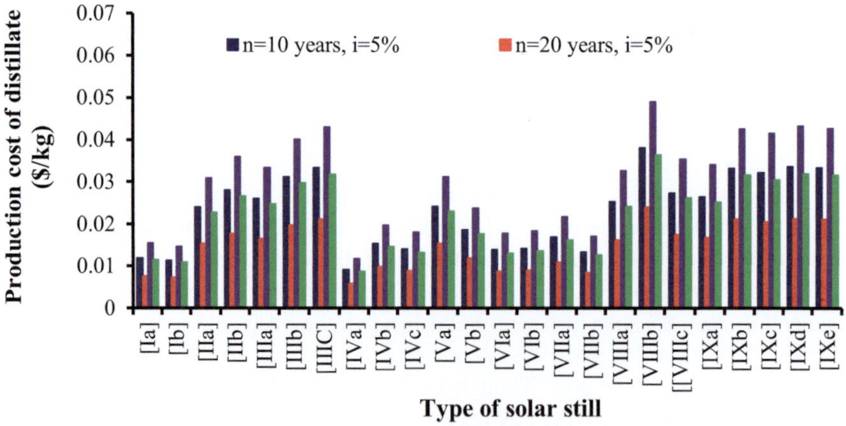

Fig. 26.3 Production cost of distillate from solar stills under consideration

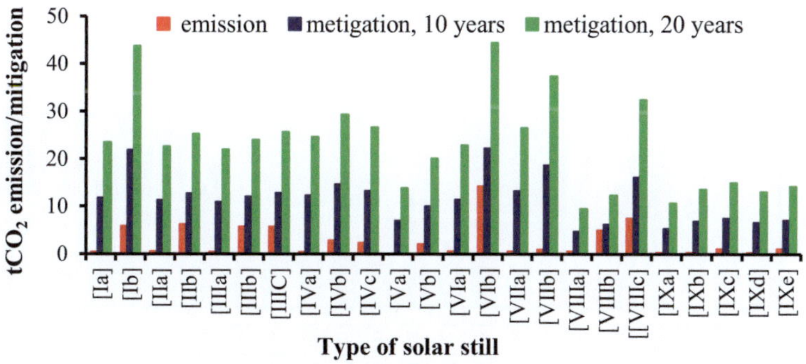

Fig. 26.4 CO_2 emission and mitigation from solar stills

and VIIb) and ~2.0 times more than the system with life period of 10 years. Considering European trading emission rate (10 \$/tCO_2), the net carbon credit earned is estimated in the range from 0.008 to 0.011 \$/kg, with and without PCM added solar stills for 20 years life. At a fixed *i*, if life of the system is doubled, the CPL is reduced ~9.0% without carbon credit. However, CPL would reduce significantly, accounting for the revenue earned due to carbon trading and help in meeting the sustainable development goal. The payback period estimated using the method proposed by Kumar and Tiwari [17] considering 5–10% interest rate, 20 years life span, and selling rate of potable water at 0.05 \$/kg show a reasonable payback time in the range of 1.4–6.6 years, which further lowered in the range of 1.1 to 5.6 years, accounting carbon credit (10 \$/tCO_2) and 5% interest rate as depicted in Fig. 26.5. The increase in interest rate from 5–10%, enhance the payback time in the range of about 5–20%, due to increase in production cost of water. The cash flow earned due to carbon credit in the global market will decrease the production cost of product water and, in some cases, be negligible, depending on the initial investment, interest rate, carbon credit earned, and system's life. Carbon credit has a significant impact on reducing payback time and is found more in PCM-loaded solar stills.

26.4 Conclusions

(a) Solar still incorporated with PCMs plays a vital role in CO_2 mitigation, and new designs would be futuristic. Carbon trading maximizes the economic benefits of solar still and will make the system economically viable in rural communities.
(b) The best production cost of 0.0072 \$/kg and 0.0108 \$/kg has been estimated at $i = 5\%$ and 10%, respectively, for 20 years life of single slope SS with PCM and maximum using FPC integrated PW added solar still (0.0174- 0.0262 \$/L).
(c) Among the passive designs, pyramid v-corrugated shaped PCM loaded solar still has the highest increase in yield (~94%) compared to its conventional design, with reasonable CPL 0.0084 \$/kg, CO_2 mitigation of about 44 tons and payback

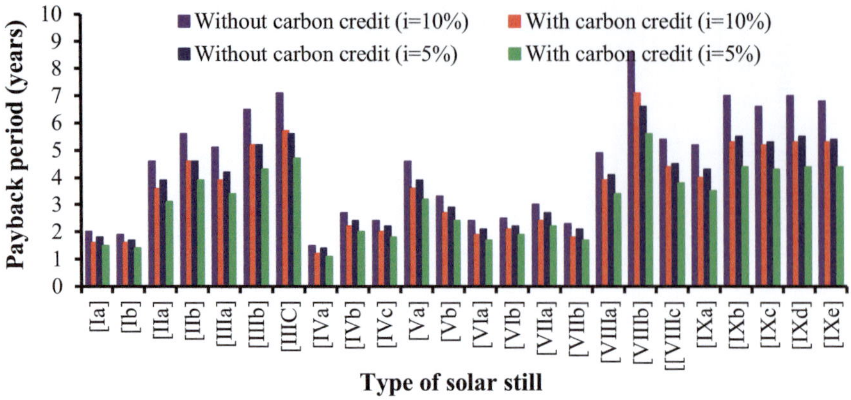

Fig. 26.5 Payback time from solar stills with and without carbon credit

period of 1.9 years accounting carbon credit for 20 years life span at 5% interest rate.

(d) With a reasonable payback time of 1.1–5.6 years by incorporating PCM and accounting for carbon credit with a selling rate of 0.05 \$/kg, the feasibility of solar stills in remote locations is clear. These solutions offer a much lower cost compared to the price of fresh water in the open market, providing practical and cost-effective option for remote areas.

Nomenclatures

i	Interest rate (%)
M_s	Maintenance cost (%)
P_s	Net present cost (\$)
S_s	Salvage value (\$)
UA	Annualized cost (\$)

References

1. M. Thimmaraju, D. Sreepada, G.S. Babu, B.K. Dasari, S.K. Velpula, N. Vallepu, Desalination of water, in *Desalination and Water Treatment* (InTech, London, 2018), pp. 333–347
2. Prabhakant, G.N. Tiwari, Evaluation of carbon credits earned by energy security in India. Int. J. Low-Carbon Technol. **4**(1), 42–51 (2009)
3. A.W. Kandeal, N.M. El-Shafai, M.R. Abdo, A.K. Thakur et al., Improved thermo-economic performance of solar desalination via copper chips, nanofluid, and nano-based phase change material. Sol. Energy **224**, 1313–1325 (2021)

4. H.E.S. Fath, M. El-Samanoudy, K. Fahmy, A. Hassabou, Thermal-economic analysis and comparison between pyramid shaped and single-slope solar still configurations. Desalination **159**(1), 69–79 (2003)
5. S.W. Sharshir, A.H. Elsheikh, E. Edreis, M.K.A. Ali et al., Improving the solar still performance by using thermal energy storage materials: a review of recent developments. Desalin. Water Treat. **165**, 1–15 (2019)
6. V.K. Chauhan, S.K. Shukla, P.K.S. Rathore, A systematic review for performance augmentation of solar still with heat storage materials: a state of art. J. Energy Storage **47**, 103578 (2022)
7. A.E. Kabeel, M. Abdelgaied, Improving the performance of solar still by using PCM as a thermal storage medium under Egyptian conditions. Desalination **383**, 22–28 (2016)
8. V.K. Sonker, J.P. Chakraborty, A. Sarkar, Development of a frugal solar still using phase change material and nanoparticles integrated with commercialization through a novel economic model. J. Energy Storage **51**, 104569 (2022)
9. S.M. Shalaby, E.E. Bialy, A.A.E. Sebaii, An experimental investigation of a v-corrugated absorber single-basin solar still using PCM. Desalination **398**, 247–255 (2016)
10. V.S. Vigneswaran, P.G. Kumar, D. Sakthivadivel et al., Energy, exergy, and economic analysis of low thermal conductivity basin solar still integrated with phase change material for energy storage. J. Energy Storage **34**, 102194 (2021)
11. M. Elashmawy, M. Alhadri, M.M.Z. Ahmed, Enhancing tubular solar still performance using novel PCM-tubes. Desalination **500**, 114880 (2021)
12. A. Hemmatian, H. Kargarsharifabad, A.A. Esfahlani, N. Rahbar, S. Shoeibi, Improving solar still performance with heat pipe/pulsating heat pipe evacuated tube solar collectors and PCM: an experimental and environmental analysis. Sol. Energy **269**, 112371 (2024)
13. S. Kumar, K.T. Kurmaji, Carbon credit earned by some designs of solar stills. Desalin. Water Treat. **51**(22–24), 4699–4708 (2013)
14. M.S. Yousef, H. Hassan, Energetic and exergetic performance assessment of the inclusion of phase change materials (PCM) in a solar distillation system. Energy Convers. Manag. **179**, 349–361 (2019)
15. A.E. Kabeel, M.A. Teamah, M. Abdelgaied, G.B. Abdel Aziz, Modified pyramid solar still with v-corrugated absorber plate and PCM as a thermal storage medium. J. Clean. Prod. **161**, 881–887 (2017)
16. A.H. Abed, H.A. Hoshi, M.H. Jabal, Experimental investigation of modified solar still coupled with high-frequency ultrasonic vaporizer and phase change material capsules. Case Stud. Therm. Eng. **28**, 101531 (2021)
17. S. Kumar, G.N. Tiwari, Life cycle cost analysis of single slope hybrid (PVT) active solar still. Appl. Energy **86**(10), 1995–2004 (2009)
18. Anon, European climate exchange (2007), www.europeanclimateexchange.com

Chapter 27
Nanomaterials and Their Applications: A Sustainable Approach to Complete Water Refinement

Bhavvya Mehta and Shubhangi Madan

Abstract Water supplies have been contaminated worldwide as a result of the fast industrialization and urbanization. Therefore, it is essential and urgent to remediate wastewater in order to mitigate the consequences on the environment and public health. In this regard, scientists have recently become interested in nanomaterials in addition to traditional therapeutic methods. The optical, electrical, chemical, magnetic and tribological, properties of matter can all be greatly tuned via nanostructuring. Because of this, nanostructured materials have been widely used in a wide range of industries and technologies that make our daily life more comfortable and convenient. Nanomaterials exhibit higher surface free energy, which enhances their surface reactivity. Despite this, they have a certain disadvantage such as: (i) They tend to aggregate in fluid systems, leading to high amount of activity loss and pressure drops. (ii) Apart from magnetic nanoparticles, separating them from treated water for reuse is difficult. While there are examples of nanostructured materials improving sustainable technologies like electrolysers, batteries, and water treatment, the full potential of nanoscale materials for sustainability has likely not yet been realised, given the nearly limitless exploration space for nanostructure- property-composition combinations. The wastewater treatment process, in particular, could see significant advancements by utilising the size-dependent properties of nanomaterials. To overcome these challenges associated with their use, extensive research has focused on modifying nanomaterials in various ways. Strategies such as designing nanocomposites with different shapes, functionalising with them with active components, and fabricating nanofibers are just a few methods that have demonstrated great promise and sustainability for water treatment. Examining the most recent developments in each of the many kinds of nanomaterials that are currently widely used for wastewater treatment is crucial. This review provides a thorough analysis of current developments in common nanomaterials used for sustainable wastewater treatment. By

B. Mehta · S. Madan (✉)
School of Engineering and Technology, Vivekananda Institute of Professional Studies-Technical Campus, Delhi 110034, India
e-mail: shubhangimadan.93@gmail.com

A. Kumar et al. (eds.), *Selected Proceedings of the 1st International Conference on Advanced Materials for Sustainable Innovation; IC-AMSI 2024; 28–30 August; New Delhi; India*, Springer Proceedings in Energy,
https://doi.org/10.1007/978-981-96-4492-6_27

addressing the current market challenges, we focus on improving the economic efficiency, sustainability and performance of new-generation nanomaterials through our assessments, ensuring the adoption of these developed technologies.

Keywords Nanomaterials · Nanotechnology · Nanocomposites · Water treatment · Sustainability

27.1 Introduction

Each year, unsafe water claims more lives than all forms of violence combined. By 2050, global freshwater demand is expected to rise by one-third. Currently, the industrial sector consumes 70% of available freshwater, which 25% is used by the domestic sector, contributing to the generation of large volumes of wastewater. Clean water is essential for human health and well-being. It refers to water that is safe for drinking, promotes sanitation and hygiene, and helps prevent waterborne diseases caused by harmful microbes. Additionally, clean water is vital for sustaining ecosystems, preserving biodiversity, and supporting crop growth and food production [1].

Water treatment has a critical role in the public healthcare system as it prevents waterborne diseases. Species on land as well as those in water suffer from untreated wastewater. Wastewater may contain organic contaminants, non-disintegrating compounds, and highly concentrated heavy metals. These pollutants need to be eliminated using efficient and safe purification techniques before they can be disposed of. Proper treatment removes pathogens, toxins, pollutants, pesticides etc. which prevents and safeguards us from chronic and other long-term diseases. Its importance ranges from irrigation to livestock, to tourism and biodiversity. Nanomaterials can be added to water in two ways to increase the effectiveness of wastewater treatment processes: (i) by adding inexpensive nano adsorbents, and (ii) by using nanofiltration. Due to their unique characteristics, nanoparticles have garnered a lot of interest over time and have become important components of water treatment processes.

Nanomaterials are defined as the particles whose size ranges from 1 to 100 nm in at least one dimension [2]. This unique property enables them to gain various characteristics such as mechanical, optical and magnetic characteristics [3–6]. This also enables them to show properties such as adsorption, high catalytic activity and high reactivity towards various elements [7–11]. This range of properties has led to development in the field of nanotechnology which has led to increase in its areas of applications such as hospitality, medicine, robotics, etc. One such notable application is water treatment. These nanoparticles have large surface areas which enhances its adsorption qualities and adsorbs all the organic pollutants, heavy metals, inorganic anions, various microbes and toxins from the surrounding matter to give clean and safe water for consumption [1].

Advanced oxidation process (AOP) systems with nanomaterials integration are a popular method of treating wastewater. Because of their high reactivity, AOPs generate hydroxyl (OH) radicals that are useful in the treatment of wastewater. Additionally, they are able to produce intermediate compounds with low toxicity. There is a research gap in wastewater treatment since only a small number of studies have explored the use of nanomaterials as photocatalysts. Some study has been conducted using micro-sized catalysts in photocatalytic AOPs. Other beneficial properties of nanomaterials include their layered structure, direct bandgap, high optical absorption coefficient, and enhanced catalysis because of their fine band edges. These properties help clean water by eliminating various pollutants and assisting in the purification of groundwater. Nano photocatalysts' quantum size contributes to both an increase in the bandgap and a decrease in the particle's overall size. Since they are inexpensive, readily available, and chemically stable, nano photocatalysts are becoming more and more popular. In this review, we aim to explore and compare new, innovative and largely unexplored industry of materials of nanotechnology or as it is called nanomaterials which are used in application of wastewater treatment to acquire safe and clean water for consumption. The application of clean water is required in every industry whether it is agriculture, domestic or robotics. The goal is to achieve sustainability by harnessing the intersection of innovation and applications of nanotechnology through the study of nanomaterials (Fig. 27.1).

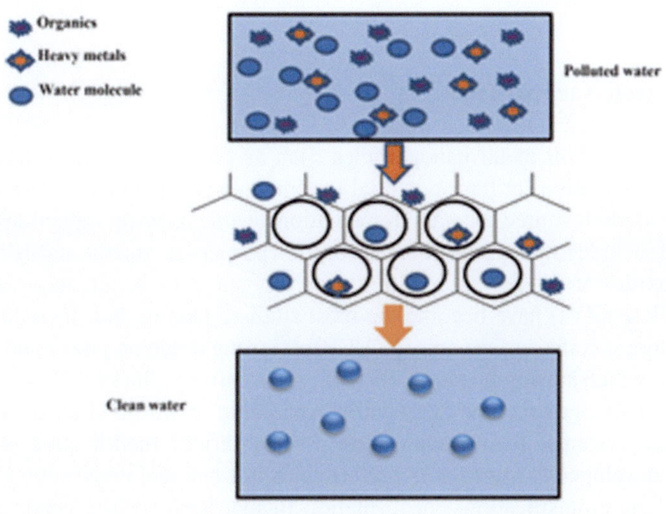

Fig. 27.1 Water treatment using nanomaterials [12]

27.2 Nanomaterials in Waste Water Treatment

27.2.1 Zero Valent Metal Nanoparticles

27.2.1.1 Silver Nanoparticles

AgNPs are a range of very highly toxic particles in nature, which makes them effective anti-bacterial against pathogens such as viruses, bacteria, fungi, etc. [11, 13 and 14]. They are so toxic in nature that they adhere themselves to the cell wall of the pathogen and then penetrate the cell wall causing structural changes within the cell and damaging its internal integrity which eventually leads to death of the pathogen [15–17]. Sometimes, they interact with the DNA of the cell by releasing anti-microbial Ag + ions which interacts with the enzymes within the cells and leads to eventual death of the cell [18]. It shows anti-bacterial properties against E. coli with minimal silver loss while maintaining clean and safe water standards set by WHO [19]. It also shows high disinfection properties which are used for domestic purposes and household wastewater treatment [20]. It promotes sustainable development by improving water quality, energy efficiency, water recycling and helping in pesticide reduction.

27.2.1.2 Iron Nanoparticles

Recently, zero-valent metal nanoparticles such as Fe, Zn, Al and Ni have drawn considerable attention for their potential to remove water pollutants. Among these, Fe is particularly favoured for water remediation despite its lower reduction capacity. Its benefits include low cost, excellent adsorption properties, and the ability to precipitate and oxidise in the presence of dissolved oxygen. As a result, zero-Valant iron nanoparticles (nZVI) have become the most studied. Due to their incredibly small size and large specific surface area, nZVI offers strong reducing power and effective adsorption, which are key factors in its ability to remove pollutants. However, nZVI also faces challenges such as aggregation, oxidation and difficulties in separation from treated systems. To overcome these issues, various modification techniques have been developed to improve its performance in water and wastewater treatment. These include emulsification, encapsulation in matrices, surface coating, doping with other metals, and binding with support materials. The main objective of these modifications is to enhance nZVI's reactivity, extending its lifespan and promoting sustainable use [21].

27.2.2 Photocatalytic Degradation Technology

Titanium oxide has high photocatalytic activity, low cost, photo stability, chemical and biological stability, thereby making it useful in electrochemical photolysis of water [22–29]. These properties of titania has drawn its attention to its potential use in wastewater treatment. In presence of light and a catalyst, contaminants are oxidised and then converted into escapable gases like CO_2, H_2O, etc. TiO_2 requires UV light for activation and have limited efficiency under visible light. To increase its efficiency, TiO_2 is doped with non-metals such as nitrogen, carbon and oxygen to enhance its photolytic activity. TiO_2 NPs with membrane technology offers easy water treatment solution by allowing easy separation of contaminants by letting the gases escape after oxidation [30–33]. It promotes sustainable development by improving water quality, energy efficiency, water recycling and improving air quality.

27.2.3 Carbon Nanomaterials

Carbon nanomaterials are available in various forms such as Carbon Nanotubes, Carbon Beads, Carbon Fibres and nonporous carbon. Amongst these, Carbon nanotubes are used in wastewater treatment plants. They have high adsorption rates, large surface areas and selectivity towards aromatic compounds which helps them trap a large variety of contaminants which are as small as 1 nm [34]. It shows mechanical, optical and electrical properties which increases its functions such as increase in dispersibility and increase in surface area which increases adsorption kinetics. Carbon nanotubes have limited production due to their high cost and they require a matrix for structural support which hinders its development [35–38]. It promotes sustainable development by improving water quality, energy efficiency, water recycling and promotes green manufacturing.

27.2.4 Nanocomposites

Nanocomposites are the materials which have dimensions in nanometer in multiple dimensions. Thin film nanocomposites and nanofiltration membranes are used in wastewater treatment. One such composite that has been reported is TiO_2 nanoparticles incorporated into a network of polyamide support [39]. It shows a unique property by increasing the stability of TiO_2, its reusability and facilitating high contaminant removal. Nanocomposites loaded with iron oxide to facilitate easy separation from water using external magnetic field has also been reported earlier [40]. This reduces the operational cost and makes it affordable for household water treatment process. These thin film nanocomposites are highly versatile in nature, making them effective for improving the overall water treatment processes [39, 40].

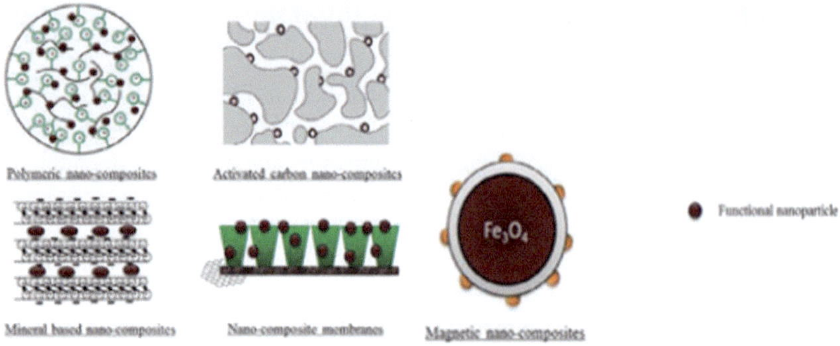

Fig. 27.2 Nanomaterials used in water treatment plants [41]

It promotes sustainable development by improving water quality, energy and resource conservation, water recycling and sustainable manufacturing (Fig. 27.2).

27.3 Benefits of Nanomaterials in Water Treatment Process

1. Removal of harmful contaminants such as microbes, inorganic pollutants, toxins, disease carrying pathogens, etc.
2. These nanomaterials have selectively permeable membranes which allows them to contain the contaminants and make the water safe for consumption.
3. These nanomaterials are chosen on the basis of their anti-microbial and biologically safe qualities. They are biologically compatible and anti-microbial within the safe amount that is approved by WHO.
4. They promote sustainable development as they require less energy than the conventional methods of water treatment.
5. They can be easily integrated within the current water treatment plants, thus they are scalable and would require very low cost of activation and promotion.
6. They act rapidly and are capable of eliminating a wide range of contaminants which are as small as 1 nm, offering greater efficiency as compared to the conventional methods.

27.4 Drawbacks of Nanomaterials in Water Treatment Process

1. These nanomaterials with size as small as 1 nm might possess high level of toxicity which might not be detected by the standards set by WHO.

2. They require high level of technology to prepare thin films and particles as small as 1 nm, which increases the cost of the nanomaterial. This ultimately leads to the high cost of equipment.
3. Due to their high toxicity levels and very thin films, they offer high level of regulatory and safety issues.

27.5 Nanomaterials in Sustainable Development

In the realm of nanomaterials, sustainable development can be promoted through following methods:

27.5.1 Removal of Heavy Metals

Nanomaterials such as iron-oxide and carbon nanotubes have very high adsorption rates and helps in removal of heavy metals such as arsenic, lead, mercury, etc. from water during waste water treatment [42].

27.5.2 Solar-Powered Water Purification

Nanomaterial such as TiO_2 and ZnO are used in photolytic degradation of organic pollutants making the water treatment process more energy efficient and sustainable [43].

27.5.3 Anti-Microbial Water Treatment

AgNPs are very famous for killing pathogens due to their highly toxic nature. This promotes sustainable development by using organic components to kill pathogens without the need for using harmful chemicals [44].

27.5.4 Energy Efficiency

Nanomaterials are developed with selectively permeable membranes which only catch pollutants of a specific size which reduces its production, cost and makes it energy efficient than the traditional methods [45].

27.5.5 *Reusability*

Nanomaterials are made from smart materials which can be used for easy and self-regeneration and thus, can be reused very easily. This plays a vital role in promoting sustainable development because it reduces the need for constant production of new materials and minimizes waste [46].

27.5.6 *Life-Cycle Assessment (LCA) of Nanomaterials Used in Waste Water Treatment*

LCA is used to evaluate the environmental impacts of a product, process or material through its entire lifecycle. LCA assess a nanomaterial on the basis of global warming potential, energy consumption, water use, toxicity and resource depletion. LCA results helps to analyse to identify which product or material has the most environmental impact. For example: If a nanomaterial is energy intensive then it could offset the environmental benefits that the material could offer [47].

27.5.7 *Cost-Effectiveness*

Sustainable nanomaterials also focus in making these materials more cost-effective so that new and efficient treatments are affordable and accessible to more consumers in the regions where clean water is scarce [45].

27.6 Conclusions

Rapid urbanization and industrialization have driven the use of nanotechnology across various industries, including wastewater treatment. Nanomaterials like Carbon Nanotubes, Ti-O_2-infused nanocomposites, FeO_2, and other biologically compatible metal oxides and highly effective in filtering contaminants as small as 1 nm and eliminating a wide range of pathogens. Despite their potential, challenges such as high production costs, toxicity concerns, and regulatory hurdles have limited their commercial adoption. Current standards for evaluating nanomaterial toxicity assessments and uniform performance evaluation criteria in wastewater treatment [1] (Fig. 27.3).

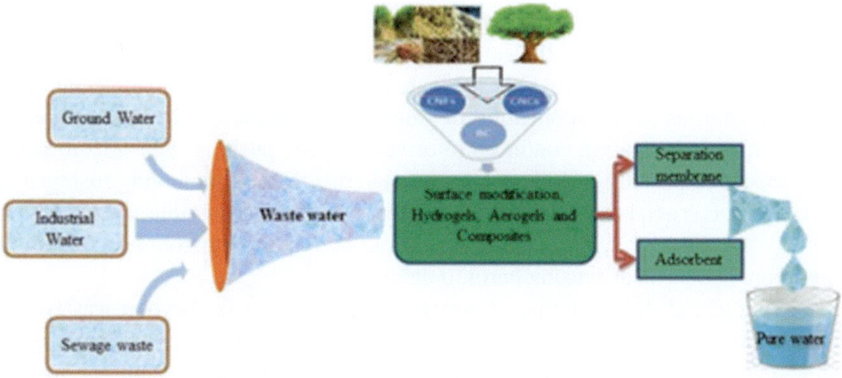

Fig. 27.3 Waste-water treatment using nanomaterials [48]

Nanomaterial	Pollutants removed	Properties
TiO_2	– Arsenic – Microbes	– High reactivity – Photocatalytic properties
FeO_2	– Chromium – Dyes – Heavy metals – Organic compounds	– High surface area – Magnetic properties
Carbon nanotubes (graphene)	– Organic contaminants – Dyes – Heavy metals – Microbes	– Large surface area – Electronic properties
Zeolites	– Manganese – Iron – Microbes – Phosphate	– High surface area – High adsorption rates – Magnetic properties – Electronic properties – Photolytic properties

From the above table, we have observed that nanomaterials for complete wastewater treatment must have the following properties:

1. High surface area
2. High adsorption rates
3. Photolytic properties.

The pollutants that are most commonly removed are:

1. Heavy Metals
2. Dyes
3. Pathogens
4. Organic Contaminants.

In conclusion, this paper tries to achieve sustainable development by enhancing the understanding and applications of nanomaterials used in wastewater treatment in the emerging field of nanotechnology. It offers products as small and thin as 1 nm to purify water and offer us the safest and cleanest water for consumption which is free of any contaminants. As we navigate the large overview of nanotechnology and nanomaterials in the application of water treatment plants, the intersection of innovation and application continues to drive progress, pushing industries towards new horizons of possibility and advancement. In this market, promoting sustainable development is crucial to ensure that new materials are both accessible and affordable for everyone. Moreover, these materials should protect the environment for future generations, rather than causing harm to them.

Looking ahead, the potential of nanomaterials extends far beyond wastewater treatment, offering transformative applications in healthcare, energy storage, environmental cleanup, and agriculture. As these advanced materials continue to find their way into new industries, a strong emphasis on sustainable practices will be vital to make these technologies widely accessible and affordable. It is equally important that their development prioritizes minimizing environmental impact, helping to protect the planet for future generations. Addressing challenges like toxicity, regulatory frameworks, and production costs will be key to realizing the full potential of nanotechnology, driving innovation across industries while promoting a more sustainable and equitable future.

Acknowledgements We are deeply grateful to all those who contributed to the successful completion of this research. First and foremost, sincere thanks to management at VSE&T and the team at IC-AMSI for their valuable guidance, insightful feedback, and unwavering support. We also express our gratitude to the School of Engineering & Technology, VIPS-TC, Delhi, for providing the resources and infrastructure necessary for conducting this study. Lastly, we would like to recognize the contributions of previous researchers and scientists whose foundational work paved the way for this study. Without their efforts, this paper would not have been possible.

References

1. V. Pavate, S. Kamble, Nanomaterials for water treatment: an overview. Int. J. Curr. Sci. (IJCSPUB) **14**(3), 876–877. Sanjay Ghodawat Group of Institutions, (2024). https://doi.org/10.1729/Journal.41337
2. C. Buzea, I.I. Pacheco, K. Robbie, Nanomaterials and nanoparticles: sources and toxicity. Biointerphases **2**(4), MR17–MR71 (2007)
3. V. Parmon, Nanomaterials in catalysis. Mater. Res. Innovations **12**(2), 60–61 (2008)
4. X.-J. Liang, A. Kumar, D. Shi, D. Cui, Nanostructures for medicine and pharmaceuticals. J. Nanomater. **2012**, 2. Article ID 921897 (2012)
5. A. Kusior, J. Klich-Kafel, A. Trenczek-Zajac, K. Swierczek, M. Radecka, K. Zakrzewska, TiO_2–SnO_2 nanomaterials for gas sensing and photocatalysis. J. Eur. Ceram. Soc. **33**(12), 2285–2290 (2013)
6. B. Bujoli, H. Roussiere, G. Montavon et al., Novel phosphate– ‘ phosphonate hybrid nanomaterials applied to biology. Prog. Solid State Chem. **34**(2–4), 257–266 (2006)
7. M.M. Khin, A.S. Nair, V.J. Babu, R. Murugan, S. Ramakrishna, A review on nanomaterials for environmental remediation. Energy Environ. Sci. **5**(8), 8075–8109 (2012)

8. W.-W. Tang, G.-M. Zeng, J.-L. Gong et al., Impact of humic/fulvic acid on the removal of heavy metals from aqueous solutions using nanomaterials: a review. Sci. Total. Environ. **468–469**, 1014–1027 (2014)
9. J. Yan, L. Han, W. Gao, S. Xue, M. Chen, Biochar supported nanoscale zerovalent iron composite used as persulfate activator for removing trichloroethylene. Biores. Technol. **175**, 269–274 (2015)
10. F. Liu, J.H. Yang, J. Zuo et al., Graphene-supported nanoscale zero-valent iron: removal of phosphorus from aqueous solution and mechanistic study. J. Environ. Sci. **26**(8), 1751–1762 (2014)
11. R.S. Kalhapure, S.J. Sonawane, D.R. Sikwal et al., Solid lipid nanoparticles of clotrimazole silver complex: an efficient nano antibacterial against Staphylococcus aureus and MRSA. Colloids Surf. B **136**, 651–658 (2015)
12. R.E. Abou-Zeid, R. Khiari, N. El-Wakil, A. Dufresne, Current state and new trends in the use of cellulose nanomaterials for wastewater treatment. Biomacromolecules 1–74 (2018). https://doi.org/10.1021/acs.biomac.8b00839
13. B. Borrego, G. Lorenzo, J.D. Mota-Morales et al., Potential application of silver nanoparticles to control the infectivity of Rift Valley fever virus in vitro and in vivo, nanomedicine: nanotechnology. Biol. Med. **12**(5), 1185–1192 (2016)
14. C. Krishnaraj, R. Ramachandran, K. Mohan, P.T. Kalaichelvan, Optimization for rapid synthesis of silver nanoparticles and its effect on phytopathogenic fungi. Spectrochim. Acta— Part A: Mol. Biomol. Spectrosc. **93,** 95–99 (2012)
15. I. Sondi, B. Salopek-Sondi, Silver nanoparticles as antimicrobial agent: a case study on E. coli as a model for Gramnegative bacteria. J. Colloid Interface Sci. **275**(1), 177–182 (2004)
16. M. Danilczuk, A. Lund, J. Sadlo, H. Yamada, J. Michalik, Conduction electron spin resonance of small silver particles. Spectrochim. Acta— Part A: Mol. Biomol. Spectrosc. **63**(1), 189–191 (2006)
17. K.I. Dhanalekshmi, K.S. Meena, DNA intercalation studies and antimicrobial activity of Ag@ZrO2 core–shell nanoparticles in vitro. Mater. Sci. Eng. C **59**, 1063–1068 (2016)
18. S. Prabhu, E.K. Poulose, Silver nanoparticles: mechanism of antimicrobial action, synthesis, medical applications, and toxicity effects. Int. Nano Lett. **2**(1), 32 (2012)
19. X. Li, J.J. Lenhart, H.W. Walker, Aggregation kinetics and dissolution of coated silver nanoparticles. Langmuir **28**(2), 1095–1104 (2012)
20. A.M. Ferreira, E.B. Roque, F.V.D. Fonseca, C.P. Borges, ´High flux microfiltration membranes with silver nanoparticles for water disinfection. Desalin. Water Treat. **56**(13), 3590–3598 (2015)
21. D. Ren, J.A. Smith, Retention and transport of silver nanoparticles in a ceramic porous medium used for point-ofuse water treatment. Environ. Sci. Technol. **47**(8), 3825–3832 (2013)
22. T. Ohsaka, K. Shinozaki, K. Tsuruta, K. Hirano, Photoelectrochemical degradation of some chlorinated organic compounds on n-TiO_2 electrode. Chemosphere **73**(8), 1279–1283 (2008)
23. M. Guo, W. Song, T. Wang, Y. Li, X. Wang, X. Du, Phenylfunctionalization of titanium dioxide-nanosheets coating fabricated on a titanium wire for selective solid-phase microextraction of polycyclic aromatic hydrocarbons from environment water samples. Talanta **144**, 998–1006 (2015)
24. Y. Lee, S. Kim, P. Venkateswaran, J. Jang, H. Kim, J. Kim, Anion co-doped Titania for solar photocatalytic degradation of dyes. Carbon Lett. **9**(2), 131–136 (2008)
25. A.T. Nguyen, C.-T. Hsieh, R.-S. Juang, Substituent effects on photodegradation of phenols in binary mixtures by hybrid H_2O_2 and TiO_2 suspensions under UV irradiation. J. Taiwan Inst. Chem. Eng. **62**, 68–75 (2016)
26. M.G. Alalm, A. Tawfik, S. Ookawara, Comparison of solar TiO_2 photocatalysis and solar photo-Fenton for treatment of pesticides industry wastewater: operational conditions, kinetics, and costs. J. Water Process. Eng. **8**, 55–63 (2015)
27. G. Moon, D. Kim, H. Kim, A.D. Bokare, W. Choi, Platinum-like behavior of reduced graphene oxide as a cocatalyst on TiO_2 for the efficient photocatalytic oxidation of arsenite. Environ. Sci. Technol. Lett. **1**(2), 185–190 (2014)

28. S.H. Kim, S.W. Lee, G.M. Lee, B.-T. Lee, S.-T. Yun, S.-O. Kim, Monitoring of TiO_2-catalytic UV-LED photo-oxidation of cyanide contained in mine wastewater and leachate. Chemosphere **143**, 106–114 (2016)
29. Z.P. Chen, Y. Li, M. Guo et al., One-pot synthesis of Mn-doped TiO_2 grown on graphene and the mechanism for removal of Cr(VI) and Cr(III). J. Hazard. Mater. **310**, 188–198 (2016)
30. A. Fujishima, K. Honda, Electrochemical photolysis of water at a semiconductor electrode. Nature **238**(5358), 37–38 (1972)
31. K. Guesh, A. Mayoral, C.M. Alvarez, Y. Chebude, I. Diaz, Enhanced photocatalytic activity of TiO_2 supported on zeolites tested in real wastewaters from the textile industry of Ethiopia. Microporous Mesoporous Mater. **225**, 88–97 (2016)
32. K. Imamura, T. Yoshikawa, K. Hashimoto, H. Kominami, Stoichiometric production of aminobenzenes and ketones by photocatalytic reduction of nitrobenzenes in secondary alcoholic suspension of titanium (IV) oxide under metal-free conditions. Appl. Catal. B **134–135**, 193–197 (2013)
33. S.B. Rawal, S. Bera, D. Lee, D.-J. Jang, W.I. Lee, Design of visible-light photocatalysts by coupling of narrow bandgap semiconductors and TiO_2: effect of their relative energy band positions on the photocatalytic efficiency. Catal. Sci. Technol. **3**(7), 1822–1830 (2013)
34. A. Chatterjee, B.L. Deopura, Carbon nanotubes and nanofibre: an overview. Fibers and Polymers **3**(4), 134–139 (2002)
35. P.Z. Ray, H.J. Shipley, Inorganic nano-adsorbents for the removal of heavy metals and arsenic: a review. RSC Adv. **5**(38), 29885–29907 (2015)
36. A.S. Adeleye, J.R. Conway, K. Garner, Y. Huang, Y. Su, A.A. Keller, Engineered nanomaterials for water treatment and remediation: costs, benefits, and applicability. Chem. Eng. J. **286**, 640–662 (2016)
37. Y.-H. Li, S. Wang, J. Wei et al., Lead adsorption on carbon nanotubes. Chem. Phys. Lett. **357**(3–4), 263–266 (2002)
38. A.S. Adeleye, A.A. Keller, Long-term colloidal stability and metal leaching of single wall carbon nanotubes: effect of temperature and extracellular polymeric substances. Water Res **49**, 236–250 (2014). [114] V.K. Gupta, S. Agarwal, T.A. Saleh, Chromium removal by combining the magnetic properties of iron oxide with adsorption properties of carbon nanotubes. Water Res **45**(6), 2207–2212 (2011)
39. M. Peyravi, M. Jahanshahi, A. Rahimpour, A. Javadi, S. Hajavi, Novel thin film nanocomposite membranes incorporated with functionalized TiO_2 nanoparticles for organic solvent nanofiltration. Chem. Eng. J. **241**, 155–166 (2014)
40. S.J. Tesh, T.B. Scott, Nano-composites for water remediation: a review. Adv. Mater. **26**(35), 6056–6068 (2014)
41. K.K. Singh, A. Singh, S. Rai, A study on nanomaterials for water purification. Mater. Today Proc. **51**, 1157–1163 (2022). https://doi.org/10.1016/j.matpr.2021.07.116
42. D. Mohan, C.U. Pittman, Arsenic removal from water/wastewater using adsorbents—a critical review. J. Hazard. Mater. **142**(1–2), 1–53 (2007). https://doi.org/10.1016/j.jhazmat.2007.01.006
43. M.N. Chong, B. Jin, C.W. Chow, C. Saint, Recent developments in photocatalytic water treatment technology: a review. Water Res. **44**(10), 2997–3027 (2010). https://doi.org/10.1016/j.watres.2010.02.039
44. M. Rai, A. Yadav, A. Gade, Silver nanoparticles as a new generation of antimicrobials. Biotechnol. Adv. **27**(1), 76–83 (2009). https://doi.org/10.1016/j.biotechadv.2008.09.002
45. A. Lee, J.W. Elam, S.B. Darling, Membrane materials for water purification: design, development, and application. Environ. Sci.: Water Res. Technol. **2**(1), 17–42 (2016). https://doi.org/10.1039/C5EW00159E
46. V.K. Upadhyayula, S. Deng, M.C. Mitchell, G.B. Smith, Application of carbon nanotube technology for removal of contaminants in drinking water: a review. Sci. Total. Environ. **408**(1), 1–13 (2009). https://doi.org/10.1016/j.scitotenv.2009.09.027
47. R. Hischier, T. Walser, Life cycle assessment of engineered nanomaterials: state of the art and strategies to overcome existing gaps. Sci. Total. Environ. **425**, 271–282 (2012). https://doi.org/10.1016/j.scitotenv.2012.03.001

48. T. Fatima, A. Mushtaq, Efficacy and challenges of carbon-based nanomaterials in water treatment: a review. Int. J. Chem. Biochem. Sci. **23**(1), 232–248 (2023), https://www.iscientific.org/wp-content/uploads/2023/05/27-IJCBS-23-23-29.pdf

Chapter 28
Thermal Characterization of Beeswax/ CuO Composites as Latent Heat Storage Media for Space Heating Applications

Arun Kumar Sharma, Prashant Sharma, Bhupendra Gupta, Anil Kumar, and Prashant Baredar

Abstract The present study is focused on the use of a phase-change material (PCM) for heat storage applications. Beeswax is investigated in this work as a PCM for heat storage because of its high latent heat. Utilizing beeswax has drawbacks because of elevated melting point and limited heat conductivity. In the present investigation, CuO nanoparticles (1–5 wt%) were added and tested based on analysis of thermal conductivity, latent heat, heat capacity characteristics and viscosity analysis. CuO nanoparticles found to give the improved result for thermal conductivity enhancement, increment in latent heat for both melting and solidifying stage. While inclusion of CuO nano particles, the specimens' heat capacity increases by 2–2.2%, with improvement in thermal conductivity upto 2.31 W/m.K. There is also in increment in viscosity.

Keywords Beewax · Nano PCM · CuO · Thermal characterization · Thermal energy storage

---

A. K. Sharma (✉)
Jabalpur Engineering College, Jabalpur, India
e-mail: arunksharma232@gmail.com

Department of Mechanical Engineering, Gyan Ganga College of Technology, Jabalpur, India

P. Sharma
Department of Mechanical Engineering, Shri Ram Institute of Technology, Jabalpur, India

B. Gupta
Department of Mechanical Engineering, Jabalpur Engineering College, Jabalpur, India

A. Kumar
Department of Mechanical Engineering, Delhi Technological University, Delhi, India

Nodal Centre of Excellence in Energy Transition (NCEET), Delhi Technological University, Delhi, India

P. Baredar
Department of Energy, MANIT, Bhopal, India

A. Kumar et al. (eds.), *Selected Proceedings of the 1st International Conference on Advanced Materials for Sustainable Innovation; IC-AMSI 2024; 28–30 August; New Delhi; India*, Springer Proceedings in Energy,
https://doi.org/10.1007/978-981-96-4492-6_28

28.1 Introduction

The amount of heat essential worldwide for domestic water heating (DHW) has increased significantly in the past few years. As such, there exists a notable reduction in the use of fossil fuels for energy [1, 2]. These days, solar energy is a great way to help lessen reliance on fossil fuels for energy. Renewable solar energy is most used in residential solar hot water systems. Though, several issues are concerned with applications. Primary is that it cannot provide sufficient warm water all day long since there is not enough solar radiation during nighttime or on days with cloud cover. Consequently, to increase the effectiveness of DHW, thermal energy storage (TES) devices must be used [3–5]. PCMs are promising for energy storage, however, they often exhibit poor thermal conductivity as a result of having an elevated concentration of energy [6, 7]. To address this issue, heat transfer improvement approaches, such as adding nanoparticles to the base PCM, can be considered [8, 9]. This will enhance thermal performance and reduce latent heat value, making PCMs more suitable for solar storage systems.

The main advantage of this inclusion would be the decrease in storage material if heat transfer improvement techniques were to improve the thermal capacity of the PCMs in thermal energy components. To address the issue of poor thermal conductivity in most phase change materials (PCMs), storage facilities need to incorporate nanoparticles into conventional PCMs as a method to boost the transmission of heat [10–12].

Experiments have shown that PCM incorporation reduces heat losses, removes temperature fluctuations in the output water, and delivers hot water for a lengthier time. Investigations on improving solar collector efficiency employing high conductivity nanofluid additions have been conducted. Additionally, several scholars have examined the impact of high thermal conductivity nanomaterial additions on PCM conductivity [13–15].

Numerous PCMs have been combined with a variety of nanoparticle kinds; nevertheless, the most appropriate nanoparticle varies depending on the circumstances. Therefore, there is a need for further research on the application of various nanoparticles with different PCM-based properties and melting points. Beeswax serves as the base material for this investigation [16, 17], to which varying percentages of CuO nanomaterial are introduced.

28.2 Methodology

28.2.1 Materials

The primary PCM in the present study is beeswax (BW). A representation of beeswax is shown in Fig. 28.1a. Beeswax was mixed with dissolving CuO nanoplatelets to improve the thermal characteristics. Copper oxide (CuO) is an oxide of copper

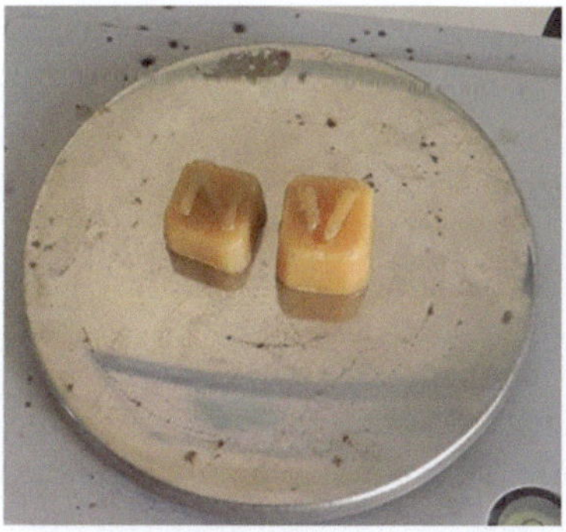

(a) Beewax

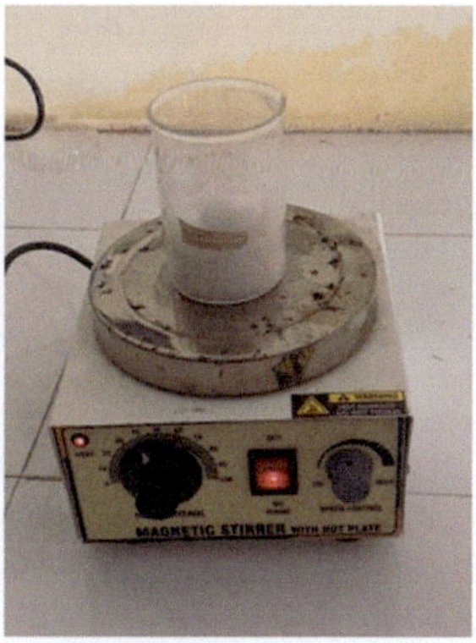

(b) Magnetic Stirring

Fig. 28.1 PCM and magnetic stirring of nano PCM

Table 28.1 Properties of CuO nanoparticle

Properties	Value/range
Appearance	Black or brownish-black powder
Size	10–100 nano-meters
Density	6.305 g/cm^3
Melting point	1201°C
Thermal conductivity	20–30 W/m·K
Specific heat capacity	530 kg·K

metal that has distinct mechanical, electronic, and magnetic characteristics. Because CuO nanoparticles have distinct physical and chemical characteristics from their bulk materials–such as diffusion, resistance to electricity, ability to conduct electricity, durability and hardness, chemical resistance, and a wide range of biological activity–research curiosity regarding these particles has risen significantly. The copper oxide nanoparticles are bought commercially from Polycam Internationals, located in Gujarat. Table 28.1 displays the CuO nanoparticles' physical and chemical characteristics.

28.2.2 Preparation of the Beeswax/CuO PCM (Nano-PCM)

A 100-ml container containing 20 g of beeswax has been utilized to generate the mixture. The container was subsequently set on an electric magnetic stirrer that was adjusted to a temperature of between 70 and 80 °C. 3–5 weight percent of nanoparticles are put into the entirely dissolved beeswax while it is continuously stirred. Nanoparticles have been heated in oven at 110 °C to remove humidity prior to being added. For thirty minutes, the beeswax containing nanoparticles was vigorously stirred. The beeswax mixture was stirred and then allowed to molt at ambient temperature in a container.

28.2.3 Characterization of Nano PCM

Thermal conductivity of the beeswax/CuO nano-PCM was measured employing a thermal conductivity detector. The sample was formed into tubular forms, and a cylinder mold was eliminated after solidifying. To stop airborne contaminants, an elevated pressure was kept, and thermal adhesive was used to stop air from escaping. Thermal loss was avoided by using polyurethane as thermal resistance. Water was pumped via the metal cylinder and cooled the specimen, and LabView software was employed for precise thermal conductivity reading.

The latent heat energy storage (LHES) and heat capacity of the materials were examined using a Perkin Elmer differential scanning calorimeter (DSC) in this investigation. The use of nitrogen as a cooling agent stabilized the heat flow. Specimens were subjected to a 5 °C/min scan speed in an aluminium DSC pan. DSC software was employed to compute the melting temperature and latent heat, and it was compared to a standard to estimate the heat capacity.

Fourier transform infrared spectroscopy (FT-IR) investigations were conducted to determine the characteristics of the beeswax/CuO nano-PCMs specimens. The CuO and base beeswax material's suitability was examined utilizing an FT-IR spectrometer. A characterization was done to find out how viscous the beeswax liquid seemed. An AP MCR viscometer was used to test the viscosity. To create the characterization graph, five distinct spindle rates were utilized to gather five beeswax viscosity measurements. While placing it into an adaptor to serve as the heating medium, the beeswax was liquefied. To maintain the liquid phase of the beeswax, water heated to approximately 70 °C was circulating around the adaptor.

28.3 Results and Discussion

To describe the characteristics of beeswax, tests were conducted first. The primary characteristics for uses involving heat transmission are density, latent heat, and melting point. Table 28.2 lists beeswax's characteristics. One benefit for thermal storage purposes is the elevated latent heat of the specimen.

The Fig. 28.2 represents the FT-IR (Fourier Transform Infrared) [18–20] spectra of pure BW and BW/CuO nano-PCM. The x-axis spans wavenumbers from approximately 500 cm^-1 to 4500 cm^-1, while the y-axis shows transmittance percentage ranging from 75 to 115%. In the beeswax spectrum, notable peaks are observed around 2900 cm^-1, indicating C-H stretching vibrations typical of alkanes. There is also a significant peak around 1700 cm^-1, which corresponds to C = O stretching,

Table 28.2 Properties of Beewax

Properties	Value/range
Melting temperature [°C]	61.2°C
Latent heat [kJ/kg]	206.8 (melting)
	203.7 (solidifying)
Specific heat [kJ/kg]	2.32 (liquid)
	1.95 (solid)
Density [kg/m^3]	790.5 (melting)
	821.5 (solidification)

a common feature of carbonyl groups. Additionally, peaks around 1500 cm$^-$1 to 1000 cm$^-$1 can be attributed to various bending and stretching vibrations associated with the molecular structure of beeswax. The beeswax/CuO spectrum shows similar peaks, but with notable differences in transmittance, particularly around the characteristic peaks. The C-H stretching region around 2900 cm$^-$1 and the C = O stretching region around 1700 cm$^-$1 both exhibits decreased transmittance in the composite compared to pure beeswax. This indicates potential interactions between the CuO nanoparticles and the beeswax matrix, suggesting successful incorporation of CuO into the beeswax. Furthermore, the overall transmittance in the beeswax/CuO spectrum is lower, which could be due to the presence of CuO particles affecting the absorption characteristics. The comparison of these spectra highlights the modifications in the chemical environment of the beeswax upon addition of CuO nanoparticles, providing insights into the compatibility and interaction between the beeswax and CuO in the composite material. The beeswax and CuO nanoplatelets just interacted physically; no undesired chemical reactions were observed.

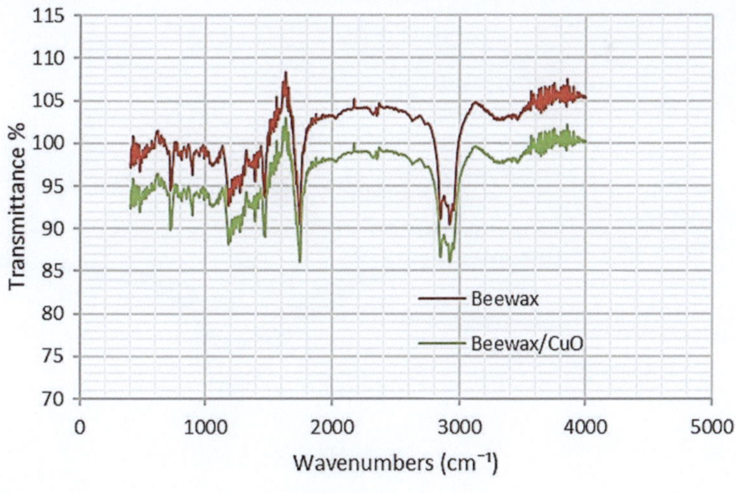

Fig. 28.2 FTIR test results of BW and BW/CuO nano-PCM

The differential scanning calorimetry (DSC) curve for beeswax and beeswax/CuO nano PCMs is displayed in Fig. 28.3. The DSC curve [21–23] displays the different thermal characteristics of the utilized beeswax and beeswax/CuO nano PCMs. The melting point, specific heat, and latent heat of the specimen's PCM are all determined using the DSC data (also showed in Table 28.3). The measurement of latent heat may be found within the region over the curve, and specific heat can be found in the gradient of the curve's dip.

Raw BW has a heat capacity of 0.52 kJ/kgK and a peak melting temperature of 61.2 °C. It also has a latent heat of 206.8 J/g. At 57.05 °C, the solidification process takes place, and the latent heat is 203.7 J/g. The peak melting temperature drops to 60.97 °C with a latent heat of 223.13 J/g and a heat capacity of 0.5304 kJ/kg·K when 1% weight of CuO is introduced.

The phase transition temperatures of beeswax were impacted by the presence of CuO nanoparticles. When the weight in percent of the CuO nanoplatelets was elevated, there was a likelihood for the melting temperatures to fall. On the other hand, there was a trend for the solidifying temperatures to rise. Beeswax/CuO nano-PCM's melting point dropped from 61.2 to 60 °C. This demonstrates that the heat transmission at which the specimen induces a phase transition in the beeswax, occurring at less temperature for dissolving and at a greater temperature for solidifying.

In terms of latent heat, the specimen's latent heat was elevated by the impact of CuO nano particles. Comparing pure beeswax to this mixture, the melting latent heat rose by 17.3%. Brownian motion and particle clustering processes account for the rise in latent heat.

According to the research, adding CuO nanoparticles to beeswax greatly enhances its capacity for heat storage. Having a heat capacity of 0.5410 kJ/kg·K and a latent

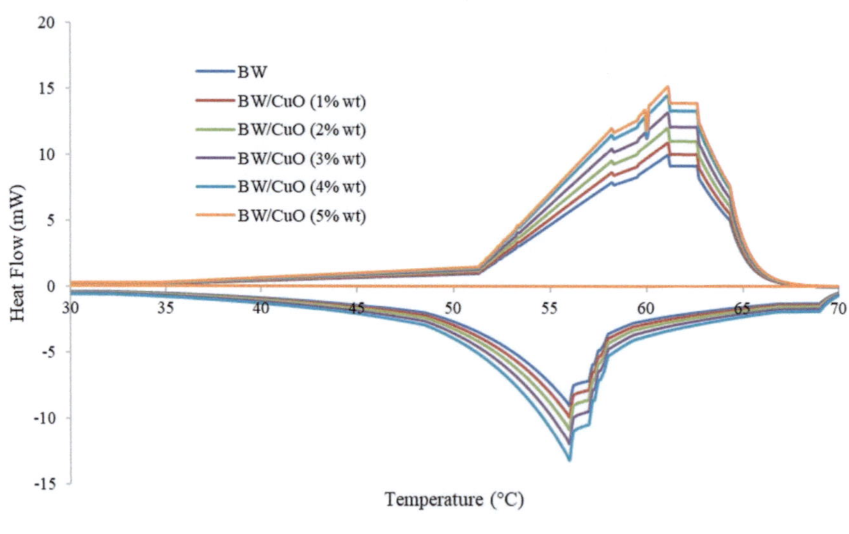

Fig. 28.3 DSC results of all the BW/CuO nano PCM's

Table 28.3 Thermal storage properties of samples

Sample	Melting process		Solidifying process		Heat capacity (KJ/KgK)
	Peak temperature (°C)	Latent heat (J/g)	Peak temperature (°C)	Latent heat (J/g)	
BW	61.2	206.8	57.05	203.7	0.52
BW/CuO (1% wt)	60.97399	223.1259	51.59942	225.6339	0.5304
BW/CuO (2% wt)	61.50462	225.4499	51.6289	229.5926	0.541008
BW/CuO (3% wt)	61.50462	238.5749	51.35376	233.8451	0.551828
BW/CuO (4% wt)	60.50231	238.8234	51.98266	234.4047	0.562865
BW/CuO (5% wt)	61.04277	263.188	51.59942	257.5276	0.574122

heat of 225.45 J/g, the highest melting temperature of 2% CuO is 61.50 °C. 51.63 °C is the highest temperature of solidification, and the latent heat is 229.59 J/g. A larger CuO concentration raises the latent heat of melting, suggesting that CuO nanoparticles improve the composite material's thermal storage effectiveness. The little differences in peak temperatures indicate that CuO nanoparticles help phase transition processes by improving heat absorption and release.

Overall, the incorporation of CuO nanoparticles in beeswax enhances its effectiveness as a PCM for TES applications. In general, adding CuO nanoparticles to beeswax improves its performance as a PCM in TES systems.

By employing the specific heat of indium as a benchmark, the specific heat capacity was determined. The finding of the heat capacity improvement in the study was unrelated to the existence of a microstructure. Since thermal capacity is not an important transmission feature, it is expected to rise because of the Brownian motion of the nano-platelets. Specimens' heat capacity rose by 2–2.2%. In comparison to the basic liquid, this layer has strong thermal characteristics.

The investigation shows how CuO nanoparticles affect the improvement of heat conductivity (Fig. 28.4). The specimen's thermal conductivity improved to 2.31 W/m.K. with the incorporation of CuO nanoparticles, suggesting that the clusters of CuO nanoparticles influence both latent heat and the speed of heat transmission. There is a direct correlation between improvement and the weight portion of nano-PCM, as evidenced by the rise in thermal conductivity of the nano-PCM with a greater weight percentage. Yet, because of nanoparticle aggregation, more nanoparticle additions might not produce a linear enhancement.

The Viscosity variations regarding CuO% growth is shown in Fig. 28.5. Because of the attraction interactions between the CuO%, the viscosity observations tended to rise as the weight percentage of the nano particles increased.

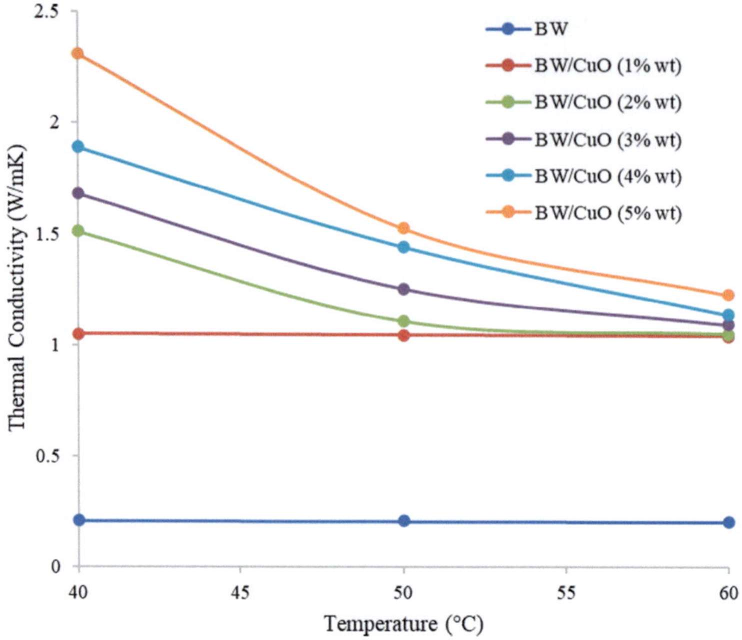

Fig. 28.4 Thermal conductivity of beeswax/CuO specimens

Fig. 28.5 Viscosity variation with respect to CuO % increment

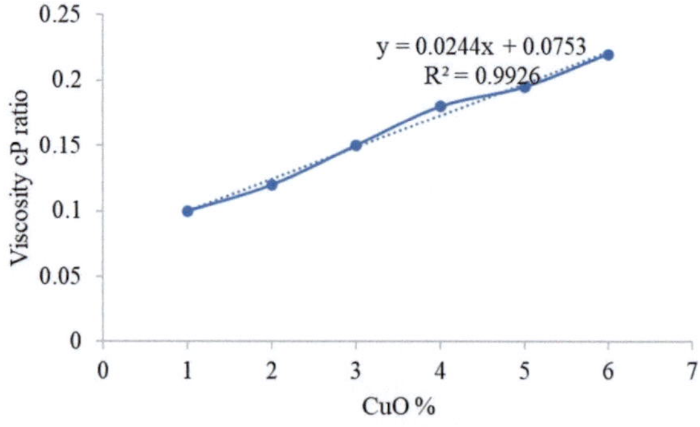

28.4 Conclusion

The present work considers the thermal characterization of Beeswax/CuO composites as LHES media for warming applications. On the analysis it was found that Beeswax's density, latent heat, and melting point make it suitable for heat transmission. FT-IR spectra show C-H stretching vibrations and C = O stretching, suggesting successful incorporation of CuO. The interaction between beeswax and CuO nanoplatelets is physical and there is not any chemical reactions. The DSC curve reveals that the addition of CuO nanoparticles to beeswax significantly reduces the peak melting temperature from 61.2 °C to 60 °C, affecting the phase transition temperatures and affecting the thermal characteristics. The study found that adding CuO nanoparticles to beeswax increased its latent heat by 17.3%, improved thermal storage effectiveness, and had a maximum melting temperature of 62.50 °C, with a specific heat capacity of 2–2.2%. The study reveals that the addition of CuO nanoparticles improves heat conductivity in a specimen, with a relationship among the weight portion of the nano-PCM and the increase in thermal conductivity. However, more nanoparticle additions may not produce a linear enhancement due to nanoparticle aggregation.

References

1. P. Sharma, B. Gupta, M. Pandey, A.K. Sharma, R. Nareliya Mishra, Recent advancements in optimization methods for wind turbine airfoil design: a review. Mater. Today Proc. **47**(xxxx), 6556–63 (2020). https://doi.org/10.1016/j.matpr.2021.02.231
2. P. Sharma, B. Gupta, M. Pandey, Hydrogen rich product gas from air–steam gasification of indian biomasses with waste engine oil as binder. Waste Biomass Valorization **13**(6), 3043–60 (2022). https://doi.org/10.1007/s12649-022-01690-4
3. S. Bouadila, S. Baddadi, T. Rehman ur, R. Ayed, Experimental investigation on the thermal appraisal of heat pipe-evacuated tube collector-based water heating system integrated with PCM. Renew. Energy **199**(March), 382–94 (2022). https://doi.org/10.1016/j.renene.2022.09.004
4. G. Sadeghi, M. Mehrali, M. Shahi, G. Brem, A. Mahmoudi, Progress of experimental studies on compact integrated solar collector-storage retrofits adopting phase change materials. Sol. Energy **237**(January), 62–95 (2022). https://doi.org/10.1016/j.solener.2022.03.070
5. S. Sadhishkumar, T. Balusamy, Performance improvement in solar water heating systems–a review. Renew. Sustain. Energy Rev. **37,** 191–8 (2014). https://doi.org/10.1016/j.rser.2014.04.072
6. Y. Zhang, J. Wang, J. Qiu, X. Jin, M.M. Umair, R. Lu et al., Ag-graphene/PEG composite phase change materials for enhancing solar-thermal energy conversion and storage capacity. Appl. Energy **237**(December 2018), 83–90 (2019). https://doi.org/10.1016/j.apenergy.2018.12.075
7. H.S. Xue, Experimental investigation of a domestic solar water heater with solar collector coupled phase-change energy storage. Renew. Energy **86**, 257–61 (2016). https://doi.org/10.1016/j.renene.2015.08.017
8. H. Kaya, M. Alkasem, K. Arslan, Effect of nanoparticle shape of Al_2O_3/pure water nanofluid on evacuated U-Tube solar collector efficiency. Renew. Energy **162**, 267–84 (2020). https://doi.org/10.1016/j.renene.2020.08.039

9. M.E. Zayed, J. Zhao, A.H. Elsheikh, Y. Du, F.A. Hammad, L. Ma et al., Performance augmentation of flat plate solar water collector using phase change materials and nanocomposite phase change materials: a review. Process. Saf. Environ. Prot.**128**, 135–57 (2019). https://doi.org/10.1016/j.psep.2019.06.002
10. Y. Zhao, X. Zhang, W. Hua Review of preparation technologies of organic composite phase change materials in energy storage. J. Mol. Liq. **336**, 115923 (2021). https://doi.org/10.1016/j.molliq.2021.115923
11. E. Douvi, C. Pagkalos, G. Dogkas, M.K. Koukou, V.N. Stathopoulos, Y. Caouris et al. Phase change materials in solar domestic hot water systems: a review. Int. J. Thermofluids **10**, 100075 (2021). https://doi.org/10.1016/j.ijft.2021.100075
12. M. Abuşka, S. Şevik, A. Kayapunar, Experimental analysis of solar air collector with PCM-honeycomb combination under the natural convection. Sol. Energy Mater. Sol. Cells **195**(February), 299–308 (2019)
13. K. Tumirah, M.Z. Hussein, Z. Zulkarnain, R. Rafeadah Nano-encapsulated organic phase change material based on copolymer nanocomposites for thermal energy storage. Energy **66**, 881–90 (2014). https://doi.org/10.1016/j.energy.2014.01.033
14. A.A. Elbrashy, F.S. Abou-Taleb, M.K. El-Fakharany, F.A. Essa, Experimental study of solar air heater performance by evacuated tubes connected in series and loaded with thermal storage material. J. Energy Storage **54**(February), 105266 (2022)
15. V.V. Tyagi, K. Chopra, R.K. Sharma, A.K. Pandey, S.K. Tyagi, M.S. Ahmad et al., A comprehensive review on phase change materials for heat storage applications: development, characterization, thermal and chemical stability. Sol. Energy Mater. Sol. Cells **2022**(234), 111392 (2021). https://doi.org/10.1016/j.solmat.2021.111392
16. E.J. D'oliveira, S.C.C. Pereira, U. Azimov Numerical analysis on hot water storage tank integrated with phase change materials. WIT Trans. Ecol. Environ. **254**, 125–35 (2021)
17. S. Rawi, M. Amin, E. Kusrini, N. Putra, Characterization of shape-stabilized phase change material using beeswax and functionalized multi-walled carbon nanotubes. IOP Conf. Ser. Earth. Environ. Sci. **105**(1) (2018)
18. X. Lu, J. Huang, W.Y. Wong, J. Qu ping, A novel bio-based polyurethane/wood powder composite as shape-stable phase change material with high relative enthalpy efficiency for solar thermal energy storage. Sol. Energy Mater. Sol. Cells **200**(March), 109987 (2019). https://doi.org/10.1016/j.solmat.2019.109987
19. A.A. Adeleke, J.K. Odusote, P.P. Ikubanni, O.O. Agboola, A.O. Balogun, O.A. Lasode, Tumbling strength and reactivity characteristics of hybrid fuel briquette of coal and biomass wastes blends. Alexandria Eng J. **60**(5), 4619–4625 (2021). https://doi.org/10.1016/j.aej.2021.03.069
20. X. Zuo, X. Zhao, J. Li, Y. Hu, H. Yang, D. Chen, Enhanced thermal conductivity of form-stable composite phase-change materials with graphite hybridizing expanded perlite/paraffin. Sol. Energy **209**(February), 85–95 (2020)
21. J. Yu, X. Chen, X. Ma, Q. Song, Y. Zhao, J. Cao, Influence of nanoparticles and graphite foam on the supercooling of acetamide. J. Nanomater. **2014** (2014)
22. O. Gencel, A. Sarı, A. Ustaoglu, G. Hekimoglu, E. Erdogmus, A. Yaras et al., Eco-friendly building materials containing micronized expanded vermiculite and phase change material for solar based thermo-regulation applications. Constr. Build. Mater. **308**(June), 125062 (2021)
23. I. Baskar, M. Chellapandian, K. Jeyasubramanian, LA-PA eutectic/ nano- SiO_2 composite phase change material for thermal energy storage application in buildings. Constr. Build. Mater. **338**(January), 127663 (2022). https://doi.org/10.1016/j.conbuildmat.2022.127663

Chapter 29
Smart Grid Management: Machine Learning for Load Balancing and Fault Detection

Ramandeep Saha, Garima Singhal, Shubhankit Sadhukar, Ajay Kumar, Aniket Singh, and Mohd Atif Wahid

Abstract The integration of renewable energy sources and the increasing complexity of power grids has led to the development of smart grids that utilize advanced sensing, communication, and control technologies. This review presents a critical discussion on the current advances in ma- chine learning (ML) techniques for load balancing and fault detection in smart grids. Frameworks for smart grid management are described, and essential sub-disciplines are reviewed, including data-intensive ML strategies and algorithms, smart grid databases and data generation approaches, and key power system descriptors used in ML models. Furthermore, an in-depth discussion on the applications of ML in load forecasting, demand response, fault detection, and self-healing is provided. Finally, how these sub-disciplines support data-driven smart grid management is outlined, and the opportunities and challenges are highlighted.

Keywords Smart grid · Machine learning · Load balancing · Fault detection · Renewable energy integration

R. Saha · S. Sadhukar · A. Kumar
Department of Artificial Intelligence (GGISPU), Delhi Technical Campus (GGISPU), Greater Noida, India
e-mail: r.saha@delhitechnicalcampus.ac.in

G. Singhal
School of Management (AKTU), NIET (AKTU), Greater Noida, India

A. Singh (✉)
Department of Computer Science and Engineering (GGISPU), Delhi Technical Campus (GGISPU), Greater Noida, India
e-mail: aniket@delhitechnicalcampus.ac.in

M. A. Wahid
School of Engineering and Technology, Vivekananda Institute of Professional Studies- Technical Campus, Delhi, India

A. Kumar et al. (eds.), *Selected Proceedings of the 1st International Conference on Advanced Materials for Sustainable Innovation; IC-AMSI 2024; 28–30 August; New Delhi; India*, Springer Proceedings in Energy,
https://doi.org/10.1007/978-981-96-4492-6_29

29.1 Introduction

The increasing use of renewable energy sources and distributed energy resources (DERs) like solar photovoltaics, wind turbines, electric vehicles, and energy storage systems is changing power networks into smart grids. Advanced sensor, communication, and control technology in smart grids enable bidirectional power flow and information exchange between utilities and consumers. Renewable energy fluctuations and dynamic loads hinder smart grid operation and control. Utilities and grid operators increasingly rely on data-driven solutions from smart meters, PMUs, and other sensors.

ML approaches have shown potential in load balancing and fault detection for smart grid control. Load balancing matches power supply with demand in real- time, taking into account grid infrastructure constraints and minimizing costs

and emissions. ML-based load forecasting models accurately predict future power demand by learning from prior load data, weather, time, and socioeconomic indicators. ML can enable dynamic demand response systems that compensate customers for adapting electricity usage patterns to price signals or grid conditions. Recent advancements in deep learning have shown promising results in short-term load forecasting, as demonstrated by Wang et al. [32]

Fault detection pinpoints abnormal grid conditions including short circuits, line outages, and equipment failures that could cause disruption or destruction. Traditional fault detection systems using circuit breakers and safety relays are slow and unreliable, causing blackouts and cascade failures. ML algorithms pro- vide proactive maintenance and grid self-healing through real-time sensor data analysis and faster, more accurate defect identification [6, 31].

This study describes recent ML approaches for smart grid defect detection and load balancing. The remaining paper is organized as follows: ML and smart grid management frameworks are explained in Section 29.2. Section 29.3 covers smart grid ML algorithms and approaches. Section 29.4 discusses smart grid databases and data generation. Section 29.5 lists the primary ML model smart grid key descriptors. ML uses in demand response, fault detection and self-healing, load and generation balancing are included in Section 29.6. Section 29.7 concludes and suggests additional research.

29.2 Frameworks for Smart Grid Management

Various frameworks have been proposed for data-driven smart grid management [30]. These systems streamline data collection, communication, processing, and control throughout the grid infrastructure by merging hardware and software. Figure 29.1 illustrates a multi-layer design for data-driven smart grid management. The key layers are described below:

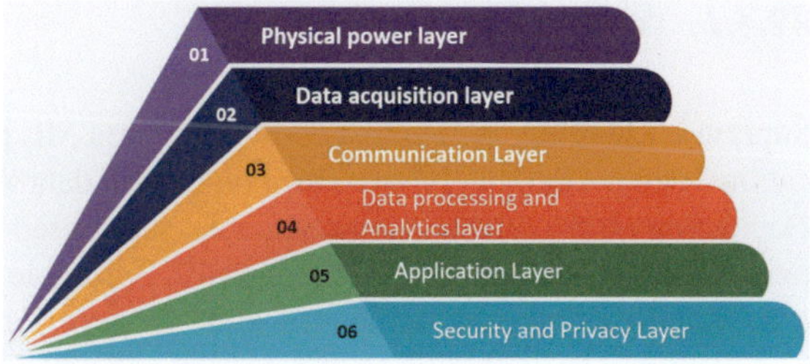

Fig. 29.1 A generic multi-layer framework for data-driven smart grid management

The physical power layer includes generators, transformers, transmission and distribution lines, and loads. DERs and renewable energy are increasingly coupled at this tier. Sensors and actuators control physical grid properties.

The data acquisition layer includes smart meters, PMUs, SCADA systems, and advanced metering infrastructure (AMI) that collect raw data from the

physical layer. The data may include voltage, current, power, frequency, device status, etc.

The communication layer provides infrastructure and procedures for sending data to higher levels for processing and analysis. Fiber optics, PLC, cellular networks, and Wi-Fi are wired and wireless technologies.

The data processing and analytics layer processes and analyzes data for smart grid management applications like load forecasting, fault detection, state estimate, optimal power flow, and dynamic pricing. Data is stored and processed utilizing big data systems and high-performance computing.

This layer provides user interfaces and decision support tools for grid operators and other stakeholders to see analytics layer findings and perform control actions. Applications include market operations, demand response, energy, and outage management.

Layer 6: Security and Privacy Smart grids rely heavily on digital technology and collect and send sensitive data, making cybersecurity and data privacy key concerns. This layer includes authentication, encryption, access control, and intrusion detection to ensure smart grid reliability.

The closed-loop data-driven smart grid management system maximizes and regulates physical grid parameters using data analytics insights, impacting future data collection. This feedback loop keeps the grid improving and adapting to changing situations.

29.3 ML Strategies and Algorithms for Smart Grid Management

This section reviews the key ML strategies and algorithms that have been applied for load balancing and fault detection in smart grids.

29.3.1 Supervised Learning

Supervised learning is the most commonly used ML paradigm for smart grid applications, where the models are trained on labeled data with known input- output pairs. Depending on the type of output variable, supervised learning can be further classified into regression and classification tasks. Ensemble methods like Random Forests and Gradient Boosting Machines have shown superior performance in various smart grid applications, including fault detection and renewable energy forecasting [2]

Regression Algorithms ML pipelines for smart grid management often include data processing and analytics [19, 30]. The model detects basic trends and relationships from training data. The validation set helps tweak model hyper parameters to prevent overfitting. The trained model is assessed generally and on a performance set. Accuracy, precision, recall, F1-score, MSE, and R-squared depend on the work. Production judgments and smart grid management estimates are available from the validated model. New data helps the model reflect grid changes. Computing resources affect ML algorithm and pipeline architecture, data volume, velocity, diversity, application complexity, scalability, interpretabil- ity, and reliability. ML models can be trained locally using subsystem or micro- grid data or centrally using grid data. Hybrid techniques combining physics- based and data-driven models aim to enhance accuracy and robustness [14].

Classification Algorithms Classification algorithms can predict discrete output variables like load type (residential, commercial, industrial), device status (normal, defective), or fault severity (low, medium, high). Smart grids use several popular classification methods: Logistic regression is a linear classifier that predicts the likelihood of an output variable belonging to a specific class based on input features. Many binary classification problems, such as defect detection, use it as a baseline model. Learning decision rules that partition feature space into groups enables categorization using decision trees and random forests [40]. They effectively handle skewed datasets and are interpretable.

29.3.2 Unsupervised Learning

Unsupervised learning finds patterns and structures in unlabeled data without knowing the output variables. It is used for data compression, anomaly detection, and exploratory analysis. Standard unsupervised learning methods used by smart grids include: A K-means clusterer separates data into K clusters based on similarity. Uses of K-means include load profiling, customer segmentation, and anomaly detection in smart meters [30].

29.3.3 Deep Learning

In deep learning, artificial neural networks with multiple layers are used to create hierarchical representations of data. Recently, deep learning has been used to smart grid management tasks such as demand forecasting, renewable energy projection, and fault detection. Popular deep learning designs for smart grids include:

CNNs use convolutional filters to handle grid-like data like images or time series, capturing local patterns and hierarchies. Sky photographs and meteorological data have been used to predict solar irradiation and wind speed.

RNNs evaluate sequential data by maintaining a hidden state that captures temporal dependencies between inputs. Application of historical time series data [18, 40] has been widely used for short-term load and renewable energy forecasts. Gated Recurrent Units (GRUs) and Long Short-Term Memory (LSTM) models can represent long-term dependencies and control vanishing gradients.

29.4 Smart Grid Databases and Data Generation

29.4.1 Smart Grid Data Sources

Smart grids generate massive amounts of data from multiple power system sources. Important sources of Advanced Metering Infrastructure (AMI) include smart meters, data management systems, and communication networks for two- way interaction between utilities and consumers [30]. After detecting and recording Smart grids generate massive amounts of data from multiple power system sources. Important sources of Advanced Metering Infrastructure (AMI) include smart meters, data management systems, and communication networks for two- way interaction between utilities and consumers [30]. After detecting and recording electricity use hourly or 15-minute intervals, smart meters send the data to the utility for monitoring and billing. AMI data can be used for load profiling, demand forecasting, and anomaly detection [30].

SCADA systems use data from RTUs and IEDs at substations and other critical sites to monitor and regulate the power grid in real-time. SCADA data includes voltage, current, and other equipment status. SCADA data can be used for state estimation, contingency analysis, and fault detection.

Phasor measurement units (PMUs) monitor high-speed voltage and current phasors (magnitude and phase angle) at multiple grid locations.

Weather and environmental sensors like temperature, humidity, wind speed, and sun irradiation substantially affect renewable energy generation and electricity demand. Installing environmental sensors and weather stations improves load and generation forecasts by collecting real-time weather data [13].

Rooftop solar panels, small wind turbines, CHP units, and energy storage devices are increasingly used off-grid. DERs generate real-time output, control settings, and performance properties. Utilizing this information can optimize the operation and integration of DERs into the grid, including demand response and other grid services.

29.4.2 Smart Grid Data Characteristics and Challenges

The "4 V" qualities of big data—volume, velocity, variety, and veracity—showcase themselves in smart grid data [30].

(1) Volume: Smart grids create enormous and fast expanding data generation capacity. One smart meter can produce several kilobytes of data per day; a utility with millions of consumers can gather terabytes of data daily. Even more data is produced by PMUs and other high-frequency sensors, which presents problems for processing, storage, and analysis.

(1) Velocity: Smart grid data are produced and sent in real-time or near realtime, therefore quick and effective processing is needed to allow prompt decision making and control activities. Some uses like fault detection and isolation call for sub-second reaction times.

(1) Variety: Smart grid data arrive in many forms and structures: time se- ries data from meters and sensors, event logs from SCADA systems, geospatial

data from GIS, textual data from maintenance records and customer complaints, and multimedia data from cameras and drones. One of the main difficulties is combining and evaluating several disparate data sources.

(1) Veracity: Smart grid data could include noise, mistakes, and inconsistencies resulting from sensor failures, poor communication, and data modification. Reliable and safe grid functioning depends on guarantees of quality, integrity, and trustworthiness of the data.

29.4.3 Smart Grid Test Systems and Synthetic Data Generation

Real-world smart grid data is frequently restricted by privacy, security, and proprietary reasons. The literature suggests several test systems and synthetic data generation methods for smart grid management solution research and development.

Birchfield et al. [5] describes synthetic power grid models as realistic representations of power networks that mimic their structural and statistical properties. Considering geography, population, economy, and technology, algorithms can simulate energy grid expansion and evolution across time. These include RT-nested ACTIVSg and SmallWorld models build synthetic power grids of various diameters and types (e.g., transmission, distribution). When real data is unavailable, synthetic grid models can test and validate smart grid management strategies.

Generative models like VAEs and GANs can provide smart grid data with real-like properties. These models can learn the data distribution and produce statistically equivalent fresh samples. Generative models have been used to create load profiles, renewable energy profiles, and PMU measurements for data augmentation and privacy preservation [15]. When real data is scarce or sensitive, generated data can be used to test and train ML models.

29.5 Key Descriptors for Smart Grid ML Models

The performance of ML models for smart grid management depends heavily on the selection of appropriate input features or descriptors that capture the relevant characteristics of the system. This section discusses some of the key descriptors used in ML models for load forecasting, renewable energy prediction, and fault detection.

29.5.1 Load Forecasting Descriptors

Load forecasting predicts electricity demand based on past load trends, weather, and other factors. Common load forecasting descriptors include: Past load data is vital for load forecasting. Historical data provides many time series elements, including hourly, daily, weekly, and seasonal load trends [23]. Forecasting models often employ lag variables, which represent prior load (e.g., previous hour, day, week) [23]. To capture temporal and cyclical trends in load data, calendar variables such as hour, day, month, and season are used [30]. Binary variables

are used to model the impact of weekends, holidays, and special events, such as sports and festivities [30]. Weather greatly impacts electricity usage, especially for heating and cooling. Load forecasting relies on meteorological elements such as temperature, humidity, wind speed, and precipitation [13]. To model the nonlinear relationship between temperature and load, [13] derived variables like HDD and CDD are used to evaluate temperature variation from a reference value. The inclusion of social media data and search engine trends has been explored to improve load forecasting accuracy, especially during special events or holidays [3]

(4) Economic and demographic variables: Identify long-term power consumption patterns and drivers using economic indicators like GDP, employment rate, and industrial output index. Demographic parameters like as population, household size, and urbanization rate are crucial for load forecasting at regional or national levels. (5) Customer Segmentation Variables: Grouping users by load profile, tariffs, or other traits improves load forecasting accuracy [30]. Customer segments can be identified using characteristics such load factor, peak-to-average ratio, customer type (residential, commercial, industrial), and load.

29.5.2 Renewable Energy Prediction Descriptors

Renewable energy forecast uses technical, topographical, and meteorological factors to estimate solar and wind power plant output. Important renewable energy prediction criteria include: Weather conditions like sun irradiation, temperature, wind speed, and direction affect solar and wind power plant production. These variables can be obtained from NWP models, satellite pictures, or weather stations [26]. Renewable energy variability and uncertainty are modeled using factors like clear skies index (ratio of real to clear sky irradiation) and turbulence intensity (wind speed standard deviation) [9]. The location and geography of the renewable energy site may affect energy output due to shade, elevation, and surface roughness. The site circumstances can be characterized using geographic variables such as latitude, longitude, altitude, slope, and aspect [21]. Land cover and usage data from satellite images or GIS can aid in understanding surrounds and problems [4]. Technical elements like configuration and standards can affect renewable energy system performance and output. ML models may be predicted using factors like solar panel capacity, efficiency, orientation, tilt angle, and wind turbine hub height, rotor diameter, and power curve [24]. Also considered [25] is equipment aging and deterioration over time. Temporal Variables: Time of day, season, and weather affect renewable energy generation. Various factors such as time of day, sun zenith angle, and season can capture solar radiation cycles [28]. Wind power temporal dependencies can be modeled using wind speed and direction persistence over 1-hour, 3-hour, and 24-hour time periods. Short-term forecasting accuracy can be improved by analyzing spatial correlation and variability of renewable energy across several sites [35]. Spatial dependencies can be determined by analyzing characteristics such as distance, direction, and terrain similarity between the target place and nearby sites [33]. Kriging and Gaussian processes can be used to interpolate and expand renewable energy data from few observations [34].

29.5.3 Fault Detection Descriptors

Renewable energy forecast uses technical, topographical, and meteorological factors to estimate solar and wind power plant output. Important renewable energy prediction criteria include: Weather conditions like sun irradiation, temperature, wind speed, and direction affect solar and wind power plant production. These variables can be obtained from NWP models, satellite pictures, or weather stations [26]. Renewable energy variability and uncertainty are modeled using factors like clear skies index (ratio of real to clear sky irradiation) and turbulence intensity (wind speed standard deviation) [9]. The location and geography of the renewable energy site may affect energy output due to shade, elevation, and surface roughness. The site circumstances can be characterized using geographic variables such as latitude, longitude, altitude, slope, and aspect [21]. Land cover and usage data from satellite images or GIS can aid in understanding surrounds and problems [4]. Technical elements like configuration and standards can affect renewable energy system performance and output. ML models may be predicted using factors like solar panel capacity, efficiency, orientation, tilt angle, and wind turbine hub height, rotor diameter, and power curve [24]. Also considered [25] is equipment aging and deterioration over time. Temporal Variables: Time of day, season, and weather affect renewable energy generation. Various factors such as time of day, sun zenith angle, and season can capture solar radiation cycles [28]. Wind power temporal dependencies can be modeled using wind speed and direction persistence over 1-hour, 3-hour, and 24-hour time periods. Short-term forecasting accuracy can be improved by analyzing spatial correlation and variability of renewable energy across several sites [35]. Spatial dependencies can be determined by analyzing characteristics such as distance, direction, and terrain similarity between the target place and nearby sites [33]. Kriging and Gaussian processes can be used to interpolate and expand renewable energy data from few observations [34].

29.6 Applications of ML in Smart Grid Management

29.6.1 Load Forecasting

Load forecasting is a critical task in smart grid management that involves predicting the future electricity demand at different spatial and temporal scales. Table 29.1 presents a survey of recent studies on ML-based load forecasting in smart grids.

Table 29.1 Survey of ML-based load forecasting studies in smart grids

Study	Methodology	Data	Best performance	Key findings
Kong et al. [18]	LSTM	University campus smart meter data	MAPE < 2% for 1–24 hour ahead	LSTM outperforms ARIMA, SVR, and ANN
Chen et al. [8]	SVR	Monthly city electricity demand	MAPE = 3.6%	Economic and demo- graphic features improve accuracy
Zheng et al. [40]	XGBoost	Hourly distribution network load	MAPE = 1.75%	XGBoost outperforms BPNN, ARIMA, and SVR
Buzau et al. [7]	LSTM-ARIMA	Hourly load of European countries	MAPE = 1.03%	Hybrid model outperforms individual models

Some key findings from the load forecasting literature include:

Deep learning models such as LSTM and CNN have shown superior performance compared to traditional models such as ARIMA, SVR, and shallow ANNs, especially for short-term load forecasting with high granularity data [18]. Incorporating additional features such as weather, economic, and demographic variables can improve the accuracy of load forecasting models, particularly for medium-term and long-term horizons [8].

Ensemble models that combine multiple ML algorithms or integrate ML with statistical models (e.g. ARIMA) have demonstrated better accuracy and robust- ness than individual models [7, 40].

Clustering and segmentation techniques can be used to group customers or load profiles into homogeneous clusters, which can then be modeled separately for improved forecasting accuracy [30].

29.6.2 Renewable Energy Forecasting

Renewable energy forecasting is another important application of ML in smart grid management, which involves predicting the output of solar and wind power plants at different time scales. Accurate renewable energy forecasts are critical for various tasks such as scheduling and dispatch, market operations, grid integration, and planning. Table 29.2 presents a survey of recent studies on ML-based renewable energy forecasting.

Some key findings from the renewable energy forecasting literature include:

Deep learning models such as CNN and LSTM have shown promising results for short-term solar and wind power forecasting by learning spatial and temporal

Table 29.2 Survey of ML-based renewable energy forecasting studies

Study	Methodology	Data	Best performance	Key findings
Sun et al. [27]	CNN-LSTM	Solar power and NWP data	RMSE = 3.92%	CNN-LSTM outperforms LSTM, GBRT, and persistence
Zhang et al. [38]	GBRT	Wind farm SCADA data	RMSE = 7.03%	GBRT out-performs SVR, ELM, and BPNN
Zeng et al. [36]	SVR	Monthly solar radiation and meteorological data	RMSE = 4.62%	SVR outper-forms ANN and empirical models
Agoua et al. [1]	Hybrid NWP-SVR	Wind speed and power data	RMSE = 11.3%	Hybrid model outperforms NWP and SVR alone

dependencies from high-dimensional data such as sky images and meteorological variables [27].

Ensemble methods such as GBRT and random forests have demonstrated superior accuracy and robustness compared to individual ML models for wind power forecasting, especially with noisy and heterogeneous data from SCADA systems [38].

Kernel-based methods such as SVR have been effective for medium-term and long-term renewable energy forecasting with limited training data, by capturing nonlinear relationships between the output and input variables such as solar radiation and meteorological factors [36].

Hybrid models that combine physical models (e.g. NWP) with data-driven models (e.g. SVR) have shown improved accuracy and computational efficiency compared to using either model alone, by leveraging the strengths of both approaches [1].

Transfer learning techniques have been applied to improve the forecasting accuracy for new renewable energy sites with limited historical data [22].

29.6.3 Fault Detection and Diagnosis

Fault detection and diagnosis is a critical application of ML in smart grid management, which involves identifying and localizing abnormal conditions or failures in power system components such as generators, transformers, transmission lines, and protective devices. Table 29.3 presents a survey of recent studies on ML- based fault detection and diagnosis in smart grids.

Some key findings from the fault detection and diagnosis literature include:

Table 29.3 Survey of ML-based fault detection and diagnosis studies in smart grids

Study	Methodology	Data	Best performance	Key findings
Jiang et al. [16]	SVM-Wavelet	Current signals	Accuracy = 98.7%	SVM outper- forms ANN and fuzzy logic
Haghigha et al. [11]	Random forest	Smart meter data	Accuracy = 97.5%	RF identifies fault location and estimates fault distance
Wang et al. [29]	CNN	Current time-frequency images	Accuracy = 99.7%	CNN outper-forms SVM, KNN, and decision tree
Nadian et al. [20]	DBSCAN	Voltage and current data	Accuracy = 95%	DBSCAN identifies fault buses and lines

SVM with wavelet-based feature extraction has been widely used for fault detection and classification in transmission lines, outperforming traditional methods such as ANN and fuzzy logic in terms of accuracy and computational efficiency [16].

Decision tree-based methods such as random forests have shown promising results for fault detection and location in distribution networks using smart meter data, by learning hierarchical rules that can handle imbalanced and noisy datasets [11].

Deep learning models such as CNN have demonstrated superior accuracy and robustness for fault detection and classification using time-frequency images of fault signals, by automatically learning discriminative features from raw data [29].

Unsupervised learning methods such as DBSCAN have been effective for fault detection and localization in distribution networks using voltage and current measurements, by identifying anomalous clusters of data points that deviate from the normal operating conditions [20].

29.6.4 *Self-Healing and Resilience*

Self-healing and resilience are emerging applications of ML in smart grid management, which aim to enable the grid to automatically detect, isolate, and recover from faults and disturbances, and maintain the continuity and quality of power supply. Some key applications of ML for self-healing and resilience include fault isolation and service restoration, microgrid control and islanding, predictive maintenance and asset management, and cyber-physical security. Table 29.4 presents a survey of recent studies on ML-based self-healing and resilience in smart grids.

Some key findings from the self-healing and resilience literature include:

Reinforcement learning has been used to learn optimal switching policies for service restoration in distribution networks, by adapting to different fault scenarios and network configurations and maximizing the amount of load restored [37].

Table 29.4 Survey of ML-based self-healing and resilience studies in smart grids

Study	Methodology	Data	Key findings
Zhang et al. [37]	Reinforcement learning	Distribution network data	Learns optimal switching policies for service restoration
Gao et al. [10]	Deep reinforcement learning	Microgrid data	Learns optimal control policies for DERs in islanded mode
Zhang et al. [39]	Deep learning	Power transformer data	Predicts remaining useful life and suggests maintenance actions
He et al. [12]	Deep learning	PMU data	Detects false data injection attacks in state estimation

Deep reinforcement learning has been applied to learn optimal control policies for DERs in islanded microgrids, by considering the stochasticity and uncertainty of renewable generation and load demand and minimizing the operating cost [10]. Deep learning has been employed for predictive maintenance of power system assets such as transformers, by learning the degradation patterns and failure modes from historical data and providing early warnings and recommendations for maintenance actions [39].

Deep learning has also been used for detecting cyber-physical attacks on smart grids such as false data injection in state estimation, by learning the normal and abnormal patterns of measurements from PMUs and identifying anomalies in real-time [12].

29.7 Conclusion

This review paper has provided a comprehensive overview of the state-of-the-art ML techniques and applications for load balancing and fault detection in smart grid management. Smart grids generate massive amounts of data from smart meters, sensors, and devices, making storage, processing, and analysis problematic. ML models' performance depends on selecting relevant and instructional features or descriptors that accurately describe the smart grid's main characteristics. Data quality and availability, scalability and computational efficiency, interoperability and standardizing, interpretability and trust, and human-machine collaboration are some of the technical and non-technical challenges of ML-based smart grid management solutions. ML could revolutionize smart grid management by providing intelligent, flexible, and robust load balancing, fault detection, and other solutions. Future research directions include the integration of physics- informed ML models to enhance the interpretability and generalization of smart grid management solutions [17].

References

1. X.G. Agoua, R. Girard, G. Kariniotakis, Probabilistic models for spatio-temporal photovoltaic power forecasting. IEEE Trans. Sustain. Energy **10**(2), 780–789 (2019)
2. A. Ahmad, A. Husain, M.N. Khan, T. Jamal, Random forests and decision tree based classification of distribution network faults using PMU measurements. Int. J. Electr. Power Energy Syst. **129**, 106802 (2021)
3. L.M. Al-Hadhrami, A. Al-Shehri, Electricity demand forecasting of smart grids using social media data. Energies **13**(9), 2383 (2020)
4. D. Assouline, N. Mohajeri, J.L. Scartezzini, Quantifying rooftop photovoltaic solar energy potential: a machine learning approach. Sol. Energy **141**, 278–296 (2017)
5. A.B. Birchfield, T. Xu, K.M. Gegner, K.S. Shetye, T.J. Overbye, Grid structural characteristics as validation criteria for synthetic networks. IEEE Trans. Power Syst. **32**(4), 3258–3265 (2017)

6. A. Bracale, P. Caramia, G. Carpinelli, A.R. Di Fazio, G. Ferruzzi, A Bayesian method for short-term probabilistic forecasting of photovoltaic generation in smart grid operation. Energies **6**(2), 733–747 (2013)
7. M.M. Buzau, J. Tejedor-Aguilera, P. Cruz-Romero, A. Gómez-Expósito, Hy-brid deep neural networks for detection of non-technical losses in electricity smart meters. IEEE Trans. Power Syst. **35**(2), 1254–1263 (2020)
8. K. Chen, K. Chen, Q. Wang, Z. He, J. Hu, J. He, Short-term load forecasting with deep residual networks. IEEE Trans. Smart Grid **10**(4), 3943–3952 (2019)
9. M. Diagne, M. David, P. Lauret, J. Boland, N. Schmutz, Review of solar irradiance forecasting methods and a proposition for small-scale insular grids. Renew. Sustain. Energy Rev. **27**, 65–76 (2013)
10. J. Gao, L. Guo, H. Xu, X. Xu, H. Wang, Fault diagnosis for distribution network based on deep reinforcement learning, in *2020 5th Asia Conference on Power and Electrical Engineering (ACPEE)* (2020), pp. 867–871
11. N. Haghighat, S. Zolriasatein, S. Eftekharnejad, M. Parsa Moghaddam, Fault location in distribution networks based on ensemble learning and distributed sensing, in *2020 IEEE Power Energy Society General Meeting (PESGM)* (2020), pp. 1–5
12. Y. He, G.J. Mendis, J. Wei, Real-time detection of false data injection attacks in smart grid: a deep learning-based intelligent mechanism. IEEE Trans. Smart Grid **8**(5), 2505–2516 (2017)
13. T. Hong, S. Fan, Probabilistic electric load forecasting: a tutorial review. Int. J. Forecast. **32**(3), 914–938 (2016)
14. W. Hu, Y. Hao, B. Zhang, Ev charging strategy optimization based on deep reinforcement learning in virtual power plant, in *2019 IEEE Innovative Smart Grid Technologies–Asia (ISGT Asia)* (2019), pp. 3531–3536
15. C. Jiang, Y. Mao, Y. Chai, M. Yu, S. Tao, Scenario generation for wind power using improved generative adversarial networks. IEEE Access **6**, 62193–62203 (2018)
16. J.F. Jiang, Z.H. Pang, S.M. Ji, S.Q. Zhao, J.W. Yang, X.H. Yang, Study on fault location for distribution network based on SVM and evidence theory, in *2014 China International Conference on Electricity Distribution (CICED)* (2014), pp. 969–973
17. A. Karpatne, W. Watkins, J. Read, V. Kumar, Physics-guided neural net-works (PGNN): an application in lake temperature modeling (2017). arXiv:1710.11431
18. W. Kong, Z.Y. Dong, Y. Jia, D.J. Hill, Y. Xu, Y. Zhang, Short-term residential load forecasting based on LSTM recurrent neural network. IEEE Trans. Smart Grid **10**(1), 841–851 (2019)
19. S. Makridakis, E. Spiliotis, V. Assimakopoulos, Statistical and machine learning forecasting methods: concerns and ways forward. PLoS ONE **13**(3), e0194889 (2018)
20. A. Naderian, S. Salemnia, Method for fault location in distribution networks based on clark transformation and anfis, in *2015 20th Conference on Electrical Power Distribution Networks Conference (EPDC)* (2015), pp. 184–189
21. Y. Noorollahi, R. Itoi, H. Yousefi, M. Mohammadi, A. Farhadi, Modeling for diversifying electricity supply by maximizing renewable energy use in Ebino city southern Japan. Sustainability **8**(11), 1139 (2016)
22. A.S. Qureshi, A. Khan, A. Zameer, A. Usman, Wind power prediction using deep neural network based meta regression and transfer learning. Appl. Soft Comput. **58**, 742–755 (2017)
23. M.Q. Raza, A. Khosravi, A review on artificial intelligence based load demand forecasting techniques for smart grid and buildings. Renew. Sustain. Energy Rev. **50**, 1352–1372 (2015)
24. N. Sharma, P. Sharma, D. Irwin, P. Shenoy, Predicting solar generation from weather forecasts using machine learning, in *2011 IEEE International Conference on Smart Grid Communications (SmartGridComm)* (2011), pp. 528–533
25. S.R. da Silva, F.L. Minku, A. Marques, T.B. Ludermir, A comprehensive search space for parameters and structures of neural networks for short-term load forecasting, in *2019 8th Brazilian Conference on Intelligent Systems (BRACIS)* (2019), pp. 576–581
26. S. Sobri, S. Koohi-Kamali, N.A. Rahim, Solar photovoltaic generation forecasting methods: a review. Energy Convers. Manag. **156**, 459–497 (2018)

27. M. Sun, C. Feng, J. Zhang, Probabilistic solar power forecasting based on weather scenario generation. Appl. Energy **266**, 114823 (2020)
28. C. Voyant, G. Notton, S. Kalogirou, M.L. Nivet, C. Paoli, F. Motte, A. Fouilloy, Machine learning methods for solar radiation forecasting: a review. Renew. Energy **105**, 569–582 (2017)
29. H. Wang, B. Zhang, W. Kong, Z. Dong, J. Ma, K. Meng, Deep learning based fault diagnosis of generators for reliable power grids, in *2019 IEEE Innovative Smart Grid Technologies–Asia (ISGT Asia)* (2019), pp. 4116–4121
30. Y. Wang, Q. Chen, T. Hong, C. Kang, Review of smart meter data analytics: applications, methodologies, and challenges. IEEE Trans. Smart Grid **10**(3), 3125–3148 (2019)
31. Y. Wang et al., Probabilistic forecasting with temporal convolutional neural network. Neurocomputing **399**, 491–501 (2020)
32. Y. Wang, Q. Chen, C. Kang, M. Zhang, K. Wang, Y. Zhao, Deep learning- based sociodemographic information identification from smart meter data. IEEE Trans. Smart Grid **10**(3), 2593–2602 (2019)
33. M. Wytock, J.Z. Kolter, Large-scale probabilistic forecasting in energy systems using sparse Gaussian conditional random fields, in *2013 IEEE 52nd Annual Conference on Decision and Control (CDC)* (2013), pp. 1019–1024
34. D. Yang, P. Jirutitijaroen, W.M. Walsh, Hourly solar irradiance time series fore- casting using cloud cover index. Sol. Energy **86**(12), 3531–3543 (2012)
35. D. Yang, H. Quan, V.R. Disfani, L. Liu, Reconciling solar forecasts: geographical hierarchy. Sol. Energy **146**, 276–286 (2017)
36. J. Zeng, W. Qiao, Short-term solar power prediction using an RBF neural network, in *2011 IEEE Power and Energy Society General Meeting (2011)*, pp. 1–8
37. H. Zhang, D. Yang, H. Xiong, Distribution network fault diagnosis and autonomous recovery approach based on deep reinforcement learning, in *2020 IEEE/IAS Industrial and Commercial Power System Asia (I CPS Asia)* (2020), pp. 662– 667
38. Y. Zhang, J. Wang, A distributed approach for wind power probabilistic forecasting considering spatio-temporal correlation without directly using off-site information. IEEE Trans. Power Syst. **33**(5), 5714–5726 (2018)
39. Y. Zhang, Z. Yuan, J. Wu, Y. Yuan, Z. Ding, Fault location in power distribution networks via deep graph convolutional networks. IEEE J. Sel. Areas Commun. **38**(1), 119–131 (2020)
40. H. Zheng, J. Yuan, L. Chen, Short-term load forecasting using EMD-LSTM neural networks with a Xgboost algorithm for feature importance evaluation. Energies **10**(8), 1168 (2017)

Chapter 30
Effect of Inclination Angle on Heat Pipe (HP) Performance with Graphene Nanofluids in Electronic Cooling Applications

Aruna Veerasamy and T. Dharini

Abstract The widespread usage of electronic devices has surged in recent decades, leading to frequent device failures primarily due to inadequate thermal management. To address these issues, heat pipes (HP) have been implemented in electronic devices. This paper investigates the use of novel graphene nanofluids in HP for electronic cooling applications. The graphene nanofluid is synthesized and injected into HP. The study discusses the effects of graphene nanofluids on improving the performance of HP compared to deionized water. The impact of an orientation on thermal performance of the HP, with input parameters including 40 W input power, inclination angles of 0°, 45°, and 90° and a 50% filling ratio. Key performance metrics such as thermal resistance, temperature difference between the evaporator and condenser, and thermal conductivity are assessed. Results demonstrates that the thermal performance of the HP improves with increasing inclination angle, exhibiting the lowest thermal resistance and highest thermal conductivity at a 90° inclination, in comparison to 45° and 0° angles. Conversely, the highest thermal resistance is achieved at a 0° orientation. These findings suggest that a 90° inclination angle is optimal for electronic cooling applications.

Keywords Heat pipes · Nanofluids · Thermal performance

A. Veerasamy (✉)
Centre of Excellence for Additive Manufacturing, Sathyabama Institute of Science and Technology, Chennai, Tamil Nadu, India
e-mail: arunaveer@gmail.com

T. Dharini
Centre for Nanoscience and Nanotechnology, Sathyabama Institute of Science and Technology, Chennai, Tamil Nadu, India

A. Kumar et al. (eds.), *Selected Proceedings of the 1st International Conference on Advanced Materials for Sustainable Innovation; IC-AMSI 2024; 28–30 August; New Delhi; India*, Springer Proceedings in Energy,
https://doi.org/10.1007/978-981-96-4492-6_30

30.1 Introduction

Heat pipes (HP) have emerged as versatile and efficient thermal management devices, finding widespread applications across various industries. As the demand for effective heat transfer solutions continues to grow, particularly in the fields of electronics [1], aerospace [2], and renewable energy [3, 4], the scientific community has dedicated significant efforts to advancing heat pipe technology. At the core of HP technology is the exceptional capacity to transfer considerable amount of heat without temperature gradient or with minimum temperature gradient. This is achieved through the evaporation and condensation of a working fluid within a sealed evacuated container. The operation of HPs is based on the principles of phase change, capillary action, and the efficient transport of latent heat. The working fluid, which can be water, ammonia, or other specialized liquids, involves a phase change mechanism from liquid to vapour in heating zone (evaporator) and from vapour to liquid in cooling zone (condenser). The hot vapour then reaches to cooling section of the heat pipe, where it condenses, releasing the latent heat and transferring it to the heat sink. One significant development is the introduction of miniature and micro HP, which have enabled the cooling of high power density electronic components in compact spaces [5, 6]. Additionally, the integration of HPs with phase change materials (PCM) and other thermal management strategies, such as vapour chambers and pulsating heat pipes [7, 8], has further expanded the capabilities of heat pipe based systems.

Over the past decades, several advancements have been made in heat pipe technology, driven by the need for more efficient and adaptable thermal management solutions. The study by Ali et al. [9] explores the use of a heat sink with (NE-PCM) and HPs for cooling of electronics parts. RT-35HC was base material, with 0.003 mass%, 0.005mass% of graphene oxide nanoparticles under different heating loads. Result show significant temperature reductions at lower heating loads and mass% concentrations. At higher heat flux of 2.5KW/m^2, the PCM/ HP combination achieved a temperature reduction of 42.81%, with peak temperature reductions of 32.95% and 37.54% for 0.003 mass% and 0.005 mass%. The TR-35HC based NE-PCM heat sinks are most effective at lower power levels, but thermal conductivity decreases at higher power levels due to nanoparticle agglomeration. Behi et al. [10–12] proposes various thermal management systems (TMS) for EVs with lithium-ion (Li-ion) batteries, focusing on heat generated during fast charging and discharging. They create a thermal model for lithium-titanate (LTO) cells to evaluate cooling startigies, natural air cooling, forced fluid cooling, and HP assisted methods. Their sandwiched HPs cooling system (SHCS) shows significant temperature reductions, with experimental results indicating that at an 8C discharging rate, SHCS lowers maximum temperatures from 56.8°C to 49°C, 38.8°C, achieving reductions of up to 33.4%. additionally, a flat copper HP's thermal efficiency is evaluated under varying inclinations (0° to 90°), revealing improved cooling efficiency and reduced thermal resistance, with a temperature difference reduction between evaporator and condenser from 5.26°C to 5.04°C, and a 1.3°C decrease in the cell's backside temperature. A CFD model in COMSOL Multiphysics validates these findings, demonstrating that single HPs handle upto 29.1% of the cooling load and 29.9% and 32.6% reduced by liquid cooling methods.

Brahim et al. [13] address the challenge of hotspots in high performance electronic chips, focusing on the potential of copper metal foam as a structure in flat plate heat pipes (FHPs). Using COMSOL multiphysics, they numerically investigate 3D FHPs with multiple heat sources. The study analyses the effects of heat flux and hotspot positions on HP performance, revealing that copper metal foam is a viable alternate wick. Visualizations indicate that hotspots near the wall are the most critical dry-out. Additionally, hotspot dimensions, distance, and effective pore radius significantly impact HP operation. Chang et al. [14] address the challenge of thermal management in increasingly miniaturized electronic devices by designing and fabricating a novel aluminium FHP with micro grooves using 3D printing technology. Aluminium powder was selectively melted and solidified to form the HP, with sintered aluminium powder enhancing the inner surface roughness and micro grooves improving capillary forces. The HP demonstrated excellent wettability with acetone and effective capillary action. The study found that a 10% working fluid filling ratio provided optimal heat transfer performance and minimal thermal resistance. The 3D printed flat HP was successfully tested for LED thermal management, maintaining LED temperature within 40°C and extending its service life.

Khalid et al. [15] review various techniques to enhance HP thermal performance and summarize advancements in the field. They discuss methods such as nanofluids, groove and fin design, different wick types, surface treatments, self-rewetting fluids, embedded HPs, and varying inclination angles. The study finds that while many techniques improve performance, each has its own limitations, including optimal parameter ranges, manufacturing challenges, economic feasibility, and commercial barriers. Ling et al. [16] propose a novel cooling method integrating PCM with leaf-shaped 3D oscillating heat pipes (OHPs) to manage heat from electronic devices. Experimental results show this method keeps device temperature below 100°C, about 35°C cooler than conventional air cooling, and reduces thermal resistance by up to 36.3%. The optimal filling ratio for the 3D-OHPs is 34–44%, and the leaf shape design lowers device surface temperatures by approximately 2°C compared to standard 3D-OHPs, offering an effective cooling solution for electronic devices. Maalej et al. [6] examine the thermal performance of a finned air cooled HPs system for CPU cooling, which includes an aluminium base, three grooved HPs, and fins. Experiments asses the system's thermal resistance across different orientations and heat loads. The thermosiphon position shows the lowest thermal resistance, with minimal impact from heat input power. The system can handle heat powers up to 150 W, 100 W, and 50 W for the thermosiphon, horizontal, and anti-gravity position, respectively, while keeping the CPU temperature below 90°C. modelling confirms that experimental temperature data closely matches predictions for heat loads below the capillary limit.

Miao et al. [17] study the thermal operation of bent copper HPs for cooling electronic components with high heat flux. They analyze both steady and transient conditions, examining the effects of cooling temperature, working conditions, and heat input on critical heat flux and thermal conductivity. Results show that increasing cooling temperature, evaporator and condenser lengths enhances critical heat flux. Additionally, extending cooling length improves critical heat flux only above 50°C. Increasing heating length shows a more significant improvement in equivalent thermal conductivity. The experiments confirm that the bent copper water HP offers rapid response to heat input changes and excellent transient heat transfer performance. Siricharoenpanich et al. [4] address challenges in thermal dissipation for electronic components, focusing on coolant heat removal limitations. They investigate the use of a nanofluid mixture of Ag and Fe_3O_4 and coolant category, and heat sink configuration, for thermal dissipation efficiency. Results show that the proposed water blocks significantly enhance coolant flow, improving thermal dissipation efficiency by 11.94%. additionally, the Ag/Fe_3O_4 nanofluid outerforms both Ag nanofluids and DI water in thermal efficiency. Wrobel et al. [8] review the use of HP technology in the thermal management of power electronics, electrical machines, and drives, which is crucial for developing high output, low emission electrified propulsion systems. They discuss the advantages of HPs, such as high heat transfer rates, low added weight, and high integration reliability. The paper covers the fundamental principles of HP construction and operation, highlighting new developments in materials and manufacturing techniques that enhance thermal management in electrical machines. Zhang et al. [1] review the challenges and advancements in cooling and thermal management for miniaturized electronic devices with increasing heat flux. Indirect cooling methods reviewed are microchannels, HPs, vapour chambers, thermoelectrics, and PCM. The review covers the effectiveness of those methods, their heat transfer capabilities. It also addresses current challenges and future prospects in electronic cooling thermal management.

In the present work, heat transfer ability of HP with graphene is evaluated. The synthesized graphene nanofluid is used and DI water is a base fluid. Thermal resistance, thermal efficiency, temperature differences is validated for same heat input power and varied inclination angles. The effect of inclination angle in HP performance is also discussed.

30.2 Experimental Setup

The experimental setup for this study is illustrated schematically in Fig. 30.1. The HP used in the experiment is fabricated from copper, a material chosen for its excellent thermal conductivity. The HP is a closed, sealed tube, where the bottom section functions as the evaporator, and the top section serves as the condenser. After the HP is vacuumed, the working fluid, which is graphene nanofluid synthesized using a two-step method, is injected into the pipe. An autotransformer is connected to the power supply and the HP to regulate the power deliver to the evaporator section. In this experiment, a constant input power of 40W is applied to the evaporator section. As the nanofluid in the evaporator absorbs heat, it undergoes a phase change from liquid to vapour. This hot vapour then reach condenser section through the adiabatic section of the HP.

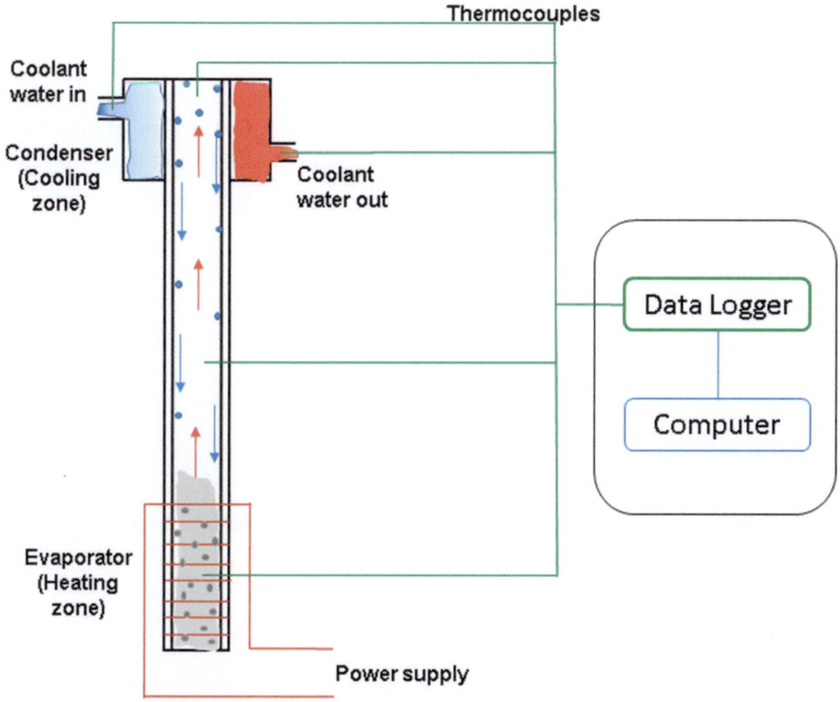

Fig. 30.1 Schematic diagram of HP

By releasing the latent heat, the hot vapor condenses back into liquid form in the condenser section. Condensates liquid then driven back to the evaporator section by capillary action and gravity, completing the cycle. Water is used as a coolant in the condenser, flowing outside the condenser section to absorb the heat released by the vapour. The flow of this coolant water is precisely controlled using a rotameter. To monitor the temperature at various points within the HP, thermocouples are strategically placed along the pipe. One end of each thermocouple is brazed to the HP, while the other end is connected to a data logger and computer, allowing for real time data acquisition and analysis. The performance of the HP is evaluated by observing its behaviour at different inclination angles, providing insights into the affects of orientation on thermal performance. This setup allows for a through investigation of the thermal behaviour of the HP filled with graphene nanofluid under varying conditions, contributing valuable data for applications in electronic cooling and other thermal management systems.

30.2.1 Data Reduction

HP is designed, fabricated and working fluid is selected, prepared and injected into the HP before starting an experiment. Filling of working fluid is a major criteria to enhance the performance of HP. Suitable filling ratio of HP is calculated from the below formula.

$$F = \frac{V_e}{V_t} \tag{30.1}$$

$$V_e = \frac{\pi}{4} d^2 L_e \tag{30.2}$$

$$V_t = \frac{\pi}{4} d^2 L_c \tag{30.3}$$

where,

F—filling ratio of working fluid in HP evaporator section
D—Diameter of the HP
V_e—Volume of evaporator section
V_t—Total volume of HP
L_e—Evaporator section length
L_c—Condenser section length

From the experimental results, the below formulas were used to calculated thermal resistance, thermal conductivity, evaporator and condenser section temperature differences and thermal efficiency.

The ratio between temperature difference between an evaporator and condenser and heat input is used to calculated the thermal resistance.

$$\Delta T = T_e - T_c \tag{30.4}$$

$$R_{th} = \frac{(T_e - T_c)}{Q} \tag{30.5}$$

Here, T_e—Evaporator temperature, T_c—condenser temperature and Q—heat input.

Thermal conductivity is calculated as,

$$k = \frac{QL}{A_c \Delta T} \tag{30.6}$$

Here, L—length of the HP, A_c—area of HP, ΔT—Temperature difference between an evaporator and condenser.

Thermal efficiency is calculated from,

$$\eta = \frac{Q_{out}}{Q} \tag{30.7}$$

Here, Q_{out}—heat output from condenser section.

Error propagation method is used to predict the uncertainty of thermal resistance. From the instruments, thermal resistance and temperature error is predicted.

$$\frac{\delta R}{R} = \sqrt{\left(\frac{\delta T_e}{T_e - T_c}\right)^2 + \left(\frac{\delta T_c}{T_e - T_c}\right)^2 + \left(\frac{\delta Q}{Q}\right)^2} \tag{30.8}$$

The maximum uncertainty of the thermal resistance, as determined from the experimental data, was calculated to be 2.03%.

30.3 Result and Discussion

The HP was designed and fabricated for the purpose of evaluating its performance when filled with a graphene nanofluid. The experiment setup was carefully constructed to assess various performance metrics under controlled conditions. The HP was charged with the graphene nanofluid, maintaining a 50% filling ratio to ensure optimal fluid distribution and capillary action within the HP. During the experiment, a constant input power of 40 W was applied to the evaporator section, simulating the thermal load typically encountered in electronic cooling applications. The inclination angle of the HP was systematically varied from 0° to 90° in incremental steps to investigate the influence of orientation on the heat transfer characteristics.

Figure 30.2 illustrates the thermal resistance of the HP filled with both graphene nanofluid and DI water. The data indicates a clear trend, as the orientation increases, thermal resistance decreasesd for both fluids. At 0° orientation, the thermal resistance is at its highest. However, as the inclination angle increases, there is a notable reduction in thermal resistance for both graphene nanofluid and DI water. Graphene nanofluids demonstrate lower thermal resistance compared to DI water across all inclination angles. This superior effect due to the combined effects of capillary action and gravity within the HP. When the HP is positioned at 0° inclination angle, the working fluid spreads across the length of the tube, reaching the condenser region. In this horizontal orientation, the evaporation and condensation processes slow down due to irregular brownian motion of the fluid particles. This irregularity hampers the efficient tranfer of heat, resulting in higher thermal resistance. As the inclination angle increases, the evaporation and condensation processes become more orderly. The improved regularity of Brownian motion at steeper angles promotes the formation of liquid slug and vapor plug within the HP. These dynamic fluid structures enhance the thermal conductivity of the system.

Additionally, the temperature differences between an evaporator and condenser plays a crucial role in reducing thermal resistance. As the orientation increases, the temperature gradient across the HP becomes more pronounced, futher contributing to the reduction in thermal resistance. From graphs, the lower thermal resistance observed with graphene nanofluids, particularly at higher inclination angles, can be

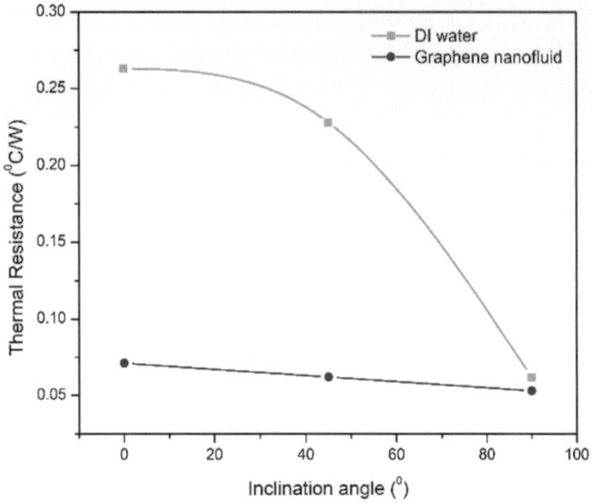

Fig. 30.2 Thermal resistance of HP with different inclination angle

primarly attributed to the enhanced capillary action, gravity effects, and more regular fluid motion within the HP.

During the evaporation process within the HP, the heat absorbs by working fluid from the internal walls of the evaporator section. As the fluid absorbs this thermal energy, it faces a phase change process from liquid to vapor. The hot vapors then travels through the adiabatic section of the HP and reaches the condenser. In the condenser, the working fluid releases its latent heat as it condenses back into a liquid, which then returns to the evaporator section due to the combined effects of capillary action and gravity. Water serves as a coolant in the condenser section, facillitating this phase change. Throughout the evaporation and condensation processes, the working fluid circulates within the HP, undergoing Brownian motion, a random movement that enhances the fluids ability to transport heat. As the fluid moves, ir carries thermal energy from the evaporator to the condenser and back again, making the temperature difference between these two sections a critical factor in improving the HP's thermal conductivity and overall thermal efficiency.

Figure 30.3 highlights this phenomenon, showing that for DI water, the temperature difference between the evaporator and condenser increases as the inclination angle of the HP is raised. However, graphene nanofluids exhibt a significant temperature difference even at a 0° inclination angle. This behavior is primarly attributed to the presence of nanoparticles within the nanofluid. The nanoparticles have a high surface area, which allows them to absorb more heat from their surrounding. As these heated nanoparticles travel through the HP, they effectively carry the absorb heat throughout the system. Although DI water shows an increase in temperature difference with higher inclination angles, graphene nanofluids demonstrate a significant increase in both temperature and temperature difference across all inclination angles. This indicates that graphene nanofluids are more effective in enhancing heat transfer within the HP, regardless of its orientation.

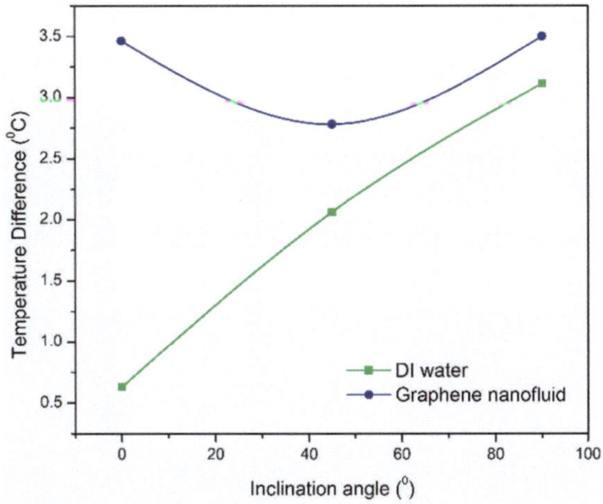

Fig. 30.3 Temperature difference between an evaporator and condenser

At a 0° inclination angle, DI water exhibits the lowest temperature difference, whereas graphene nanofluids achieve the highest temperature difference at a 90° orientation. The temperature differences of evaporator and condenser sections is crucial for improving heat transfer rate and thermal conductivity within the HP. As the heated nanoparticles in the evaporator section reach boiling point, they form bubbles. These bubbles, upon further heating, break into smaller bubbles, which travel through the HP. The movement of these bubbles, which carry both the nanoparticles and the absorbed heat, acts a vital role in enhancing heat transfer efficiency. The interaction of the nanoparticvles with the bubbles ensures that heat is effectively distributed throughout the HP, contributing to the overall improvement in thermal performance.

Figure 30.4 presents the thermal conductivity of the HP filled with either DI water or graphene nanofluids at various inclination angles. The results clearly show that the thermal conductivity is significantly high for graphene nanofluid across all tested angles compared to DI water. This enhanced thermal conductivity can be credited to the presence of graphene nanoparticles in the working fluid. These nanoparticles posses a high heat absorbing capacity, which allows them to absorb and transfer thermal energy more effectively than conventional fluids. The nanoparticles within the graphene nanofluid move through the HP due to Brownian motion, effectively carrying heat from the evaporator section to the condenser section. The intrinsic properties of graphene, including its superior thermal conductivity, contribute to the overall enhancement of heat transfer within the HP. Even when the HP is positioned horizontally, where thermal conductivity typically diminishes, the graphene nanofluid still demonstrates considerable thermal conductivity. This indicates that the graphene nanofluids can maintain efficient heat transfer regardless of the HPs orientation. As the inclination angle increases, there is a further improvement in thermal conductivity for both DI water and graphene nanofluid. However, the increase is more pronounced in the case of graphene nanofluid, which continues to outerform

Fig. 30.4 Thermal conductivity of HP with different inclination angle

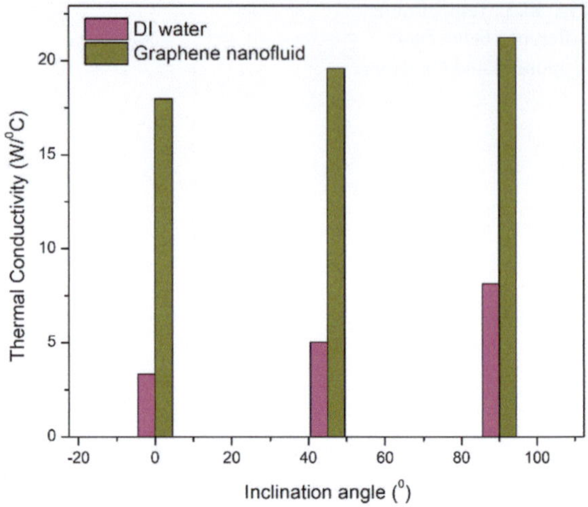

DI water across all angles. This can be explained by the synergistic effect of gravity and capillary action in facilitating better circulation and more effective heat transfer in the HP filled with graphene nanofluid.

Figure 30.5 illustrates the thermal efficiency of the HP with different working fluids at various inclination angles. The data reveals a trend of increasing thermal effcicency with a rise in inclination angle, which is consistent across both DI water and graphene nanofluids. The presence of nanoparticles in the base fluid is critical for enhancing the themal efficiency of the HP. As shown in the figure, the thermal efficiency increases more significantly for graphene nanofluid than for DO water when the inclination angle is increased. Specifically, the percentage increase in thermal efficiency for DI water at a 90° inclination compared to a 0° inclination is about 20%. In contrast, the increase for graphene nanofluid is approximately 32.1%.

The higher thermal efficiency observed in the HP with graphene nanofluid is a direct consequence of the improved heat transfer capabilities provided by the nanoparticles. The nanoparticles not only increase the fluid thermal conductivity but also enhance the boiling process within the evaporator section. The heated nanoparticles reach the boiling point and form bubbles, which subsequently break into smaller bubbles, further facilitating efficient heat transfer as these bubbles move through the HP. The nanoparticles, along with the heat they carry, are transported more effictively, leading to higher thermal efficiency. The experimental result demonstrate that graphene nanofluids significantly improve both the thermal conductivity and thermal efficiency of heat pipes across various inclination angles compared to DI water. The addition of graphene nanoparticles to the working fluid enhance the overall thermal performance of the HP, making it a promising candidate for advanced thermal management applications.

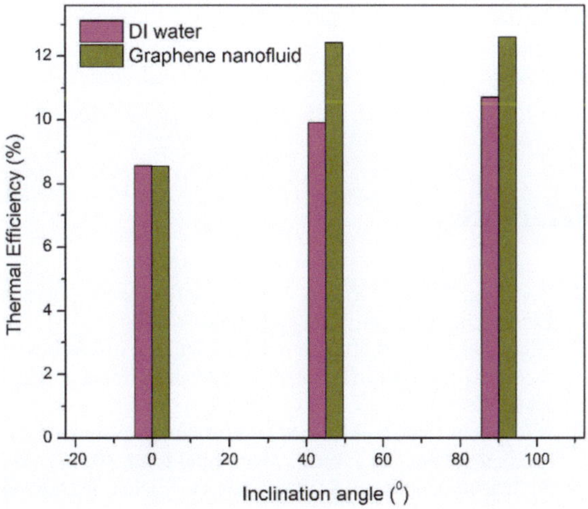

Fig. 30.5 Thermal efficiency of HP with different inclination angle

30.4 Conclusion

The HP was designed and fabricated, and working fluids such as DI water and graphene nanofluids were injected and evaluated. In this work, the heat input power was fixed at 40 W, and the working fluid filling ratio in the evaporator section was set at 50%. Thermal resistance, thermal conductivity, thermal efficiency, and temperature difference were evaluated by varying the inclination angle from 0° to 90°. During the experimentation, temperatures were observed, and the output parameters were calculated. From the validation of the results, it was observed that graphene nanofluid exhibits lower thermal resistance than DI water at all inclination angles. Since thermal resistance is directly propotional to thermal conductivity and thermal efficiency, there was a remarkable increase in thermal conductivity at all positions, especially with hraphene nanofluids. The increase in thermal conductivity also led to improvement in thermal efficiency of the graphene nanofluid. The temperature difference between the evaporator and condenser sections is a key factor in improving the performance of the HP.

Graphene nanofluid demonstrated 12.6% lower thermal resistance at 45° and 14.5% lower resistance at 90°, which is 51% lower than that of DI water. Similarly, the thermal conductivity of graphene nanofluid was 8.31% higher at 45° and 7.67% higher at 90°, which is 43.85% higher than DI water. The improvement in thermal efficiency was 31.2% at 45° and 1.34% at 90°. This work demonstrates the influence of orientation on the performance of HPs. The same input heat power produced different results depending on the working fluid and inclination angles. From this experimentation, it is suggested that graphene nanofluid with HO is well suited for all electronic cooling applications, regardless of position.

Acknowledgements The authors thank the Chancellor, Vice Chancellor, Centre of Excellence for Additive Manufacturing, International Research Centre, Sathyabama Institute of Science and Technology for providing infrastructure and instrumrntal facilities.

References

1. Z. Zhang, X. Wang, Y. Yan, A review of the state-of-the-art in electronic cooling. E-Prime-Adv. Electr. Eng. Electron. Energy **1**, 100009 (2021)
2. Nikolaenko Y.E. et al., Thermal performance of low-cost cooling systems for transmit/receive modules of phased array antennas with and without gravity heat pipes. Front. Heat Mass Transf. **18** (2022)
3. R.R. Riehl, S.M.S. Murshed, Performance evaluation of nanofluids in loop heat pipes and oscillating heat pipes. Int. J. Thermofluids **14**, 100147 (2022)
4. A. Siricharoenpanich, S. Wiriyasart, A. Srichat, P. Naphon, Thermal cooling system with Ag/ Fe_3O_4 nanofluids mixture as coolant for electronic devices cooling. Case Stud. Therm. Eng. **20**, 100641 (2020)
5. H. Mbulu, Y. Laoonual, S. Wongwises, Experimental study on the thermal performance of a battery thermal management system using heat pipes. Case Stud. Therm. Eng. **26**, 101029 (2021)
6. S. Maalej, A. Zayoud, I. Abdelaziz, I. Saad, M.C. Zaghdoudi, Thermal performance of finned heat pipe system for central processing unit cooling. Energy Convers. Manag. **218**, 112977 (2020)
7. G.F. Smaisim, H. Al-Madhhachi, A.M. Abed, Study the thermal management of Li-ion batteries using looped heat pipes with different nanofluids. Case Stud. Therm. Eng. **37**, 102227 (2022)
8. R. Wrobel, R.J. McGlen, Opportunities and challenges of employing heat-pipes in thermal management of electrical machines, in *2020 International Conference on Electrical Machines (ICEM)*, (IEEE, Gothenburg, Sweden, 2020), pp. 961–967
9. H.M. Ali, Analysis of heat pipe-aided graphene-oxide based nanoparticle-enhanced phase change material heat sink for passive cooling of electronic components. J. Therm. Anal. Calorim.Calorim. **146**(1), 277–286 (2021)
10. H. Behi et al., Heat pipe air-cooled thermal management system for lithium-ion batteries: high power applications. Appl. Therm. Eng. **183**, 116240 (2021)
11. H. Behi et al., Thermal management analysis using heat pipe in the high current discharging of lithium-ion battery in electric vehicles. J. Energy Storage **32**, 101893 (2020)
12. H. Behi, D. Karimi, J. Jaguemont, M. Berecibar, J.V. Mierlo, Experimental study on cooling performance of flat heat pipe for lithium-ion battery at various inclination angels **1**(1) (2020)
13. T. Brahim, A. Jemni, CFD analysis of hotspots copper metal foam flat heat pipe for electronic cooling applications. Int. J. Therm. Sci. **159**, 106583 (2021)
14. C. Chang, Z. Han, X. He, Z. Wang, Y. Ji, 3D printed aluminum flat heat pipes with micro grooves for efficient thermal management of high power LEDs. Sci. Rep. **11**(1), 8255 (2021)
15. S.U. Khalid, H. Babar, H.M. Ali, M.M. Janjua, M.A. Ali, Heat pipes: progress in thermal performance enhancement for microelectronics. J. Therm. Anal. Calorim.Calorim. **143**(3), 2227–2243 (2021)
16. Y.-Z. Ling, X.-S. Zhang, F. Wang, X.-H. She, Performance study of phase change materials coupled with three-dimensional oscillating heat pipes with different structures for electronic cooling. Renew. Energy **154**, 636–649 (2020)
17. S. Miao, J. Sui, Y. Zhang, F. Yao, X. Liu, Experimental study on thermal performance of a bent copper-water heat pipe. Int. J. Aerosp. Eng. **2020**, 1–10 (2020)

Chapter 31
Enhancing Home Comfort and Efficiency: A Comprehensive Study of Smart Home Automation Systems

Himanshu Lohia, Nupur, Sunidhi Prasad, Rishi Singhal, Aniket Mehra, and Divya Agarwal

Abstract Smart Home Automation Systems (SHASs) have transformed the way we interact with our living spaces, without living spaces, offering convenience, efficiency, and enhanced adaptability. This paper provides a broad area for smart homes, targeting the integration of innovative technologies such as IoT, data analytics, mobile application development, wireless connectivity, cloud computing, machine learning, and security measures. Through these technologies, users can remotely control and monitor devices such as smart fans, smart lights, and smart meters via intuitive mobile apps. The Internet of Things (IoT) forms the foundation of smart automation, enabling devices to connect to the Internet and communicate with each other. Wireless connectivity technologies such as Bluetooth and Wi-Fi enable communication between devices hosting applications, and processing commands. Data analytics and machine learning techniques analyze device data to offer information about usage trends and optimize device operations. Security measures ensure the safety and integrity of user data and device communications.

Keywords Data analytics · Machine learning · Wireless connectivity · Living spaces · Smart homes · Mobile app development

31.1 Introduction

Technology has become an integral part of daily life, profoundly reshaping human experiences. Among these advancements, Smart Home Automation Systems (SHASs) have emerged as transformative innovations that redefine how individuals interact with their living environments. This paper explores the expansive realm of

H. Lohia · Nupur · S. Prasad · R. Singhal · A. Mehra · D. Agarwal (✉)
Vivekananda Institute of Professional Studies – Technical Campus, New Delhi, India
e-mail: divya.agarwal@vips.edu

A. Kumar et al. (eds.), *Selected Proceedings of the 1st International Conference on Advanced Materials for Sustainable Innovation; IC-AMSI 2024; 28–30 August; New Delhi; India*, Springer Proceedings in Energy,
https://doi.org/10.1007/978-981-96-4492-6_31

smart home automation, focusing on the integration of advanced technologies such as voice commands, sensor-based automation, and app-controlled functionalities. These technologies are designed to enhance convenience, improve energy efficiency, and elevate overall quality of life.

A smart home functions as a network of interconnected appliances and devices capable of performing tasks automatically or upon command. Key components like the Arduino Uno board, Bluetooth module, and ESP32 microcontroller facilitate essential functions such as data processing, communication, and system control within this ecosystem. These components empower users to tailor solutions to their specific needs, whether it involves managing lighting, regulating temperature, or enhancing security systems. Speech recognition technology plays a pivotal role in modern smart homes, enabling users to interact seamlessly with their environment using natural language commands. Voice modules process spoken instructions, translating them into actionable commands for connected devices. This capability allows users to manage multiple tasks effortlessly, from adjusting room temperatures to controlling lighting levels.

Sensor-based automation further augments the efficiency and functionality of smart home devices. By deploying sensors such as temperature, light, and motion detectors, automated systems can respond dynamically to environmental changes without human intervention. For instance, motion sensors activate lights in unoccupied rooms to conserve energy, while light sensors adjust interior lighting based on outdoor daylight levels to optimize comfort and reduce energy consumption. The introduction of app-based control interfaces has revolutionized user interaction with smart home appliances. Intuitive mobile applications, developed using technologies like HTML and CSS, provide homeowners with remote access to their home automation network from anywhere with internet connectivity. These applications serve as centralized platforms for real-time notifications, device management, and automated scheduling, enabling users to seamlessly manage their home environment and personalize their living experience.

In conclusion, while SHASs hold immense potential to enhance energy efficiency, bolster security measures, and improve overall quality of life through advanced technologies and intelligent algorithms, challenges such as interoperability and concerns over data privacy and cybersecurity must be effectively addressed. This paper provides a comprehensive analysis of SHASs, examining their technologies, functionalities, and future implications, aiming to foster the development of smarter, more efficient, and sustainable living environments. Recent advancements in IoT-based SHASs facilitate internet-enabled connectivity and control of household appliances, allowing homeowners to remotely monitor and manage their homes, optimize energy usage, and enhance convenience through voice and gesture commands.

Fig. 31.1 Smart home connected features

31.1.1 Fan Control

This study investigates a voice-controlled SHAS for fan speed regulation, emphasizing its hands-free operation and benefits in comfort and convenience across user demographics. The system integrates voice control technology with other smart devices, utilizing multidisciplinary design principles encompassing hardware, software, voice recognition algorithms, and user interfaces. Evaluation includes user testing, feedback surveys, and energy efficiency metrics, demonstrating its potential to enhance home comfort and sustainability (Figs. 31.1 and 31.2).

31.1.2 Light Control

Smart lighting technologies have transformed the lighting sector, granting users unprecedented control and adaptability. Through intuitive smartphone applications, these systems facilitate remote adjustment of lighting settings from anywhere, whether at home, work, or during travel. Integration with Bluetooth and Wi-Fi distinguishes smart lighting, enabling instant modifications via mobile devices. Wi-Fi-enabled apps offer remote access for seamless on/off, brightness adjustments, and color changes, catering to diverse settings and enhancing energy efficiency (Table 31.1).

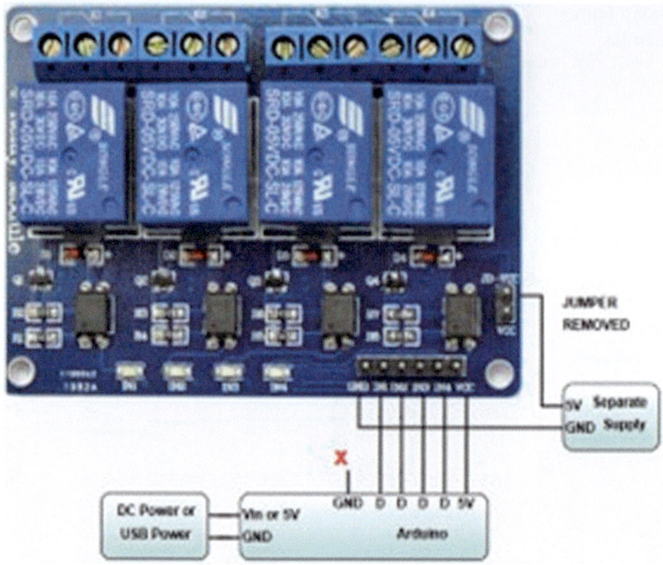

Fig. 31.2 Block diagram

31.1.3 *Smart Meter*

IoT-enabled smart electricity meters revolutionize energy management by empowering consumers with unprecedented control and visibility over their energy usage. Through these advanced devices and smartphone applications, property owners can monitor and oversee their energy consumption in real-time, regardless of their location—whether at home, work, or away. These smart meters, equipped with Bluetooth and Wi-Fi connectivity, offer direct control via Bluetooth proximity and remote adjustments through Wi-Fi. The intuitive mobile app interface provides access to historical data and real-time consumption patterns, facilitating informed decisions for energy conservation and cost savings. Users can also remotely operate connected appliances, further enhancing overall energy efficiency. For utility companies, IoT-based smart meters enhance billing accuracy and enable proactive maintenance. Ultimately, these innovations empower consumers to reduce energy consumption and contribute to sustainable practices through advanced monitoring and control capabilities (Fig. 31.3).

Table 31.1 Summary of recent research

Authors	Description	Applications
[1]	Aims to integrate advanced smart home applications into an IoT-enabled environment for seamless functionality	Renewable energy generation with wind turbines and solar panels at home
[2]	Evaluate benefits and risks of smart home technologies comprehensively	Enhanced monitoring and control functions for household management
Chakraborty et al.	Study homeowners' adoption and adjustment to SHTs for energy efficiency	Smart home technologies for efficient energy usage and savings
Menachem Domb	Develop low-cost, secure IoT-based SHAS using Arduino and sensors	Smartphone-based control with robust password protection for security
Soliman et al.	Implement IoT SHAS with Arduino-Uno, MATLAB-GUI, and FPGA for reliability	Wireless communication for cellular control and efficient energy management
Zhou et al.	Focus on smart home energy management and renewable integration	Energy management solutions for both residential and industrial sectors
Farooq et al.	Propose blockchain-based security for SHAS data privacy	Enhanced security using tamper-proof transaction records
[3]	Design IoT-based SHAS monitoring comfort levels with Emon CMS and NodeMCU	Interactive electricity use to improve energy efficiency for users
Benammar et al.	Create a sensor-based smart home for elderly care and activity monitoring	Home monitoring, remote control, and real-time hygiene comfort sensing
Kastrati et al.	Investigate IoT protocols and security in smart home automation	Smartphone control with emphasis on privacy and motion detection
Gaglio et al.	Implement SVM and blockchain in Raspberry Pi-controlled smart home	Home automation, security systems, and smart lighting solutions
Ashraf et al.	Explore IoT-driven smart home automation functionalities and user interactions	Wireless communication and cellular control for efficient energy management
Satyarthi et al.	Design voice-controlled smart home system with NLP and UI advancements	Hands-free control and accessibility enhancements for convenience
Okour et al.	Develop SHAS using IoT and Edge Computing on Raspberry Pi for efficiency	Residential applications with privacy preservation and edge computing
[18]	Integrate qToggle API for flexible home automation in proposed system	Automation, monitoring, and security for smart home appliances

(continued)

Table 31.1 (continued)

Authors	Description	Applications
[4]	Study IoTSHT applications in Malaysia for remote control and security	Entertainment, surveillance, and health monitoring functionalities
Chung et al.	Overview smart home automation benefits, technologies, and protocols	Core functionalities and potential applications understanding in smart homes
Chaudhary et al.	Explore IoT applications in SHASs, new uses, and challenges	Appliance automation, environmental monitoring, and voice control integration
[5]	Prototype solar-powered smart home with MPPT and healthcare tracking	Promotion of sustainable living practices through smart home technologies
Behera et al.	Implement gesture recognition for human–computer interaction using RNN	Proactive healthcare services and emergency alert systems in smart homes
[12]	Use image processing and Raspberry for device control in virtual environments	Research-oriented smart home prototypes for various sectors
Gurumurthy et al.	Compare existing smart home automation technologies	Security applications in education, retail, marketing, and industrial control
Petra et al.	Enhance energy consumption decisions with TensorFlow-based learning	Wireless communication systems for cost reduction and risk prevention
Chawda et al.	Integrate devices and sensors for efficient, sustainable IoT-based SHASs	Providing affordable solutions with wide coverage and low data rates
Hauxwell et al.	Analyze SHAuto use cases and security vulnerabilities	Annotating smart home appliance data based on climate information
Kiah et al.	Enable global appliance control via IoT in SHASs	Enhancing home occupants' quality of life with potential security considerations
Aldrich et al.	Evaluate IoT-based home security systems with PIR sensors and automation	Implementation possibilities in sensitive areas like homes, schools, hospitals, etc
[15]	Develop low-cost, secure IoT-based SHAS using Arduino and sensors	Smart home security enhancements with PIR sensors and improved UI for automation

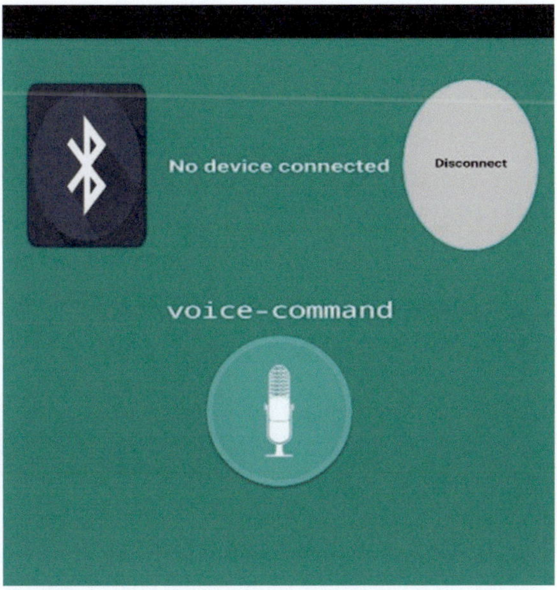

Fig. 31.3 Interface for developed voice module app

31.2 Background Study

31.3 Methodology Adopted

The automation of devices like smart fan regulators, lights, and meters via mobile apps integrates various technologies for seamless control and monitoring. Here are the key components involved:

- **Wireless Connectivity**: Utilizes Wi-Fi and Bluetooth for internet access, enabling communication and data exchange with devices remotely.
- **Internet of Things (IoT)**: Connects devices to the internet, incorporating actuators, sensors, and communication modules for smart functionality.
- **Mobile Application Development**: Interfaces with users to control and monitor devices using APIs, developed with frameworks like Kotlin and Swift for iOS and Android platforms.
- **Cloud Computing**: Platforms such as AWS and Google Cloud provide scalable infrastructure for command processing, application hosting, and data storage, ensuring real-time communication between mobile apps and devices.
- **Security and Privacy Measures**: Implements HTTPS, OAuth, and encryption protocols to secure communications. User authentication, access controls, and data encryption safeguard confidentiality and prevent unauthorized access.

Fig. 31.4 Circuit of voice module

- **Data Analytics and Machine Learning**: Utilizes machine learning to automate tasks, optimize device settings, and predict user preferences. Data analytics techniques analyze smart device data, providing insights into energy usage, trends, and user behavior.

These systems enhance comfort, efficiency, and convenience in homes and offices by enabling remote monitoring and operation through robust technological integration (Figs. 31.4, 31.5 and 31.6).

31.4 Results and Discussion

Integrating intelligent home automation technologies significantly enhances home interactions, offering greater convenience, reduced energy expenses, and improved quality of life. Key components like the ESP32 microcontroller, Bluetooth module, and Arduino Uno Board quickly integrate to create a unified and responsive home environment. This approach leverages advanced features such as voice commands, sensor-based automation, and app-controlled functionalities, simplifying daily tasks and enhancing overall user experience. The amalgamation of sophisticated hardware and technology in intelligent home automation systems provides substantial benefits including enhanced efficiency, security, and usability (Figs. 31.7 and 31.8).

Fig. 31.5 Process flow chart

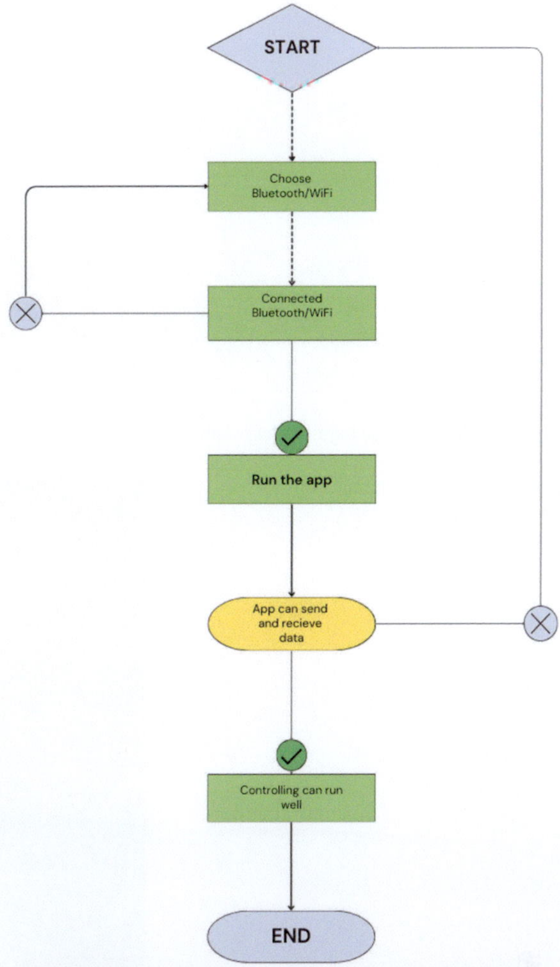

Fig. 31.6 Circuitry for fan and light controlling unit

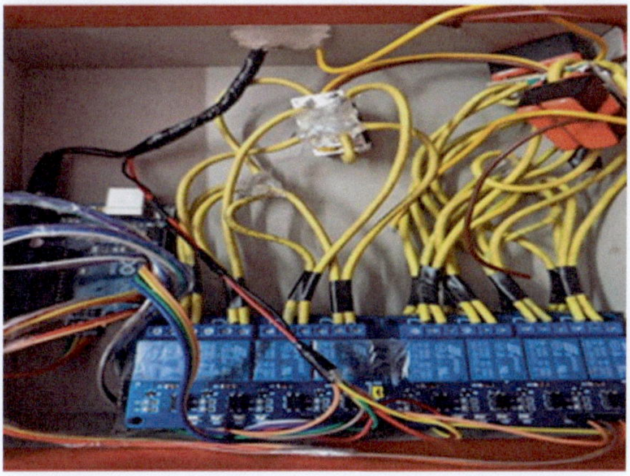

Fig. 31.7 Smart home automation prototype

Fig. 31.8 Voice command prompt for app

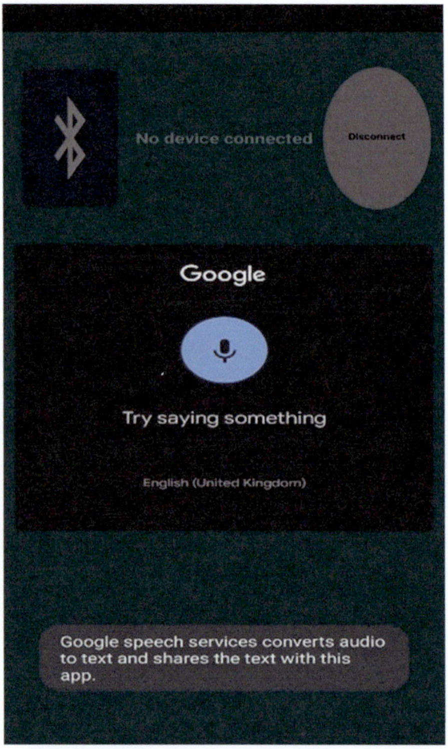

31.5 Conclusion

In summary, integrating intelligent home automation systems marks a significant advancement in how we interact with our living spaces, offering unmatched ease of use, reduced energy consumption, and improved quality of life. These systems integrate hardware components such as the Arduino Uno board, Bluetooth module, and ESP32 microcontroller to create a cohesive and responsive home environment. They employ advanced technologies like voice commands, sensor-based automation, and app-controlled features. Voice recognition technology enables natural language interaction, enhancing accessibility and reducing manual tasks. Sensor-based automation further optimizes energy efficiency and security by autonomously adjusting devices based on environmental changes, such as adjusting lighting according to daylight levels or activating security measures upon detecting motion. App-based interfaces revolutionize user interaction by providing centralized platforms for remote monitoring and management in real-time, enhancing convenience and control over the home environment. Despite the evident benefits, challenges such as interoperability issues and concerns about cybersecurity and data privacy remain to be addressed. Nonetheless, smart home automation holds immense potential to transform energy management, enhance security, and elevate overall living standards through continual innovation and integration of cutting-edge technologies.

31.6 Future Scope

In the future, smart home automation systems (SHAS) will revolutionize living spaces with advanced customization and connectivity. They will ensure seamless device interoperability through standardized protocols like Matter, enhance responsiveness with AI and machine learning, and improve security while optimizing energy usage with renewable sources.

References

1. Trivodaliev et al., A review of internet of things for smart home: challenges and solutions. J. Clean. Prod. **140**(3), 1454–1464 (2017)
2. Wilson et al., Benefits and risks of smart home technologies **103**, 72–83 (2017)
3. Zhnag et al., Smart home: architecture, technologies and systems. Procedia Comput. Sci. **131**, 393–400 (2018)
4. Butunoi et al., An IoT-based smart home automation system, in *International Symposium on Electronics and Telecommunications ISETC 2020* (2021)
5. Hasan et al., Smart home system: a comprehensive review. J. Electr. Comput. Eng. (2023)
6. Kumar, Chawda, A research paper on smart home. ResearchGate (2020)
7. Baldwin et al., Learning to live in a smart home. Taylor & Francis (2017)
8. Kiah et al., A review of smart home automation based on internet of things **97**, 48–65 (2017)
9. F.K. Aldrich, Smart homes: past, present and future. Springer Link, pp. 17–39

10. Petra et al., State of the art of smart homes. Eng. Appl. Artif. Intell. **25**, 1313–1321 (2012)
11. Carlsson et al., A risk analysis of a smart home automation system. Futur. Gener. Comput. Syst. **56**, 719–733 (2016)
12. Jalal et al., Smart home automation-based hand gesture recognition using feature fusion and recurrent neural network **23**(17), 7523 (MDPI, AI and Big Data for Smart Construction, 2023). https://doi.org/10.3390/s23177523
13. Anandha et al., Augmented reality-based smart home automation using dynamic vision sensor. IEEE (2023)
14. Behra et al., Solar powered smart home automation and smart health monitoring with IoT. Springer Link, pp. 169–182 (2023)
15. Klongdee et al., Development and assessment of Internet of things- driven smart home security and automation with voice commands. MDPI **5**(1), 79–99 (2024). https://doi.org/10.3390/iot5010005
16. Umer et al., An intelligent, secure, and smart home automation system. Sci. Program. (2020). https://doi.org/10.1155/2020/4579291
17. Benammaar et al., Smart Home automation using IoT-based sensing and monitoring platform, IEEE Xplore. Doha (Qatar) (2018). https://doi.org/10.1109/CPE.2018.8372548
18. Kastrati et al., Towards smart home automation using IoT-enabled edge-computing paradigm. MDPI **21**(14), 4932 (2021). https://doi.org/10.3390/s21144932
19. Chung et al., Usage and Impact of the internet-of-things based smart home technology: a quality-of-life perspective. Springer Link **23**, 345–364 (2024)
20. Almalki et al., Towards the design and implementation, a smart home automation system based on internet of things approach. Int. J. Appl. Eng. Res. **12**, 2731–2737 (2017)

Chapter 32
Estimation of Optimum Condensing Temperature for Three Stage Vapor Compression Refrigeration System Using Alternative Refrigerants

Saquib Khan, Vaibhav Jain, and Kanchan Mugil

Abstract Three-stage vapor compression refrigeration systems (TSVCRS) are extensively preferred for very low temperature refrigeration applications due to the limitation of low volumetric efficiency of compressors in a single stage. In this paper, the thermodynamic analysis of TSVCRS is presented using R744, R717, and R410A in the LTC (low-temperature cycle), MTC (medium-temperature cycle), and HTC (high-temperature cycle) at the base case. It is observed that the overall COP of TSVCRS increases with an increase in cascade condensing temperatures of LTC and MTC up to a certain limit of −43.33°C at LTC and −7.22°C at MTC, then starts decreasing. Thus, the optimum condensing temperature is a subtle design parameter for the performance of TSVCRS. The maximum COP at these optimum condensing temperatures of LTC and MTC are estimated to be 0.716. These optimum values are further estimated using different combinations of refrigerants in LTC, MTC, and HTC. The potential candidates considered in present study are R744, R170, R1150 (LTC); R717, R290, R717 (MTC) and R410A, R1234yf, R600 (HTC). These alternative refrigerants are selected based on their commercial viability under the given temperature and pressure range. The system predicts the highest COP of 0.763 using R744, R290, and R410A in LTC, MTC, and HTC, respectively. Further, the optimum condensing temperature of LTC and MTC at this operating point is −47.22°C and −6.22°C, respectively.

Keywords Three-stage vapor compression · COP · Optimum condensing temperature · Alternative refrigerants

S. Khan (✉) · V. Jain · K. Mugil
Maharaja Agrasen Institute of Technology, New Delhi, India
e-mail: saquib.k765@gmail.com

A. Kumar et al. (eds.), *Selected Proceedings of the 1st International Conference on Advanced Materials for Sustainable Innovation; IC-AMSI 2024; 28–30 August; New Delhi; India*, Springer Proceedings in Energy,
https://doi.org/10.1007/978-981-96-4492-6_32

32.1 Introduction

Fundamentally based on the concepts of thermodynamics and heat transfer, refrigeration is used in many different industrial and commercial applications. The most common refrigeration system in use today is the vapor compression refrigeration system (VCRS), which operates by compressing a refrigerant, condensing it, expanding it through an expansion valve, and then evaporating it to absorb heat from the desired area. The VCRS is widely used due to its efficiency and reliability [1]. Despite its widespread use, the simple VCRS faces significant limitations, particularly when lower temperatures are required. The load on a single compressor increases as the target temperature drops because the compressor's volumetric efficiency declines. This inefficiency becomes critical when extremely low temperatures are necessary, rendering the simple VCRS less effective and economically viable for such applications. To overcome these limitations and achieve ultra-low temperatures, cascade refrigeration systems have emerged as a viable solution [2]. By employing multiple stages of compression and condensation, cascade systems can effectively operate at lower temperatures while maintaining reasonable efficiency. Nicola et al.'s investigation [3] of the cascade system's performance using various mixes of natural refrigerants revealed that the cascade cycle's COP approaches respectable levels. Udroiu et al. [4] explore the integration of ejectors in cascade refrigeration systems of two levels for ultra low temperature refrigeration, particularly relevant due to the increased demand for vaccine storage during the COVID-19 pandemic. Walid et al. [5] performed the thermodynamic analysis of a triple cascade vapor refrigeration system using hydrocarbon refrigerants for ultra-low temperature application. Different hydrocarbon refrigerant pairs were used to obtain the most suitable refrigerant combination by mathematical modeling of the system. A three-stage cascade vapor compression refrigeration system is necessary for these ultra-low temperature applications. This advanced system further improves volumetric efficiency and enhances the coefficient of performance (COP), making it suitable for extremely demanding conditions. Literature indicates that three-stage systems can achieve significantly lower temperatures with better energy efficiency compared to their two-stage counterparts, thereby extending the practical limits of vapor compression refrigeration technology. For cryogenic applications between 110 and 150 K, Rogala and Kwiatkowski [6] investigate the viability of a three-stage cascaded refrigeration system using conventional refrigeration compressors.

In this paper, a comprehensive study for performance enhancement of a three stage cascade VCR system is presented using alternative refrigerants employed in LTC, MTC and HTC system. The alternative refrigerants considered in the present study are R744, R170, R1150 (LTC); R717, R290, R717 (MTC) and R410A, R1234yf, R600 (HTC). These alternative refrigerants are selected based on their thermophysical properties at the temperature and pressure. The selected refrigerants also confirm their industrial suitability. The optimum condensing temperature is presented for different combinations of working pairs. In order to provide useful insights for both academic research and industrial implementation, this study intends to enhance refrigeration technology for very low temperature applications.

32.2 Working of a Three-Stage Vapor Compression Refrigeration System

A TSVCRS is engineered to achieve ultra-low temperatures by integrating three distinct vapor compression refrigeration cycles, each utilizing a different refrigerant and operating at varying temperature levels. This intricate design enhances efficiency and allows the system to reach temperatures as low as −80 °C.

A TSVCRS consists of three cycles, each connected by a cascade condenser (CHX). Each cycle operates in different temperature ranges. The cascade condenser acts as a condenser for the lower cycle and an evaporator for the upper cycle. The three cycles are named low-temperature cycle, medium-temperature cycle, and high-temperature cycle. The components in each cycle comprise a compressor, an expansion valve, an evaporator, and a condenser as shown in Fig. 32.1. The lower temperature cycle is connected to the medium temperature cycle via cascade condenser 1. The MTC is linked to the high-temperature cycle via another cascade condenser 2, and the condenser of HTC is in direct contact with the surroundings. The cascade condenser's degree of overlap must be defined to get the optimized performance parameters. Natural refrigerants, with their 0 ODP and small GWP, have regained attention as alternative refrigerants for many applications due to the negative environmental impacts of synthetic refrigerants. This paper thus presents a thermodynamic analysis of hydrofluorocarbons (HFCs) and hydrocarbons (HC) refrigerants with zero ODP and very low GDP. The base refrigerants used in the LTC, MTC, and HTC are R744, R717, and R410A, respectively at the base case.

The evaporator extracts heat from the space to be cooled and evaporates into gaseous state form liquid state and then the refrigerator goes into the compressor where it's pressure and temperature is increased and then it passes through the first cascade condenser where the heat is exchanged between the refrigerant in medium temperature cycle (R717) and lower temperature cycle (R744) and the refrigerant in lower temperature cycle condenses into liquid state and then passes through the expansion valve where it's temperature and pressure decreases and the cycle repeats. In MTC the refrigerant (R717) absorbs heat from the refrigerant of LTC (R744) the heat absorbed ideally is the sum of heat extracted from the space to be cooled and work done by the compressor, due to which the refrigerant of MTC changes phase and turns into a gas from liquid, then it goes into the compressor where it's temperature and pressure is increased and then passes through the second cascade condenser where heat is exchanged between refrigerant of HTC (R410A) and MTC (R717) and due to heat rejection phase change happens and the refrigerant of MTC (R717) changes from a gaseous state to liquid state and then passes through the expansion valve and the MTC cycle repeats.

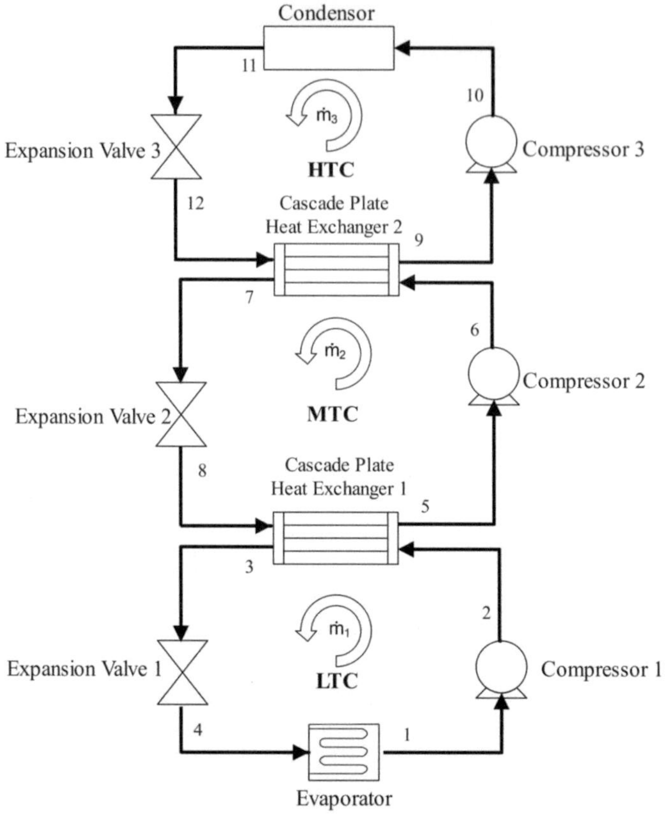

Fig. 32.1 Schematic diagram of the proposed TSVCRS

In HTC the refrigerant (R410A) absorbs heat from the refrigerant of MTC (R717) in the second cascade condenser and this heat ideally is the addition of heat extracted by the evaporator and the actual work done by the compressors of MTC and LTC due to which phase change of refrigerant in HTC takes place from liquid state to gaseous state, then the refrigerant passes through the compressor of HTC where it's temperature and pressure is increased and the refrigerant (R410a) is then passed on to the condenser where the heat is rejected to the surrounding and the heat rejected ideally is the addition of heat extracted from space to be cooled and work done by the compressors of all the three cycles (LTC, MTC and HTC) due to the heat rejection phase change happens of refrigerant of HTC (R410A) from gaseous state to liquid state, then the refrigerant passes through the expansion valves and the HTC cycle. In this way the TSVCRS is used for achieving ultra-low temperature while maintaining good COP. The cycle's P–h and T-s diagrams are shown in Fig. 32.2.

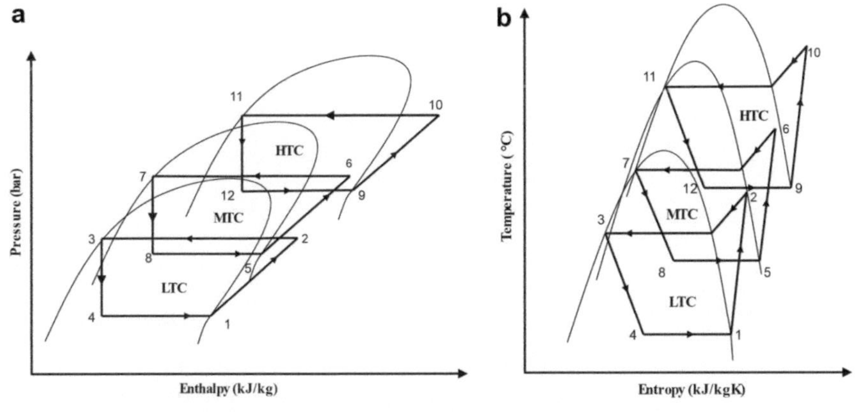

Fig. 32.2 **a** P–h graph and **b** T-s graph of the TSVCRS

32.3 Thermodynamic Modelling of TSVCRS

This section includes the formulation of a mathematical model and associated equations for each system component based on the relevant state points. All of the system's parts must have their mass and energy conservation equations applied to conduct a thermodynamic analysis of the system from an energy perspective. In this regard, Table 32.1 displays the regulating formulas of the TSVCR. Furthermore, Table 32.2 displays the basic operational parameters to replicate the mathematical model of the system. Initial parameters taken into account for the thermodynamic analysis are displayed in Table 32.3 (base case). A few presumptions are made to streamline the mathematical model's simulation. These are as follows:

(i) It is anticipated that every component will be in a stable state with constant flow.

Table 32.1 Governing equations of components

S. no	Component	Equations
1	Evaporator	$\dot{Q} = \dot{m}_1(h_1 - h_4) = mc_p\Delta T$
2	Compressor LTC	$\dot{W}_1 = \dot{m}_1(h_2 - h_1)$
3	Cascade condenser 1	$\dot{Q}_{c1} = \dot{m}_2(h_5 - h_8) = \dot{m}_1(h_2 - h_3)$
4	Expansion valve LTC	$h_3 - h_4$
5	Compressor MTC	$\dot{W}_2 = \dot{m}_2(h_6 - h_5)$
6	Cascade condenser 2	$\dot{Q}_{c2} = \dot{m}_3(h_9 - h_{12}) = \dot{m}_2(h_6 - h_7)$
7	Expansion valve MTC	$h_7 - h_8$
8	Compressor HTC	$\dot{W}_3 = \dot{m}_3(h_{10} - h_9)$
9	Condenser	$\dot{Q}_c = \dot{m}_3(h_{10} - h_{11})$
10	Expansion valve HTC	$h_{11} - h_{12}$

Table 32.2 Governing formulas for the TSVCRS performance analysis

Parameters	Equation
COP of LTC	$COP_{LTC} = \dot{Q}_e/\dot{W}_1$
COP of MTC	$COP_{MTC} = \dot{Q}_{C1}/\dot{W}_2$
COP of HTC	$COP_{HTC} = \dot{Q}_{C2}/\dot{W}_3$
Total compressor work	$\dot{W}_{Total} = \dot{W}_1 + \dot{W}_2 + \dot{W}_3$
Overall COP	$COP = \dot{Q}_e/\dot{W}_{Total}$

Table 32.3 Initial values for the examination of thermodynamics analysis

Parameters	Values
Ambient temperature, T_o	40 °C
Ambient pressure, P_{amb}	101.325 kPa
Cooling load, $\dot{Q}_e$	50 kW
Range temperature in evaporator, T_{evp}	−70°C to −80°C
Degree of overlap in CHX, ΔT	5°C
Condenser temperature, T_{cond}	40°C

(ii) The pipelines connecting the various parts of the system experience very small pressure drops and heat losses.
(iii) The condenser and evaporator fans' use of power is not taken into account.
(iv) Heat exchangers are thought to have minimal environmental heat loss.

The EES program has developed a mathematical model to simulate the performance characteristics, such as the coefficient of performance (COP) and total compressor work, in order to calculate the system's performance. The EES software's built-in package is used to calculate the thermodynamic properties of the refrigerants. The conversion of mass, or the mass of refrigerant entering or leaving a system component that is equal, is known as mass balance. The mass balance is given by Eq. 32.1:

$$\sum \dot{m}_{in} = \sum \dot{m}_{out} \tag{32.1}$$

Energy balance is the conversion of energy that is energy entering the system is totally converted into work used and shown in Eq. 32.2:

$$\dot{Q} + \sum \dot{m}_{in}h_{in} = \dot{W} + \sum \dot{m}_{out}h_{out} \tag{32.2}$$

32.3.1 Performance Indicators

The COP is defined as the energy conversion efficiency in the system, the net COP is the ratio of total heat transfer to the total work done by the compressor (Table 32.2).

32.4 Results and Discussions

The TSVCRS is examined in the current study using a thermodynamic performance evaluation methodology. In this work, a parametric investigation is conducted to find the best operating parameters and analyze the system's performance. Initially, three refrigerant pairs, carbon dioxide (R744), ammonia (R717), and R410A were used in LTC, MTC, and HTC respectively as the base pairs. Two more alternate pairs for each refrigerant are taken in consideration for the study. For carbon dioxide, R170 and R1150 were considered as the alternate pairs in the LTC. In the MTC, R290 and R32 were taken as the alternatives to ammonia (R717). For the alternative of refrigerant R410a in the HTC, R1234yf and R600 were taken. All these alternates for each refrigerant in different cycles are taken for optimization of the overall performance of TSVCRS and have similar physical properties as of the base pairs considering the literature done. Consequently, an investigation is conducted into how the cascade condensing temperature at CHX affects the change in overall COP, and COPs at each temperature cycle. The shift in COP about the evaporation temperature (T_{evp}), LTC (Tc_1), and MTC (Tc_2) intermediate condensation temperatures must be taken into account first, though, before examining those parameters.

Refrigerants at MTC and HTC–R717 and R410A, respectively–are held constant during the simulation for the first instance, while two additional refrigerants are used in the LTC. The variation in the optimum condensing temperature at LTC and overall COP is being analyzed with alternate refrigerants. Furthermore, MTC serves as a link between LTC and HTC, which makes MTC crucial to the system's operation. Therefore, choosing MTC refrigerants over HTC or LTC is more important.

LTC and MTC cascade condensers function as both a condenser and an evaporator in this system. They function within the TSVCRS as a connection. As a result, one crucial parameter for the TSVCRS's optimization performance is the cascade condensation temperature. This implies that for different refrigerant combinations, the ideal MTC cascade condensing temperature (T_{c2}) and the ideal LTC cascade condensing temperature (T_{c1}) are distinct.

Variation of COP corresponding to T_{c1}

Figure 32.3 shows the effect of cascade condensing temperature on the COP_{LTC}, COP_{MTC}, COP_{HTC}, and $COP_{overall}$ using base pair refrigerants. By altering the condenser temperature at the cascade condenser (CHX_1), the parameters were tuned. At MTC (T_{c2}), the cascade condensing temperature and overlap degree were determined to be 5 °C and 0 °C, respectively.

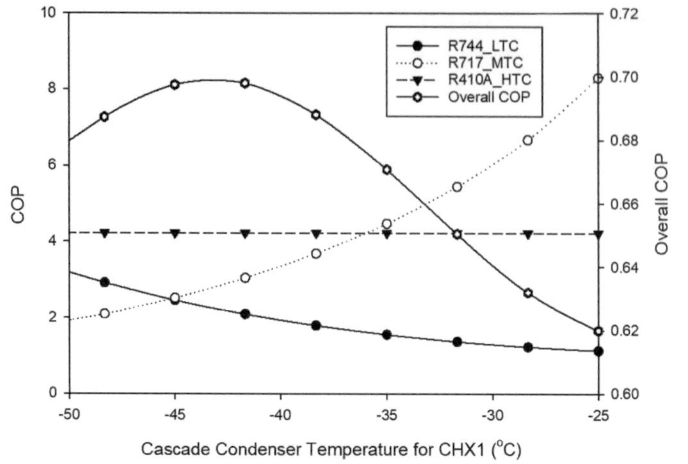

Fig. 32.3 Change in COP with base pair refrigerants

The impact of T_{c1} on COP_{LTC}, COP_{MTC}, and COP_{HTC} when TSVCRS uses R744, R717, and R410A with an evaporation temperature (T_{evp}) of −80 °C is shown in Fig. 32.3. As T_{c1} (the condensing temperature at CH_{X1}) rises, COP_{MTC} increases while COP_{LTC} and COP_{HTC} fall, as illustrated in Fig. 32.3. The system's overall COP reaches its maximum value at the optimal cascade condensing temperature, after which it gradually decreases as it rises. The cascade condenser CHX_1's ideal overall COP, with the values displayed in Fig. 32.3, is 0.699 at an ideal coupling temperature of −43.33 °C.

Figure 32.3 depicts the influence of T_{c1} on COP_{LTC}, COP_{MTC}, and COP_{HTC} when TSVCRS employs R744, R717, and R410a with the temperature at evaporator (T_{evp}) of −80 °C. Figure 32.3 shows that the COP_{MTC} increases as T_{c1} (the condensing temperature at CHX_1) increases whereas COP_{LTC} and COP_{HTC} decrease as T_{c1} increases. At the ideal cascade condensing temperature, the system's overall COP increases to its maximum value and then progressively declines as it grows. Under the parameters shown in Fig. 32.3, the cascade condenser CHX_1's optimal overall COP is 0.699 at an optimum coupling temperature of −43.33 °C.

Variation of COP corresponding to T_{c1} with alternate refrigerants at LTC

Figure 32.4 depicts the influence of T_{c1} on COP_{LTC}, COP_{MTC}, and COP_{HTC} when TSVCRS employs R170, R717, and R410a respectively with the temperature at evaporator (T_{evp}) of −80 °C. Figure 32.4 shows that the graph follows the same pattern as seen in Fig. 32.3. Here again the COP_{MTC} increases as T_{c1} (the condensing temperature at CHX_1) increases whereas COP_{LTC} and COP_{HTC} decreases as T_{c1} increases. COP increases to its maximum value and then progressively declines as it grows. Under the parameters shown in Fig. 32.4, the cascade condenser CHX_1's optimal overall COP is 0.7519 at an optimum coupling temperature of −40.56 °C.

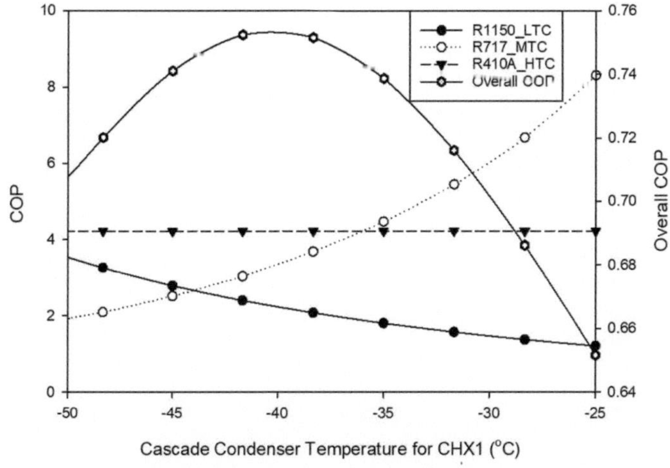

Fig. 32.4 Change in COPs corresponding to T_{C1} with R1150, R717 and R410A

Figure 32.5 depicts the influence of T_{c1} on COP_{LTC}, COP_{MTC}, and COP_{HTC} when TSVCRS employs R1150, R717, and R410a respectively with the temperature at evaporator (T_{evp}) of −80 °C. Figure 32.5 shows that the graph follows the same pattern as seen in Fig. 32.4. Here again the COP_{MTC} increases as T_{c1} (the condensing temperature at CHX_1) increases whereas COP_{LTC} and COP_{HTC} decrease as T_{c1} increases. COP increases to its maximum value and then progressively declines as it grows. Under the parameters shown in Fig. 32.5, the cascade condenser CHX_1's optimal overall COP is 0.7536 at an optimum coupling temperature of −40.22 °C.

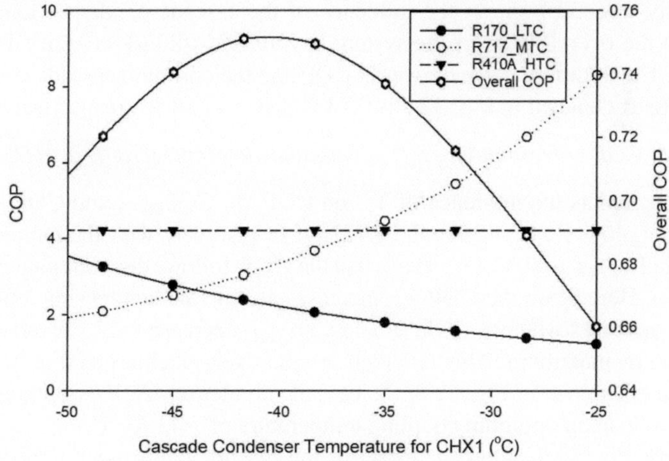

Fig. 32.5 Change in COPs corresponding to T_{C1} with R170, R717 and R410A

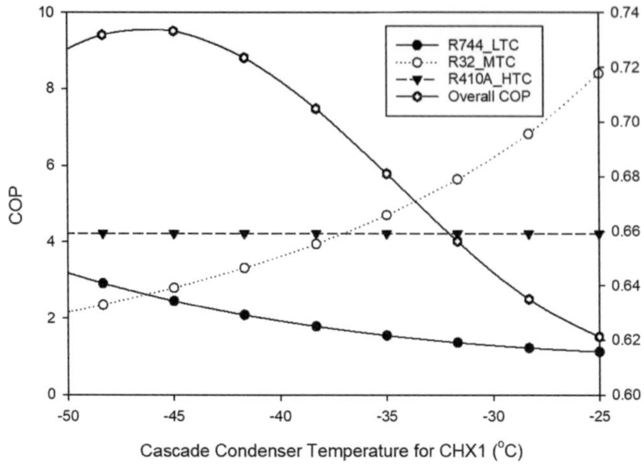

Fig. 32.6 Change in COPs corresponding to T_{C1} with R744, R32 and R410A

Variation of COP corresponding to T_{c1} with alternate refrigerants at MTC

Figure depicts the influence of T_{c1} on COP_{LTC}, COP_{MTC}, and COP_{HTC} when TSVCRS employs R744, R290, and R410a respectively with the temperature at evaporator (T_{evp}) of −80 °C. Figure 32.6 shows that the graph follows the same pattern as seen in Fig. 32.4. Here again the COP_{MTC} increases as T_{c1} (condensing temperature at CHX_1) increases whereas COP_{LTC} and COP_{HTC} decrease as T_{c1} increases. COP increases to its maximum value and then progressively declines as it grows. Under the parameters shown in Fig. 32.6, the cascade condenser CHX_1's optimal overall COP is 0.7519 at an optimum coupling temperature of −47.22 °C.

Similarly, Fig. 32.7 shows the influence of the cascade condenser temperature of LTC on the overall COP of the system having R744/R32/R410a at LTC, HTC, and MTC. The optimum value of overall COP and the optimum cascade condensing temperature as depicted in Fig. 32.7 is 0.7338 and −45.56 °C, respectively.

Variation of COP corresponding to T_{c1} with alternate refrigerants at HTC

Figure 32.8 depicts the influence of T_{c1} on COP_{LTC}, COP_{MTC}, and COP_{HTC} when TSVCRS employs R744, R290, and R1234yf respectively with the temperature at evaporator (T_{evp}) of −80 °C. Fig. shows that the graph follows the same pattern as seen in Fig. 32.4. Here again the COP_{MTC} increases as T_{c1} (the condensing temperature at CHX_1) increases whereas COP_{LTC} and COP_{HTC} decrease as T_{c1} increases. COP increases to its maximum value and then progressively declines as it grows. Under the parameters shown in Fig. 32.8, the cascade condenser CHX_1's optimal overall COP is 0.7006 at an optimum coupling temperature of −43.33 °C.

Similarly, Fig. 32.9 shows the effect of the cascade condensing temperature of LTC on the total COP of the system having R744/R32/R600 at LTC, HTC, and MTC.

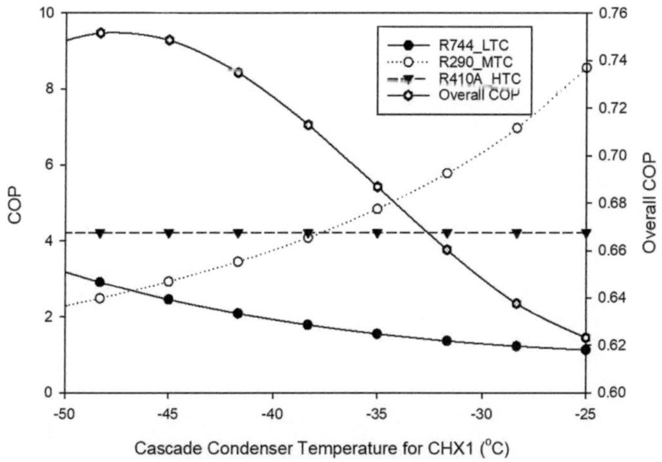

Fig. 32.7 Change in COPs corresponding to T_{C1} with R744, R290 and R410A

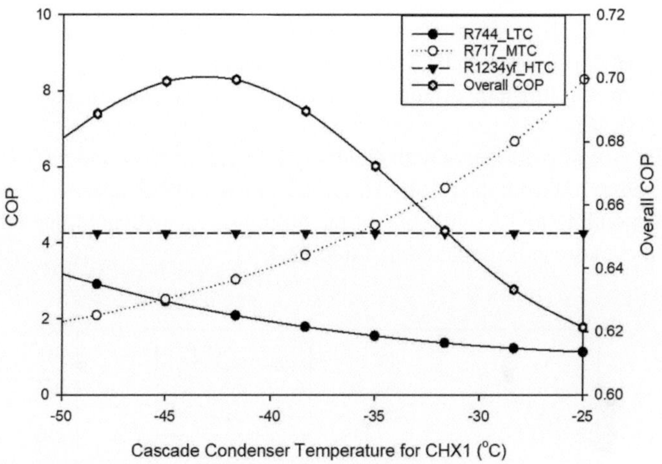

Fig. 32.8 Change in COPs corresponding to T_{C1} with R744, R717 and R1234yf

The optimum value of overall COP and the optimum cascade condensing temperature as depicted in Fig. 32.9 is 0.7136 and −43.33 °C, respectively.

Variation of COP corresponding to T_{c2} while taking the optimum T_{c1} with alternate refrigerants in different cycles

Figure 32.10 depicts the influence of T_{c2} on COP_{LTC}, COP_{MTC}, and COP_{HTC} when TSVCRS employs R744, R717, and R410A respectively with the temperature at evaporator (T_{evp}) of −80 °C. Figure 32.10 shows that the graph follows the same pattern as seen in Fig. 32.4. Here, the COP_{HTC} increases as T_{c2} (the condensing

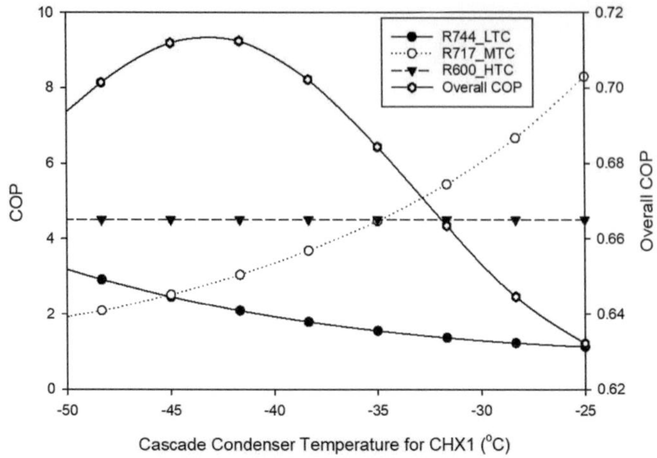

Fig. 32.9 Change in COPs corresponding to T_{C1} with R744, R717 and R600

temperature at CHX_2) increases whereas COP_{LTC} and COP_{MTC} decrease as T_{c2} increases. The optimum overall COP at the stated conditions of Fig. 32.10 is 0.7161 at an optimum coupling temperature of −7.222 °C for cascade condenser CHX_2 and T_{c1} at −43.33 °C.

The behavior of parameters with changing refrigerant pairs and varying T_{c1} and T_{c2} can be observed from the results discussed above. Furthermore, the variations of COP corresponding to T_{c2} while taking the optimum T_{c1} with alternate refrigerants in different cycles can be depicted in Table 32.4.

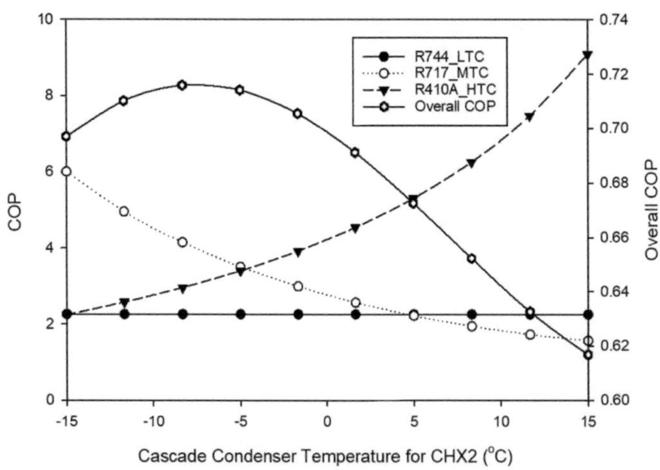

Fig. 32.10 Change in COPs corresponding to T_{C2} with R744, R717 and R410A

Table 32.4 Overall COP at optimum cascade condensing temperatures

LTC	MTC	HTC	Opt temp T_{c1}	Opt temp T_{c2}	Overall COP
R744	R717	R410A	−43.33 °C	−7.22 °C	0.7161
R170	R717	R410A	−40.56 °C	−5.33 °C	0.7616
R1150	R717	R410A	−40.22 °C	−5.33 °C	0.7623
R744	R290	R410A	−47.22 °C	−6.22 °C	0.7634
R744	R32	R410A	−45.56 °C	−6.66 °C	0.7454
R744	R717	R1234yf	−43.33 °C	−6.66 °C	0.7129
R744	R717	R600	−43.33 °C	−6.11 °C	0.7286

The system predicts the highest COP of 0.763 using R744, R290, and R410A in LTC, MTC, and HTC, respectively. Further, the optimum condensing temperature of LTC and MTC at this operating point is −47.22°C and −6.22°C, respectively.

32.5 Conclusions

Present study on TSVCRS revels that selection of refrigerant pairs and cascade condensing temperatures must be optimized to obtain the maximum COP of the system for very low-temperature applications. The study offers a detailed observation and indicates that cascade condensing temperature at LTC (T_{c1}) and MTC (T_{c2}) has a significant effect on the COP and performance of TSVCRS. Based on the refrigerant qualities, low global warming potential, and ODP, the study also illustrates the possible application of natural refrigerants on the three-stage cascade VCR system. Hydrocarbon refrigerants need to be thoroughly studied in order to guarantee safety concerns as well.

References

1. M. Fadaeefath Abadi, M. Hosseini Rahdar, F. Nasiri, F. Haghighat, Fault identification and fault impact analysis of the vapor compression refrigeration systems in buildings: a system reliability approach. Energies **15**, 5774 (2022). https://doi.org/10.3390/en15165774
2. V. Rangel, A. Almeida, F. Almeida, L. Duarte, Cascade refrigeration system for low temperatures using natural fluids. Revista Foco. **15**, e295 (2022). https://doi.org/10.54751/revistafoco.v15n1-013
3. G. Nicola, F. Polonara, R. Stryjek, A. Arteconi, Performance of cascade cycles working with blends of CO_2 + natural refrigerants. Int. J. Refrig. -revue Internationale Du Froid **34**, 1436–1445 (2011). https://doi.org/10.1016/j.ijrefrig.2011.05.004
4. C.M. Udroiu, A. Mota-Babiloni, P. Giménez-Prades, Á. Navarro-Esbrí, Two-stage cascade configurations based on ejectors for ultra-low temperature refrigeration with natural refrigerants. Int. J. Thermofluids 100287 (2023). https://doi.org/10.1016/j.ijft.2023.100287

5. F.M. Walid, M.H. Nabil, M. Uddin, M. Ehsan, S. Salehin, Thermodynamic assessment of a triple cascade refrigeration system utilizing hydrocarbon refrigerants for ultra-low temperature applications. Energy Convers. Manag.: X. **14**, 100207 (2022). https://doi.org/10.1016/j.ecmx.2022.100207
6. Z. Rogala, A. Kwiatkowski, Modeling of a three-stage cascaded refrigeration system based on standard refrigeration compressors in cryogenic applications above 110 K. Modelling **3**, 255–271 (2022). https://doi.org/10.3390/modelling3020017

Chapter 33
Design and Experimental Analysis of Free Piston Stirling Cooling System: An Environment Friendly Cooling Technique

Ehsas Srivastava, Kavya Taneja, Vaibhav Jain, and Kanchan Mudgil

Abstract This work presents an experimental investigation on the implementation of free piston Stirling system, which is an energy efficient cooling technique, for low temperature refrigeration applications. The system has compact size and reduced weight compared to recuperative refrigeration systems and uses Helium as refrigerant (environmentally friendly) unlike CFCs or HCFCs used in conventional vapor compression refrigeration system. In present work, an experimental test facility for the free piston Stirling cooling system has been designed and developed. The experiments were done for the cold head temperature of −40 °C to −70 °C, however the present test rig can reach upto −90 °C and from the experiments it has been found that as the cold head temperature decreases from −40 °C to −70 °C, the power consumption increases from 75.54 W to 83.17 W, while the refrigeration capacity decreases from 153.69 W to 12.69 W. As expected, the COP of the system decreases from 2.03 to 0.15 with reducing refrigeration temperature. Present result shows that implementing Stirling techniques-based cooling system can lead to enhanced energy efficiency and overall performance, contributing to the development of more sustainable cooling solutions.

Keywords Free piston Stirling cooling system · Low temperature applications · Environmentally friendly · COP

E. Srivastava (✉) · K. Taneja · V. Jain · K. Mudgil
Department of Mechanical Engineering, Maharaja Agrasen Institute of Technology, Delhi, India
e-mail: ehsassrivastava.work@gmail.com

A. Kumar et al. (eds.), *Selected Proceedings of the 1st International Conference on Advanced Materials for Sustainable Innovation; IC-AMSI 2024; 28–30 August; New Delhi; India*, Springer Proceedings in Energy,
https://doi.org/10.1007/978-981-96-4492-6_33

33.1 Introduction

The growing demand for ultra-low temperature freezers, essential for preserving special foods and biological materials, has highlighted the need for efficient and environmentally friendly cooling technologies. The free-piston Stirling cooler, known for its an energy efficiency and eco-friendly refrigerants, is gaining attention as a viable solution for extreme low temperature applications. This paper details the development and assembly of a free-piston Stirling cooling system capable of operating between −70°C and −10°C.

Ultra-low temperature freezing, defined as cooling below −60°C, has become increasingly important for various applications, including the cryo-preservation of seafood and prolonged scarcity of biological and medical products, including vaccines and plasma. Traditional household refrigerators, operating above −20°C, utilize single-stage vapor compression refrigeration cycles, whereas ultra-low temperatures, around −80°C, often require two-stage cascade refrigeration systems. Notable advancements in this field include SANYO and thermo Scientific's commercial freezers. The free-piston Stirling cooler emerges as a promising solution due to its compact design, high efficiency, and eco-friendly refrigerants.

Chakravarthy et al. [1] review refrigeration methods from 4 to 300 K, including steady-state and periodic systems, highlighting selection factors like cost, reliability, and application needs. The paper provides a comprehensive overview for refrigeration designers, emphasizing current technology and future challenges. Kim et al. [2] numerically investigated the cooling performance and pressure loss of a free-piston Stirling cooler (FPSC) using ANSYS Fluent. They discovered that expanding the heat transfer area of the regenerator reduces the temperature of the cold head but leads to a higher pressure drop, requiring more input power and reducing efficiency. Zhu et al. [3] investigated a FPSC for liquid nitrogen applications, focusing on key parameters like displacer mass, regenerator porosity, and operating frequency. They found that increasing displacer mass reduced exergy loss, while higher regenerator porosity decreased efficiency, and operating frequency significantly affected acoustic performance. Chaudhari et al. [4] introduced a free piston Stirling cooler operating on the reversed Stirling cycle, presenting a simplified numerical model. They assessed the system's performance indicators, such as the coefficient of performance, heat absorption, and work input, using both theoretical and graphical analyses. Yu et al. [5] designed an FPSC with a cooling capacity of 300 W at 80 K using thermo acoustic theory. They analyzed the phase shift effect, internal energy loss, and attained a relative Carnot efficiency of approximately 0.5 at 80 K through simulation and theoretical modeling. E. Oguz et al. [6] experimentally investigated two prototype FPSCs, assessing their cooling capacity and thermal efficiency. They tested the coolers at varying input voltages and cold head temperature ranging from − 40 °C to 0 °C, correlating the performance data to predict refrigeration capacity changes based on applied voltage. S. Celik et al. [7] implemented a FPSC system in a 27-L compact refrigerator, using water and a nano-fluid with aluminum oxide particles. They tested a custom jet-impingement heat exchanger and recorded temperatures with T-type thermocouples. Both fluids were tested and simulated, showing close agreement between experimental and numerical results. Song et al. [8] investigated the FPSC system for capturing CO_2 at cryogenic temperatures, focusing on enhancing efficiency by examining the coefficient of performance (COP). They found that using a copper cold head with specific dimensions significantly increased the COP, achieving 0.82 for the FPSC and 0.70 for the CO2 capture system. Zhu et al. [9] designed a FPSC utilizing the thermoacoustic principle, delivering 350 W of cooling power at 80 K with a relative Carnot efficiency of 26.8% and an input power of 3.57 kW. Sun et al. [10] examined a reverse Stirling cycle for cooling applications, constructing a free piston Stirling cooler and applying it to butter churning. They found that a larger displacer reduced cooler temperature, and increased working fluid volume or higher compression ratios improved cooling. The Stirling cooler also produced butter with lower water and higher fat content more quickly than conventional methods. Park et al. [11] evaluated the FPSC with a lubricated-free linear compressor, achieving 1.03 kW of heat absorption at 110 K with an 8 kW electric input and a cooler efficiency of 0.13. The study, using a numerical model, demonstrates the FPSC's suitability for applications like LNG, semiconductor fabrication, and biological material preservation, and its integration into a small-scale liquefier for natural gas liquefaction. Sun et al. [12] examined the performance of CCHP with a triple effect FPSC. In present paper, the experimental results demonstrating the efficiency of a free-piston Stirling cooling system, highlighting its potential in ultra-low temperature freezing applications are reported.

33.2 Working of Free Piston Stirling Engine

The P–V diagram of Stirling cooling cycle is shown in Fig. 33.1.

Process 1–2: Isothermal compression;

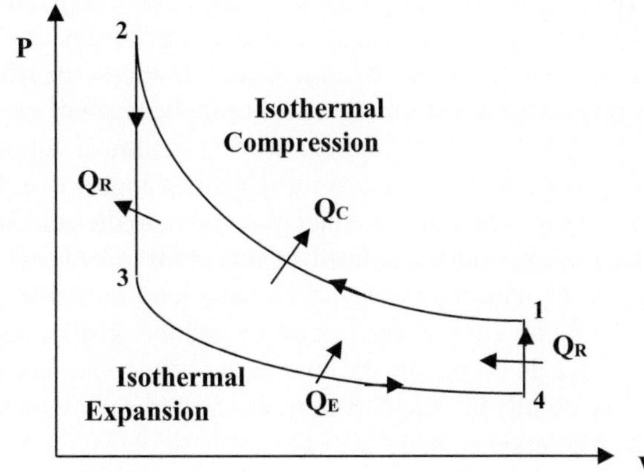

Fig. 33.1 P–v Diagram of Stirling cooling cycle

The piston in the apparatus moves towards regenerator with the motion given by linear motor which compresses the working fluid, helium, within the compression chamber. At this point displacer remains stationary. Heat gets transferred from helium to external sink at ambient temperature therefore heat from helium is rejected by rejectors.

Process 2–3: Constant volume heat release;

The compression and expansion chambers are linked via a regenerator. After piston movement compressing the working fluid, the piston and displacer moves simultaneously keeping the volume between them constant. It promotes the thermal transfer from helium to regenerator matrix. Now this fluid is transferred to expansion space during which fluid is cooled.

Process 3–4: Isothermal expansion;

In this process, with the piston stationary, the displacer continuously moves away from the regenerator. As the expansion occurs in the expansion chamber, the pressure and temperature drop to the refrigeration temperature. Following this, heat is transferred to the working fluid, and heat is absorbed from the surroundings of the expansion chamber.

Process 4–1: Constant volume heat absorption;

Finally in the last process, again the piston and displacer move simultaneously due to expansion of helium gas, to move the working fluid, which has absorbed heat from the surroundings, back through the regenerative matrix from the expansion space to the compression space. This process also involves transferring heat from the regenerator matrix back to the working fluid.

33.3 Experimental Setup and Procedure

The block diagram and photo of FPSC system employed in this investigation are shown in Fig. 33.2. These coolers function as pressure vessels, wherein approximately 1 g of Helium is cyclically displaced through the coordinated movements of two primary components: the piston and the displacer. The piston, which compresses the gas, is powered by a linear motor; whereas, the displacer's movement is driven by pressure differentials. Due to the relatively small surface areas of the compression and expansion spaces and the need to minimize temperature differentials between the gas and the heat sink/source to achieve higher coefficients of performance, additional heat exchangers are incorporated onto the cold and warm heads of the cooler. The Stirling coolers utilized in this study operate at a frequency of 50 Hz with alternating current (AC). A schematic representation delineating the configuration of the thermally insulated chamber and the positioning of the FPSC is provided in Fig. 33.2. In this setup, the cold head of the cooler is enclosed in a chamber that has been carefully insulated to minimize heat transfer with the surrounding environment. The insulation strategy incorporates multiple layers to optimize thermal resistance. Firstly, an inner layer comprising aluminium composite plastic (ACP), characterized by a thermal conductivity of 0.15 W/m–K, is used. Subsequently, polyurethane (PU), possessing a thermal conductivity of 0.022 W/m–K, acts as the intermediate insulation layer. The outer layer incorporates thermocol and mild steel, with respective thermal conductivities of 0.033 W/m–K and 45 W/m–K, to further attenuate heat transfer. The collective design yields an estimated overall thermal conductivity for the insulation system to be 0.3547 W/m–K, as calculated theoretically. Additionally, a layer of thermocol is introduced to safeguard the PU layers from potential damage, followed by encasement within a mild steel outer shell. Within the adiabatic chamber, an aluminum fin is affixed to the piston head, facilitating enhanced heat dissipation. Furthermore, a fan, rated at 0.24 A and powered by a 12 V DC supply, is mounted on the fin within the adiabatic chamber to augment convective heat transfer processes. In the experimental setup, various parameters were assessed to evaluate the performance of a FPSC system. These parameters included cold head temperature, warm head temperature, fin head temperature, adiabatic chamber temperature, and ambient air temperature. Temperature measurements were conducted using Pt100RTD thermocouples and a data logger that has an accuracy of ± 0.1 °C. Electrical variables, such as voltage input and current input were also measured during the experiments using voltmeter (panel mount size 96 × 46 (mm), range: 0–500 V AC, accuracy: 1 V) and ammeter (panel mount size 96 × 46 (mm), range: 0–10 A AC, accuracy: 0.5 A). The experiments were conducted utilizing an uninterruptible power supply providing 240 V. The experimental setup comprised six thermocouples strategically placed within the test rig. One thermocouple was positioned on the inner surface of the adiabatic chamber, while another was placed above and below the fan within the chamber to measure temperature differentials. Additional thermocouples were located at the piston head, heat sink, and outer surface of the adiabatic chamber. At the start of each test, the input voltage of the cooler was adjusted. Steady-state conditions were observed to take some time to stabilize, particularly at certain cold head temperatures. After a suitable period, the system achieved a steady state, and the coolers were operated for about an additional hour under these conditions.

COP is a measure of the efficiency of a cooling system. In the context of a FPSC system, the COP represents the ratio of the heat extracted from the cooling chamber to the power input required to achieve that cooling. Mathematically, COP is expressed as COP = Q_c/W, where Q_c is the heat extracted from the cooling chamber, and W is the power input to the system. A higher coefficient of performance signifies greater efficiency within the cooling system. In case of a FPSC system, achieving a high COP involves optimizing the heat transfer processes and minimizing losses during the reverse Stirling cycle.

To calculate the mass flow rate across the fan, parameters considered are:

Density (p) = 1.293 kg/m^3

Area of the fan (A) = 12.56 × 10^{-4} m^2

Velocity (v) = 2 m/s.

Power factor (cos(q)) = 0.97

$$\mathrm{COP} = \frac{Q_C}{W} \tag{33.1}$$

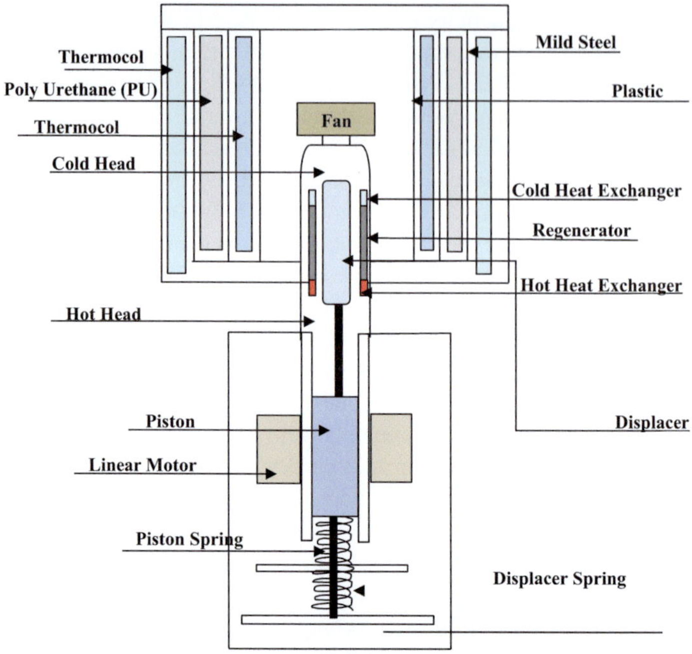

(a)

(b)

Fig. 33.2 Free piston Stirling cooling system. **a** Block diagram and **b** Experimental setup

$$Q_C = m \times C_p \times dT \tag{33.2}$$

$$m = p \times A \times v \tag{33.3}$$

$$W = V \times I \times \cos(q) \tag{33.4}$$

In the above equations, m represents the mass flow rate across fan, C_p represent the specific heat of air, v is the velocity of air flowing across the fan, V represent the voltage input, I is the current consumed, p is the density of air.

33.4 Result and Discussion

In this experimental investigation of FPSC system, readings of experimental setup are taken at four different cold head temperatures ranging from −40 °C to −70 °C with an interval of 10 °C, by operating the control unit. To conduct the experiment, the cold head temperature in being set to a particular value by using the switch on control unit.

This allowed for observing real changes and obtaining a better understanding of the temperature change in chamber, hot end and cold head during the experiment. When the temperature starts to decrease to the particular set value, the temperature values measured by the thermocouple are recorded in data logger, which we can access at any point of time using the Prolog software.

Figure 33.3 shows the variation of temperatures measured by various thermocouples over time at an input voltage of 240 V. It is evident that cold head temperature decreases from 28.3 °C to −73.2 °C, where the initial temperature drop was at a faster rate, and then the rate of temperature drop became slower with time. This is because initially the load on the system was less as compared to the later condition.

The variation of heat sink temperature from 28.3 °C to 45.3 °C can also be observed in Fig. 33.3. This is because with the compression of helium being undergoing in the compression chamber, heat is produced, excess heat is taken out from the compression chamber with the help of hot heat exchanger.

First the system started at ambient temperature and current consumption was 0.09 A, the cold head temperature was set at −40 °C, then system took 209 min to reach −40 °C at cold head, while consuming power of 75.54 W and producing a refrigeration capacity of 153.69 W. The current consumption at this temperature was 0.33 A, the voltage consumption was 236 V and the COP came out to be 2.03.

Then after analyzing the system at −40 °C, the temperature of cold head was set at −50 °C, now system took 48 min to reach the steady state at −50 °C, the system was made to operate at this state for 30 min during which the current consumption was found to be 0.33 A, the voltage consumption was found to be 239 V and the power consumption was 76.5 W. The refrigeration capacity came out to be 72.61 W

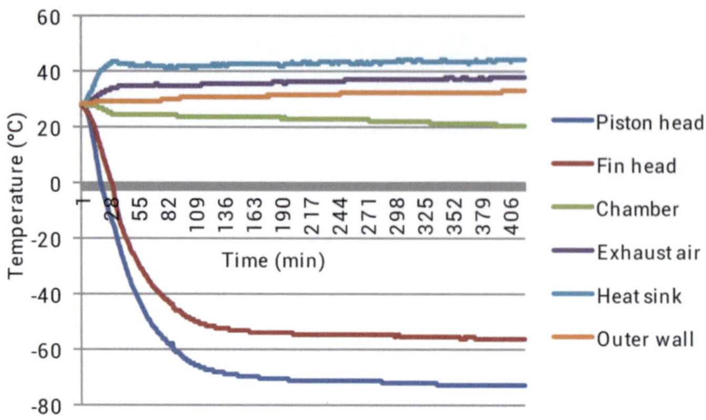

Fig. 33.3 Variation of temperature (°C) of piston head, fin head, chamber, heat sink, exhaust air, outer wall with time (min)

at −50 °C, using the refrigeration capacity and power consumed the COP was found be to 0.94.

Then after analyzing the system at −50 °C, the temperature of cold head was set at −60 °C, now system took 55 min to reach the steady state at −60 °C, the system was made to operate at this state for 30 min during which the current consumption was found to be 0.34 A, the voltage consumption was found to be 243 V and the power consumption was 80.14 W. The refrigeration capacity came out to be 34.5 W at −60 °C, using the refrigeration capacity and power consumed the COP was found be to 0.43.

Then after analyzing the system at −60 °C, temperature of cold head was set at −70 °C, now system took 73 min to reach the steady state at −70 °C, the system was made to operate at this state for 30 min during which the current consumption was found to be 0.35 A, the voltage consumption was found to be 245 V and the power consumption was 83.17 W. The refrigeration capacity came out to be 12.699 W at −70 °C, and the COP was found be to 0.152. The refrigeration capacity of the FPSC system being tested is given in Fig. 33.4. The cooler was tested at different cold head temperature ranging from −40 °C to −70 °C with an interval of 10 °C, meanwhile the warm head temperature varied form 41.6 °C to 42.4 °C. Referring to Fig. 33.4, it is easily seen that with the decrease in cold head temperature, the refrigeration capacity decreases. This can be due to increase in load on the system, also the system is taking more time to reach steady state at every consecutive set temperature.

Referring to Fig. 33.5, it is easily seen that the power consumption is increasing as the cold head temperature decreases, this is because as the temperature of system goes down, its current consumption increases and hence the power consumption increases.

The coefficient of performance of the FPSC system being tested is shown in Fig. 33.6. The cooler was tested at different cold head temperature ranging from −

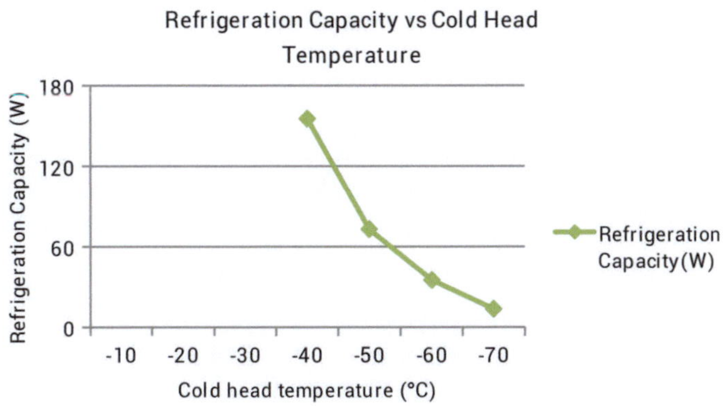

Fig. 33.4 Variation of refrigeration capacity with cold head temperature

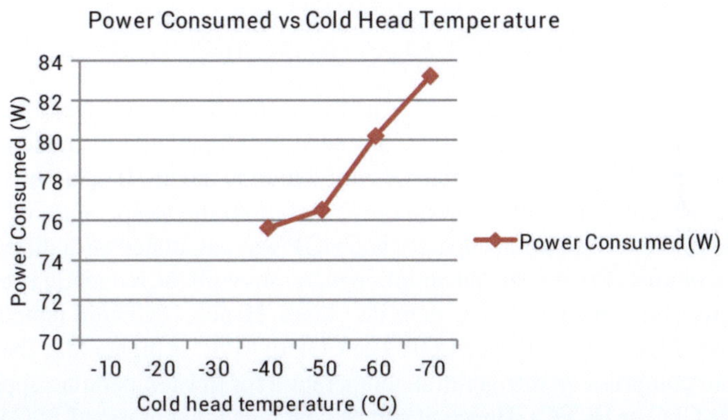

Fig. 33.5 Variation of power consumption with cold head temperature

40 °C to −70 °C with an interval of 10 °C, meanwhile the warm head temperature varied from 41.6 °C to 42.4 °C. Referring to Fig. 33.6, it is easily seen that the COP is decreasing as the cold head temperature decreases, this is because as the temperature of system goes down, its refrigeration capacity decreases and the power consumption increases because of high load on the system. The present trends in the results are inline with the previous work on FPSC system which provides confidence in the findings reported in the paper.

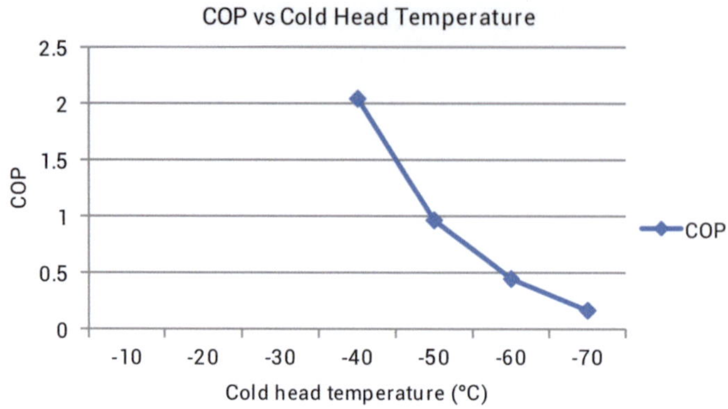

Fig. 33.6 Variation of cop with cold head temperature

33.5 Conclusions

In the research experimental test facility of free piston Stirling refrigeration system, an energy efficient cooling technique, has been developed and examined. Helium is employed as refrigerant in present system which is having 0 value of ODP and GWP. The present test facility can quickly reach upto the temperature of −90 °C. The effect on the refrigeration capacity and COP was calculated with different cold head temperatures. The power consumption increases with the reduction in cold head temperature; hence the COP of system decreases. Hence, high cold temperature is recommended for higher value of COP. The COP of FPSC is higher than the conventional vapor compression refrigeration temperature for low temperature applications which uses CFC or HCFCs. Hence, it can be concluded that present FPSC is energy efficient and environmentally friendly.

Acknowledgements We are thankful to ASHRAE for providing us the grant to develop the test facility of FPSC.

References

1. V.S. Chakravarthy, R.K. Shah, G. Venkatarathnam, A review of refrigeration methods in the temperature range 4–300 K. J. Therm. Sci. Eng. Appl. **3**(2), 1–19 (2011)
2. H.S. Kim, I.C. Gwak, S.H. Lee, Numerical analysis of heat transfer area effect on cooling performance in regenerator of free-piston Stirling cooler. Case Stud. Therm. Eng. **2022**(32), 101875 (2021)
3. S. Zhu, G. Yu, X. Li, M. Ying, C. Yan, W. Dai et al., Acoustic field characteristics of a free-piston Stirling cryocooler with large cooling capacity at liquid nitrogen temperature. Appl. Therm. Eng. **147**, 324–335 (2019)
4. P. Chaudhari, D.D. Souza, S. Borkar, V. Haribhakta, S. Trimbake, Mathematical formulation of free piston stirling cooler for domestic refrigeration using loss factor. **7**(11), 1505–10 (2016)

5. G.Y. Yu, K. Li, W. Dai, E.C. Luo, Z.H. Wu, Investigation on a free piston Stirling cryocooler with large cooling capacity. AIP Conf. Proc. **2014**(1573), 1626–1632 (2004)
6. E. Oguz, F. Ozkadi, An experimental study on the refrigeration capacity and thermal performance of free piston Stirling coolers, in *International Refrigeration and Air Conditioning Conference*, (2000), 497–504
7. S. Celik, Experimental analysis of a stirling refrigerator employing jet-impingement heat exchanger and nanofluids, in *International Refrigeration and Air Conditioning Conference Paper 1098*, (2010)
8. C. Song, J. Lu, Y. Kitamura, Study on the COP of free piston Stirling cooler (FPSC) in the anti-sublimation CO_2 capture process. Renew. Energy **74**, 948–954 (2015)
9. S. Zhu, G. Yu, X. Li, W. Dai, E. Luo, Parametric study of a free-piston Stirling cryocooler capable of providing 350 W cooling power at 80 K. Appl. Therm. Eng. **174**(February), 115101 (2020)
10. J.F. Sun, Y. Kitamura, T. Satake, Application of Stirling cooler to food processing: feasibility study on butter churning. J. Food Eng. **84**(1), 21–27 (2008)
11. J. Park, H. Kim, J. Ko, S. In, Y. Hong, H. Yeom et al., Free-piston Stirling cryocooler (FPSC) for low-temperature applications down to 110 K. Int. J. Refrig. **117**, 218–229 (2020)
12. Y. Sun, K. Luo, J. Wang, E. Luo, J. Hu, Z. Wu, L. Zhang, Y. Zeng, J. Chi, Integrated combined heating, cooling and electricity system based on triple-effect free-piston Stirling configuration. Int. J. Refrig. **164**(August 2024), 117–121

Chapter 34
Turning Trash into Treasure: The Role of Plastic Waste Recycling in Decarbonization

Mudita Nagpal, Sania, Khushi Garg, and Nidhi Sharma

Abstract Our dependence on disposable plastics has created a massive problem. Plastic waste, especially single-use plastics, presents a substantial threat to ecosystems, marine life, and human health. Improper waste disposal methods and a lack of recycling facilities has led to the buildup of plastic waste in rivers, landfills, and seas, where they remain for hundreds of years, posing a threat to wildlife by being ingested or causing entanglement. In this study, the increasing plastic waste generation trend in India in the last decade has been analyzed and the waste generation in the next five years has been projected. Recycling has been proposed as an answer to solve the issue of plastic waste, aiming to transform this challenge into an opportunity. Recycling plastic isn't just good for the environment by conserving resources, reducing energy use, and cutting greenhouse gases, it also creates valuable new products. Innovative ways to transform plastic waste into valuable products like fuel, construction materials, and even advanced carbon-based materials have been discussed. The article delves into various methods for tackling plastic waste, including established techniques like pyrolysis and thermal processing, and explores promising new approaches like upcycling and catalytic conversion. The conclusion suggests that recycling plastic not only helps alleviate negative environmental impacts but also opens up avenues for development and innovation by exploring novel products and designs.

Keywords Plastic waste · Recycling · Fuels · Fabrics · Carbon-based nanomaterials · Circular economy · Data analysis

---

M. Nagpal · Sania (✉) · K. Garg · N. Sharma
School of Engineering & Technology, Vivekananda Institute of Professional Studies-Technical Campus, Delhi, India
e-mail: saniagarg.info@gmail.com

A. Kumar et al. (eds.), *Selected Proceedings of the 1st International Conference on Advanced Materials for Sustainable Innovation; IC-AMSI 2024; 28–30 August; New Delhi; India*, Springer Proceedings in Energy,
https://doi.org/10.1007/978-981-96-4492-6_34

34.1 Introduction

Plastic, a ubiquitous material in our lives, is synthetic polymers prized for their affordability, light weight, enduring nature, ease of manufacture, and remarkable chemical versatility. From packaging to clothing to medical equipment, a diverse range of polymers, encompassing polypropylene (PP), polyethylene terephthalate (PET), polyethylene (PE), polystyrene (PS) and polyvinyl chloride (PVC) serves as the building blocks for commercially available plastics we rely on today [1]. However, this widespread use of plastic presents a significant environmental challenge, as its degradation can take hundreds of years, and it contains highly toxic elements. The production of plastic has surged dramatically in recent decades, now peaking at approximately 350 million tons annually [2]. The exponential rise in plastic production has led to widespread pollution of our water, air, and land ecosystems. This surge has resulted in vast quantities of plastic accumulating in our seas, leading to the tragic deaths of marine animals. Furthermore, improper disposal of plastic on land has contributed to the outbreak of diseases such as dengue and malaria [3]. The burning of plastic exacerbates the problem by emitting toxic gases into the air, posing significant health risks to humans and making it difficult to breathe (Burning one ton of plastic waste results in the creation of 2.9 tons of carbon emissions). Additionally, plastic waste plays a substantial role in exacerbating global warming, further compounding the environmental challenges we face [4] (Fig. 34.1).

Despite significant progress, plastic recycling lags behind most other materials. In the EU, only about 33% of plastic is recycled overall. However, it was estimated that the amount of packaging waste which was recycled was 41.5% in the EU in 2018 [6]. However, there's a glimmer of hope. Recycling rates are rising globally. Sustainability concerns are driving the development of innovative recycling technologies, which in turn is creating a thriving market for recycled plastics. There are various plastic recycling techniques, categorized as primary, secondary, tertiary, and quaternary [4]. Progression in technology has made it possible to recycle a wider range of polymers and metals.

Recycling efforts are important in reducing the environmental impact of plastic waste and extending its life cycle [4]. The recycling process for most plastics often results in downcycling, meaning the recycled material is converted into products with a lower quality compared to original plastic [6]. Indeed, numerous emerging technologies offer promising avenues for converting plastic waste into valuable resources (bricks, fabrics, nanomaterial and H_2 gas production, fuels and many more), thereby contributing to eco-friendly development. This research paper explores several of these technologies, highlighting their potential to recycle and repurpose plastic effectively. By harnessing innovative approaches, we can transform plastic waste into valuable resources, mitigating environmental harm and promoting sustainability.

Fig. 34.1 Plastic pollution and the impact of plastic on human health [5]

34.2 Analyzing the Growth in Plastic Waste Generation in India

According to CPCB data, the quantity of plastic waste produced in India over the past decade is represented in Table 34.1 for the years, 2013–2021 [7]. The volume of plastic waste generated in India has increased from 1,508,597.87 TPA in 2013 to 4,126,997 TPA in 2021, which is 2.7 times the plastic waste generated in 2013. This

Table 34.1 Plastic waste generated in India [7]

Year ending	Plastic waste generated (TPA)
2013	1,508,597.87
2014	1,829,059.00
2015	788,999.62
2016	1,589,418.00
2017	1,568,714.00
2018	660,787.00
2019	3,360,043.00
2020	3,469,780.00
2021	4,126,997.00

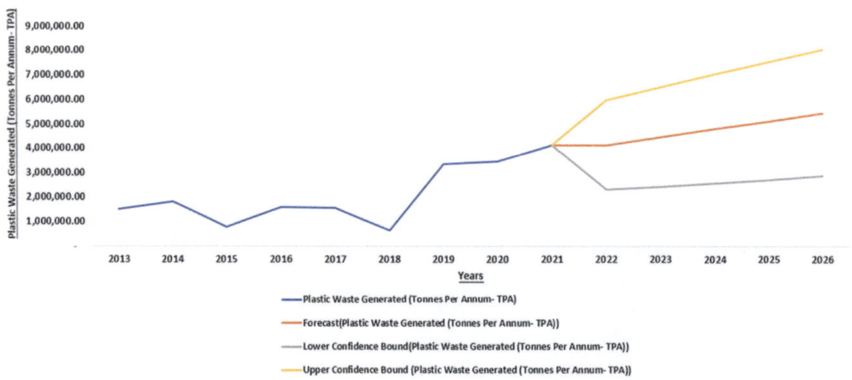

Fig. 34.2 Plastic waste trends in India: a look at past generation and future projections (2013–2026)

has also been depicted graphically in Fig. 34.2. According to the graph, there was a notable surge in plastic waste generation from 2013 to 2021.

However, there were intermittent declines in 2015 and 2018. Several factors could explain these temporary downturns:

- **Awareness Campaigns:** Heightened awareness regarding plastic pollution and its environmental consequences might have led to decreased plastic usage.
- **Regulations and Policies:** Various regulations and policies were implemented to address plastic waste, such as, Plastic Waste Management Rules, 2016, which emphasized waste minimization, source segregation, and recycling, alongside the adoption of the polluter's pay principle [8].
- Legislative Guide for Plastic Pollution Reduction, offering policy and legal approaches to tackle plastic pollution.
- Ban on Plastic Sachets, targeting single-use plastic sachets used for gutkha, tobacco, and pan masala due to their environmental impact [9].
- State-Specific Stringent Norms, where several states and union territories enforced partial and complete bans on plastic carry bags and other items [10].

Table 34.2 Projected data for plastic waste generated in India

Year ending	Forecast (plastic waste generated (TPA)	Lower confidence bound (plastic waste generated (TPA)	Upper confidence bound (plastic waste generated (TPA)
2022	2,318,103.90	5,952,591.88	4,135,347.89
2023	2,429,009.38	6,494,117.16	4,461,563.27
2024	2,559,890.14	7,015,667.17	4,787,778.65
2025	2,705,879.01	7,522,109.07	5,113,994.04
2026	2,863,802.18	8,016,616.66	5,440,209.42

- The Plastic Waste Management (Amendment) Rules, 2018, include provisions for phasing out non-recyclable materials. Multilayered Plastic (MLP) was introduced to promote sustainable alternatives [11].
- **Economic Factors:** Fluctuations in the economy could influence production and consumption patterns. Efforts promoting a circular economy are aimed at reducing the environmental impact of plastic packaging [12].
- **Urbanization Trends:** Urbanization and lifestyle changes may have impacted plastic consumption.

We have also projected the plastic waste generation in India for the next five years (2022–2026) (Table 34.2). In the graph (Fig. 34.2), the forecast line illustrates the predicted values derived from historical data. The upper confidence line represents a range above the forecast line and indicates the upper limit within which the actual data points are likely to fall, given a certain level of confidence. The lower confidence line shows the lower limit within which the actual data points are likely to fall. The projected data for plastic waste generation in India for the years, 2022–2026 indicates a continued rapid increase. Research conducted by the Federation of Indian Chambers of Commerce and Industry (FICCI) and Accenture forecasts significant projected losses in the value of plastic materials by 2030, attributed to unsustainable packaging practices [13]. Implementing effective circular economy practices is crucial to mitigate this environmental impact. India faces a significant challenge in managing plastic waste, necessitating sustainable practices, recycling promotion, and heightened awareness for a cleaner and greener future.

34.3 Transforming Plastic Waste into Useful Products

34.3.1 Fuel

Today, the dearth of adequate energy supplies and efficient transformation systems remains a significant obstacle to widespread human and socio-economic advancement. Globally, the vast quantities of plastic waste generated could potentially be harnessed to meet global energy needs if managed effectively. The conversion of plastics to fuel is a leading research subject around the globe. Hydrocracking, thermochemical conversion and catalytic cracking methods are popular plastic-to-fuel technologies [14]. Of these ways, thermochemical conversions, especially the pyrolysis process seem to be the most attractive alternative. Thermochemical processes break the large polymers down into small hydrocarbons with a wide range of carbon numbers and boiling points. This transformation occurs in an environment devoid of oxygen, typically under controlled conditions with anaerobic heating ranging from 350 to 600 °C [15]. Pyrolysis of polyvinyl chloride (PVC) can yield liquid oil with a high calorific value, reaching up to 40 MJ/kg. This process involves heating PVC without oxygen, leading to the polymer decomposition into liquid hydrocarbons [16].

Sivagami and colleagues investigated the performance of ZSM-5 zeolite in processing different combinations of polyolefin waste and post-consumer solid waste (PSW), particularly from flexible packaging and municipal sources. They found that the gross calorific value of the oil derived from various blends of multilayer packaging waste ranged from 10,789 to 7,156 kcal kg^{-1}. By employing a synthesized ZSM-5 zeolite catalyst, they achieved significant results with low-density polyethylene (LDPE) plastic waste. LDPE yielded a maximum oil yield of 70%, along with 16% gas and 14% char production during the pyrolysis process [17]. Based on the properties of distilled plastic oils and waste plastic oil, it is clear that these fuels are far better for a compression ignition engine (diesel engines) than spark ignited engine.

The composition of waste plastic oil typically includes approximately 84% carbon, 0.5% oxygen and 15.5% hydrogen, by mass. These proportions contribute to the fuel's suitability for compression ignition engines, where the high carbon content facilitates combustion efficiency [18]. The conversion of plastics into fuel depends on factors like thermal degradation conditions, temperature, type of reactor and reaction time, which vary with the type of plastic. Adding catalysts during pyrolysis can increase the yield of oil from plastics. Batch-type reactors have been extensively studied due to their ease of control, fabrication and design. In contrast, fluidized bed reactors are noted for their ability to perform uniform mixing of feedstock and catalyst, enhancing the efficiency of the pyrolysis process [19]. Additional research is required to optimize the proportion of catalysts and additives added to different types of plastics. This optimization aims to enhance the efficiency of conversion, improving both the quality and quantity of the resulting products. Plastic waste is being converted into useful fuels successfully in India. The process to convert polyethylene and polypropylene–the most common types of plastic that constitute approximately 60% of plastic waste—into either diesel or gasoline has been developed by CSIR-Indian Institute of Petroleum. A similar endeavor is also underway at Rudra Environmental Solutions in Pune. In Goa, which produces 66 metric tons' plastic waste per day, plants have been established at Bicholim and Sonsoddo through a public–private partnership with Bangalore-based M K Aromatics Ltd for fuel conversion from this waste. Ventana, a Hyderabad based company, plans to open several plants for turning waste into energy with the support of Ramky Environment [20].

34.3.2 Construction Materials (Bricks)

Bricks, a common construction material, particularly in residential projects, have traditionally been manufactured from red clay. However, this conventional approach to brick production poses environmental challenges, including topsoil erosion, greenhouse gas emissions from coal and biomass combustion, and consequent fertility loss. Consequently, there is an imperative to explore alternative methods of brick production, with plastic bricks emerging as a promising solution [21]. Plastic brick manufacturing offers an optimal means to solve the problem of plastic waste accumulation, while also presenting cost-effective solutions for building materials. Plastic possesses several advantageous properties, including low density, ease of processing, robust mechanical properties, chemical resistance, as well as excellent thermal and electrical insulating properties, all of which are pivotal for the construction of durable bricks [20]. Materials utilized in the manufacturing process of plastic bricks include, manufactured sand, cement, flash, polypropylene fiber and water [23]. The process for manufacturing bricks from plastic waste has been shown in Fig. 34.3.

Numerous studies have documented the correlation between the plastic waste amount in the brick and its compressive strength. In a study by Raunak shah and group, plastic waste was used to synthesize bricks. The findings reveal that LDPE plastic bricks could withstand a maximum compressive load of 13.69 N/mm^2. Following closely are LDPE fly ash composite bricks, exhibiting a strength ranging from 11.48 to 10.42 N/mm^2, surpassing the typical compressive strength of conventional clay bricks available in the market, which ranges from 3 to 5 N/mm^2 [24]. Mahajan and group [25] and Alighiri and group [26] have identified a direct relationship between the amount of plastic waste (PW) and the compressive strength of bricks, indicating that as PW content increases, so does compressive

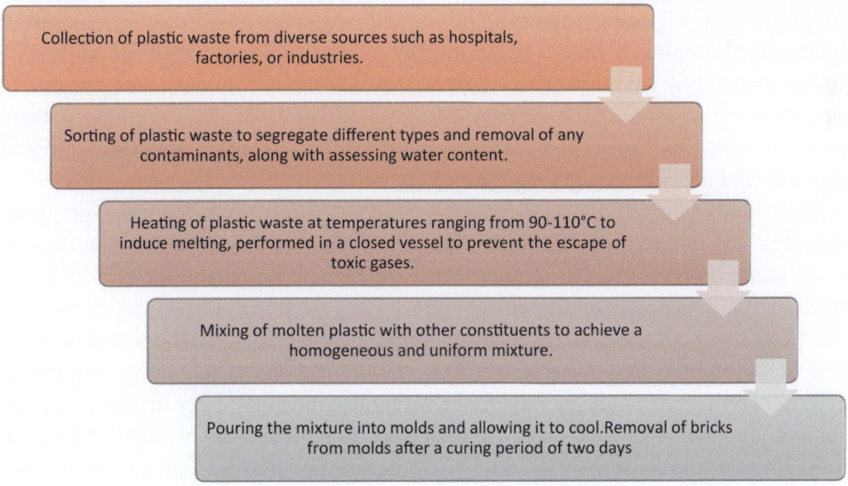

Fig. 34.3 Step by step process for manufacturing bricks from plastic waste [32]

strength. Singhal and Netula (2018) [27] conducted a study revealing that bricks manufactured from melted plastic waste mixed with stone dust achieved the highest compressive strength of 5.6 N/mm2, underscoring the potential for increased strength with higher PW content. Similarly, [28] demonstrated that the highest compressive strength of 11 N/mm^2 was attained with a PW content of 70%, further emphasizing the influence of PW quantity on brick strength [28].

Plastic waste has been shown to impact the rate at which water is absorbed. Thirugnanasambantham et al. (2017) [29] demonstrated that plastic bricks exhibit reduced water absorption rates compared to fly ash bricks [29]. Thus, plastic bricks offer a compelling alternative to clay bricks due to their affordability, sustainability, and satisfactory quality characteristics. Plastic waste is being converted into bricks successfully in India. Rhino Machines converts plastics into bricks using foundry dust and waste plastic in India. The company got orders for 650 paver blocks and 1,000 bricks since February 2020, attracting interest from Chile, Colombia, and China [30] Meanwhile, Abhishek Banerjee and his team at Jadavpur University developed "Plastiqube," a brick made from recycled plastic, addressing India's plastic waste problem with a sustainable construction alternative [31].

34.3.3 Carbon Nanomaterials

The buildup of plastic waste has led to significant environmental concerns, exacerbated by their non-biodegradable nature and the presence of harmful substances. Conversion of waste into various carbon nanomaterials (CNMs) holds great promise in reducing plastic debris and promoting the production of CNMs. Examples of carbon materials derived from plastics include carbon filaments such as graphene, carbon nanosheets, carbon nanotubes, carbon spheres, porous carbons and carbon nanofibers [33]. RIL, which operates the world's largest integrated and petrochemical and refining complex, is India's first company to achieve this great feat of converting plastic waste into an innovative, high-quality Circular Polymer. The process of converting pyrolysis oil derived from plastic waste into ISCC-Plus certified as feedstock for production plant, accords well with Reliance's approach towards circular economy and chemical recycling vision [34].

An appealing factor is that lots of research investigated the production of graphene with high quality from plastic waste. Yao et al. reported in a research that decomposed waste plastics from local food courts to carbon nanomaterials (CNMs), and byproducts composed of hydrogen-rich gases using the pyrolysis process in the presence of a catalyst. Bimetallic catalysts (Co-Fe, Co–Ni and Fe–Ni) supported on MgO were prepared until they were tested. Compared with the pH-increase precipitation method-synthesized catalysts, Fe–Ni-Mg was one of the most active in plastic conversion among all investigated catalytic materials. It achieved the highest CNM yield of 30.25 wt% and hydrogen yield of 31.52 mmol/g of plastic. Conversely, Co–Ni/MgO, prepared by impregnation, exhibited the lowest activity among the catalysts studied [35].

The synthesis of multi-layer, few-layer and monolayer graphene materials, including flash graphene, has been predominantly reported using four main methods [36] (Fig. 34.4):

1. Thermal decomposition of plastics directly on a metal substrate.
2. Initial thermal decomposition of plastics, with the resultant hydrocarbon gases directed into a Chemical Vapor Deposition (CVD) system housing the metal substrate.
3. Thermal decomposition, subsequent ball milling, and microwave sintering.
4. Flash Joule heating (FJH).

34.3.4 Fabrics

Sustainable fashion presents a comprehensive approach to environmental responsibility, social justice, and cultural preservation within the clothing industry. Through industry-wide collaboration, technological innovation, and consumer awareness, the concept of sustainable fashion could transform how clothing is produced and consumed, ultimately fostering a more positive and sustainable future for the industry [37]. Researchers have found that plastic waste can be transformed into polyester textiles through processes such as spinning, washing, sterilization, weaving and high temperature melting [38]. The resulting textiles, including carpets and clothing, are more durable and flexible, and resistant to insects compared to those made from plant and animal fibers [39].

Recycled PET is an environmentally friendly alternative to virgin polyester, produced by breaking down used plastic bottles into small chips. These chips are then melted and spun into polyester yarn, significantly reducing the reliance on new petroleum-based materials and helping to minimize waste (Fig. 34.5) [40]. Although the production of Recycled PET (R-PET) has increased considerably in recent years, only 30% of PET bottles are currently recycled. Aditya Banger, a 17-year-old from Rajasthan, founded Trash to Treasure in Bhilwara district. His company turns PET-grade plastics into durable fabrics, sourced locally and cleaned rigorously before processing into fibers mixed with cotton for diverse product use [41]. Raw Pressery, a Mumbai-based startup, has launched RawCycle to tackle this issue. They've collected 1.2 million bottles and transformed them into 1,500 t-shirts. These t-shirts, part of the Street to Studio collection, are made from 95% recycled plastic and 5% spandex [42]. Significantly, R-PET fibers are 20% more cost-effective compared to fibers possessing comparable physical characteristics [43].

Various studies have reported innovative ways of converting plastic waste into useful fabrics. Ho and group reported a novel methodology for transforming polymeric waste into potentially valuable materials for the textile industry. The method introduced the conversion of reclaimed PET (polyethylene terephthalate) waste films into chemically transformed thermoplastic polyester elastomers (TPEE), specifically PBT-co-PTMEG. This process involves an esterification process followed by condensation polymerization. Subsequently, the upcycled TPEE undergoes solubilization

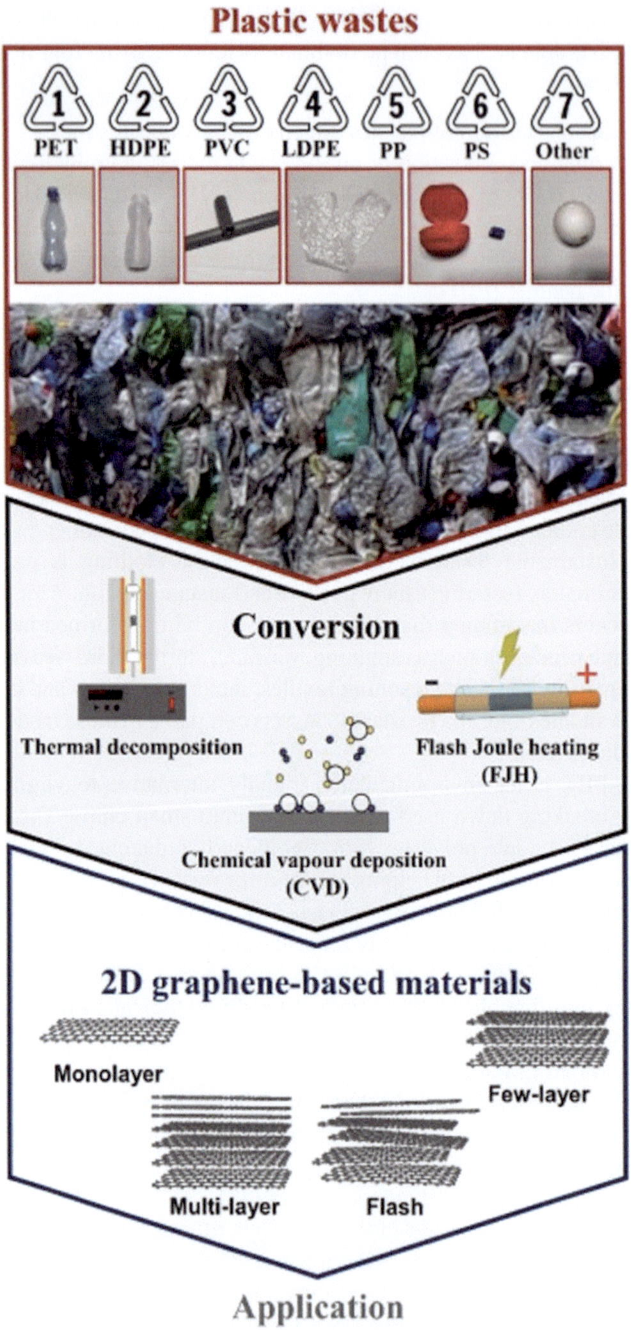

Fig. 34.4 Methods for production of different types of graphene from plastic waste [36]

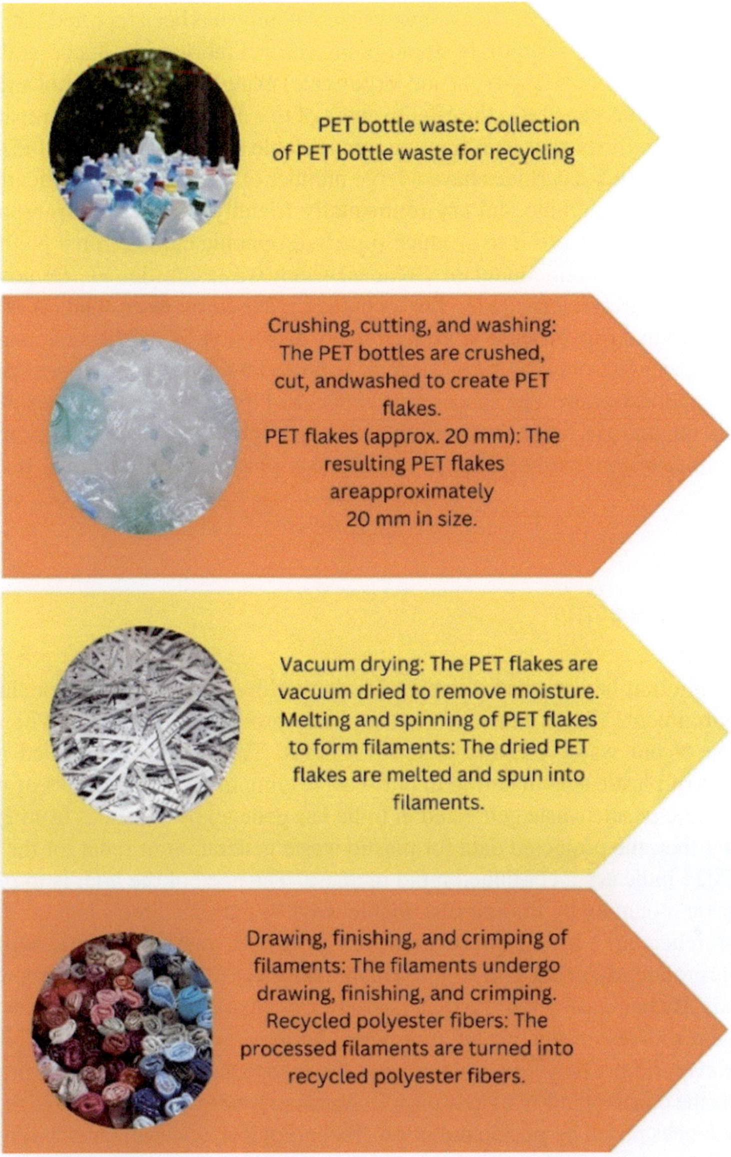

Fig. 34.5 Steps for manufacturing fabric from waste PET bottles [35]

in a TFA/DCM solvent system before being electrospun into aligned nanofibrous scaffolds using a rotating drum [44].

Traditional aerogel fabrication suffers from lengthy processing times, hazardous chemicals, and complex implementation, hindering cost-effective production and widespread adoption. Recently, a nonwoven technology has been reported for continuous fiber generation, potentially streamlining aerogel fabrication. Successful fabrication of rPET aerogels (lab-scale and large-scale) using polyvinyl alcohol and recycled PET fibers demonstrates the effectiveness of this approach for fiber preparation, paving the way for further investigation into scalable aerogel production [45].

Superhydrophobic carbides have a large number of technological applications as they are versatile, durable and environmentally friendly. Ali and group reported a novel technique in an effort to produce superhydrophobic carbides on a plethora of substrates (metals, semiconductors, oxides) using waste plastics via femtosecond laser-assisted carburization [11]. The technique is based on laser ablation of waste plastic in toluene to produce a concentrated dispersion. This dispersion is then deposited dropwise onto the target surface, followed by femtosecond laser scanning to induce carburization. The successful synthesis of hydrogenated carbides of tungsten, silicon, and zirconium with exceptional hydrophobic properties demonstrates the potential of this method for upcycling waste plastics into functional materials [46].

34.4 Conclusions

Plastic pollution has reached alarming levels worldwide, posing a major threat to our environment. The exponential rise in plastic production has led to widespread pollution of our water, air, and land ecosystems. This surge has resulted in vast quantities of plastic accumulating in our seas, leading to the tragic deaths of marine animals. The plastic waste generated in India has gone up by 2.7 times from 2013 to 2021. Further, the projected data for plastic waste generation in India for the years, 2022–2026 indicates a continued rapid increase. The need of the hour is to manage this plastic mountain by finding sustainable ways to reuse and recycle it.

Researchers have been exploring innovative ways to transform plastic waste into valuable products like fuel, construction materials, fabrics and advanced carbon nanomaterials. By harnessing innovative approaches, we can transform plastic waste into valuable resources, mitigating environmental harm and promoting sustainability.

However, simply recycling isn't enough. We need to tackle the root of the problem, which is the constant influx of new plastic. Awareness needs to be increased to tackle the challenges posed by plastic pollution. Technological advancements will be ineffective unless the movement of plastic waste is curtailed or significantly diminished. It is crucial for people to alter their behavior, attitudes, and mindset regarding plastic use. Accountability, traceability, and openness are crucial to ensure that waste is managed responsibly and utilized as a valuable resource whenever feasible.

References

1. K. Zheng, Y. Wu, Z. Hu, S. Wang, X. Jiao, J. Zhu, Y. Sun, Y. Xie, Progress and perspective for conversion of plastic wastes into valuable chemicals. Chem. Soc. Rev. **52**(1), 8–29 (2023)
2. L.M. Heidbreder, I. Bablok, S. Drews, C. Menzel, Tackling the plastic problem: a review on perceptions, behaviors, and interventions. Sci. Total. Environ. **668**, 1077–1093 (2019)
3. R. Proshad, T. Kormoker, M.S. Islam, M.A. Haque, M.M. Rahman, M.M.R. Mithu, Toxic effects of plastic on human health and environment: a consequences of health risk assessment in Bangladesh. Int. J. Health **6**(1), 1–5 (2018)
4. N. Singh, D. Hui, R. Singh, I.P.S. Ahuja, L. Feo, F. Fraternali, Recycling of plastic solid waste: a state of art review and future applications. Compos. B Eng. **115**, 409–422 (2017)
5. R. Verma, K.S. Vinoda, M. Papireddy, A.N.S. Gowda, Toxic pollutants from plastic waste–a review. Procedia Environ. Sci. **35**, 701–708 (2016)
6. L. Shen, E. Worrell, Plastic recycling, in *Handbook of recycling*, pp. 497–510. Elsevier (2024)
7. https://cpcb.nic.in/status-of-implementation-of-plastic-waste/. Accessed 26 Apr 2024
8. https://www.hspcb.org.in/content/laws/pwm/PWM_Rules.pdf. Accessed 26 Apr 2024
9. Press Information Bureau (pib.gov.in). Accessed 26 Apr 2024
10. https://cpcb.nic.in/displaypdf.php?id=cGxhc3RpY3dhc3RlL0FubnVhbF9SZXBvcnRfMjAxNC0xNV9QV00ucGRm. Accessed 26 Apr 2024
11. https://pib.gov.in/newsite/PrintRelease.aspx?relid=178707. Accessed 27 Apr 2024
12. OECD, Towards a circular economy: key drivers, in *The Circular Economy in Cities and Regions: Synthesis Report*. OECD Publishing, Paris (2020). https://doi.org/10.1787/7bf512c1-en. Accessed 27 Apr 2024
13. R. Shanker, D. Khan, R. Hossain, M.T. Islam, K. Locock, A. Ghose, V. Sahajwalla, H. Schandl, R. Dhodapkar, Plastic waste recycling: existing Indian scenario and future opportunities. Int. J. Environ. Sci. Technol. **20**(5), 5895–5912 (2023)
14. O. Awogbemi, D.V. Von Kallon, Achieving affordable and clean energy through conversion of waste plastic to liquid fuel. J. Energy Inst. **106**, 101154 (2023)
15. B. Kunwar, H.N. Cheng, S.R. Chandrashekaran, B.K. Sharma, Plastics to fuel: a review. Renew. Sustain. Energy Rev. **54**, 421–428 (2016)
16. Z. Li, S. Niu, J. Liu, Y. Wang, Solid fuel production from co-hydrothermal carbonization of polyvinyl chloride and corncob: higher dechlorination efficiency and process water recycling. Sci. Total. Environ. **843**, 157082 (2022)
17. K. Sivagami, K.V. Kumar, P. Tamizhdurai, D. Govindarajan, M. Kumar, I. Nambi, Conversion of plastic waste into fuel oil using zeolite catalysts in a bench-scale pyrolysis reactor. RSC Adv. **12**(13), 7612–7620 (2022)
18. M. Chandran, S. Tamilkolundu, C. Murugesan, Characterization studies: waste plastic oil and its blends. Energy Sources Part A: Recover. Util. Environ. Eff. **42**(3), 281–291 (2020)
19. K. Murthy, R.J. Shetty, K. Shiva, Plastic waste conversion to fuel: a review on pyrolysis process and influence of operating parameters. Energy Sources Part A: Recover. Util. Environ. Eff. **45**(4), 11904–11924 (2023)
20. S. Dutta, Recycling plastic in India: converting plastic waste to fuel. The unrealised potential. NDTV-Dettol Banega Swasth India (2017). https://swachhindia.ndtv.com/recycling-plastic-in-india-converting-plastic-waste-to-fuel-the-unrealised-potential-9436/. Accessed 25 July 2024
21. P.L. Paihte, A.C. Lalngaihawma, G. Saini, Recycled aggregate filled waste plastic bottles as a replacement of bricks. Mater. Today: Proc. **15**, 663–668 (2019)
22. S.S. Chauhan, J.K. Singh, H. Singh, S. Mavi, V. Singh, M.I. Khan, An overview on recycling plastic wastes in bricks. Mater. Today: Proc. **47**, 4067–4073 (2021)
23. R. Kumar, M. Kumar, I. Kumar, D. Srivastava, A review on utilization of plastic waste materials in bricks manufacturing process. Mater. Today: Proc. **46**, 6775–6780 (2021)
24. R. Shah, H. Garg, P. Gandhi, R. Patel, A. Daftardar, Study of plastic dust brick made from waste plastic. Int. J. Mech. Prod. Eng. **5**(10), 2320–2092 (2017)
25. P. Mahajan, D. Girase, A.R. Lokhande, Plastic soil bricks substituiting traditional bricks. Int. J. Res. Appl. Sci. Eng. Technol. **6**(1), 3380–3386 (2018)

26. D. Alighiri, M.N. Yasin, B. Rohmawati, A. Drastisianti, Processing of recycled waste PET (polyethylene terephthalate) plastics bottle into for the lightweight and reinforcement bricks. J. Phys. Conf. Ser. **1321**(2), 022023 (2019)
27. A. Singhal, O. Netula, Utilization of plastic waste in manufacturing of plastic sand bricks, in *3rd International Conference on New Frontiers of Engineering, Science, Management and Humanitie*, pp.206–210 (2018)
28. C.G. Mohan, J. Mathew, J.N. Kurian, J.T. Moolayil, C. Sreekumar, Fabrication of plastic brick manufacturing machine and bricks analysis. Int. J. Innov. Res. Sci. Technol. **2**, 455–462 (2016)
29. N. Thirugnanasambantham, P.T. Kumar, R. Sujithra, R. Selvaraman, P. Bharathi, Manufacturing and testing of plastic sand bricks. Int. J. Sci. Eng. Res. **5**, 1150–1155 (2017)
30. G. Angarika, Made using 100% waste, these bricks cost just rs 10 & are 250% Stronger. Thebetterindia (2020). https://thebetterindia.com/233190/plastic-waste-bricks-cheap-low-cost-purchase-features-recycle-engineer-architect-anand-gujarat-ang136/. Accessed 25 July 2024
31. GCR Staff, Young engineer invents brick made from waste plastic, global construction review (2019). https://www.globalconstructionreview.com/young-engineer-invents-brick-made-waste-plastic/. Accessed 25 July 2024
32. R. Bhushaiah, S. Mohammad, D.S. Rao, Study of plastic bricks made from waste plastic. Int. Res. J. Eng. Technol. **6**(4), 6 (2019)
33. S. Ren, X. Xu, K. Hu, W. Tian, X. Duan, J. Yi, S. Wang, Structure-oriented conversions of plastics to carbon nanomaterials. Carbon Res. **1**(1), 15 (2022)
34. RIL Communication, Reliance industries becomes first in india to use chemical recycling for circular polymers. Reliance Industries Limited (2023). https://rilstaticasset.akamaized.net/sites/default/files/2023-12/Reliance-Industries-Becomes-First-in-India-to-Use-Chemical-Recycling-for-Circular-Polymers.pdf. Accessed 25 July 2024
35. D. Yao, H. Li, B.C. Mohan, A.K. Prabhakar, Y. Dai, C.H. Wang, Conversion of waste plastic packings to carbon nanomaterials: investigation into catalyst material, waste type, and product applications. ACS Sustain. Chem. Eng. **10**(3), 1125–1136 (2022)
36. O. Vieira, R.S. Ribeiro, J.L.D. de Tuesta, H.T. Gomes, A.M. Silva, A systematic literature review on the conversion of plastic wastes into valuable 2D graphene-based materials. Chem. Eng. J. **428**, 131399 (2022)
37. R. Gupta, V.K. Shukla, P. Agarwal, Sustainable transformation in modest fashion through "RPET technology" and "dry-dye" process, using recycled PET plastic. Int. J. Recent. Technol. Eng. **8**(3), 5415–5421 (2019)
38. G.P. Karayannidis, A.K. Nikolaidis, I.D. Sideridou, D.N. Bikiaris, D.S. Achilias, Chemical recycling of PET by glycolysis: polymerization and characterization of the dimethacrylated glycolysate. Macromol. Mater. Eng. **291**(11), 1338–1347 (2006)
39. C.A. Tshifularo, A. Patnaik, Recycling of plastics into textile raw materials and products, in *Sustainable Technologies for Fashion and Textiles*, pp. 311–326. Woodhead Publishing (2020)
40. B. Sadeghi, Y. Marfavi, R. AliAkbari, E. Kowsari, F. Borbor Ajdari, S. Ramakrishna, Recent studies on recycled PET fibers: production and applications: a review. Mater. Circ. Econ. **3**, 1–18 (2021)
41. Buzz Staff, Rajasthan Teen's startup converts 1,000 Kg plastic waste to fabric every day. News-18 (2021). https://www.news18.com/news/buzz/rajasthan-teens-startup-converts-1000-kg-plastic-waste-to-fabric-every-day-4205924.html. Accessed 25 July 2024
42. R. Balaji, From recycling to reusing: how this Mumbai startup is going green by turning plastic into wearable fabric (2019). https://yourstory.com/socialstory/2019/07/mumbai-startup-raw-pressery-recycling-plastic-clothing. Accessed 25 July 2024
43. M. Abbasi, M.R. Mohades Mojtahedi, R. Kotek, Experimental study on texturability of filament yarns produced from recycled PET. Text. Res. J. **90**(23–24), 2703–2713 (2020)
44. J.H. Ho, K.C. Kuo, T.Y. Lo, C.T. Chang, Y.H. Tseng, B. Gautam, C.T. Lin, J.T. Chen, Upcycling fabrics: valorization of recycled polyethylene terephthalate (r-PET) plastic waste into thermoplastic polyester elastomers (TPEE) for fiber spinning. ACS Appl. Polym. Mater. **6**(1), 552–560 (2024)

45. X.Y. Goh, X. Deng, W.S. Teo, R.H. Ong, L.T. Nguyen, T. Bai, H.M. Duong, Scale-up fabrication and advanced properties of recycled polyethylene terephthalate aerogels from plastic waste. Polym. Eng. Sci. (2024)
46. A. Ali, A.S. Alnaser, Laser-based fabrication of super hydrophobic carbide surfaces from waste plastics. Surf. Interfaces **44**, 103602 (2024)

Chapter 35
Smart Waste Segregation System for Urban Sustainability

Ajeet Pathak, Shivani, Preeti Goyal, Amit Das, and Swati Sharma

Abstract As India's population grows, the amount of waste produced by each person every day is substantial. Some waste can be recycled, some cannot. Therefore, a system that takes into account various types of waste should be developed. To address this problem, we propose a smart waste machine that separates metal waste, dry waste and wet waste, and also automatically converts wet waste into compost. This waste can be used in horticulture, urban agriculture and organic farming. By developing such a system, we aim to contribute to a cleaner Bharat and promote sustainable practices. This initiative not only helps in managing the waste but also decreases the landfill pressure and nurtures greener surroundings for coming generations.

Keywords Rocker-waste management · Internet of Things (IoT) · Real time monitoring · Sensors

35.1 Introduction

The rising population of India poses significant challenges regarding the availability of living space, natural resource utilization, education, and employment. In addition to this, one more critical issue is, the escalating amount of waste generated on daily basis. Our country produces a staggering 100,000 tons of waste every day, yet only 5% of this colossal amount is recycled. In India, the collection, transportation,

A. Pathak · Shivani · P. Goyal (✉) · S. Sharma
Department of Electronics and Communication Engineering, Maharaja Agrasen Institute of Technology, New Delhi, India
e-mail: preetigoyal.s@gmail.com

A. Pathak
e-mail: pathakajeet64@gmail.com

A. Das
Department of Electronics and Communication Engineering, GLA University, Mathura, Uttar Pradesh, India

A. Kumar et al. (eds.), *Selected Proceedings of the 1st International Conference on Advanced Materials for Sustainable Innovation; IC-AMSI 2024; 28–30 August; New Delhi; India*, Springer Proceedings in Energy,
https://doi.org/10.1007/978-981-96-4492-6_35

and disposal of Municipal Solid Waste (MSW) is characterized by unscientific and chaotic practices. Unregulated dumping at the peripheries of urban areas has created overflowing landfills. The mounting waste crisis accelerated the pollution levels, significantly contribute to severe air, soil, and water contamination, which in turn affects public health. These sites are not only environmentally damaging but also impossible to rehabilitate due to the chaotic dumping practices.

Previous research on waste management highlights the importance of segregating waste at the disposal level itself. For instance, a study by Abuga et al. [1] proposed a smart garbage management system that uses ZigBee and GSM technology for efficient waste collection and segregation. Similarly, research by Wang et al. [3] introduced an IoT-based smart garbage system for managing food waste efficiently. These studies have laid the groundwork for developing more advanced systems for waste segregation and management. Our project builds upon these previous works by proposing an automatic waste segregator that segregates waste into wet, dry, or metal categories and also includes a provision which helps in converting wet waste into compost, which can further be used in horticulture, urban agriculture, and organic farming. This initiative aims to contribute to the 'Swachh Bharat' mission by reducing waste and promoting recycling and composting.

35.2 Literature Survey

Effective waste management is a growing global issue, leading to extensive research on smart systems for waste management. This section reviews significant research and technological developments in this field. A smart waste management system using Zigbee and GSM technology is utilized for waste management and classification. The system employs sensors to monitor the level of waste in the bin, enabling rapid disposal, and reducing spillage. These principles emphasize the importance of communication technology in streamlining the waste management process [1–5]. Furthermore, a smart waste management system based on the Internet of Things (IoT) is proposed to efficiently manage food waste through a combination of sensors and communication modules. Research indicates that IoT can improve collection times and thereby reducing the environmental impact of waste disposal. Anagnostopoulos et al. [5] introduced a garbage collection scheduling in smart cities based on the IoT, focusing on optimizing garbage trucks routes to reduce fuel consumption and operating costs. Collectively, these studies illustrate that IoT integration can lead to smarter, more efficient waste management. Researchers are also investigating the use of smart devices and IoT systems to separate waste from the environment [6–8]. Various sensors, such as metal detectors, infrared detectors, and moisture detectors, are used in waste separation processes, increasing recycling efficiency and reducing the need for manual processing. This technology increases the accuracy and efficiency of waste classification, which is an important element of waste management [9, 10]. The discussion also addresses the challenges of managing sensor data and reviews actions taken on the Internet. The importance of integrating IoT technology with traditional waste management processes to improve data collection, analysis, and decision-making is emphasized. Such integration is essential for addressing the transportation and operational challenges faced by urban cities in managing municipal solid waste. Some researchers provide a comprehensive review of IoT applications in waste management, highlighting the revolutionary potential of real-time monitoring and control systems. Their research underscores the role of IoT in ensuring safety and efficiency [11, 12]. A recent study explores an innovative method for waste segregation to enhance recycling and disposal efficiency, leveraging a deep learning approach. The YOLOv3 algorithm was employed within the Darknet neural network framework to train a custom dataset. The model was trained to recognize six categories of waste: cardboard, glass, metal, paper, plastic, and organic material. However, significant disparities exist, particularly in developing countries [13]. Future research should focus on overcoming these challenges by developing effective, large-scale solutions that can be widely implemented. Additionally, integrating machine learning with IoT systems could improve the accuracy and flexibility of waste management systems, enabling them to better respond to diverse situations.

35.3 Proposed System

The proposed system discussed in this work is an advanced garbage segregation system designed to efficiently classify and manage waste. This system aims to automate the process of waste segregation at the source, thereby reducing manual intervention and improving recycling rates. The system employs a combination of sensors, microcontrollers, and mechanical components to segregate waste into three categories: metallic waste, wet waste, and dry waste. The segregation process is managed by an Arduino Uno microcontroller, which processes the sensor data and controls the movement of waste to the appropriate bins. The complete block diagram of the proposed system is presented in Fig. 35.1.

35.4 Components

35.4.1 Metal Detector (Fig. 35.2)

Detects metal waste using electromagnetic induction principles. This sensor generates alternating magnetic field, which alters the original field of nearby conducting material. When metal waste is detected, the sensor sends a signal to the Arduino Uno.

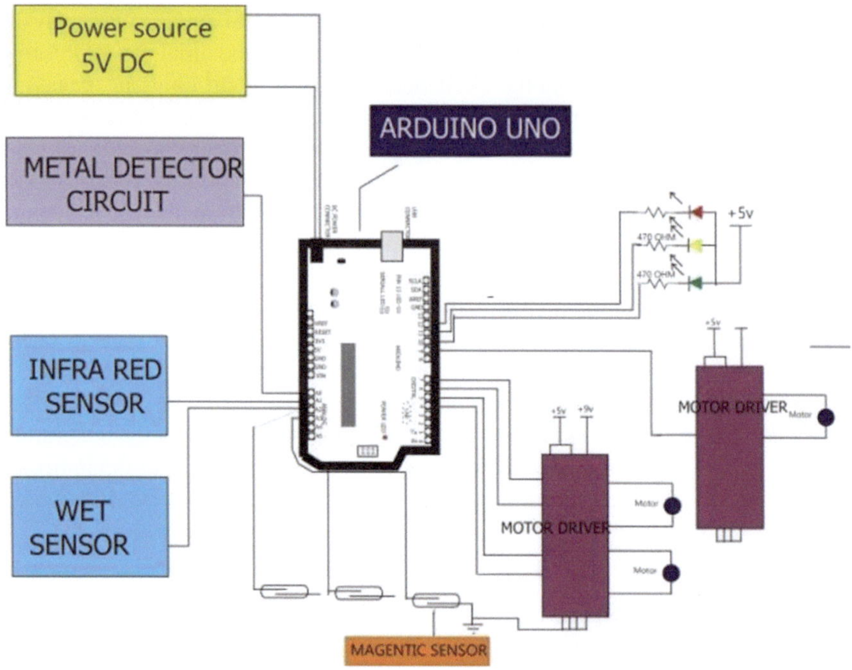

Fig. 35.1 Block diagram of smart garbage segregator

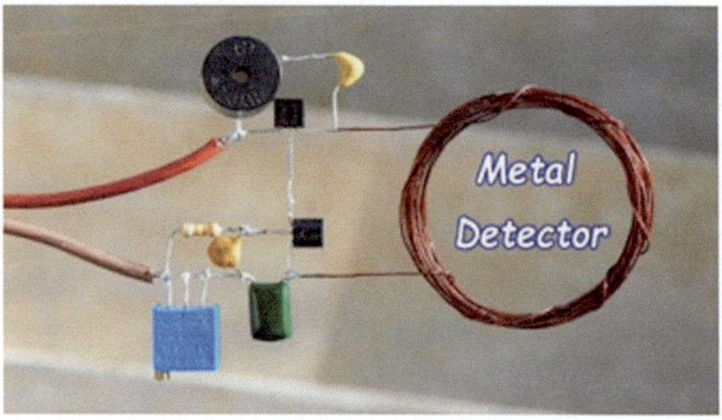

Fig. 35.2 Metal detector

35.4.2 Infrared Sensor (Fig. 35.3)

This sensor identifies dry waste such as plastics and paper. It uses an infrared LED and photodiode to detect the presence of dry waste materials.

Fig. 35.3 Infrared sensor

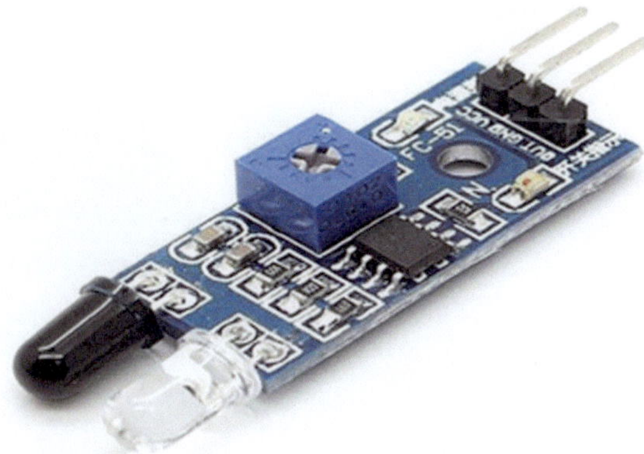

Fig. 35.4 Moisture sensor

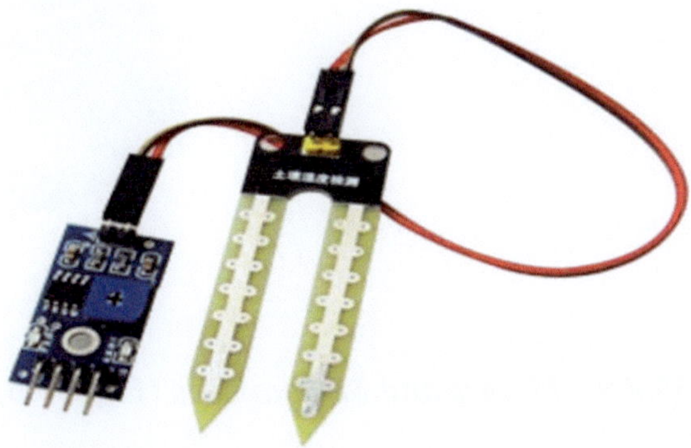

35.4.3 Moisture Sensor (Fig. 35.4)

Detects wet waste by measuring the conductivity between two metal probes. The presence of moisture completes the circuit, indicating wet waste.

35.4.4 Arduino UNO (Fig. 35.5)

The central microcontroller that processes inputs from all sensors and controls the actuators. It features 14 digital I/O pins, 6 analog inputs, and operates at 5 V.

Fig. 35.5 Arduino Uno

Fig. 35.6 Servo motor

35.4.5 Motors and Mechanisms (Fig. 35.6)

DC Gear Motors: Controlled by an L293 motor driver circuit, these motors facilitate the movement of waste into the designated bins.

35.4.6 Motor Driver (L293D)

This H-bridge IC controls the direction and speed of the DC gear motors based on commands from the Arduino.

35.4.7 Power Supply

The Arduino Uno and associated components are powered via USB or an external AC-to-DC adapter, providing the necessary voltage and current for stable operation.

35.5 Technical Specification and Working

The technical details of the proposed device are outlined in this section to specify the key technical specifications. The device is built around the Microcontroller ATmega328, operating at a voltage of 5 V. It has a recommended input voltage range of 7–12 V, with an allowable limit of 6–20 V. The device features 14 digital I/O pins, 6 of which provide PWM output, and 6 analog output pins. Each I/O pin can handle a DC current of up to 40 mA, while the 3.3 V pin can handle up to 50 mA. Additionally, the device includes 32 KB of flash memory.

The proposed smart garbage segregation system operates through a series of systematic steps that integrate sensor inputs and microcontroller processing to automate waste segregation. Initially, the system is powered by the Arduino Uno microcontroller, which can draw power from a USB connection or an external AC-to-DC adapter, ensuring stable operation. Upon startup, the Arduino Uno initializes all connected sensors, preparing them to detect three types of waste: metal, wet, and dry. When waste is deposited, a metal detector sensor scans for metallic objects using electromagnetic induction and sends an analog signal to the Arduino Uno if metal is detected. For dry waste, an infrared sensor module consisting of an infrared LED and a photodiode, detects materials like plastic and paper, sending a digital signal to the Arduino Uno. The moisture sensor uses two metal probes to detect the presence of wet waste. When the probes come into contact with moisture, a circuit is completed, indicating presence of wet waste. The Arduino Uno processes these signals, using pre-programmed logic to determine the waste type. It then generates control signals to operate DC gear motors via an L293D motor driver IC, directing the waste to the appropriate bin. The Arduino Uno sends control signals to the motor driver, which in turn controls the direction and movement of the motors. A motor-driven shutter mechanism guides the waste into the correct bin—metal, dry, or wet—based on the sensor inputs. The detailed operation of this device is illustrated in a flowchart presented in Fig. 35.7.

35.6 Conclusion

The proposed system reduces the time required to segregate and collect garbage from various locations. It also lowers the overall expenditure associated with garbage collection. Additionally, it promotes cleanliness and hygiene in the areas where it is implemented. Therefore, the IoT-based garbage management system makes garbage collection more efficient. Although the smart dustbin is more expensive than a regular dustbin, the overall cost such as manpower needed for manual segregation and the total number of trips required for garbage collection, will be reduced with the proposed system. The system also contributes to maintaining cleanliness in the area.

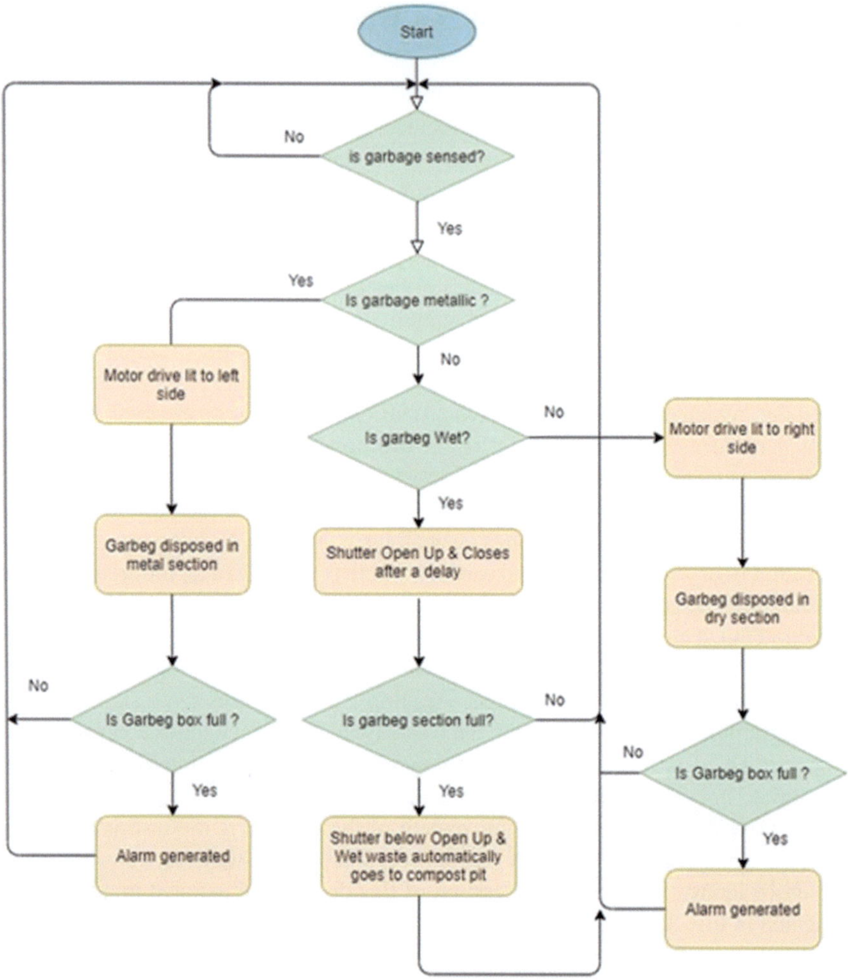

Fig. 35.7 Flow chart for working of proposed model

35.7 Future Scope

The future scope of the smart garbage segregation system includes integrating advanced sensor technologies for enhanced detection capabilities and machine learning for improved accuracy. IoT integration will enable real-time data analytics and smart city applications. Energy-efficient components and solar power can increase sustainability. Expanded use cases in industrial settings and customizable segregation options will broaden the system's applicability. Enhanced user interfaces and accessibility features will improve user interaction. Direct integration with recycling and composting facilities, along with automated logistics, will streamline waste management processes. Ongoing research, pilot programs, and collaborations with policymakers will drive continuous improvements and broader adoption, contributing to environmental sustainability and cleaner communities.

References

1. D. Abuga, N.S. Raghava, Sustain. Cities Soc. **75**, 103347 (2021)
2. V. Bhor, P. Morajkar, M. Gurav, D. Pandya, A. Deshpande. Int. J. Eng. Res. Technol. (IJERT) **4**(03) (2015)
3. C. Wang, J. Qin, C. Qu, X. Ran, C. Liu, B. Chen, Waste Manag. **135**, 20–29 (2021)
4. I. Saukenova, M. Oliskevych, I. Taranic, A. Toktamyssova, D. Aliakbarkyzy, R. Pelo. East. -Eur. J. Enterp. Technol. **1**(3), 115 (2022)
5. T. Anagnostopoulos, A. Zaslavsy, A. Medvedev, S. Khoruzhnicov, in *16th IEEE International Conference on Mobile Data Managagement*, vol. 2, pp. 50–55. IEEE (2015)
6. A.A. Laghari, K. Wu, R.A. Laghari, M. Ali, A.A. Khan. Arch. Comput. Methods Eng. 1–19 (2021)
7. M. Look, Natl. Conf. Prod. Des. (NCPD 2016). In earth 911B (2016)
8. Md A. Al Rakib, Md S. Rana, Md M. Rahman, F.I. Abbas. Eur. J. Eng. Technol. Res. **6**(5), 129–133 (2021)
9. R. Chatterjee, S. Sharma, Procedia Comp. Sci **160**, 238–246 (2019)
10. M. Mojoudi, M. Kalantari, D. Brito, R. IEEE Internet of Things J. **7**(5), 4320–4330
11. S. Sahoo, B. Ray. Mater. Today: Proc. **46**(10), 7654–7661
12. P. Gupta, R. Prasad, J. Environ. **305**, 114355
13. S. Kumar, D. Yadav, H. Gupta, O.P. Verma, I.A. Ansari, C.W. Ahn. Electronics **10**(1), 14 (2020)

Chapter 36
Consumer Insights on E-Waste, Energy, and the Circular Economy: A Case Study of Delhi

Urvashi Chawla, Jayati Ahuja, and Sonal Khurana

Abstract Electronic waste or E-waste is the reference to discarded electronic devices or equipment or their constituents. These often contain complex valuable materials like gold or hazardous substances like mercury which need to be recycled effectively. Here, the concept of circular economy, i.e. a reuse, repair, recycle economy is ideal. The beneficial effects of a circular economy are manifold. It minimizes waste, protects the environment, by increasing the life of materials. By promoting reuse, repair, and recycling, it minimizes the quantity of garbage transported to landfills, lowering methane emissions. Economically, it benefits society by lowering total demand for raw materials. In the current work, a group of individuals have been surveyed to examines the level of awareness among Delhi's middle-class population, finding high familiarity with e-waste but lower understanding of the circular economy concept. Barriers to e-waste recycling include lack of convenient facilities, uncertainty about recycling methods, and perceived inconvenience. Motivators for recycling include environmental awareness and financial incentives. It develops a crucial understanding of how people can be more motivated to participate in recycling and embrace the new concept. It aids inconceptualising why the current population may dodge recycling or purchasing products which are a result of the circular approach.

Keywords E-waste · Circular economy · Recycling · Consumer behaviour · Awareness · Environmental sustainability

---

U. Chawla · J. Ahuja (✉) · S. Khurana
School of Engineering and Technology, VIPS-TC, Delhi, India
e-mail: 06417702723_cse@vipstc.edu.in

U. Chawla
e-mail: 35817702723_cse@vipstc.edu.in

S. Khurana
e-mail: sonal.khurana@gmail.com

A. Kumar et al. (eds.), *Selected Proceedings of the 1st International Conference on Advanced Materials for Sustainable Innovation; IC-AMSI 2024; 28–30 August; New Delhi; India*, Springer Proceedings in Energy,
https://doi.org/10.1007/978-981-96-4492-6_36

36.1 Introduction

Every year, millions of electrical and electronic devices are discarded globally as they break or are no longer required, constituting what is known as electronic waste or e-waste. This category includes a wide range of products such as computers, mobile phones, household appliances, and medical equipment [1]. E-waste is categorized into hazardous and non-hazardous materials, comprising ferrous and non-ferrous metals, plastics, glass, and more. Steel and Iron make up about half of e-waste, followed by plastics (21%) and non-ferrous metals like Copper and Aluminium (13%). Hazardous elements, including lead, mercury, and flame retardants, present in e-waste can cause significant environmental and health risks upon disposal [2].

India, one of the world's largest producers of e-waste, generated 3.2 million tonnes of it in 2019, ranking third globally after China and the United States [3]. Despite this massive volume, the country only collected a small fraction for recycling, highlighting a significant gap between e-waste generation and proper disposal practices.

According to studies, developing countries will produce e-waste double than that of developed countries within next 6–8 years. It is also estimated that the developing and developed countries will discard 400–700 million and 200–300 million obsolete computers by 2030, respectively [4]. The volume of e-waste in India has witnessed a significant increase to 1.6 million tonnes in 2021–22 [5]. This trend underscores the urgent need for sustainable practices in resource management, particularly in the production and consumption of electronics.

36.1.1 Impact of E-waste on the Environment

Improper management of e-waste poses significant environmental and health risks, exacerbated by informal recycling practices that release hazardous substances into the environment. E-waste contains materials such as mercury and plastics with brominated flame retardants, which are released during improper disposal processes.

It also contributes to soil contamination due to high levels of lead and mercury leaching from discarded electronics. These toxic substances can persist in the soil for years, affecting both terrestrial and aquatic ecosystems [6]. When it is incinerated or left in landfills, toxic compounds like lead, cadmium, and beryllium are released into the air, which further pollutes the environment. It also contributes to greenhouse gas emissions, for instance, manufacturing alone potentially emitting 10 tonnes of CO_2 per tonne of laptops produced [7]. Furthermore, these substances can also contaminate groundwater, posing risks to human health and biodiversity. Exposure to e-waste has been linked to adverse health effects, including negative birth outcomes, respiratory issues, and changes in lung function etc.

Proper e-waste disposal faces challenges like lack of awareness, accessibility issues, and mistrust in recycling processes. Addressing these is crucial for shifting from conventional waste management to a circular economy approach, requiring changes in consumer perspectives and behaviour.

The economic potential of circularity is substantial. The FICCI Circular Economy Report, 2017, highlighted that the business opportunity for extracting gold from e-waste is estimated at $0.7 to $1 billion. Despite this potential, only 17.4 percent of e-waste was collected and recycled globally in 2019, leading to a loss of nearly $47 billion in recoverable materials, including gold, silver, copper, platinum, and other high-value materials [8]. Embracing a circular economy not only promises environmental benefits but also unlocks economic value from materials that would otherwise be lost in the linear economy model.

36.1.2 Objectives

The primary objective is to analyse and understand the behavioural habits of consumers in the context of e-waste collection and their acceptance to the concept of circular economy. The research focuses on:

1. Investigating the levels of awareness and understanding among consumers regarding the environmental impact of e-waste.
2. Identifying the barriers and motivations that influence consumer behaviour towards e-waste recycling.
3. Exploring the role of convenience, accessibility, and incentives in encouraging consumer participation in e-waste collection programs.

36.2 Literature Review and Related Works

36.2.1 Circular Economy

Our current economic model operates on a linear basis, where raw materials are extracted, transformed into products, and ultimately discarded as waste, with only a small fraction being recycled back into the economy. This linear approach not only strains natural resources but also worsens environmental problems such as climate change, biodiversity loss, and pollution crises. Globally, only 7.2 percent of used materials are currently cycled back into our economies after use, highlighting the inefficiencies of our current system [9].

On the other hand, the idea of circular economy presents the society with a progressive model of change. A circular economy is based on the framework that will help to sustain resources and reduce waste at the same time. Some of circular economy strategies include designing for longevity and for ease of recyclability means that products should be designed to have long durability, to be easily repaired, and at the end of their useful life they should be recycled [10]. Another principle is resource

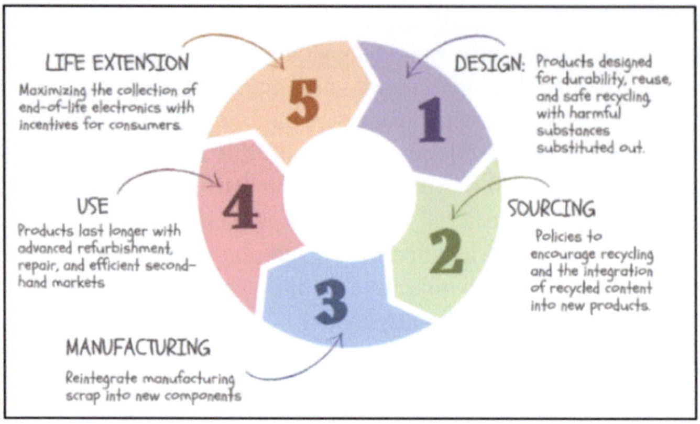

Fig. 36.1 Flowchart on circular approach to e-waste management (author's own contribution)

intensity, which has to do with using resources in the most efficient manner possible: by avoiding waste, recycling and where possible using renewable inputs, especially energy and the least emitting throughout the production-consumption chain [11]. Also, utilizing circular consultancy that signifies the reusing considerations originating from one cycle as inputs for another cycle is useful as it reduces virgin material usage and the company's overall effect on the environment [12]. It does not only diminish the pollution related to the processes of production and consumption but also opens vast possibilities for resolving more extensive tasks, such as climate change and the decrease in biodiversity [9] (Fig. 36.1).

Circular economy principles can be applied across various sectors, including textiles, construction, and electronics, at every stage of a product's lifecycle—from design and manufacturing to distribution and disposal. For instance, in the textile industry, initiatives are underway to promote regenerative agriculture and produce garments using organic materials and natural dyes. By prioritising durability and repairability, these practices extend the lifespan of clothing and facilitate recycling [9].

The 2017 report released by the Ellen MacArthur Foundation [13] looks ahead to consumer electronics designed to last, through either ongoing utilization by the original owner or passing to other users. Following their lifecycle, specialists either update, reuse, or recycle useful components and materials.

36.2.2 Advantages of Circular Economy

The circular economy presents numerous advantages that extend beyond environmental sustainability to economic growth and social well-being. By 2030, India is projected to become the world's third-largest economy, comprising approximately 8.5% of the global GDP. Embracing circular economy principles can leverage this economic expansion while delivering significant environmental benefits, establishing a sustainable and resilient framework [14].

Circular practices in e-waste recycling significantly mitigate environmental harm by reducing pollution from traditional disposal methods and landfills, thereby decreasing greenhouse gas emissions and aligning with global climate goals. These methods enhance resource efficiency by extending product life spans through repair, refurbishment, and recycling, minimising new material extraction and conserving natural resources. Economically, they create opportunities across sectors, fostering markets for refurbished goods and benefiting informal workers in tier II and III cities with formalised roles. This shift drives innovation in sustainable technologies, promoting job growth in recycling, remanufacturing, and digital resale platforms. Socially, integrating informal waste sectors into formal circular initiatives enhances community resilience, broadens access to end-of-life solutions, and promotes inclusive economic participation.

Moreover, the study by Parajuly et al. (2024) emphasises the significant impact of consumer behaviour on the circular economy, particularly in e-waste management within the EU. The authors review behavioural theories relevant to sustainable consumption and propose interventions for improving e-waste management. Key findings suggest that understanding the techno-economic aspects alone is insufficient; integrating behavioural insights is crucial. The study highlights community-based social marketing and nudging as effective strategies, emphasising the importance of creating favourable environments for behaviour change. The authors call for targeted interventions designed to achieve specific outcomes, rather than attempting to permanently alter general consumer attitudes and behaviours [15].

Additionally, adopting circular practices boosts India's global competitiveness by meeting international sustainability standards and consumer preferences for eco-friendly products, supporting long-term economic stability.

Another important study was conducted in Delhi by [16] that found a considerable fraction of the middle-class population lacks awareness about e-waste management. While some respondents understood the health impacts of improper e-waste handling, many were unaware of correct recycling practices. Internet and print media were identified as primary sources of information. A notable consumer behaviour revealed that 12–26% replace major electronic goods within three years of purchase, contributing significantly to e-waste generation. Respondents expressed a strong preference for efficient recycling facilities and supported the idea of shared responsibility between consumers and producers for managing e-waste. These findings underscore the urgent need for targeted awareness campaigns and improved infrastructure for sustainable e-waste management in urban India [16].

The purpose of our research is to check the awareness level of people in Delhi after 10 years of this research, especially in light of the rapid technological advancements. Over the past decade, technology has evolved dramatically, increasing both the volume and complexity of e-waste. This transformation necessitates an evaluation of how awareness and behaviours regarding e-waste management have changed.

36.2.3 Japan's Approach [17]

The case study by Satoshi Yoshida, Ministry of Japan, explains the transition of a linear economy model to a circular approach using society aspects as one of its pillars. It uses Circular Economy and Resource efficiency Principles (CEREP) as a guidance on CERE businesses. It uses 6 principles in reforming of decision-making affecting leadership, identification of risks and opportunities, changes in product design, enhancing monitoring and reporting, and inclusion of varied stakeholder partnerships.

Examples of past reforms included are Japan's resource circulation strategy of 2019 and plastic resource circular act, April 2022. Both focus on stakeholders' promoting the "3Rs + Renewable" as a primary part and work on a circular approach for the business industry.

The Plan of building a well-working Material-Cycle Society was developed under the Basic Act on Establishing a Sound Material-Cycle Society, which was implemented in January 2001. This plan aims to promote measures for creating a sustainable material-cycle society in a comprehensive and systematic manner and is reviewed roughly every five years. The latest iteration, the 4th plan, was established in June 2018.

The forthcoming 5th plan will focus on transitioning to a circular economy, which will support achieving a net-zero, nature-positive economy, enhance industrial competitiveness, boost circular business models, revitalize local areas, and ensure economic security.

36.3 Study Area

The Study area analysed for implementation of circular economy is Delhi, the capital of India.

- Delhi is the second largest city as producer of E-waste in India, after Mumbai. Delhi generates about 55,000 Metric tonnes of e-waste per year according to the ASSOCHAM 2014.
- Delhi is the leading centre for e-waste processing in India. A study by GTZ in 2007 revealed that around 25,000 workers were refurbishing 10,000–20,000 tonnes of e-waste annually. This work is carried out in small, illegal units without adherence to regulations or environmental and health protections. The absence of proper facilities for the storage and disposal of e-waste leads to incidents like the radiation exposure in Mayapuri and more [18].
- Delhi is one of the most developed cities in India, and thus, understanding how awareness can be spread about the circular economy on such a sample of population is more efficient.

Table 36.1 Datasets from survey conducted

1	Age (71 Participants)	
2	Age Group	No. Of responses
3	Under 18	7
4	18-24	50
5	25-34	3
6	35-44	4
7	44-65	7
8	Above 65	None
9	Gender (71 Participants)	
10	Gender	No. of responses
11	Male	32
12	Female	39
13	Others	None
14	Are participating aware where to recycle e-waste	
15	Yes	20
16	No	35
17	Maybe	16
18	Feedback on E-waste and circular economy	
19	Positive	12
20	Negative	6
21	No feedback	52

36.4 Results and Discussion

36.4.1 Dataset

In total 71 responses were collected in the survey. The data also analyses the gender and age group in accord with the answers submitted. It also reflects people's awareness and participation in the circular economy and e-waste management.

The Sample population of the survey conducted is majorly middle class citizens that reside in Delhi. Most participants do not have any history of working towards the environment or the concept of circular economy professionally and thus, give us a fair idea of the conceptual and practical awareness of the topic (Table 36.1).

36.4.2 Observation and Interpretation

The survey included analysis of varied responses that illustrate how the general public or the sample population reacts to e-waste and the Circular economy. The Circular Economy model brings up various issues and as well as requires understanding of consumer patterns. Following is some of the consumer patterns observed in a circular economic structure.

As observed (see Fig. 36.2), the said model is supposed to bring the product's focus from acquisition to retaining, repairing and reselling. This can only be achieved

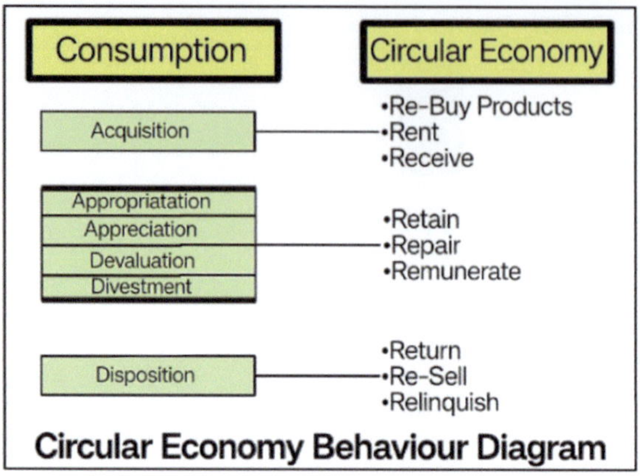

Fig. 36.2 Circular Economy Consumer Behaviour (author's own contribution)

by understanding the consumer's behaviour, response and awareness to the circular economy model.

Awareness of Circular Economy

The majority of respondents (94.4%) are familiar with e-waste to some extent, with a significant portion being very familiar (59.2%). This suggests that there is a high level of awareness about electronic waste among the respondents. In contrast, the awareness of the circular economy is lower. While a combined 69.1% of respondents are familiar with the concept to some degree, only 25.4% are very familiar. A notable 31% of respondents are either not very familiar or not at all familiar with the circular economy concept (Fig. 36.3).

This indicates that although e-waste is a well-known issue, the broader concept of the circular economy—which is vital for the sustainable management of resources and waste reduction—is less understood. To address this gap, it is essential to implement targeted educational programs and awareness campaigns that highlight the principles and advantages of the circular economy. Improving knowledge in this area can promote more sustainable practices and enhance the management of electronic waste, contributing to broader environmental objectives.

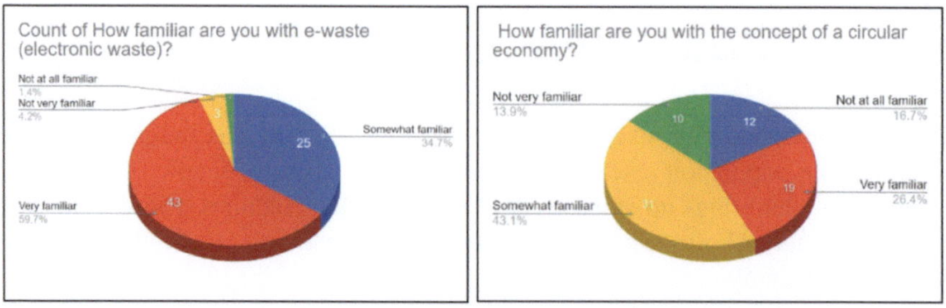

Fig. 36.3 Pie charts of count of familiarity with e-waste and count of familiarity with circular economy from the survey conducted (author's own contribution)

People's Concept of E-waste Management and How They Act

Most respondents understand key e-waste management practices like recycling, reuse/refurbishing, and proper disposal. However, fewer are aware of donating or selling old electronics as viable options. Most respondents took appropriate actions for disposing of electronic devices, such as returning them to manufacturers or giving them to friends/family. However, a significant portion still disposed of them with regular trash, indicating room for improvement in proper e-waste management.

The data indicates that while there is a solid understanding of fundamental e-waste management practices such as recycling, reuse/refurbishing, and proper disposal, fewer people are aware of alternatives like donating or selling old electronics.

In practice, many respondents engage in responsible actions like returning devices to manufacturers or recycling centres. However, a significant minority still dispose of electronics improperly with regular trash.

This underscores the need for enhanced awareness and education about all e-waste management options, especially those that prevent improper disposal through directly throwing them in the trash (Figs. 36.4 and 36.5).

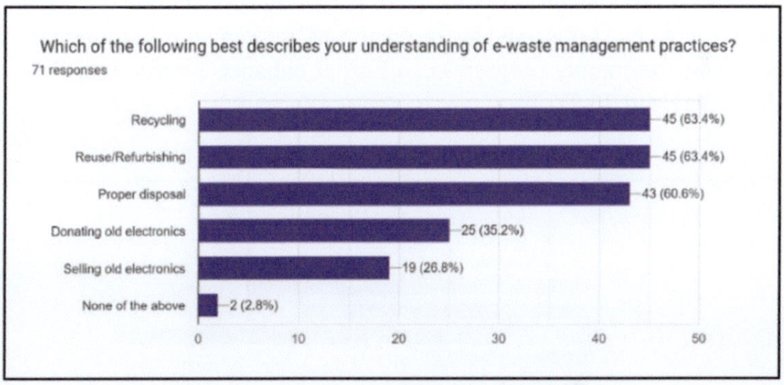

Fig. 36.4 Survey responses to understanding of e-waste management (author's own contribution)

When disposing of electronic devices, which of the following actions have you taken in the past year?
71 responses

Action performed	Amount of Responses	Percentage
Used a designated e-waste recycling service	27	38%
Returned to the manufacturer or retailer	32	45.10%
Participated in a community e-waste collection event	12	16.90%
Gave it to a friend or family member	32	45.10%
Disposed of it with regular trash	13	18.30%
Never Disposed	4	5.60%
sold to recyclers	1	1.40%

Fig. 36.5 Survey responses to actions performed last year (author's own contribution)

Why People Don't Recycle E-waste and How to Promote it

The most significant barrier is the absence of nearby facilities. This suggests that many people do not have convenient access to places where they can dispose of their e-waste properly. Nearly half of the respondents are unsure about how or where to recycle their e-waste. This indicates a substantial gap in public knowledge and the need for better information dissemination. Many respondents also find the process inconvenient, as observed the reasons to be time and cost. Cost is a barrier for about a quarter of the respondents. This might include the cost of transportation to recycling facilities or fees associated with e-waste disposal. Some varied responses reasons such as already recycling, not seeing a benefit, or relying on their parents for help.

People also responded to what may be more motivational to promote recycling to e-waste. The most compelling motivator is the availability of convenient drop-off locations. This reinforces the finding that accessibility is a major issue. Over half of the respondents would be motivated by greater environmental awareness. This includes awareness of consequences of not recycling the said waste and what hazards it may cause.

Financial incentives and convenient options are strong motivators. Introducing monetary rewards, discounts, or convenient recycling services can encourage more people to participate. As observed by the responses, implementing stricter regulations and supporting community programs can further enhance e-waste recycling efforts (Figs. 36.6, 36.7 and 36.8).

Fig. 36.6 Survey responses to reasons for not recycling (author's own contribution)

Fig. 36.7 Survey responses to methods to motivate e-waste recycling (author's own contribution)

Fig. 36.8 Pie chart of survey responses to location awareness of recycling centres (author's own contribution)

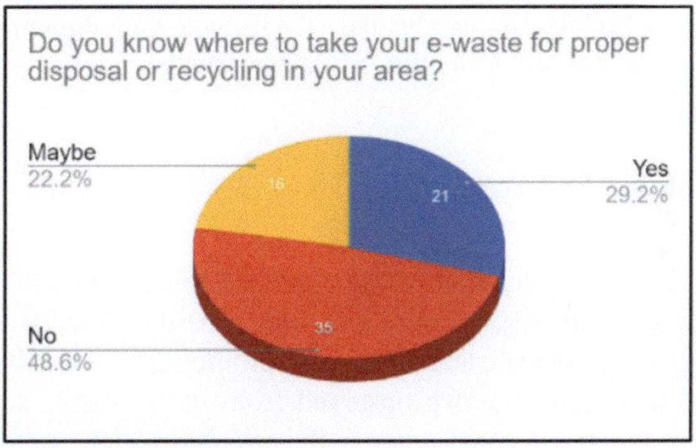

People's Response to Recycled Products

As we know, circular economy is based on the sole concept of recycling, thus it is important to understand the consumer's to recycled products as a whole. The largest group of respondents (46.5%) expressed that they might consider purchasing refurbished or recycled electronics. This indicates that a notable portion of the population is open to the idea but may require additional information or reassurance about the products' quality and reliability. A quarter of the respondents (25.4%) are willing to purchase refurbished or recycled electronics, who are already supportive of sustainable purchasing practices.

Although, over a quarter of the respondents (28.2%) are not willing, indicating the group might have concerns about the performance, lifespan, or stigma associated with refurbished products (Fig. 36.9).

Thus, in case of application of a circular economy, the following practices might be a necessity, especially as most responded with maybe:

1. Consumer Education and Awareness to end stigma
2. Quality Assurance of the recycled products

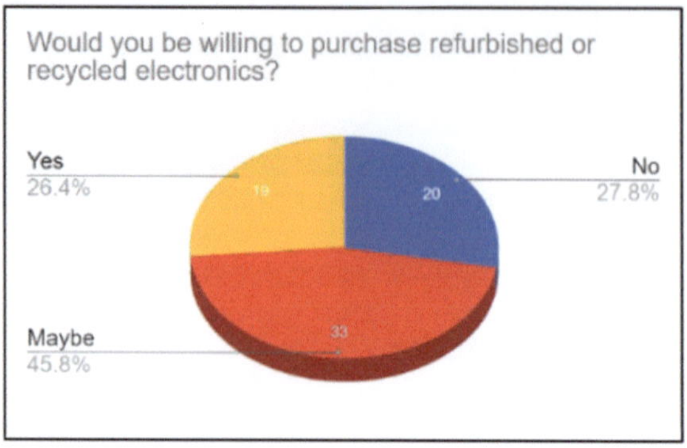

Fig. 36.9 Pie chart of survey responses to opinion on purchasing recycled electronics (author's own contribution)

3. Emphasising the environmental benefits and cost savings associated with purchasing refurbished electronics can appeal to the "maybe" group and potentially sway the "no" respondents.

36.5 Future Scope

The future potential of the circular economy is vast, impacting numerous sectors. It aims to revolutionise the lifecycle of products—from design and production to consumption and disposal—promoting a more sustainable and robust global economy. But as we observed from the research, some areas need to be worked on to achieve consistency in the circular model.

1. First and foremost, it includes the importance of awareness. A lot of individuals care for the environment and would like to contribute but lack even the basic knowledge on how to handle the e-waste produced. By spreading awareness in the future, we work directly on the root cause or root production of waste before handling other steps in the process.
2. Accessibility improvement and pick-up services in the future may boost the model greatly as time and distance are leading factors of ignorance of the existing curriculum for the concept.
3. Government intervention and creation of rules and fines may prove to be an easy-to-implement change in the linear system that may boost the circular economy exponentially in the future.
4. The advancement and incorporation of technologies like digital platforms, Artificial Intelligence, etc., facilitate more efficient tracking, sharing, and management of resources. AI can even be used to predict and pattern waste production, making it easier to manage.

36.6 Limitations of Current Research

The current survey-based research also has some possible drawbacks stated below:

1. **Sample population's representation**: Since the sample population of the research is mostly adults and belongs to the middle class, some parts of the society remain inaccessible in the research. Thus, the opinions of the said research survey are based on the middle-class population of Delhi.
2. **Limited Depth**: A survey does not provide an in-depth look into each and every individual's opinion, giving us limited knowledge on each subject or issue as the answers may have a missed context to them.
3. **Fixed Responses in some questions**: Closed-ended questions limit the ability of respondents to express all their problems fully. The survey may not fully tell us about how people tackle their e-waste in a deep conceptual matter.
4. **Result Interpretation**: A survey's result may conclude with varied opinions based on person to person. This can be overcome by having a general and unbiased outlook while coming to a conclusion from the research survey.

36.7 Conclusion

This study reviews the impact of E-waste not only the environment, but its adverse effects on biodiversity and health risks it poses to humanity. Thus, the implementation of circular economy model is crucial. Research and the data collected demonstrate the severity of lack of education about circular model of production. The research also statistically elucidates that accessibility to recycling centers, along with clear and comprehensive information on e-waste management, is crucial for enhancing public participation and promotion of the same. Additionally, government interventions, including stricter regulations and financial incentives, could significantly bolster recycling efforts.

Use of some present day technologies can also be made to improve the inclusion of circular economy into the lives of many. These include:

- Use of digital platforms and IoT devices to analyse participation, track E-waste's life cycles and maintaining material records in the said electronics.
- Predictive analysis to understand which resources' addition is economically beneficial as well as more biodegradable than the current subsets of Electronic production.
- Use of 3-D printing to substitute for materials that cannot be replaced completely.
- Inclusion of Artificial intelligence and machine learning to ameliorate current recycling practices for E-waste.

Many more methods can be used for addressing e-waste management practices in Delhi, promoting environmental sustainability and economic benefits while also supporting the broader goals of a circular economy.

References

1. World Health Organization, Electronic waste (e-waste) (2023). https://www.who.int/news-room/fact-sheets/detail/electronic-waste-(e-waste)
2. Research Unit (LARRDIS), E-waste in India. Rajya Sabha Secretariat, New Delhi (2011)
3. Down to Earth, (2020). India collected just 3% e-waste generated in 2018, 10% in 2019: CPCB report (2023). https://www.downtoearth.org.in/news/waste/india-collected-just-3-e-waste-generated-in-2018-10-in-2019-cpcb-report-75072
4. S. Arya, S. Kumar, E-waste in India at a glance: Current trends, regulations, challenges and management strategies. J. Clean. Prod. **271**, 122707 (2020). https://doi.org/10.1016/j.jclepro.2020.122707
5. Press Information Bureau, Ministry of environment, forest and climate change: generation of e-waste (2023). https://pib.gov.in/PressReleasePage.aspx?PRID=1943201
6. Earth.org., Environmental impact of e-waste (2023). https://earth.org/environmental-impact-of-e-waste/
7. M. Adam, How we think about e-waste is in need of repair. Anthropocene (2016). www.anthropocenemagazine.org/ewaste-repair
8. Ministry of Electronics and Information Technology, Government of India, Circular economy in electronics and electrical sector: action plan (version 7, May 2021). New Delhi (2021)
9. UNDP Climate Promise. What is circular economy and how it helps fight climate change (2023). https://climatepromise.undp.org/news-and-stories/what-is-circular-economy-and-how-it-helps-fight-climate-change. Accessed 15 July 2024
10. F. Blomsma, M. Tennant, Circular economy: preserving materials or products? Introducing the resource states framework. Resour. Conserv. Recycl. **156**, 104698 (2020). https://doi.org/10.1016/j.resconrec.2020.104698
11. B. Corona, L. Shen, D. Reike, J.R. Carreon, E. Worrell, Towards sustainable development through the circular economy—a review and critical assessment on current circularity metrics. Resour. Conserv. Recycl. **151**, 104498 (2019). https://doi.org/10.1016/j.resconrec.2019.104498
12. S. Kara, M. Hauschild, J. Sutherland, T. McAloone, Closed-loop systems to circular economy: a pathway to environmental sustainability? CIRP Ann. **71**, 505–528 (2022). https://doi.org/10.1016/j.cirp.2022.05.008
13. Ellen MacArthur Foundation, Circular consumer electronics: an initial exploration (2018)
14. India Brand Equity Foundation, Circular economy for sustainable development in India (2023). https://www.ibef.org/blogs/circular-economy-for-sustainable-development-in-india. Accessed 15 July 2024
15. K. Parajuly, C. Fitzpatrick, O. Muldoon, R. Kuehr, Behavioral change for the circular economy: a review with focus on electronic waste management in the EU. Resour. Conserv. Recycl.: X **6**, 100035 (2020). https://doi.org/10.1016/j.rcrx.2020.100035
16. S. Kwatra, S. Pandey, S. Sharma, Understanding public knowledge and awareness on e-waste in an urban setting in India: a case study for Delhi. Manag. Environ. Qual. **25**(5), 480–495 (2014)
17. S. Yoshida, Transition to a circular economy as a national strategy. Ministry of the environment, Japan (2023). https://gj-eedf.org/system/files/document/04.-moej_mr.-yoshida.pdf
18. E. Green. E-waste in India (2018). https://greene.gov.in/wp-content/uploads/2018/01/e-waste_in_india-Document.pdf

MIX
Papier aus verantwortungsvollen Quellen
Paper from responsible sources
FSC® C105338

If you have any concerns about our products,
you can contact us on
ProductSafety@springernature.com

In case Publisher is established outside the EU,
the EU authorized representative is:
Springer Nature Customer Service Center GmbH
Europaplatz 3, 69115 Heidelberg, Germany

Printed by Libri Plureos GmbH
in Hamburg, Germany